T0394950

Advances in Experimental Medicine and Biology

Volume 1281

Advances in Experimental Medicine and Biology provides a platform for scientific contributions in the main disciplines of the biomedicine and the life sciences. This series publishes thematic volumes on contemporary research in the areas of microbiology, immunology, neurosciences, biochemistry, biomedical engineering, genetics, physiology, and cancer research. Covering emerging topics and techniques in basic and clinical science, it brings together clinicians and researchers from various fields.

Advances in Experimental Medicine and Biology has been publishing exceptional works in the field for over 40 years, and is indexed in SCOPUS, Medline (PubMed), Journal Citation Reports/Science Edition, Science Citation Index Expanded (SciSearch, Web of Science), EMBASE, BIOSIS, Reaxys, EMBiology, the Chemical Abstracts Service (CAS), and Pathway Studio.

2018 Impact Factor: 2.126.

More information about this series at http://www.springer.com/series/5584

Bernardino Ghetti
Emanuele Buratti
Bradley Boeve • Rosa Rademakers
Editors

Frontotemporal Dementias

Emerging Milestones of the 21st Century

 Springer

Editors
Bernardino Ghetti
Department of Pathology and
Laboratory Medicine
Indiana University
Indianapolis, IN, USA

Bradley Boeve
Department of Neurology
Mayo Clinic
Rochester, MN, USA

Emanuele Buratti
International Centre for Genetic
Engineering and Biotechnology
(ICGEB)
Trieste, Italy

Rosa Rademakers
Department of Neuroscience
Mayo Clinic
Jacksonville, FL, USA

VIB Center for Molecular Neurology
University of Antwerp-CDE
Antwerpen, Belgium

ISSN 0065-2598 ISSN 2214-8019 (electronic)
Advances in Experimental Medicine and Biology
ISBN 978-3-030-51139-5 ISBN 978-3-030-51140-1 (eBook)
https://doi.org/10.1007/978-3-030-51140-1

This Springer imprint is published by the registered company Springer Nature Switzerland AG
The registered company address is: Gewerbestrasse 11, 6330 Cham, Switzerland

The image intended to represent visually the mathematical formula for Pi (π). The artwork was created by a patient with non-fluent Primary Progressive Aphasia (PPA)

Preface

The desire of giving a voice to the International Society for Frontotemporal Dementias provided the inspiration for bringing together modern pioneers and their trainees to share their awareness and vision about the diseases of the nervous system that destroy language and most human qualities. As the heritage of the founders of Frontotemporal Dementia research is legendary, nothing matters more than what we can promise and do next.

We have made the effort of identifying what we call "emerging milestones of the twenty-first century," building a bridge between the work of the founders and that of today's pioneers. Each chapter of the volume not only illustrates the present state of the art, but also reveals the challenges ahead. The original clinical and neuropathologic observations made in individuals affected by frontotemporal dementia are now investigated with the most advanced methods and continuously evolving instrumentation of microscopy and in vivo imaging that allow us to interrogate the biology of frontotemporal dementia at the molecular and soon at the atomic level. Discoveries made during the last two decades of the twentieth century have provided the foundations for new molecular investigations in the twenty-first century; in fact, protein chemistry and genetics have contributed to the exploration of unmapped territories, revealing how the words "frontotemporal dementia" includes a growing multiplicity of disorders.

Thus, the emerging milestones are meant to remind us that many miles are ahead for the International Society for Frontotemporal Dementias, before we reach the end of a journey which is challenging for science and perilous for the patients and the families that we, the members of the International Society for Frontotemporal Dementias, wish to help.

Indianapolis, IN, USA — Bernardino Ghetti
Trieste, Italy — Emanuele Buratti
Rochester, MN, USA — Bradley Boeve
Antwerpen, Belgium — Rosa Rademakers

Contents

Behavioural Variant Frontotemporal Dementia: Recent Advances in the Diagnosis and Understanding of the Disorder

Rebekah M. Ahmed, John R. Hodges, and Olivier Piguet

Introduction

Over the past 30 years, the understanding of the clinical phenomenology, neuroimaging, genetics and pathology of frontotemporal dementia (FTD) has undergone a metamorphosis. This has, in turn, opened the door to potential treatment trials, which would have been thought to be out of reach not that long ago. Since the original descriptions of FTD, originally known as Pick's disease, our ability to accurately diagnose and differentiate patients presenting with predominantly behavioural changes (so-called behavioural variant FTD) and with forms of primary progressive aphasias has improved considerably. Recently,

R. M. Ahmed (✉)
Memory and Cognition Clinic, Department of Clinical Neurosciences, Royal Prince Alfred Hospital, Sydney, NSW, Australia

Central Sydney Medical School and Brain & Mind Centre, Faculty of Medicine and Health, The University of Sydney, Sydney, NSW, Australia
e-mail: rebekah.ahmed@sydney.edu.au

J. R. Hodges
Central Sydney Medical School and Brain & Mind Centre, Faculty of Medicine and Health, The University of Sydney, Sydney, NSW, Australia
e-mail: john.hodges@sydney.edu.au

O. Piguet
School of Psychology and Brain & Mind Centre, The University of Sydney, Sydney, NSW, Australia
e-mail: olivier.piguet@sydney.edu.au

the concept of frontotemporal lobar degeneration spectrum disorders has evolved to encompass the overlap between FTD and amyotrophic lateral sclerosis (ALS), as well as conditions such as progressive supranuclear palsy and corticobasal degeneration.

FTD primarily refers to a group of neurodegenerative brain disorders characterised by atrophy of the frontal and anterior temporal lobes. Prevalence studies suggest that FTD is the second most common cause of younger onset dementia [1, 2]. Three main clinical syndromes of FTD are generally recognised, based on their clinical presentations: a behavioural variant FTD (bvFTD) in which deterioration in social function and personality is most prominent and two language presentations, classified under primary progressive aphasia (PPA), in which an insidious decline in language skills is the primary feature. These PPAs are divided based on the pattern of language breakdown into semantic dementia (SD, also labelled semantic variant PPA) and progressive nonfluent aphasia (PNFA, also labelled nonfluent variant PPA) [3, 4]. Each of these syndromes has distinct clinical symptoms, imaging and pathological characteristics, although considerable heterogeneity and overlap exist in clinical practice, particularly as the disease progresses.

This chapter specifically focuses on bvFTD and on the recent advances in our understanding of the clinical features of this syndrome, its diag-

B. Ghetti et al. (eds.), *Frontotemporal Dementias*, Advances in Experimental Medicine and Biology 1281, https://doi.org/10.1007/978-3-030-51140-1_1

nosis, including its overlap with ALS. Other chapters of this special issue will cover the genetic, pathological and imaging advances.

A major advance in the field of bvFTD was the publication, in 2011, of international consensus diagnostic criteria with increasing levels of diagnostic certainty (Table 1). At the lowest level of diagnostic certainty is possible bvFTD: a pure clinical diagnosis requiring the presence of three of six behavioural changes, namely disinhibition, apathy, loss of empathy, perseverative/compulsive behaviours, hyperorality and a dysexecutive neuropsychological profile. A diagnosis of prob-

Table 1 Key diagnostic symptoms of bvFTD, forming part of diagnostic criteria [5]

A. Early *behavioural disinhibition [one of the following symptoms (A.1–A.3) must be present]:
A.1. Socially inappropriate behaviour
A.2. Loss of manners or decorum
A.3. Impulsive, rash or careless actions
B. Early apathy or inertia [one of the following symptoms (B.1–B.2) must be present]:
B.1. Apathy
B.2. Inertia
C. Early loss of sympathy or empathy [one of the following symptoms (C.1–C.2) must be present]:
C.1. Diminished response to other people's needs and feelings
C.2. Diminished social interest, interrelatedness or personal warmth
D. Early perseverative, stereotyped or compulsive/ritualistic behaviour [one of the following symptoms (D.1–D.3) must be present]:
D.1. Simple repetitive movements
D.2. Complex, compulsive or ritualistic behaviours
D.3. Stereotypy of speech
E. Hyperorality and dietary changes [one of the following symptoms (E.1–E.3) must be present]:
E.1. Altered food preferences
E.2. Binge eating, increased consumption of alcohol or cigarettes
E.3. Oral exploration or consumption of inedible objects
F. Neuropsychological profile: executive/generation deficits with relative sparing of memory and visuospatial functions [all of the following symptoms (F.1–F.3) must be present]:
F.1. Deficits in executive tasks
F.2. Relative sparing of episodic memory
F.3. Relative sparing of visuospatial skills

* Refers to symptom presentation within first 3 years

able bvFTD is based on the clinical syndrome, plus demonstrable functional decline and structural or functional changes in the frontotemporal regions on neuroimaging. A diagnosis of definite bvFTD is limited to those patients with the clinical syndrome and evidence of a pathogenic mutation or FTLD histopathology [5]. It has been shown that the probable level is robust and consistent when cases are followed over a number of years while only a half of those with possible bvFTD progress to clear-cut FTD over a 3-year follow-up period [6], and a proportion of such cases have the phenocopy syndrome discussed below.

While in the early 2000s research understandably focused on cognition, it has become apparent that tests of executive dysfunction have limited specificity in detecting bvFTD. More recent studies have examined other aspects which are not included in the current diagnostic guidelines [5] or have attempted to get at the core changes in social cognition and emotion processing. (Fig. 1). In addition, there has been the realisation that the effects of bvFTD are more widespread and affect physiological functioning.

Physiological Functioning

It is increasingly recognised that the changes in bvFTD are not simply restricted to behaviour, cognition and motor function, but that fundamental alterations in bodily functions including satiety and metabolism, as well as autonomic function occur. These changes have been linked to the disruption of large-scale neural networks linked to the hypothalamus with associated neuroendocrine changes [7].

Central to our understanding of physiological disturbances in bvFTD are changes in hypothalamic volume which have been shown in a number of neurodegenerative conditions including FTD and ALS [8], with abnormalities in eating and metabolism in bvFTD linked to potential connections between the hypothalamus and reward pathways [9]. Two studies have examined hypothalamic volumes in bvFTD. In the first, posterior hypothalamic atrophy was associated

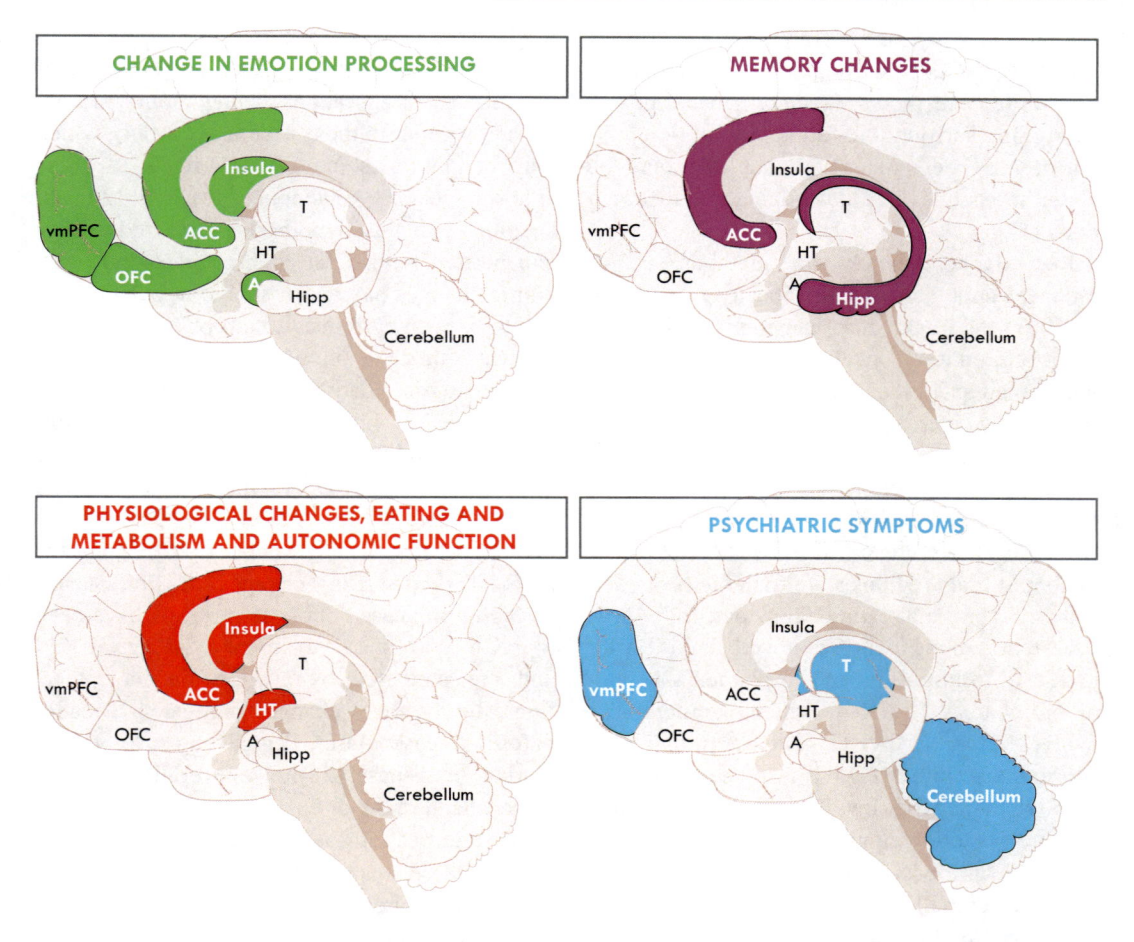

Fig. 1 Key neural structures implicated in each of the emerging symptom groups in bvFTD. (*vmPFC* ventral medial prefrontal cortex; *OFC* orbitofrontal cortex; *ACC* anterior cingulate cortex; *HT* hypothalamus; *T* thalamus; *Hipp* hippocampus; *A* amygdala)

with feeding abnormalities [10]. This relationship was observed within 2 years of disease onset, with continuing atrophy over the course of the disease. Importantly, atrophy was more pronounced in cases with transactive response DNA binding protein 43 kDa (TDP-43) inclusion pathology than in those with tau inclusions, pointing to a potential in vivo biomarker [10]. A second study reported a 17% reduction in hypothalamic volume on neuroimaging in bvFTD compared to controls, again particularly involving the posterior hypothalamus [11].

Eating and Metabolism in Behavioural Variant Frontotemporal Dementia

Hyperorality and dietary changes, which form one of the six core criteria for the diagnosis of bvFTD [5], are reported in over 60% of patients at initial presentation [12]. Such changes discriminate FTD from other dementias, notably Alzheimer's disease [13]. The changes in eating habits vary across the clinical subtypes of FTD. Alterations in bvFTD patients have been characterised by hyperphagia, indiscriminate eating, increased preference for sweet foods and other oral behaviours compared to patients with Alzheimer's disease [14]. In SD, changes are also present but take a different flavour. In this syndrome, patients show prominent changes in food preference including increased selectivity and rigidity surrounding food consumption [14–16]. It has been suggested that this may be related to changes in knowledge about different foods [17].

Recently, ecologically valid methods, such as test meal approach used in obesity research to measure food intake, have been applied in FTD. When offered a test meal of breakfast after fasting, Ahmed and colleagues (2016) demonstrated a markedly increased total caloric intake in bvFTD patients compared to both AD and control subjects and a preference for sugar. In addition, they also revealed rigid eating behaviour and a strong sugar preference in SD patients [18]. A number of brain regions were found to be associated with abnormal eating behaviour. In bvFTD, consistent regions identified have been a distributed set of frontoinsular and anteromedial temporal brain areas [19, 20], which parallel those involved early in bvFTD [21, 22]. Increased caloric intake in bvFTD has also been related to atrophy of a network involving the bilateral anterior and posterior cingulate gyri, the thalamus, bilateral lateral occipital cortex, lingual gyri and the right cerebellum. These structures are also implicated in the control of cognitive reward, autonomic, neuroendocrine and visual modulation of eating behaviour [18]. Changes in eating behaviour have also been linked to hypothalamic atrophy and changes in key neuroendocrine peptides (Fig. 2) including agouti-related peptide, neuropeptide Y (NPY) and leptin [8, 23]. How hypothalamic changes and changes in neuroendocrine peptides control eating behaviour in bvFTD and interact with cortical networks controlling eating behaviour requires further investigation.

Given the prominent changes in eating behaviour in bvFTD, it is not surprising that patients also exhibit changes in metabolism including changes in body mass index (BMI), insulin and cholesterol levels. Both bvFTD and SD patients have modestly increased BMI and waist circumference compared to normal controls [16], although the degree of change is less than one

might predict, given the level of eating abnormalities found in bvFTD, raising the question of whether other alterations in metabolic rate are present, similar to those seen in ALS [24], which may counteract some of the effects of these abnormal eating behaviours on BMI [20]. In keeping with this hypothesis, increased energy expenditure with a raised heart rate and autonomic changes have been shown in bvFTD [25] and have been correlated to atrophy in structures known to mediate autonomic function including the anterior cingulate cortex and insula.

Changes in insulin levels and lipid levels including insulin resistance have been identified in both bvFTD and SD with increased insulin and triglycerides and lower HDL cholesterol (reflecting a state of insulin resistance) [26]. Along the ALS-FTD spectrum, changes in lipid levels including increased cholesterol levels have been found to correlate with improved survival [27] and are mediated by changes in fat intake. Interestingly, these changes may occur decades before disease onset, suggesting a potential marker of disease [28]. The overall impact of these changes on disease progression and survival requires further exploration, including whether these changes are the result of atrophy in specific brain areas or actually modify the neurodegenerative process.

Autonomic Functions in Behavioural Variant Frontotemporal Dementia

In addition to changes in eating and metabolism, autonomic dysfunction has been identified in both bvFTD and SD [29]. Anecdotally, many carers report episodes of dizziness, as well as changes in thermoregulation in patients. Carer-based surveys have reported a high rate of symptoms related to blood pressure control, gastrointestinal function, thermoregulation, sweating and urinary symptoms [29, 30]. Objective measures of autonomic processing show abnormal responsiveness to emotion stimuli in FTD using physiological measures such as skin conductance [31, 32]. Changes in pain perception have been reported with bvFTD potentially associated with blunted pain and temperature responsiveness, while heightened

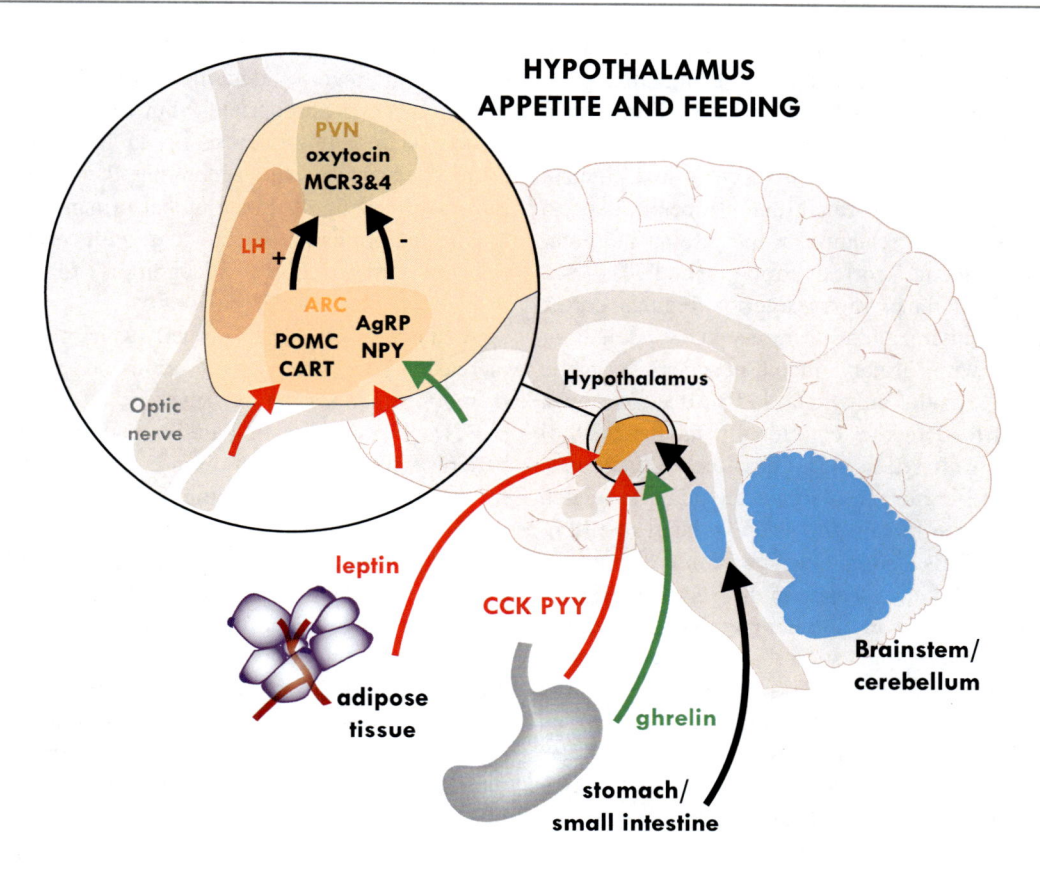

Fig. 2 Eating behaviour and the hypothalamus. Structures implicated in eating behaviour in FTD and pathways controlling eating behaviour in healthy individuals. Structures implicated in FTD include orbito-frontal cortex, right-sided reward structures including putamen, pallidum and striatum and posterior hypothalamus. Normal eating behaviour is controlled by an appetite stimulating pathway (shown in green) which results from ghrelin being released peripherally and targeting neurons of the arcuate nucleus (ARC) of the hypothalamus that contain neuropeptide Y (NPY) and agouti-related peptide (AgRP). An appetite suppressing pathway involves leptin (shown in red) being released from peripheral adipocytes, which then acts on pro-opiomelanocortin (POMC) and the cocaine and amphetamine-related transcript (CART) neu-rons in the hypothalamus. Peptide tyrosine tyrosine (PYY) and cholecystokinin (CCK), released peripherally, also suppress appetite. AgRp, NPY, POMC and CART neurons in the hypothalamus project to and act on mela-nocortin receptors (MCR). POMC is cleaved into alpha and beta-melanocyte-stimulating hormones that act on melanocortin receptor subtypes 3 and 4 (MCR 3 and 4) to decrease food intake. AgRP stimulates food intake by antagonism of MCR 3 and 4 receptors. In bvFTD, ele-vated levels of AgRP have been found. Increased leptin levels have also been found likely secondary to increased adipose stores. Autonomic pathways (black arrow) are also involved in food intake through projections via the brainstem and cerebellum to the hypothalamus. (*PVN* paraventricular nucleus)

responses are observed in SD and PNFA [33]. Recent studies using heart rate monitoring have shown increased heart rate and decreased heart rate variability in bvFTD [34]. Abnormalities in autonomic dilation of pupils in response to auditory stimuli are considered a physiological signature of neurodegeneration in FTD [35].

It is well established that autonomic changes may result from damage to cortical structures including the anterior and mid-cingulate cortices, prefrontal cortex, insula, ventral striatum, amygdala and hypothalamus [36, 37] regions known to undergo marked changes in FTD. Atrophy in the amygdala, ventral striatum, insula and anterior cingulate cortices has been reported in FTD [21,

36]. In bvFTD, pathological changes in these structures have traditionally being linked to disturbance of behaviour and social-emotional functioning [38–41]; however, their role in autonomic function has been recently investigated. Decreased cardiac vagal tone has been linked to left-lateralised structural frontoinsular and anterior cingulate cortex atrophy in FTD [34]. Atrophy in the premotor/anterior cingulate cortex and the putamen/claustrum/insula has been associated with urinary incontinence [42], while changes in the amygdala and insula have been linked to defective emotionally mediated autonomic dysfunction [43]. Pathology in the mesial temporal cortex, insula and amygdala is related to increased resting and sleeping heart rate [25]. The insula is also involved early in the course of bvFTD [21] and atrophy in this region correlates with altered pain and temperature perception [33], with the suggestion that the insula forms a network hub for sensory homeostatic signaling together with the thalamus [44]. Further research is required to examine how atrophy in these key regions regulates changes in autonomic function in FTD, how these changes are reflected in the different clinical phenotypes of FTD and how they could be harnessed as markers of disease progression.

Memory Function in Behavioural Variant Frontotemporal Dementia

Historically, memory functions have been reported to be preserved in bvFTD, with integrity of memory a key feature distinguishing Alzheimer's disease and bvFTD. Indeed, in clinical practice, it is not uncommon to read in clinical letters that a diagnosis of bvFTD is unlikely because the patient is exhibiting impaired memory function on cognitive testing. This position is further reflected in the consensus diagnostic criteria, where the cognitive profile in bvFTD (symptom F) is defined as one of executive/generation deficits, *with relative sparing of memory and visuospatial functions* [5]. Indeed, when present, memory deficit was thought to reflect a disturbance in retrieval efficiency, rather than a true episodic memory deficit, whereby information is encoded appropriately but recall performance is impaired because of an inability to retrieve efficiently and accurately the relevant information. Improvement in performance following the provision of cues (e.g. with recognition or forced-choice recognition formats) provides support for this position.

In the past decade, however, it has become increasingly apparent that various aspects of memory function can be severely affected in bvFTD, to a degree comparable to that seen in patients with Alzheimer's disease. Impaired performance is observed on common tasks of verbal and nonverbal episodic memory, such as short stories, word list learning or design recall, as well as on autobiographical memory and future thinking/prospective memory tasks [45–47], that correlates with the integrity of the hippocampus and other brain regions known to participate in memory functions [40, 48]. Deficits are also observed on tasks that rely on intact episodic and semantic memory systems, such as scene construction [49]. Further, evidence indicates that over time, episodic memory tends to worsen more rapidly in bvFTD than in AD [50]. Performance on topographical memory may, however, differentiate these two groups, where patients in AD tend to experience greater spatial orientation disturbance compared with patients with bvFTD [51].

Arguably, a differential diagnosis is not based solely on the presence/absence of a single clinical feature but is made within the context of multiple indices of clinical phenomenology, background and clinical history and ancillary investigations (e.g. brain MRI). Given the prominence of episodic memory deficit towards a clinical diagnosis of AD, it is important to emphasise that the presence of impaired memory, either on testing or clinical history, should not rule out a diagnosis of bvFTD.

Social Cognition in Behavioural Variant Frontotemporal Dementia

As its name indicates, disturbance in various aspects of social cognition is at the core of the prototypical presentation of bvFTD. In the current diagnostic criteria, these changes are covered by Symptom A (Early behavioural disinhibition) and Symptom C (Early loss of sympathy or empathy), both of which comprise additional subcategories. As is the case with the other symptoms, these symptom lack clear definitions, apart from the fact that they need to be persistent and recurrent, rather than one off or rare events [5]. Nearly 280 peer-reviewed articles have been published investigating social cognition in frontotemporal dementia to date. Of these, over 200 were published in the last decade, denoting the increasing interest in this topic. This should not come as a surprise for at least two reasons. First, social cognition forms a central block of interpersonal relationships. Humans are essentially social beings that have evolved because of their capacity to live in increasingly complex social environments. As such, disturbance in the capacity to engage or respond socially will have an impact not just for the affected individuals but for their broad social structure as well. In addition, evidence from epidemiological studies has shown that social interactions and social networks are protective risk factors against dementia in later life [52].

Second, the increasing availability of novel technologies in recent years, such as functional MRI, eye tracking or virtual reality, has opened the door to a variety of investigations, not possible until then, to understand the phenomenology of social cognition in healthy and clinical populations and their biological substrates (see for example [53] for a review). These investigations in healthy individuals and in clinical – stroke, tumour, neurodevelopmental and neurodegenerative – populations have identified a number of brain regions that play a central role in supporting social cognition. These regions are widespread and include frontal (anterior insula, anterior cingulate, orbitofrontal, medial frontal), temporal (temporal pole, superior temporal sulcus, amygdala) and parietal (temporo-parietal junction) brain regions [53–55].

Investigations in bvFTD have further confirmed the presence of pervasive changes in social cognition, which take many forms including emotion processing (recognition, expression), empathy, theory of mind, moral reasoning, reward sensitivity and understanding of social rules [32, 56–59]. These deficits can occur in isolation or in various combinations. Importantly, these findings suggest that single-test investigations of social cognition integrity are unlikely to be sufficient for ascertaining the presence of positive Symptoms A and C in the clinic.

While remarkable in its phenomenology and variability, the emergence of social cognition deficits in bvFTD is consistent with the pattern of brain atrophy observed in this syndrome. Indeed, the regions most susceptible to neuropathological changes and atrophy are the same that have consistently implicated in social cognition [60]. Importantly, these investigations have also identified that, in addition to brain atrophy, social cognition deficits also arise from global system disturbance, in particular in the autonomic system, leading to inaccurate integration of internal signals with external stimuli, resulting in inadequate or inappropriate responses [32, 61–63].

Importantly, the work of the past couple of decades has also demonstrated that disturbance in social cognition in FTD is not confined to its behavioural variant. Indeed, although beyond the scope of this chapter, it is important to note the emergence of such deficits in the language presentations of FTD, semantic dementia and progressive nonfluent aphasia. The characteristics of these deficits appear to differ from those in bvFTD, in their severity and quality. As such, and similar to what was discussed in the memory section, disturbance of social cognition capacity in the presence of a co-existing language disturbance should not necessarily rule out a diagnosis of language variant of FTD.

Overlap of Behavioural Variant Frontotemporal Dementia and Psychiatric Conditions

The current diagnostic criteria for bvFTD state that behavioural disturbances may not be better explained by a psychiatric condition [5]. The early clinical diagnosis of bvFTD, however, is often made difficult by the overlap with late onset psychiatric conditions. Patients often initially present with apathy and inertia and changes in empathy, which is mistakenly diagnosed as late onset depression. Not uncommonly, patients are placed on anti-psychotic medication, which can lead to changes in eating behaviour and weight gain, often blurring the presence of hyperorality changes. Once patients develop the florid behavioural changes including psychotic features, obsessive compulsive features, they are often misdiagnosed as schizophrenia, schizoaffective disorder or bipolar disorder [64].

Compounding the overlap between bvfTD and psychiatric conditions is the finding of high rates of psychiatric features in bvFTD patients with the chromosome 9 open reading frame 72 (C9orf72) gene expansion and the fact that such symptoms may be present for many years before the emergence of more characteristic FTD features. In a recent study of 56 bvFTD cases [65], a third showed psychotic features, with C9orf72 expansion cases more likely to exhibit psychotic symptoms than non-carriers (64% vs. 26%). Delusions, which comprise of persecutory, somatic, jealous and grandiose types, were more likely to occur in C9orf72 expansion carriers (57% vs. 19%), as were hallucinations (36% vs. 17%). Increased psychotic symptoms in C9orf72 expansion carriers correlated with atrophy in a distributed cortical and subcortical network that included discrete regions of the frontal, temporal and occipital cortices, as well as the thalamus, striatum and cerebellum. These structures are similar to structures involved in psychiatric conditions such as schizophrenia [65]. The situation is further confounded by the findings of a large study of 1414 family members of patients with bvFTD that found that relatives of patients with the C9orf72 gene expansion have an increased incidence of young onset schizophrenia and autism spectrum disorder [66]. Further research is needed to understand the overlap between bvFTD and psychiatric conditions and predisposition to psychiatric conditions as this may aid in earlier detection and treatment targeting.

Behavioural Variant Frontotemporal Dementia Phenocopy Syndrome

Along the bvFTD-psychiatric spectrum are patients that initially present with behavioural and neuropsychiatric features; yet, they do not show frontotemporal atrophy or hypometabolism on imaging and do not progress to develop cognitive decline or functional impairment [67]. It has been proposed that these patients may represent a late onset decompensation of life-long personality disorders or a neuropsychiatric condition, rather than true bvFTD [68]. Two patients with this disorder that went to autopsy showed no evidence of FTD pathology [69]. Caution, however, should be taken when classifying patients with the phenocopy syndrome in the absence of genetic testing for the C9orf72 gene expansion. Indeed, a recent meta-analysis on the phenocopy syndrome reported 7 cases of slowly progressive FTD that were associated with the C9orf72 gene expansion, out of a total of 292 reported phenocopy cases [67]. This finding is in keeping with a very long-term follow up of 16 cases from Cambridge, UK, all of whom were tested for the C9orf72 gene expansion found in 1 case only (6.25%). Reports showing the phenotypic variability in patients with the C9orf72 gene expansion are also increasing, with reports of patients within the same family having a rapid course in their 40s and death within 3 years and a much more indolent course in their 70s [70]. Further studies are required to ascertain the difference in penetrance and the underlying pathological mechanisms responsible for this. Studies of the effect of repeat size have produced discordant findings, and the contribution of repeat size to penetrance and phenotype remain uncertain and require further investigation [71–73].

Current Areas of Research Development

Amyotrophic Lateral Sclerosis: Frontotemporal Dementia Overlap

Since the mid-2000s, FTD and amyotrophic lateral sclerosis (ALS) have been increasingly conceptualised as representing the opposite ends of a disease spectrum [74, 75], with mounting evidence pointing towards an aetiological overlap between ALS and FTD and a multitude of studies showing behavioural and cognitive changes across the spectrum [27, 76–78]. This has largely been driven by genetics, with the discovery of the *C9orf72* expansion causing both bvFTD and ALS [79, 80]. In contrast to FTD, patients diagnosed with ALS typically exhibit limb or bulbar symptoms at initial presentation [81–83]. Much debate continues over the incidence of cognitive changes in ALS (behavioural, cognitive, language), with most large and community-based surveys reporting some cognitive changes in around 40–50% of cases [84, 85], while up to 15% of patients may satisfy the criteria for a diagnosis of concomitant FTD [86]. Conversely, 10–15% of FTD patients develop ALS, with varying estimates of motor neuron dysfunction in FTD insufficient to reach criteria for ALS, at between 25% and 30% [74, 87]. Further confirmation of the aetiological overlap between FTD and ALS is the finding of TDP-43 pathology in virtually all ALS cases and around half of those with bvFTD, although only 25% of bvFTD patients have similar motor neuron-like neuronal TDP-43 inclusion pathology [88, 89].

Recent research has suggested that bvFTD and ALS with TDP-43 inclusions may potentially result from the regional spreading ('prion like') of TDP-43 in the brain and spinal cord [90–92], with different initiating regions of pathology involved. In ALS, the pathology begins in the motor neocortex, progressing rapidly to the spinal cord and brainstem, prior to the involvement of nearby frontal and parietal regions, and then finally involving the temporal lobes [93]. Such a pattern of spread may potentially explain the late development of cognitive symptoms in ALS. In bvFTD, the disease process is thought to begin in the frontal lobe prior to spreading into the premotor, primary motor, parietal and temporal cortices, and eventually into the spinal cord [94].

In contrast to a suggested spectrum of disease, recent evidence indicates that the overlap between ALS and FTD is far more complex [95], with debate focusing on the cognitive and behavioural differences between ALS-FTD and bvFTD (i.e. are ALS-FTD and bvFTD part of the same disease). Previous studies have shown greater language involvement in ALS-FTD than in bvFTD [96, 97] including reduced sentence comprehension and grammatical difficulties with a language presentation of ALS-FTD with progressive non-fluent aphasia associated with anterior temporal and frontal language area atrophy, while that with prominent semantic problems is associated with temporal lobe and orbitofrontal cortex atrophy [98]. Currently, many studies are focusing on the longitudinal progression of behavioural and cognitive changes in ALS and ALS-FTD. These will help delineate the true nature of the progression and allow us to better clinically phenotype patients, which will aid in clinical trial development.

Predictors of Clinical Progression

One of the most common questions asked in clinical practice is 'how will bvFTD progress' and 'what is a patient's predicted survival'. Longitudinal large-scale follow-up studies of bvFTD patients are limited, but a number of cohort studies including genetic mutation carriers are currently underway around the world. Patients with combined ALS-FTD tend to show more rapid progression to death than those with either pure ALS or bvFTD [99]. It has also been shown that survival in those with both ALS and FTD may be dependent on initial phenotypic presentation, with those with initial motor symptoms having a shorter survival than those with initial cognitive or behavioural symptoms [100]. A recent study examined predictors of progression and survival in a cohort of 75 bvFTD patients. Median survival time from disease onset was

10.8 years and median survival prior to transition to nursing home was 8.9 years. Shorter survival was predicted by shorter disease duration at presentation, greater atrophy in the anterior cingulate cortex, older age and a higher burden of behavioural symptoms. In terms of disease progression, presence of a known pathogenic frontotemporal dementia genetic mutation was the strongest predictor of progression. Deficits in letter fluency and greater atrophy in the motor cortex were also associated with faster progression [101]. Research is now focusing on variables that can aid in early diagnosis including potential markers that develop prior to cognitive change to aid in early diagnosis. These aspects have particularly focused on imaging analyses including examining cerebral blood flow patterns [102], showing abnormalities up to 12 years prior to disease onset and grey matter atrophy patterns between those affected mutation carriers and asymptomatic mutation carriers, with different atrophy patterns visible presymptomatically, between *C9orf72, MAPT* and *GRN* genetic abnormality carriers, but also a common network of atrophy involving the insula, orbitofrontal lobe and anterior cingulate cortex [103]. As discussed above, these regions potentially mediate a number of physiological changes, potentially offering potential physiological markers that could be developed to facilitate earlier diagnosis and monitoring of disease progression.

Treatment and Intervention

Disease-modifying treatments do not currently exist for FTD and recent efforts have yielded disappointing results, for example the double-blind, placebo-controlled trial of LMTM (leuco-methylthioninium bis(hydromethanesulphonate)), a derivative of methylthioninium chloride, a drug targeting tau protein aggregation, in bvFTD [104]. A few clinical trials are, however, in the pipeline, but mostly targeting the familial forms of the disease or symptomatic management (see clinicaltrials.gov). Drugs used in Alzheimer's disease, such as acetylcholinesterase inhibitors, or NMDA receptor antagonists, provide no benefits to bvFTD patients and may even have a negative impact on cognition. Similarly, symptomatic treatments of challenging behaviours (e.g. disinhibition, agitation, aggression) with selective serotonin reuptake inhibitors (SSRIs) or antipsychotics have had mixed results.

A number of non-pharmacological approaches targeting behavioural difficulties, such as apathy or aggression have shown promise. For example, a subset of patients will develop repetitive behaviours over time (e.g. lining up objects, jigsaw puzzles), which can negatively impact on the patient's level of independence and interpersonal relationship. Interventions, such as the Tailored Activities Program (TAP), that directly target a specific behaviour and redirect it into personalised and relevant activities (selected by the carer) have demonstrated positive results, in reducing the disruption associated with the behaviour, increased meaningful activity engagement and reduction in carer stress [105]. Unlike in mild cognitive impairment and in Alzheimer's disease, targeted cognitive retraining has not been widely investigated in bvFTD and its suitability remains to be established. The prominent and early lack of insight, common in this population, complicates direct patient interventions [106], and carer-based interventions may therefore be more suitable.

Supporting families by providing education and coping skills is an avenue with demonstrated success in other clinical populations, such as traumatic brain injury [107]. A pilot study in FTD reported positive findings, but these will need to be replicated on a larger scale to determine their applicability in FTD [108].

Concluding Remarks

It is clear that much has been learnt about bvFTD. In this review, we focus on topics which have been of particular interest to FRONTIER, our frontotemporal dementia clinical research group based in Sydney, Australia. We have shown that the effects of bvFTD extend to fundamental aspects of physiology and metabolism, and that, contrary to clinical opinion, episodic memory is affected in bvFTD and reflects involvement of the hippocampus. Work on social cognition has emphasised the importance of breakdown in interpreting and expressing emotions, while the overlap between psychiatric disorders and bvFTD has been brought into focus by the finding of high rates of psychotic features in carriers of the *C9orf72* gene expansion and of psychiatric disorders in their family members. We have progressed knowledge on predictors of rapid versus slow decline in bvFTD, yet the holy grail for all researchers in the field – an effective therapy which can modify the clinical course of FTD – still remains beyond our grasp. We will certainly be ready when it comes, and there is some hope given the raft of new drugs under development, at least for use in those with known gene abnormalities who are still symptom free.

Acknowledgements Work reported in this article was supported in part by funding to Forefront, a collaborative research group dedicated to the study of frontotemporal dementia and motor neurone disease, from the National Health and Medical Research Council of Australia (NHMRC) program grant (#1037746 to OP and JRH) and the Australian Research Council Centre of Excellence in Cognition and its Disorders Memory Program (#CE110001021 to OP and JRH) and other grants/sources (NHMRC project grant #1003139 to OP), and Royal Australasian College of Physicians, MND Research Institute of Australia. RA is a NHMRC Early Career Fellow (GNT1120770). OP is an NHMRC Senior Research Fellow (GNT1103258). We are grateful to the research participants involved with our research studies over the years.

References

1. Ratnavalli E, Brayne C, Dawson K, Hodges JR (2002) The prevalence of frontotemporal dementia. Neurology 58(11):1615–1621
2. Rosso SM, Donker Kaat L, Baks T, Joosse M, de Koning I, Pijnenburg Y et al (2003) Frontotemporal dementia in The Netherlands: patient characteristics and prevalence estimates from a population-based study. Brain J Neurol 126(Pt 9):2016–2022
3. Hodges JR, Patterson K (2007) Semantic dementia: a unique clinicopathological syndrome. Lancet Neurol 6(11):1004–1014
4. Gorno-Tempini ML, Hillis AE, Weintraub S, Kertesz A, Mendez M, Cappa SF et al (2011) Classification of primary progressive aphasia and its variants. Neurology 76(11):1006–1014
5. Rascovsky K, Hodges JR, Knopman D, Mendez MF, Kramer JH, Neuhaus J et al (2011) Sensitivity of revised diagnostic criteria for the behavioural variant of frototemporal dementia. Brain J Neurol 134(Pt 9):2456–2477
6. Devenney E, Bartley L, Hoon C, O'Callaghan C, Kumfor F, Hornberger M et al (2015) Progression in behavioral variant frontotemporal dementia: a longitudinal study. JAMA Neurol 72(12):1501–1509
7. Ahmed RM, Ke YD, Vucic S, Ittner LM, Seeley W, Hodges JR et al (2018) Physiological changes in neurodegeneration – mechanistic insights and clinical utility. Nat Rev Neurol 14(5):259–271
8. Ahmed RM, Latheef S, Bartley L, Irish M, Halliday GM, Kiernan MC et al (2015) Eating behavior in frontotemporal dementia: peripheral hormones vs hypothalamic pathology. Neurology 85(15):1310–1317
9. Perry DC, Sturm VE, Seeley WW, Miller BL, Kramer JH, Rosen HJ (2014) Anatomical correlates of reward-seeking behaviours in behavioural variant frontotemporal dementia. Brain J Neurol 137(Pt 6):1621–1626
10. Piguet O, Petersen A, Yin Ka Lam B, Gabery S, Murphy K, Hodges JR et al (2011) Eating and hypothalamus changes in behavioral-variant frontotemporal dementia. Ann Neurol 69(2):312–319
11. Bocchetta M, Gordon E, Manning E, Barnes J, Cash DM, Espak M et al (2015) Detailed volumetric analysis of the hypothalamus in behavioral variant frontotemporal dementia. J Neurol 262(12):2635–2642
12. Piguet O, Hornberger M, Shelley BP, Kipps CM, Hodges JR (2009) Sensitivity of current criteria for the diagnosis of behavioral variant frontotemporal dementia. Neurology 72(8):732–737
13. Mendez MF, Licht EA, Shapira JS (2008) Changes in dietary or eating behavior in frontotemporal dementia versus Alzheimer's disease. Am J Alzheimers Dis Other Dement 23(3):280–285
14. Ikeda M, Brown J, Holland AJ, Fukuhara R, Hodges JR (2002) Changes in appetite, food preference, and eating habits in frontotemporal dementia and Alzheimer's disease. J Neurol Neurosurg Psychiatry 73(4):371–376
15. Snowden JS, Bathgate D, Varma A, Blackshaw A, Gibbons ZC, Neary D (2001) Distinct behavioural profiles in frontotemporal dementia and semantic dementia. J Neurol Neurosurg Psychiatry 70(3):323–332
16. Ahmed RM, Irish M, Kam J, van Keizerswaard J, Bartley L, Samaras K et al (2014) Quantifying the eating abnormalities in frontotemporal dementia. JAMA Neurol 71(12):1540–1546
17. Vignando M, Rumiati RI, Manganotti P, Cattaruzza T, Aiello M (2019) Establishing links between abnormal eating behaviours and semantic deficits in dementia. J Neuropsychol. Online ahead of print
18. Ahmed RM, Irish M, Henning E, Dermody N, Bartley L, Kiernan MC et al (2016) Assessment of eating behavior disturbance and associated neural

networks in frontotemporal dementia. JAMA Neurol 73(3):282–290

19. Whitwell JL, Sampson EL, Loy CT, Warren JE, Rossor MN, Fox NC et al (2007) VBM signatures of abnormal eating behaviours in frontotemporal lobar degeneration. NeuroImage 35(1):207–213

20. Woolley JD, Gorno-Tempini ML, Seeley WW, Rankin K, Lee SS, Matthews BR et al (2007) Binge eating is associated with right orbitofrontal-insular-striatal atrophy in frontotemporal dementia. Neurology 69(14):1424–1433

21. Seeley WW, Crawford R, Rascovsky K, Kramer JH, Weiner M, Miller BL et al (2008) Frontal paralimbic network atrophy in very mild behavioral variant frontotemporal dementia. Arch Neurol 65(2):249–255

22. Irish M, Piguet O, Hodges JR (2011) Self-projection and the default network in frontotemporal dementia. Nat Rev Neurol 8(3):152–161

23. Ahmed RM, Phan K, Highton-Williamson E, Strikwerda-Brown C, Caga J, Ramsey E et al (2019) Eating peptides: biomarkers of neurodegeneration in amyotrophic lateral sclerosis and frontotemporal dementia. Ann Clin Transl Neurol 6(3):486–495

24. Ahmed RM, Dupuis L, Kiernan MC (2018) Paradox of amyotrophic lateral sclerosis and energy metabolism. J Neurol Neurosurg Psychiatry 89(10):1013–1014

25. Ahmed RM, Landin-Romero R, Collet TH, van der Klaauw AA, Devenney E, Henning E et al (2017) Energy expenditure in frontotemporal dementia: a behavioural and imaging study. Brain J Neurol 140(Pt 1):171–183

26. Ahmed RM, MacMillan M, Bartley L, Halliday GM, Kiernan MC, Hodges JR et al (2014) Systemic metabolism in frontotemporal dementia. Neurology 83(20):1812–1818

27. Ahmed RM, Highton-Williamson E, Caga J, Thornton N, Ramsey E, Zoing M et al (2018) Lipid metabolism and survival across the frontotemporal dementia-amyotrophic lateral sclerosis spectrum: relationships to eating behavior and cognition. J Alzheimers Dis 61(2):773–783

28. Mariosa D, Hammar N, Malmstrom H, Ingre C, Jungner I, Ye W et al (2017) Blood biomarkers of carbohydrate, lipid, and apolipoprotein metabolisms and risk of amyotrophic lateral sclerosis: a more than 20-year follow-up of the Swedish AMORIS cohort. Ann Neurol 81(5):718–728

29. Ahmed RM, Iodice V, Daveson N, Kiernan MC, Piguet O, Hodges JR (2015) Autonomic dysregulation in frontotemporal dementia. J Neurol Neurosurg Psychiatry 86(9):1048–1049

30. Diehl-Schmid J, Schulte-Overberg J, Hartmann J, Forstl H, Kurz A, Haussermann P (2007) Extrapyramidal signs, primitive reflexes and incontinence in fronto-temporal dementia. Eur J Neur 14(8):860–864

31. Hoefer M, Allison SC, Schauer GF, Neuhaus JM, Hall J, Dang JN et al (2008) Fear conditioning in

32. Kumfor F, Hazelton JL, Rushby JA, Hodges JR, Piguet O (2019) Facial expressiveness and physiological arousal in frontotemporal dementia: phenotypic clinical profiles and neural correlates. Cogn Affect Behav Neurosci 19(1):197–210

33. Fletcher PD, Downey LE, Golden HL, Clark CN, Slattery CF, Paterson RW et al (2015) Pain and temperature processing in dementia: a clinical and neuroanatomical analysis. Brain J Neurol 138(Pt 11):3360–3372

34. Guo CC, Sturm VE, Zhou J, Gennatas ED, Trujillo AJ, Hua AY et al (2016) Dominant hemisphere lateralization of cortical parasympathetic control as revealed by frontotemporal dementia. Proc Natl Acad Sci U S A 113(17):E2430–E2439

35. Fletcher PD, Nicholas JM, Downey LE, Golden HL, Clark CN, Pires C et al (2016) A physiological signature of sound meaning in dementia. Cortex 77:13–23

36. Jones SE (2011) Imaging for autonomic dysfunction. Cleve Clin J Med 78(Suppl 1):S69–S74

37. Critchley HD, Nagai Y, Gray MA, Mathias CJ (2011) Dissecting axes of autonomic control in humans: insights from neuroimaging. Auton Neurosci 161(1–2):34–42

38. Kumfor F, Irish M, Hodges JR, Piguet O (2013) Discrete neural correlates for the recognition of negative emotions: insights from frontotemporal dementia. PLoS One 8(6):e67457

39. Galton CJ, Gomez-Anson B, Antoun N, Scheltens P, Patterson K, Graves M et al (2001) Temporal lobe rating scale: application to Alzheimer's disease and frontotemporal dementia. J Neurol Neurosurg Psychiatry 70(2):165–173

40. Irish M, Piguet O, Hodges JR, Hornberger M (2014) Common and unique gray matter correlates of episodic memory dysfunction in frontotemporal dementia and Alzheimer's disease. Hum Brain Mapp 35(4):1422–1435

41. Kipps CM, Nestor PJ, Acosta-Cabronero J, Arnold R, Hodges JR (2009) Understanding social dysfunction in the behavioural variant of frontotemporal dementia: the role of emotion and sarcasm processing. Brain J Neurol 132(Pt 3):592–603

42. Perneczky R, Diehl-Schmid J, Forstl H, Drzezga A, May F, Kurz A (2008) Urinary incontinence and its functional anatomy in frontotemporal lobar degenerations. Eur J Nucl Med Mol Imaging 35(3):605–610

43. Rosen HJ, Levenson RW (2009) The emotional brain: combining insights from patients and basic science. Neurocase 15(3):173–181

44. Craig AD (2009) How do you feel – now? The anterior insula and human awareness. Nat Rev Neurosci 10(1):59–70

45. Hornberger M, Piguet O, Graham AJ, Nestor PJ, Hodges JR (2010) How preserved is episodic memory in behavioral variant frontotemporal dementia? Neurology 74(6):472–479

46. Irish M, Hornberger M, Lah S, Miller L, Pengas G, Nestor PJ et al (2011) Profiles of recent autobiographical memory retrieval in semantic dementia, behavioural-variant frontotemporal dementia, and Alzheimer's disease. Neuropsychologia 49(9):2694–2702

47. Dermody N, Hornberger M, Piguet O, Hodges JR, Irish M (2016) Prospective memory impairments in Alzheimer's disease and behavioral variant frontotemporal dementia: clinical and neural correlates. J Alzheimers Dis 50(2):425–441

48. Hornberger M, Wong S, Tan R, Irish M, Piguet O, Kril J et al (2012) In vivo and post-mortem memory circuit integrity in frontotemporal dementia and Alzheimer's disease. Brain J Neurol 135(Pt 10):3015–3025

49. Wilson NA, Ramanan S, Roquet D, Goldberg ZL, Hodges JR, Piguet O et al (2020) Scene construction impairments in frontotemporal dementia: evidence for a primary hippocampal contribution. Neuropsychologia 137:107327

50. Schubert S, Leyton CE, Hodges JR, Piguet O (2016) Longitudinal memory profiles in behavioral-variant frontotemporal dementia and Alzheimer's disease. J Alzheimers Dis 51(3):775–782

51. Tu S, Wong S, Hodges JR, Irish M, Piguet O, Hornberger M (2015) Lost in spatial translation – a novel tool to objectively assess spatial disorientation in Alzheimer's disease and frontotemporal dementia. Cortex 67:83–94

52. Livingston G, Frankish H (2015) A global perspective on dementia care: a lancet commission. Lancet 386(9997):933–934

53. Kennedy DP, Adolphs R (2012) The social brain in psychiatric and neurological disorders. Trends Cogn Sci 16(11):559–572

54. Frith CD, Frith U (1999) Interacting minds – a biological basis. Science 286(5445):1692–1695

55. Saxe R, Kanwisher N (2003) People thinking about thinking people. The role of the temporo-parietal junction in "theory of mind". NeuroImage 19(4):1835–1842

56. Sturm VE, Ascher EA, Miller BL, Levenson RW (2008) Diminished self-conscious emotional responding in frontotemporal lobar degeneration patients. Emotion 8(6):861–869

57. Sturm VE, Perry DC, Wood K, Hua AY, Alcantar O, Datta S et al (2017) Prosocial deficits in behavioral variant frontotemporal dementia relate to reward network atrophy. Brain Behav 7(10):e00807

58. Shdo SM, Ranasinghe KG, Gola KA, Mielke CJ, Sukhanov PV, Miller BL et al (2018) Deconstructing empathy: neuroanatomical dissociations between affect sharing and prosocial motivation using a patient lesion model. Neuropsychologia 116(Pt A):126–135

59. Synn A, Mothakunnel A, Kumfor F, Chen Y, Piguet O, Hodges JR et al (2018) Mental states in moving shapes: distinct cortical and subcortical contributions to theory of mind impairments in dementia. J Alzheimers Dis 61(2):521–535

60. Kumfor FHJ, De Winter FL, Cleret de Langavant L, van den Stock J (2017) Clinical studies of social neuroscience: a lesion model approach. In: Ibanez AGA (ed) Neuroscience and social science: the missing link. Springer, Cham, pp 255–296

61. Garcia-Cordero I, Sedeno L, Babino A, Dottori M, Melloni M, Martorell Caro M et al (2019) Explicit and implicit monitoring in neurodegeneration and stroke. Sci Rep 9(1):14032

62. Ibanez A, Manes F (2012) Contextual social cognition and the behavioral variant of frontotemporal dementia. Neurology 78(17):1354–1362

63. Eckart JA, Sturm VE, Miller BL, Levenson RW (2012) Diminished disgust reactivity in behavioral variant frontotemporal dementia. Neuropsychologia 50(5):786–790

64. Lanata SC, Miller BL (2016) The behavioural variant frontotemporal dementia (bvFTD) syndrome in psychiatry. J Neurol Neurosurg Psychiatry 87(5):501–511

65. Devenney EM, Landin-Romero R, Irish M, Hornberger M, Mioshi E, Halliday GM et al (2017) The neural correlates and clinical characteristics of psychosis in the frontotemporal dementia continuum and the C9orf72 expansion. NeuroImage Clin 13:439–445

66. Devenney EM, Ahmed RM, Halliday G, Piguet O, Kiernan MC, Hodges JR (2018) Psychiatric disorders in C9orf72 kindreds: study of 1,414 family members. Neurology 91(16):e1498–ee507

67. Valente ES, Caramelli P, Gambogi LB, Mariano LI, Guimaraes HC, Teixeira AL et al (2019) Phenocopy syndrome of behavioral variant frontotemporal dementia: a systematic review. Alzheimers Res Ther 11(1):30

68. Hornberger M, Shelley BP, Kipps CM, Piguet O, Hodges JR (2009) Can progressive and nonprogressive behavioural variant frontotemporal dementia be distinguished at presentation? J Neurol Neurosurg Psychiatry 80(6):591–593

69. Devenney E, Forrest SL, Xuereb J, Kril JJ, Hodges JR (2016) The bvFTD phenocopy syndrome: a clinicopathological report. J Neurol Neurosurg Psychiatry 87(10):1155–1156

70. Foxe D, Elan E, Burrell JR, Leslie FVC, Devenney E, Kwok JB et al (2018) Intrafamilial phenotypic variability in the C9orf72 gene expansion: 2 case studies. Front Psychol 9:1615

71. van Blitterswijk M, DeJesus-Hernandez M, Niemantsverdriet E, Murray ME, Heckman MG, Diehl NN et al (2013) Association between repeat sizes and clinical and pathological characteristics in carriers of C9ORF72 repeat expansions (Xpansize-72): a cross-sectional cohort study. Lancet Neurol 12(10):978–988

72. Gijselinck I, Van Mossevelde S, van der Zee J, Sieben A, Engelborghs S, De Bleecker J et al (2016) The C9orf72 repeat size correlates with onset age of disease, DNA methylation and transcriptional downregulation of the promoter. Mol Psychiatry 21(8):1112–1124

73. Nordin A, Akimoto C, Wuolikainen A, Alstermark H, Jonsson P, Birve A et al (2015) Extensive size variability of the GGGGCC expansion in C9orf72 in both neuronal and non-neuronal tissues in 18 patients with ALS or FTD. Hum Mol Genet 24(11):3133–3142

74. Lomen-Hoerth C, Anderson T, Miller B (2002) The overlap of amyotrophic lateral sclerosis and frontotemporal dementia. Neurology 59(7):1077–1079

75. Clark CM, Forman MS (2006) Frontotemporal lobar degeneration with motor neuron disease: a clinical and pathological spectrum. Arch Neurol 63(4):489–490

76. Strong MJ, Abrahams S, Goldstein LH, Woolley S, McLaughlin P, Snowden J et al (2017) Amyotrophic lateral sclerosis – frontotemporal spectrum disorder (ALS-FTSD): revised diagnostic criteria. Amyotroph Lateral Scler Frontotemporal Degener 18(3–4):153–174

77. Lillo P, Garcin B, Hornberger M, Bak TH, Hodges JR (2010) Neurobehavioral features in frontotemporal dementia with amyotrophic lateral sclerosis. Arch Neurol 67(7):826–830

78. Lillo P, Mioshi E, Burrell JR, Kiernan MC, Hodges JR, Hornberger M (2012) Grey and white matter changes across the amyotrophic lateral sclerosis-frontotemporal dementia continuum. PLoS One 7(8):e43993

79. Hodges J (2012) Familial frontotemporal dementia and amyotrophic lateral sclerosis associated with the C9ORF72 hexanucleotide repeat. Brain J Neurol 135(Pt 3):652–655

80. Mitsuyama Y, Inoue T (2009) Clinical entity of frontotemporal dementia with motor neuron disease. Neuropathology 29(6):649–654

81. Turner MR, Hardiman O, Benatar M, Brooks BR, Chio A, de Carvalho M et al (2013) Controversies and priorities in amyotrophic lateral sclerosis. Lancet Neurol 12(3):310–322

82. Kiernan MC, Vucic S, Cheah BC, Turner MR, Eisen A, Hardiman O et al (2011) Amyotrophic lateral sclerosis. Lancet 377(9769):942–955

83. Vucic S, Rothstein JD, Kiernan MC (2014) Advances in treating amyotrophic lateral sclerosis: insights from pathophysiological studies. Trends Neurosci 37(8):433–442

84. Strong MJ (2008) The syndromes of frontotemporal dysfunction in amyotrophic lateral sclerosis. Amyotroph Lateral Scler 9(6):323–338

85. Montuschi A, Iazzolino B, Calvo A, Moglia C, Lopiano L, Restagno G et al (2015) Cognitive correlates in amyotrophic lateral sclerosis: a population-based study in Italy. J Neurol Neurosurg Psychiatry 86(2):168–173

86. Ringholz GM, Appel SH, Bradshaw M, Cooke NA, Mosnik DM, Schulz PE (2005) Prevalence and patterns of cognitive impairment in sporadic ALS. Neurology 65(4):586–590

87. Burrell JR, Kiernan MC, Vucic S, Hodges JR (2011) Motor neuron dysfunction in frontotemporal dementia. Brain J Neurol 134(Pt 9):2582–2594

88. Rohrer JD, Lashley T, Schott JM, Warren JE, Mead S, Isaacs AM et al (2011) Clinical and neuroanatomical signatures of tissue pathology in frontotemporal lobar degeneration. Brain J Neurol 134(Pt 9):2565–2581

89. Josephs KA, Hodges JR, Snowden JS, Mackenzie IR, Neumann M, Mann DM et al (2011) Neuropathological background of phenotypical variability in frontotemporal dementia. Acta Neuropathol 122(2):137–153

90. Ludolph AC, Brettschneider J (2015) TDP-43 in amyotrophic lateral sclerosis – is it a prion disease? Eur J Neurol 22(5):753–761

91. Braak H, Brettschneider J, Ludolph AC, Lee VM, Trojanowski JQ, Del Tredici K (2013) Amyotrophic lateral sclerosis – a model of corticofugal axonal spread. Nat Rev Neurol 9(12):708–714

92. Tan RH, Kril JJ, Fatima M, McGeachie A, McCann H, Shepherd C et al (2015) TDP-43 proteinopathies: pathological identification of brain regions differentiating clinical phenotypes. Brain J Neurol 138(Pt 10):3110–3122

93. Brettschneider J, Del Tredici K, Toledo JB, Robinson JL, Irwin DJ, Grossman M et al (2013) Stages of pTDP-43 pathology in amyotrophic lateral sclerosis. Ann Neurol 74(1):20–38

94. Brettschneider J, Del Tredici K, Irwin DJ, Grossman M, Robinson JL, Toledo JB et al (2014) Sequential distribution of pTDP-43 pathology in behavioral variant frontotemporal dementia (bvFTD). Acta Neuropathol 127(3):423–439

95. Lule DE, Aho-Ozhan HEA, Vazquez C, Weiland U, Weishaupt JH, Otto M et al (2019) Story of the ALS-FTD continuum retold: rather two distinct entities. J Neurol Neurosurg Psychiatry 90(5):586–589

96. Saxon JA, Thompson JC, Jones M, Harris JM, Richardson AM, Langheinrich T et al (2017) Examining the language and behavioural profile in FTD and ALS-FTD. J Neurol Neurosurg Psychiatry 88(8):675–680

97. Long Z, Irish M, Piguet O, Kiernan MC, Hodges JR, Burrell JR (2019) Clinical and neuroimaging investigations of language disturbance in frontotemporal dementia-motor neuron disease patients. J Neurol 266(4):921–933

98. Vinceti G, Olney N, Mandelli ML, Spina S, Hubbard HI, Santos-Santos MA et al (2019) Primary progressive aphasia and the FTD-MND spectrum disorders: clinical, pathological, and neuroimaging correlates. Amyotroph Lateral Scler Frontotemporal Degener 20:1–13

99. Hodges JR, Davies R, Xuereb J, Kril J, Halliday G (2003) Survival in frontotemporal dementia. Neurology 61(3):349–354

100. Hu WT, Seelaar H, Josephs KA, Knopman DS, Boeve BF, Sorenson EJ et al (2009) Survival profiles of patients with frontotemporal dementia and motor neuron disease. Arch Neurol 66(11):1359–1364

101. Agarwal S, Ahmed RM, D'Mello M, Foxe D, Kaizik C, Kiernan MC et al (2019) Predictors of survival

and progression in behavioural variant frontotemporal dementia. Eur J Neurol 26(5):774–779

102. Mutsaerts H, Mirza SS, Petr J, Thomas DL, Cash DM, Bocchetta M et al (2019) Cerebral perfusion changes in presymptomatic genetic frontotemporal dementia: a GENFI study. Brain J Neurol 142(4): 1108–1120

103. Cash DM, Bocchetta M, Thomas DL, Dick KM, van Swieten JC, Borroni B et al (2018) Patterns of gray matter atrophy in genetic frontotemporal dementia: results from the GENFI study. Neurobiol Aging 62:191–196

104. Gauthier S, Feldman HH, Schneider LS, Wilcock GK, Frisoni GB, Hardlund JH et al (2016) Efficacy and safety of tau-aggregation inhibitor therapy in patients with mild or moderate Alzheimer's disease: a randomised, controlled, double-blind, parallel-arm, phase 3 trial. Lancet 388(10062):2873–2884

105. O'Connor CM, Clemson L, Brodaty H, Gitlin LN, Piguet O, Mioshi E (2016) Enhancing caregivers' understanding of dementia and tailoring activities in frontotemporal dementia: two case studies. Disabil Rehabil 38(7):704–714

106. Scherling CS, Zakrzewski J, Datta S, Levenson RW, Shimamura AP, Sturm VE et al (2017) Mistakes, too few to mention? Impaired self-conscious emotional processing of errors in the behavioral variant of frontotemporal dementia. Front Behav Neurosci 11:189

107. Fisher A, Bellon M, Lawn S, Lennon S, Sohlberg M (2019) Family-directed approach to brain injury (FAB) model: a preliminary framework to guide family-directed intervention for individuals with brain injury. Disabil Rehabil 41(7):854–860

108. O'Connor CMC, Mioshi E, Kaizik C, Fisher A, Hornberger M, Piguet O (2020) Positive behaviour support in frontotemporal dementia: a pilot study. Neuropsychol Rehabil 1–24. Online ahead of print

The Neuropsychiatric Features of Behavioral Variant Frontotemporal Dementia

Bradley T. Peet, Sheila Castro-Suarez, and Bruce L. Miller

Introduction

Frontotemporal dementia (FTD) is a clinically and neuropathologically heterogenous neurodegenerative disorder characterized by disturbances in behavior, personality, and language associated with degeneration of frontal and temporal brain regions [1]. FTD consists of three clinical variants distinguished by the predominant presenting symptoms: behavioral variant FTD (bvFTD), semantic variant primary progressive aphasia (svPPA), and nonfluent/agrammatic variant primary progressive aphasia (nfvPPA) [2, 3]. Additionally, there are several related disorders which share features with FTD, including FTD with motor neuron disease (FTD-MND), corticobasal syndrome (CBS), and progressive supranuclear palsy (PSP).

B. T. Peet (✉) · B. L. Miller
Memory and Aging Center, Department of Neurology, Weill Institute for Neurosciences, University of California, San Francisco School of Medicine, San Francisco, CA, USA
e-mail: Bradley.Peet@ucsf.edu

S. Castro-Suarez
Memory and Aging Center, Department of Neurology, Weill Institute for Neurosciences, University of California, San Francisco School of Medicine, San Francisco, CA, USA

Atlantic Fellow for Equity in Brain Health at UCSF and clinical researcher Instituto Nacional de Ciencias Neurológicas (INCN), Lima, Peru

Behavioral variant FTD is the most common form of FTD and comprises over 50% of all FTD cases [4]. The syndrome has an early age of onset with a mean of 58 years of age. The time between disease onset and the initial evaluation is approximately 3 years and the duration of the illness is approximately 8 years [2]. Behavioral variant FTD is characterized by a set of core diagnostic criteria proposed by Rascovsky et al. in 2011, which include behavioral disinhibition; apathy or inertia; loss of sympathy or empathy; perseverative, stereotyped, or compulsive/ritualistic behavior; hyperorality; and executive dysfunction (Table 1) [2]. All but one of these criteria (executive dysfunction) are behavioral in nature and have overlapping features with many psychiatric syndromes, including schizophrenia, obsessive-compulsive disorder (OCD), bipolar disorder (BPD) and major depressive disorder (MDD). Not surprisingly, many patients with bvFTD are mistakenly diagnosed with a primary psychiatric condition [5].

The neuroanatomical correlates of bvFTD typically show a pattern of atrophy in the frontal and anterior temporal lobes, with the right hemisphere being predominately affected. The earliest structures involved include the anterior cingulate cortex (ACC), anterior insula (AI), and orbitofrontal cortex (OFC). As the disease progresses, atrophy is found in the dorsolateral prefrontal cortex (dlPFC), frontal poles, dorsal insula (DI), striatum, thalamus, and anterior hippocampus. In later stages of the disease, atrophy becomes more

B. Ghetti et al. (eds.), *Frontotemporal Dementias*, Advances in Experimental Medicine and Biology 1281, https://doi.org/10.1007/978-3-030-51140-1_2

Table 1 International consensus criteria for behavioral variant frontotemporal dementia

I. Neurodegenerative disease

The following symptom must be present to meet criteria for bvFTD:

A. Shows progressive deterioration of behavior and/or cognition by observation or history (as provided by a knowledgeable informant)

II. Possible bvFTD

Three of the following behavioral/cognitive symptoms (A–F) must be present to meet criteria. Ascertainment requires that symptoms be persistent or recurrent rather than single or rare events

A. Early* behavioral disinhibition (one of the following symptoms [A.1–A.3] must be present):

A.1. Socially inappropriate behavior

A.2. Loss of manners or decorum

A.3. Impulsive, rash, or careless actions

B. Early apathy or inertia (one of the following symptoms [B.1–B.2] must be present):

B.1. Apathy

B.2. Inertia

C. Early loss of sympathy or empathy (one of the following symptoms [C.1–C.2] must be present):

C.1. Diminished response to other people's needs and feelings

C.2. Diminished social interest, interrelatedness, or personal warmth

D. Early perseverative, stereotyped, or compulsive/ritualistic behavior (one of the following symptoms [D.1–D.3] must be present):

D.1. Simple repetitive movements

D.2. Complex, compulsive, or ritualistic behaviors

D.3. Stereotypy of speech

E. Hyperorality and dietary changes (one of the following symptoms [E.1–E.3] must be present):

E.1. Altered food preferences

E.2. Binge eating, increased consumption of alcohol or cigarettes

E.3. Oral exploration or consumption of inedible objects

F. Neuropsychological profile: executive/generation deficits with relative sparing of memory and visuospatial functions (all of the following symptoms [F.1–F.3] must be present):

F.1. Deficits in executive tasks

F.2. Relative sparing of episodic memory

F.3. Relative sparing of visuospatial skills

III. Probable bvFTD

All of the following symptoms (A–C) must be present to meet criteria.

A. Meets criteria for possible bvFTD

B. Exhibits significant functional decline (by caregiver report or as evidenced by clinical dementia rating scale or functional activities questionnaire scores)

C. Imaging results consistent with bvFTD (one of the following [C.1–C.2] must be present):

C.1. Frontal and/or anterior temporal atrophy on MRI or CT

C.2. Frontal and/or anterior temporal hypoperfusion or hypometabolism on PET or SPECT

IV. Behavioral variant FTD with definite FTLD pathology

Criterion A and either criterion B or C must be present to meet criteria:

A. Meets criteria for possible or probable bvFTD

B. Histopathological evidence of FTLD on biopsy or at postmortem

C. Presence of a known pathogenic mutation

V. Exclusionary criteria for bvFTD

Criteria A and B must be answered negatively for any bvFTD diagnosis. Criterion C can be positive for possible bvFTD but must be negative for probable bvFTD

A. Pattern of deficits is better accounted for by other non-degenerative nervous system or medical disorders

B. Behavioral disturbance is better accounted for by a psychiatric diagnosis

C. Biomarkers strongly indicative of Alzheimer's disease or other neurodegenerative processes

*As a general guideline 'early' refers to symptom presentation within the first 3 years. Adapted from Rascovsky et al. [2]. Used with permission

widespread and involves the posterior hippocampi, posterior insula (PI), and parietal lobes, regions that are prominently involved in Alzheimer's disease (AD) [6]. For that reason, in the late stages of bvFTD, there is considerable overlap with AD.

Neuropsychiatric Features of bvFTD

The bvFTD syndrome has substantial overlap with the symptomology of multiple primary psychiatric conditions, which presents a significant diagnostic challenge for clinicians. The difficulty in diagnostic accuracy is delineated in a study by Woolley et al. [5], who performed a systematic, retrospective, blinded chart review of 252 patients at the University of California, San Francisco Memory and Aging Center in order to identify the rate of psychiatric diagnoses which precede that of a neurodegenerative disorder. Of this population, 71 patients (28.2%) received a psychiatric diagnosis prior to ultimately being diagnosed with a neurodegenerative disorder, and approximately 50% with bvFTD were first diagnosed with a primary psychiatric disorder [5]. This study highlights the importance of understanding the key features of bvFTD in order to make an accurate diagnosis.

In this section, we will outline the neuropsychiatric features of bvFTD followed by a brief review of the overlap between bvFTD and primary psychiatric conditions, namely, MDD and BPD.

Disinhibition

Disinhibition is an early symptom of bvFTD and is present in 76% of cases at the time of the initial evaluation [2]. Disinhibition is often the most salient feature of bvFTD as patients frequently exhibit impulsivity, socially inappropriate behavior, and loss of social decorum attributable to aberrant reward processing and a lack of regard for potential consequences of inappropriate actions [2, 7, 8].

The most common manifestation of behavioral disinhibition is impulsivity [9], such as new-onset gambling or substance use, excessive spending, reckless behavior, or oversharing of

personal information [2, 9, 10]. Violation of social norms often includes overfamiliarity, inappropriate touching, and inappropriate sexual acts [2, 11, 12]. Additionally, criminal behaviors occur in approximately 50% of cases [13]. A general lack of etiquette may be demonstrated by inappropriate laughing at a serious event, touching others, unbridled profanity, or making offensive jokes [2, 12].

Studies exploring the neuroanatomical correlates of disinhibition have largely implicated dysfunction of the right subgenual cingulate cortex (SGC) and the posteromedial aspect of the right OFC [14–16]. In addition to SGC and OFC involvement, Franceschi et al. also found that the bilateral inferior temporal cortex, hippocampus, amygdala, and nucleus accumbens were hypometabolic on fluorodeoxyglucose positron emission tomography (FDG-PET) in those with bvFTD who predominately exhibited disinhibition [15].

Sturm et al. found that atrophy of the right pregenual ACC in subjects with bvFTD was associated with a lower degree of self-conscious emotional reactivity [17], which suggests that this brain region may play a role in disinhibited behavior. In another study, Perry et al. found that the inability of participants to subjectively differentiate between pleasant and unpleasant odors was correlated with atrophy of the right ventral mid-insula and right amygdala, suggesting that the lack of aversion to negative stimuli may be a component of reward-seeking behavior seen in those with disinhibited behavior [8].

Pharmacological treatments targeting disinhibited behavior are limited, though some studies demonstrate that selective serotonin reuptake inhibitors (SSRI) may be effective. In a small open-label study by Swartz et al., 11 subjects with FTD were treated with fluoxetine, sertraline, or paroxetine for three months and were found to have a reduction in the degree of disinhibition [18]. In another open-label study by Herrmann et al., 15 subjects with FTD were treated with citalopram 30 mg daily for six weeks and were found to have significant decreases on the neuropsychiatric inventory (NPI) questionnaire total score and on the disinhibition subscore of the NPI [19].

Apathy

Apathy is the most common presenting symptom of bvFTD, occurring in 84% of cases at the time of the initial evaluation [2]. It is characterized by the reduction of goal-directed behavior, diminished emotional reactivity, and a decrease in social engagement [20]. Apathy initially manifests as a reduction of spontaneous activity and a general sense of indifference [20–22]. In later stages of the disease process, apathy may result in substantial functional impairment with limited ability to perform instrumental and basic activities of daily living (ADL) [23]. The functional impairment resulting from apathy is often very difficult for those caring for the patient and can lead to substantial emotional distress [24].

A diagnostic framework for the diagnosis of apathy was proposed by Marin in 1991 [21]. Marin described apathy as a distinct neuropsychiatric syndrome defined by a lack of motivation, a decrease in goal-directed behavior, and a loss of interest [21]. Despite Marin's proposed diagnostic criteria describing the apathy syndrome, the concept of apathy has varied throughout the literature leading to confusion due to lack of a formal definition. Levy and Dubois defined apathy as a "quantitative reduction of voluntary, goal-directed behaviors" and defined three subtypes, emotional-affective, cognitive, and autoactivation. This definition was important for delineating three different behavioral syndromes associated with three different anatomical correlates [25]. A more recent set of diagnostic criteria describing the apathy syndrome has been proposed by a consensus panel of 23 experts (Table 2) in order to formally define the concept of apathy [20, 22].

The neuroanatomical correlates of apathy in those with bvFTD have been largely associated with dysfunction of the frontal lobes. Studies utilizing voxel-based morphometry (VBM) demonstrate predominately right-sided atrophy of the medial prefrontal cortex (mPFC) [14, 16, 26] and the ACC [14, 16, 26, 27]. Other areas of involvement include the dlPFC [16, 26], OFC [14, 16, 27], insula [26, 28], and the caudate [28].

Pharmacological treatments targeting apathy are limited. Psychostimulants may result in

Table 2 Apathy diagnostic criteria

Criterion A: A quantitative reduction of goal-directed activity either in behavioral, cognitive, emotional, or social dimensions in comparison to the patient's previous level of functioning in these areas. These changes may be reported by the patient himself/herself or by observation of others
Criterion B: The presence of at least two of the three following dimensions for a period of at least four weeks and present most of the time:
B1. Behavior and cognition Loss of, or diminished, goal-directed behavior or cognitive activity as evidenced by at least one of the following:
General level of activity: The patient has a reduced level of activity either at home or work, makes less effort to initiate or accomplish tasks spontaneously, or needs to be prompted to perform them **Persistence of activity:** He/she is less persistent in maintaining an activity or conversation, finding solutions to problems, or thinking of alternative ways to accomplish them if they become difficult **Making choices:** He/she has less interest or takes longer to make choices when different alternatives exist (e.g., selecting TV programs, preparing meals, choosing from a menu, etc.) **Interest in external issue:** He/she has less interest in or reacts less to news, either good or bad, or has less interest in doing new things **Personal well-being:** He/she is less interested in his/her own health and well-being or personal image (general appearance, grooming, clothes, etc.)
B2. Emotion Loss of, or diminished, emotion as evidenced by at least one of the following:
Spontaneous emotions: The patient shows less spontaneous (self-generated) emotions regarding their own affairs or appears less interested in events that should matter to him/her or to people that he/she knows well **Emotional reactions to environment:** He/she expresses less emotional reaction in response to positive or negative events in his/her environment that affect him/her or people he/she knows well (e.g., when things go well or bad, responding to jokes or events on a TV program or a movie, or when disturbed or prompted to do things he/she would prefer not to do) **Impact on others:** He/she is less concerned about the impact of his/her actions or feelings on the people around him/her **Empathy:** He/she shows less empathy to the emotions or feelings of others (e.g., becoming happy or sad when someone is happy or sad or being moved when others need help) **Verbal or physical expressions:** He/she shows less verbal or physical reactions that reveal his/her emotional states

(continued)

Table 2 (continued)

B3. Social interaction
Loss of, or diminished, engagement in social interaction as evidenced by at least one of the following:
Spontaneous social initiative: The patient takes less initiative in spontaneously proposing social or leisure activities to family or others **Environmentally stimulated social interaction:** He/she participates less or is less comfortable or more indifferent to social or leisure activities suggested by people around him/her **Relationship with family members:** He/she shows less interest in family members (e.g., to know what is happening to them, to meet them or make arrangements to contact them) **Verbal interaction:** He/she is less likely to initiate a conversation, or he/she withdraws soon from it **Homebound:** He/she prefers to stays at home more frequently or longer than usual and shows less interest in getting out to meet people
Criterion C: These symptoms (A–B) cause clinically significant impairment in personal, social, occupational, or other important areas of functioning
Criterion D: The symptoms (A–B) are not exclusively explained or due to physical disabilities (e.g., blindness and loss of hearing), to motor disabilities, to a diminished level of consciousness, to the direct physiological effects of a substance (e.g., drug of abuse, medication), or to major changes in the patient's environment

Adapted from Robert et al., 2018 [20]. Used with permission

some improvement, though careful patient selection is essential due to the potential for significant side effects, including insomnia, hypertension, irritability, and psychosis. Methylphenidate has shown benefit in multiple small studies [29]. In several randomized controlled trials (RCT), methylphenidate was associated with significant improvement in apathy in those with AD [30, 31, 32]. One small RCT of eight patients diagnosed with bvFTD found that treatment with dextroamphetamine resulted in an improvement in neuropsychiatric inventory (NPI) subscales of apathy by 2.8 points [33]. Despite some improvement in symptoms, these small studies do not justify treating FTD with stimulants in most cases given the potential for significant adverse events.

Loss of Empathy or Sympathy

Loss of empathy is a presenting symptom in 73% of bvFTD cases [2]. Symptoms manifest as emotional detachment and a decrease in social interest, as well as a lack of concern for the feelings of others [2, 34]. This indifference can profoundly impact relationships early in the disease course and often results in substantial caregiver burden by disruption the emotional connection between the caregiver, often the patient's spouse, and the patient [35, 36]. In extreme cases, lack of empathy may manifest as sociopathic behavior [37]. When coupled with disinhibition and impulsivity commonly demonstrated by those with bvFTD, loss of empathy may lead to criminal behavior, ranging from petty theft to homicide [38].

Empathy is a complex construct in which an observer is able to identify with the feelings, thoughts, or emotions of another individual, leading to a change in the observer's affective state [39]. The ability to empathize is a fundamental aspect of social interaction and involves affective perspective taking and affect sharing [39]. The affective perspective taking and affect sharing of empathy are broken down into cognitive components and affective components, each of which has multiple subcomponents that are beyond the scope of this chapter [40, 41]. The cognitive components of empathy involve the observer understanding what others may be thinking or feeling, while the affective components involve sharing and responding to the emotional experience of others [40]. This paradigm allows for a better understanding of the empathy deficits demonstrated in bvFTD.

The underlying neuroanatomy implicated in the loss of empathy is largely related to the widespread dysfunction of structures that are associated with both cognitive and affective empathy. A lesional study by Shamay-Tsoory et al. demonstrated that cognitive empathy and affective empathy were associated with distinct neuroanatomical substrates. Deficits in cognitive empathy were found to be associated with lesions of the right ventromedial PFC (vmPFC), while deficits in affective empathy were found to be associated with lesions of the left inferior frontal gyrus (IFG)

[42]. A by Dermody et al. in subjects with bvFTD correlated well with the earlier study by Shamay-Tsoory et al. Dermody et al. found that the diminished cognitive empathy in bvFTD was associated with predominately right-sided atrophy of the mPFC, OFC, insular cortices, and lateral temporal lobes. Diminished affective empathy was associated with predominately left-sided atrophy of the OFC, (IFG), insula, thalamus, putamen, and the bilateral mid-cingulate gyrus [43].

Pharmacological treatments targeting loss of empathy in bvFTD are limited, though studies evaluating the therapeutic effects of oxytocin have shown positive results. Hurlemann et al. assessed the effect of intranasal oxytocin in a randomized, double-blind, placebo-controlled trial of 48 healthy male volunteers and demonstrated that intranasal oxytocin enhanced emotional empathy but not cognitive empathy [44]. Subsequent randomized, double-blind, placebo-controlled trials regarding the effects of oxytocin were performed by Jesso et al. (n = 20) and Finger et al. (n = 46) in subjects with bvFTD. Both studies demonstrated significant improvement in measures of empathy [45, 46].

Perseverative, Stereotyped, or Compulsive/Ritualistic Behaviors

Repetitive behaviors occur in bvFTD at the time of initial evaluation in 71% of cases [2] and may be related to deficits in suppressing urges to perform an action [47]. These behaviors can present as simple repetitive movements or vocalizations, such as eye blinking, throat clearing, or tapping, among others. More complex behaviors are also frequently observed, including collecting and hoarding behavior, repetitive storytelling, and frequent unnecessary trips to the bathroom [2, 48, 49, 50].

Some of the repetitive behaviors seen in bvFTD have features related to deficits in impulse control. Compulsions, as typically seen in OCD, are characterized in the Diagnostic and Statistical Manual of Mental Disorders (DSM-5) as purposeful repetitive motor acts that are associated with obsessive thoughts and performed to reduce anxiety or distress [51]. In those with OCD, per-forming the behavior results in relief from distress, though the action is not intrinsically pleasurable [52, 53]. In contrast to that seen in OCD, impulsivity typically seen in bvFTD is characterized by the inability to resist urges due to deficits in response inhibition and delayed gratification, as well as a lack of consideration for potential consequences. This leads to an act performed due to the need to immediately satisfy a desire rather than to alleviate anxiety or distress [54, 55]. In a study by Moheb et al., it was demonstrated that typical repetitive behaviors seen in bvFTD were complex and included stereotypic speech, hoarding, and frequent unnecessary trips to the restroom. In contrast, symptoms typical of OCD, such as checking, counting, and ordering, were infrequent. Furthermore, repetitive behaviors in bvFTD were not associated with anxiety, and were able to be stopped on command without causing distress [50]. In contrast to the study by Moheb et al., other studies have found that symptoms typical of OCD are prominent in bvFTD [47, 56, 57, 58].

The neuroanatomical correlates of repetitive behaviors in bvFTD are largely related to the dysfunction of structures in the left temporal lobe and striatum. A study by Rosso et al. performed computed tomography (CT) and/or MRI on 87 subjects and found that left temporal lobe atrophy was associated with complex compulsive behaviors, which included preoccupation with ideas or activities, strict adherence to a fixed schedule, frugality, arranging items in a particular order, and cleaning rituals [59]. Similarities were found in a functional imaging study using single-photon emission computed tomography (SPECT), which correlated hypoperfusion of the left temporal lobe with compulsive behavior, and hypoperfusion of the right frontal lobe with stereotypical behavior [60]. The role of the striatum in stereotypical behavior associated with bvFTD was further described in a study utilizing VBM by Josephs et al., who demonstrated disproportionate atrophy of the putamen and caudate head bilaterally [61]. Additionally, atrophy of the dorsal ACC and right supplementary motor area was found to be associated with repetitive behavior in a study by Rosen et al. [14]

The pharmacological treatments specifically targeting repetitive behaviors in those with bvFTD are limited, though some studies have demonstrated modest improvement in symptoms. In a small open-label study published by Swartz et al., previously described in this chapter, a reduction in compulsive behaviors was noted after treatment with an SSRI [18]. A case series of three patients reported improvement in compulsive behavior associated with bvFTD with the use of clomipramine [62]. A case report indicated that topiramate was effective in helping to reduce alcohol use in an individual with bvFTD [63]. Lastly, a small study by Mendez et al. found that those with FTD who were treated with donepezil were more likely to experience worsening of disinhibition and compulsive behavior, which returned to baseline after donepezil was discontinued [64].

Hyperorality and Dietary Changes

Hyperorality and dietary changes have long been recognized as a feature of neurodegenerative disease [65] and have been considered a core feature of bvFTD since the diagnostic criteria proposed in 1998 by Neary et al. [66]. Significant changes in eating behavior are present in 59% of those with bvFTD at the initial evaluation [2] and increase with the progression of disease [67]. A range of disordered eating behavior has been described, including increased appetite, excessive eating regardless of satiety, alterations in food preferences often with a preference for carbohydrates, and, in severe cases, oral exploration, chewing, or ingestion of inedible objects [2, 68, 69, 70]. The potential degree of insatiability that may be seen in those with bvFTD was clearly demonstrated in a study by Woolley et al., where participants were given sandwiches for lunch and allowed to eat as many as they wished for up to one hour. Sandwiches were continuously brought to the participants, regardless of their requests, in order to maintain a constant volume of sandwiches in front of the participant. Those with bvFTD were much more likely to eat more sandwiches than controls. In some cases, participants with bvFTD requested that sandwiches stop being brought to them, despite continuing to eat the sandwiches [70].

The underlying neuroanatomical correlates of eating behavior are likely multifactorial, though multiple studies have implicated orbitofrontal-insular-striatal networks as a mediator of eating behavior and satiety [70, 71, 72]. A study by Woolley et al. utilizing VBM demonstrated that atrophy of the right ventral insular cortex, striatum, and anterior OFC was associated with increased food consumption [70]. A similar study by Whitwell et al. demonstrated that increased food consumption was associated with atrophy of the anterolateral OFC bilaterally [71]. Whitwell et al. correlated the increase in carbohydrate craving associated with bvFTD with the right AI and the posterolateral OFC bilaterally [71]. There is also evidence that changes in the hypothalamus may lead to disturbance in eating behavior. In a study by Piguet et al., significant atrophy of the hypothalamus was present on structural MRI as well as on postmortem analyses in patients with bvFTD who demonstrated a significant disturbance in eating behavior [72].

Treatment modalities targeting hyperorality and dietary changes are limited, though there have been a small number of medications that have resulted in improvement. Serotonergic medications are the most widely studied in bvFTD. In a small open-label study performed by Swartz et al., previously described in this chapter, treatment with an SSRI resulted in a decrease in carbohydrate craving [18]. Additional studies found that fluvoxamine and trazodone were also effective at improving eating behaviors [73–75]. Given the limited number of available pharmacological treatment options, close caregiver supervision is often necessary in order to mitigate the potential for overeating, weight gain, and possible attempts to ingest inedible objects [11].

Overlapping Characteristics of bvFTD and Psychiatric Disorders

Neurodegenerative disorders, particularly bvFTD, and primary psychiatric disorders have many features in common, which presents a significant diagnostic challenge for clinicans. Recognizing these overlapping features is an essential first step in uncovering the etiology of the patient's presenting symptoms. A thorough evaluation, including an exhaustive history, neuroimaging, and neuropsychological testing, can aid in narrowing the differential diagnoses and may result in findings that would have otherwise be missed, such as severe executive dysfunction noted on neuropsychological testing or frontotemporal atrophy demonstrated on MRI of the brain. In this section, we will briefly discuss some of the similarities between bvFTD, MDD, and BPD.

Overlapping Symptomatology

Early symptoms of bvFTD can mimic those found in late-life primary mood disorders, and are often experienced by those with MDD and BPD [5].

Major depressive disorder is a common psychiatric disorder characterized by intermittent episodes of depressed mood with several clinical features with bvFTD, which can potentially congtribute to diagnostic uncertainty, particularly in complicated or atypical cases. The diagnostic criteria for MDD are defined in DSM-5 by nine characteristic features: (1) depressed mood, (2) diminished interest or pleasure in activities, (3) a significant change in weight or appetite, (4) insomnia or hypersomnia, (5) psychomotor agitation or retardation, (6) fatigue or loss of energy, (7) feelings of worthlessness or guilt, (8) diminished concentration, and (9) recurrent thoughts of death or suicide [51]. Anhedonia is often present in those with MDD, though it can appear very similar to that of apathy associated with bvFTD as both often present clinically as decreased motivation. In this case, other symptoms would likely aid in clarifying the underlying diagnosis, however, differentiating bvFTD and MDD can become increasingly difficult in those with atypical or severe cases of depression. In those with atypical depression, appetite is often signficantly increased and may appear similar to the increase in appetite commonly associated with bvFTD. Severe cases of depression may present as emotional disengagement and social withdrawal which can be mistaken for lack of empathy associated with bvFTD. [5, 76–79]. Despite the overlapping symptomatology between bvFTD and MDD, clinicians generally consider MDD much more prevalent than bvFTD [80, 81]. Clinician's familiarity with similarities and differences between bvFTD and MDD, as well as a thorough evaluation, can improve diagnostic accuracy and help to avoid a delay in care.

Bipolar disorder is a psychiatric disorder characterized by intermittent episodes of mania and depression. The diagnostic criteria for BPD are defined in DSM-5 as a distinct period of abnormally and persistently elevated or irritable mood in addition to seven characteristic features: (1) inflated self-esteem or grandiosity, (2) decreased need for sleep, (3) more talkative than usual or pressure to keep talking, (4) flight of ideas or racing thoughts, (5) distractibility, (6) increased goal-directed activity or psychomotor agitation, and (7) excessive involvement in activities that have a high potential for negative consequences. As in MDD, the symptoms of BPD have a large overlap with those of bvFTD. One of the most salient features in both BPD and bvFTD is excessive impulsivity, however, in BPD it is predominantly associated with manic episodes, though this is not the case in bvFTD. Psychomotor agitation (i.e., engaging in purposeless movements) is another common feature of BPD that may be mistaken for repetitive movements which may be seen in bvFTD [5, 76–79].

Neuroimaging Correlates Between bvFTD and Psychiatric Disorders

Neuroimaging plays an important role in helping to distinguish bvFTD from a primary psychiatric disorder, but it has also been instrumental in helping to establish the underlying neuroanatomical correlates of psychiatric symptoms. In those with MDD and BPD, evidence suggests that structural changes in the brain impact regions involved in bvFTD [6, 82]. Two large meta-analyses examined gray matter abnormalities in MDD and BPD via VBM [82, 83]. Redlich et al.

found that in subjects with MDD, there was a reduction in gray matter volume of the vmPFC, dorsomedial PFC (dmPFC), hippocampus, caudate, and precuneus. In subjects with BPD, there was a reduction in the dlPFC, insula, bilateral hippocampi, amygdala, caudate, thalamus, and putamen [84]. Lu et al. found that in subjects with MDD, there was a reduction of gray matter volume in the vmPFC, ACC, anterior superior temporal gyrus, left caudate, and left hippocampus when compared to healthy controls. In subjects with BPD, there was a reduction in the bilateral insula, superior temporal gyrus, mPFC, ACC, bilateral medial frontal gyrus, and the right medial and inferior temporal gyrus [83].

Regions of volume loss described in these meta-analyses have overlapped with regions involved with bvFTD, described elsewhere in this chapter. The overlapping brain regions of bvFTD, MDD, and BPD may explain the commonality between specific features of these conditions.

Conclusions

Behavioral variant FTD often presents predominately as a neuropsychiatric syndrome early in the disease course. The symptomology of bvFTD and underlying neuroanatomical correlates have overlap with those of primary psychiatric conditions. The commonalities between bvFTD in the early stages, and primary psychiatric conditions, may lead to a misdiagnosis of bvFTD. Clinicians should be aware of this pitfall and be diligent in the evaluation of patients who present with complaints that appear to be psychiatric in nature, especially those who present in late life.

Cases 1

Ms. BH is a right-handed woman with 16 years of education who was employed by the United States government. At baseline, BH was very involved with family and friends and enjoyed volunteering for local community organizations. In her mid-50s, BH began consuming an excessive amount of alcohol at a far greater extent than she had previously. Around this time, she began performing simple repetitive movements which manifested as constantly rubbing her hands together. More complex compulsive behaviors were also noted as she began collecting discarded aluminum cans and other items from the roadside. BH's hygiene progressive worsened and she began showering only once every two weeks.

One year after symptom onset, she was noted by her husband to be increasingly withdrawn socially and not meeting her obligations in the community organizations which she was involved. She became less interested in socializing with family and friends and was less engaged with her husband. On one notable occasion, after the death of a close friend, she did not reach out to her friend's family or attend the funeral. BH stopped completing household chores and began spending most of her day on the sofa watching television. Additionally, she began exhibiting some degree of disinhibition as she began engaging strangers in conversation and disclosing personal details.

Neurological exam revealed difficulty with the Luria sequence, mild postural tremor, and global, symmetric hyperreflexia without clonus.

Neuropsychological testing revealed impairment in verbal and visual memory, visuospatial ability, confrontation naming, working memory, set switching, and response inhibition.

Laboratory studies and electroencephalogram were unremarkable.

FDG-PET was notable for decreased glucose metabolism in the right anterior temporal lobe. Florbetapir [^{18}F] PET was negative for amyloid aggregates. Structural MRI of the brain demonstrated disproportionate volume loss in the right frontotemporal region (Figs. 1 and 2).

Given BH's symptoms of disinhibition, apathy, loss of empathy, repetitive/compulsive behavior, hyperorality, and executive dysfunction in addition to atrophy of the frontal and anterior temporal lobes, BH was ultimately diagnosed with bvFTD.

Case 2

Mr. KC is a right-handed man with 16 years of education who was employed as a photographer.

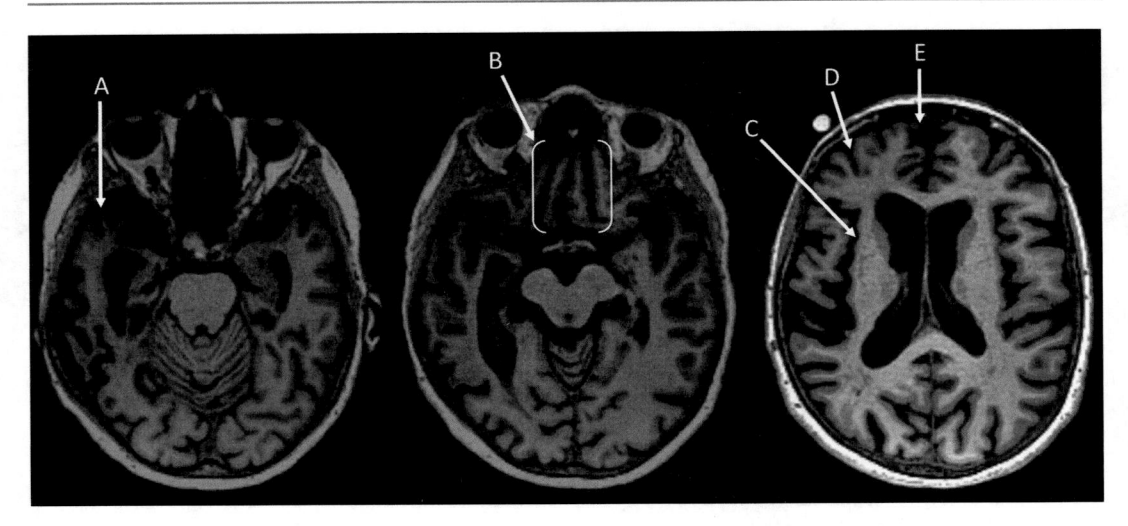

Fig. 1 Axial view in radiological orientation showing asymmetric atrophy of the right anterior temporal lobe (**a**) and the insular cortex (**c**). Other areas of degeneration include the bilateral orbitofrontal cortex with prominence of the orbital fissures bilaterally and the longitudinal cerebral fissure (**b**), the lateral prefrontal cortex (**d**), and the medial prefrontal cortex (**e**)

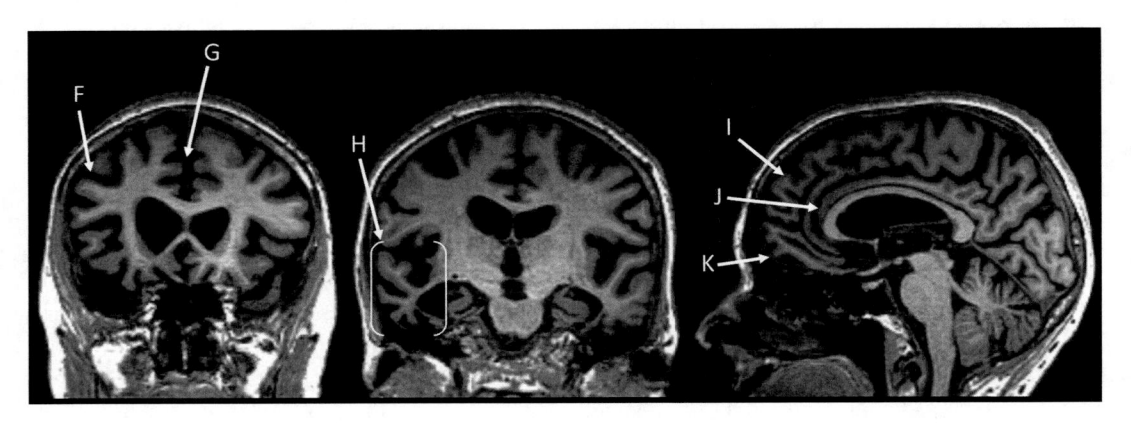

Fig. 2 Coronal view (left and middle) and sagittal view (right) in radiological orientation showing significant areas of atrophy in the lateral prefrontal cortex (**f**), medial prefrontal cortex and anterior cingulate gyrus (**g**), right anterior temporal lobe (**h**), medial prefrontal cortex (**i**), subgenual cingulate gyrus (**j**), and orbitofrontal cortex (**k**)

In his early-50s, KC began experiencing difficulty performing tasks related to his job, and was noted to be repeatedly purchasing incorrect items for his camera. Additionally, the quality of KC's photography began to decline and he became increasingly rigid, often arguing with clients and insisting that his ideas were better.

Approximately 1 year after KC began to experience difficulty at work, he began to experience word-finding difficulty. His speech became increasingly generalized and non-specific, often referring to objects as a "thing" rather than the specific name of the object. Over the following year, KC experienced substantial progressive of word-finding difficulty. He began taking pictures of objects that he was unable to name and used the pictures to communicate. For example, he took pictures of multiple types of fruit and would send the pictures to his wife via text message in order to let her know that they needed more fruit from the store. Further progression of symptoms led to KC being unable to recall the meaning of words or conceptual knowledge making it difficult to use objects. For example, he was unable to identify what broccoli was or what to do with it. He also had been found performing activities inappropriately on mulitple occasions, such as washing his hands in the toilet rathger than the sink. In addition to language deficits, KC also exhibited memory impairment, visuospatial impairment, and behavioral changes, which included apathy and disinhibition. Notably, his cognition was impaired to the degree that he was unable to work and his wife had to help with most chores.

On neurological examination, KC did not engage in conversation unless directly questioned. He demonstrated echolalia and laughed inappropriately. He was perseverative and stimulus-bound, demonstrating utilization behaviors throughout the evaluation. His thought process was tangential, and he rarely answered questions directly. He could not describe his mood, however, his affect was borderline euphoric. He did not demonstrate appropriate concern or emotion given the nature of some of the topics of discussion. His speech was fluent and grammatically intact but generalized and vague. He was unable to name any items on a task of confrontational naming but was able to correctly describe the function of some of the words. He made semantic paraphasic errors. He was unable to read short, simple sentences. He was unable to write words or draw an animal when asked.

Neuropsychological testing revealed profound impairment in language, most notably impaired confrontation naming, impaired single-word comprehensionm impaired object knowledge. He was also noted to have impaired verbal and visual memory and executive functioning. Structural MRI demonstrated profound asymmetric left temporal atrophy (Fig. 3).

KC was ultimately diagnosed with semantic variant primary progressive aphasia (svPPA) based on the presence of impaired confrontation naming, impaired single word comprehension, impaired object knowledge, alexia, and agraphia. Additionally, MRI demonstrated left anterior temporal lobe atrophy consistent with svPPA.

Case 3

Ms. KA is a right-handed woman with 14 years of education who worked as an accountant. In her early-50s, KA began experiencing slowed speech and was noted to be omitting prepositions and conjunctions when communicating with her children via text message, though there were no spelling errors. The following year, KA began to have difficulty with grammar and using appropriate sentence structure predominantly when writing, resulting in shorter, more simple sentences. KA also began experiencing word-finding difficulty which impacted her ability to communicate.

Fig. 3 Axial (left) and sagittal (right) views in radiological orientation showing profound asymmetric left temporal lobe atrophy (**a**)

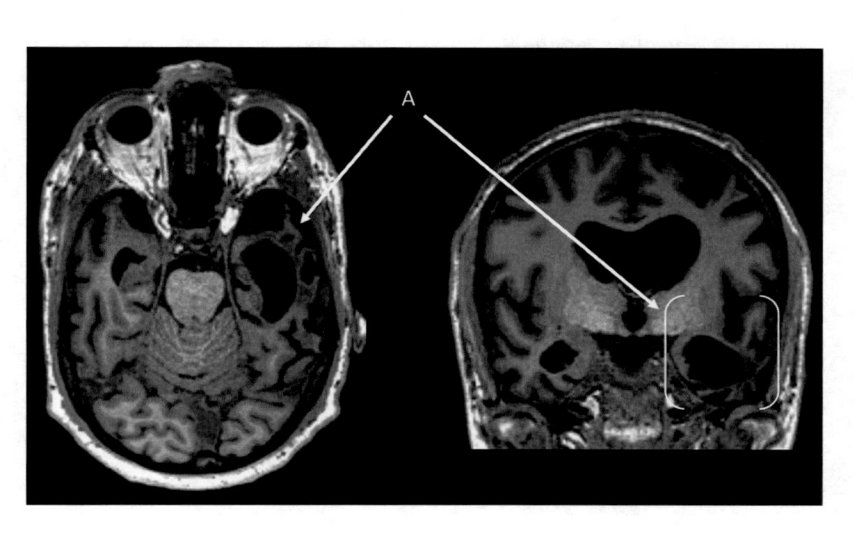

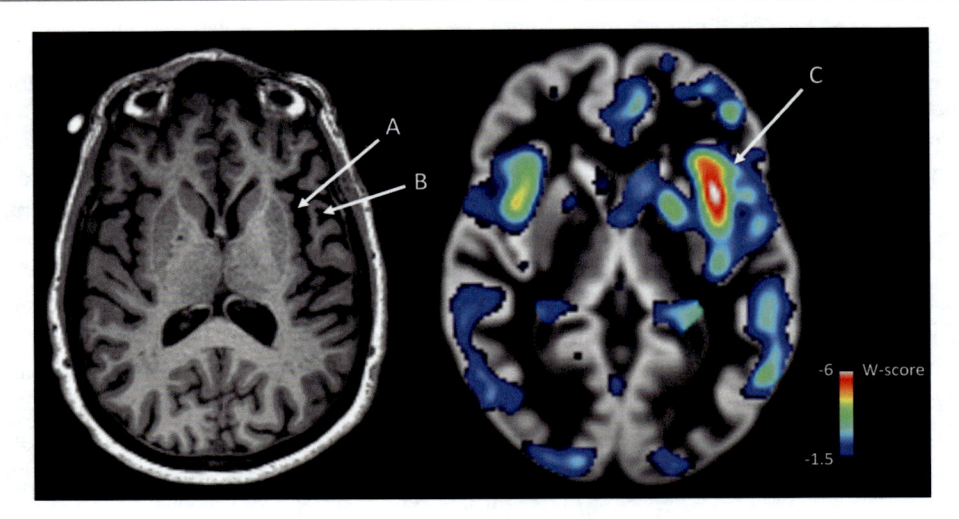

Fig. 4 Axial structural MRI (left) view in radiological orientation showing asymmetric left insular cortex (**a**) and left operculum (**b**) atrophy. Voxel-based morphometry* (VBM) (right) showing the greatest degree of atrophy in the left insular/opercular region (**c**). *Map of this patient's brain volume compared to 534 healthy older controls (age range 44–99 y.o., M ± SD: 68.7 ± 9.1; 220 male/302 female) from the UCSF MAC Hillblom Cohort, adjusted for age, sex, total intracranial volume, and magnet strength. W-scores are interpreted like z-scores, with mean = 0/standard deviation = 1. Negative W-scores represent below-average volume. <−1.50 are below 7th %ile compared to healthy controls and might be considered clinically abnormal. VBM analyses were performed using the open-source Brainsight system, developed at the University of California, San Francisco, Memory and Aging Center by Katherine P. Rankin, Cosmo Mielke, and Paul Sukhanov, and powered by the VLSM script written by Stephen M. Wilson, with funding from the Rainwater Charitable Foundation and the UCSF Chancellor's Fund for Precision Medicine

Three years after the onset of symptoms, she began exhibiting halting, effortful speech. Reading and verbal comprehension was impaired, particularly to long, syntactically complex sentences. Writing was much more difficult than speaking, though she was still able to compose letters and emails with the help of her family. Despite her language impairment, she remained fully independent and continued to work as an accountant without additional difficulty.

On neurological examination, KA was exhibited halting, effortful speech, with agrammatism and frequent phonological errors. When she spoke, her sentences were short and grammatically simplistic. Comprehension of syntactically complex sentences was impaired. Single-word comprehension and object knowledge were spared. She exhibits subtle nondominant limb apraxia and orobuccal apraxia.

On neuropsychological testing, she demonstrated severely impaired sentence repetition, diminished verbal agility, and agrammatism with relatively preserved naming, comprehension, and semantic knowledge. Additionally, she exhibited impaired lexical fluency, markedly impaired design fluency, executive dysfunction, and mild visual memory deficits.

Ms. KA was ultimately diagnosed with non-fluent variant primary progressive aphasia (nfvPPA) on the basis of her effortful, halting speech with agrammatism, spared single-word comprehension and object knowledge, and neuroimaging demonstrating predominant left fronto-insular atrophy (Fig. 4).

References

1. Rabinovici GD, Miller BL (2010) Frontotemporal lobar degeneration. CNS Drugs 24(5):375–398
2. Rascovsky K, Hodges JR, Knopman D et al (2011) Sensitivity of revised diagnostic criteria for the behavioural variant of frontotemporal dementia. Brain 134(9):2456–2477

3. Gorno-Tempini ML, Hillis AE, Weintraub S et al (2011) Classification of primary progressive aphasia and its variants. Neurology 76(11):1006–1014

4. Johnson JK, Diehl J, Mendez MF et al (2005) Frontotemporal lobar degeneration: demographic characteristics of 353 patients. Arch Neurol 62(6):925–930

5. Woolley JD, Khan BK, Murthy NK et al (2011) The diagnostic challenge of psychiatric symptoms in neurodegenerative disease: rates of and risk factors for prior psychiatric diagnosis in patients with early neurodegenerative disease. J Clin Psychiatry 72(2):126–133

6. Seeley WW, Crawford R, Rascovsky K et al (2008) Frontal paralimbic network atrophy in very mild behavioral variant frontotemporal dementia. Arch Neurol 65(2):249–255

7. Eckart JA, Sturm VE, Miller BL et al (2012) Diminished disgust reactivity in behavioral variant frontotemporal dementia. Neuropsychologia 50(5):786–790

8. Perry DC, Datta S, Sturm VE et al (2017) Reward deficits in behavioural variant frontotemporal dementia include insensitivity to negative stimuli. Brain J Neurol 140(12):3346–3356

9. Le Ber I, Guedj E, Gabelle A et al (2006) Demographic, neurological and behavioural characteristics and brain perfusion SPECT in frontal variant of frontotemporal dementia. Brain J Neurol 129(Pt 11):3051–3065

10. Lanata SC, Miller BL (2016) The behavioural variant frontotemporal dementia (bvFTD) syndrome in psychiatry. J Neurol Neurosurg Psychiatry 87(5):501–511

11. Manoochehri M, Huey ED (2012) Diagnosis and management of behavioral issues in frontotemporal dementia. Curr Neurol Neurosci Rep 12(5):528–536

12. Miller B, Llibre Guerra JJ (2019) Frontotemporal dementia. Handb Clin Neurol 165:33–45

13. Diehl-Schmid J, Perneczky R, Koch J et al (2013) Guilty by suspicion? Criminal behavior in frontotemporal lobar degeneration. Cogn Behav Neurol Off J Soc Behav Cogn Neurol 26(2):73–77

14. Rosen HJ, Allison SC, Schauer GF et al (2005) Neuroanatomical correlates of behavioral disorders in dementia. Brain J Neurol 128(11):2612–2625

15. Franceschi M, Anchisi D, Pelati O et al (2005) Glucose metabolism and serotonin receptors in the frontotemporal lobe degeneration. Ann Neurol 57(2):216–225

16. Massimo L, Powers C, Moore P et al (2009) Neuroanatomy of apathy and disinhibition in frontotemporal lobar degeneration. Dement Geriatr Cogn Disord 27(1):96–104

17. Sturm VE, Sollberger M, Seeley WW et al (2013) Role of right pregenual anterior cingulate cortex in self-conscious emotional reactivity. Soc Cogn Affect Neurosci 8(4):468–474

18. Swartz JR, Miller BL, Lesser IM et al (1997) Frontotemporal dementia: treatment response to serotonin selective reuptake inhibitors. J Clin Psychiatry 58(5):212–216

19. Herrmann N, Black SE, Chow T et al (2012) Serotonergic function and treatment of behavioral and psychological symptoms of frontotemporal dementia. Am J Geriatr Psychiatry Off J Am Assoc Geriatr Psychiatry 20(9):789–797

20. Robert P, Lanctôt KL, Agüera-Ortiz L et al (2018) Is it time to revise the diagnostic criteria for apathy in brain disorders? The 2018 international consensus group. Eur Psychiatr 54:71–76

21. Marin RS (1991) Apathy: a neuropsychiatric syndrome. J Neuropsychiatry Clin Neurosci 3(3):243–254

22. Robert P, Onyike CU, Leentjens AFG et al (2009) Proposed diagnostic criteria for apathy in Alzheimer's disease and other neuropsychiatric disorders. Eur Psychiatry 24(2):98–104

23. Clarke DE, Ko JY, Lyketsos C et al (2010) Apathy and cognitive and functional decline in community-dwelling older adults: results from the Baltimore ECA longitudinal study. Int Psychogeriatr 22(5):819–829

24. Merrilees J, Dowling GA, Hubbard E et al (2013) Characterization of apathy in persons with frontotemporal dementia and the impact on family caregivers. Alzheimer Dis Assoc Disord 27(1):62–67

25. Levy R, Dubois B (2006) Apathy and the functional anatomy of the prefrontal cortex-basal ganglia circuits. Cereb Cortex 16(7):916–928

26. Sheelakumari R, Bineesh C, Varghese T et al (2019) Neuroanatomical correlates of apathy and disinhibition in behavioural variant frontotemporal dementia. Brain Imaging Behav.

27. Zamboni G, Huey ED, Krueger F et al (2008) Apathy and disinhibition in frontotemporal dementia: insights into their neural correlates. Neurology 71(10):736–742

28. Eslinger PJ, Moore P, Antani S et al (2012) Apathy in frontotemporal dementia: behavioral and neuroimaging correlates. Behav Neurol 25(2):127–136

29. Dolder CR, Nicole Davis L, McKinsey J (2010) Use of psychostimulants in patients with dementia. Ann Pharmacother 44(10):1624–1632

30. Rosenberg PB, Lanctôt KL, Drye LT et al (2013) Safety and efficacy of methylphenidate for apathy in Alzheimer's disease: a randomized, placebo-controlled trial. J Clin Psychiatry 74(8):810–816

31. Padala PR, Padala KP, Lensing SY et al (2018) Methylphenidate for apathy in community-dwelling older veterans with mild Alzheimer's disease: a double-blind, randomized, placebo-controlled trial. Am J Psychiatry 175(2):159–168

32. Scherer RW, Drye L, Mintzer J et al (2018) The apathy in dementia methylphenidate trial 2 (ADMET 2): study protocol for a randomized controlled trial. Trials 19(1):46

33. Huey ED, Garcia C, Wassermann EM et al (2008) Stimulant treatment of frontotemporal dementia in 8 patients. J Clin Psychiatry 69(12):1981–1982

34. Carr AR, Mendez MF (2018) Affective empathy in behavioral variant frontotemporal dementia: a meta-analysis. Front Neurol 9

35. Hsieh S, Irish M, Daveson N et al (2013) When one loses empathy: its effect on carers of patients with dementia. J Geriatr Psychiatry Neurol 26(3):174–184

36. Pomponi M, Ricciardi L, La Torre G et al (2016) Patient's loss of empathy is associated with caregiver burden. J Nerv Ment Dis 204(9):717–722

37. Mendez MF (2010) The unique predisposition to criminal violations in frontotemporal dementia. J Am Acad Psychiatry Law 38(3):318–323

38. Liljegren M, Naasan G, Temlett J et al (2015) Criminal behavior in frontotemporal dementia and Alzheimer disease. JAMA Neurol 72(3):295–300

39. Singer T, Lamm C (2009) The social neuroscience of empathy. Ann N Y Acad Sci 1156(1):81–96

40. Rankin KP, Kramer JH, Miller BL (2005) Patterns of cognitive and emotional empathy in frontotemporal lobar degeneration. Cogn Behav Neurol 18(1):28–36

41. Baez S, Manes F, Huepe D et al (2014) Primary empathy deficits in frontotemporal dementia. Front Aging Neurosci 6:262

42. Shamay-Tsoory SG, Aharon-Peretz J, Perry D (2009) Two systems for empathy: a double dissociation between emotional and cognitive empathy in inferior frontal gyrus versus ventromedial prefrontal lesions. Brain 132(3):617–627

43. Dermody N, Wong S, Ahmed R et al (2016) Uncovering the neural bases of cognitive and affective empathy deficits in Alzheimer's disease and the behavioral-variant of frontotemporal dementia. J Alzheimers Dis 53(3):801–816

44. Hurlemann R, Patin A, Onur OA et al (2010) Oxytocin enhances amygdala-dependent, socially reinforced learning and emotional empathy in humans. J Neurosci 30(14):4999–5007

45. Jesso S, Morlog D, Ross S et al (2011) The effects of oxytocin on social cognition and behaviour in frontotemporal dementia. Brain J Neurol 134(Pt 9):2493–2501

46. Finger EC, MacKinley J, Blair M et al (2015) Oxytocin for frontotemporal dementia: a randomized dose-finding study of safety and tolerability. Neurology 84(2):174–181

47. Mendez MF, Perryman KM, Miller BL et al (1997) Compulsive behaviors as presenting symptoms of frontotemporal dementia. J Geriatr Psychiatry Neurol 10(4):154–157

48. Olney NT, Spina S, Miller BL (2017) Frontotemporal Dementia. Neurol Clin 35(2):339–374

49. Mateen FJ, Josephs KA (2009) The clinical spectrum of stereotypies in frontotemporal lobar degeneration. Mov Disord: Off J Mov Disord Soc 24(8):1237–1240

50. Moheb N, Charuworn K, Ashla MM et al (2019) Repetitive behaviors in frontotemporal dementia: compulsions or impulsions? J Neuropsychiatry Clin Neurosci 31(2):132–136

51. American Psychiatric Association (2013) Diagnostic and statistical manual of mental disorders, 5th edn. American Psychiatric Association

52. Stein DJ (2002) Obsessive-compulsive disorder. Lancet 360(9330):397–405

53. Veale D, Roberts A (2014) Obsessive-compulsive disorder. BMJ 348

54. Robbins T, Curran H, de Wit H (2012) Special issue on impulsivity and compulsivity. Psychopharmacology 219(2):251–252

55. Berlin GS, Hollander E (2014) Compulsivity, impulsivity, and the DSM-5 process. CNS Spectr 19(1):62–68

56. Ames D, Cummings JL, Wirshing WC et al (1994) Repetitive and compulsive behavior in frontal lobe degenerations. J Neuropsychiatry Clin Neurosci 6(2):100–113

57. Perry DC, Whitwell JL, Boeve BF et al (2012) Voxel-based morphometry in patients with obsessive-compulsive behaviors in behavioral variant frontotemporal dementia. Eur J Neurol 19(6):911–917

58. Mitchell E, Tavares TP, Palaniyappan L et al (2019) Hoarding and obsessive-compulsive behaviours in frontotemporal dementia: clinical and neuroanatomic associations. Cortex 121:443–453

59. Rosso SM, Roks G, Stevens M et al (2001) Complex compulsive behaviour in the temporal variant of frontotemporal dementia. J Neurol 248(11):965–970

60. McMurtray AM, Chen AK, Shapira JS et al (2006) Variations in regional SPECT hypoperfusion and clinical features in frontotemporal dementia. Neurology 66(4):517–522

61. Josephs KA, Whitwell JL, Jack CR (2008) Anatomic correlates of stereotypies in frontotemporal lobar degeneration. Neurobiol Aging 29(12):1859–1863

62. Furlan JC, Henri-Bhargava A, Freedman M (2014) Clomipramine in the treatment of compulsive behavior in frontotemporal dementia: a case series. Alzheimer Dis Assoc Disord 28(1):95–98

63. Cruz M, Marinho V, Fontenelle LF et al (2008) Topiramate may modulate alcohol abuse but not other compulsive behaviors in frontotemporal dementia: case report. Cogn Behav Neurol 21(2):104–106

64. Mendez MF, Shapira JS, McMurtray A et al (2007) Preliminary findings: behavioral worsening on donepezil in patients with frontotemporal dementia. Am J Geriatr Psychiatry 15(1):84–87

65. Sourander P, Sjögren H (2008) The concept of Alzheimer's disease and its clinical implications. In: Ciba foundation symposium – Alzheimer's disease and related conditions. Wiley, p 11–36

66. Neary D, Snowden JS, Gustafson L et al (1998) Frontotemporal lobar degeneration: a consensus on clinical diagnostic criteria. Neurology 51(6):1546–1554

67. Piguet O, Hornberger M, Shelley BP et al (2009) Sensitivity of current criteria for the diagnosis of behavioral variant frontotemporal dementia. Neurology 72(8):732–737

68. Miller BL, Darby AL, Swartz JR et al (1995) Dietary changes, compulsions and sexual behavior in frontotemporal degeneration. Dementia 6(4):195–199

69. Mendez MF, Licht EA, Shapira JS (2008) Changes in dietary or eating behavior in frontotemporal dementia versus Alzheimer's disease. Am J Alzheimers Dis Other Demen 23(3):280–285
70. Woolley JD, Gorno-Tempini M-L, Seeley WW et al (2007) Binge eating is associated with right orbitofrontal-insular-striatal atrophy in frontotemporal dementia. Neurology 69(14):1424–1433
71. Whitwell JL, Sampson EL, Loy CT et al (2007) VBM signatures of abnormal eating behaviours in frontotemporal lobar degeneration. NeuroImage 35(1):207–213
72. Yi DS, Bertoux M, Mioshi E et al (2013) Fronto-striatal atrophy correlates of neuropsychiatric dysfunction in frontotemporal dementia (FTD) and Alzheimer's disease (AD). Dement Neuropsychol 7(1):75–82
73. Piguet O, Petersén Å, Yin Ka Lam B et al (2011) Eating and hypothalamus changes in behavioral-variant frontotemporal dementia. Ann Neurol 69(2):312–319
74. Ikeda M, Shigenobu K, Fukuhara R et al (2004) Efficacy of fluvoxamine as a treatment for behavioral symptoms in frontotemporal lobar degeneration patients. Dement Geriatr Cogn Disord 17(3):117–121
75. Lebert F, Stekke W, Hasenbroekx C et al (2004) Frontotemporal dementia: a randomised, controlled trial with trazodone. Dement Geriatr Cogn Disord 17(4):355–359
76. Block NR, Sha SJ, Karydas AM et al (2016) Frontotemporal dementia and psychiatric illness: emerging clinical and biological links in gene carriers. Am J Geriatr Psychiatry 24(2):107–116
77. Banks SJ, Weintraub S (2008) Neuropsychiatric symptoms in behavioral variant frontotemporal dementia and primary progressive aphasia. J Geriatr Psychiatry Neurol 21(2):133–141
78. Gossink F, Vijverberg E, Pijnenburg Y et al (2018) Neuropsychiatry in clinical practice: the challenge of diagnosing behavioral variant frontotemporal dementia. J Neuropsychiatry 02(01)
79. Gossink FT, Dols A, Kerssens CJ et al (2016) Psychiatric diagnoses underlying the phenocopy syndrome of behavioural variant frontotemporal dementia. J Neurol Neurosurg Psychiatry 87(1):64–68
80. Hasin DS, Sarvet AL, Meyers JL et al (2018) Epidemiology of adult DSM-5 major depressive disorder and its specifiers in the United States. JAMA Psychiat 75(4):336–346
81. Knopman DS, Roberts RO (2011) Estimating the number of persons with frontotemporal lobar degeneration in the US population. J Mol Neurosci 45(3):330–335
82. Charney DS, Sklar PB, Buxbaum JD et al (2018) Charney & Nestler's neurobiology of mental illness. Oxford University Press, New York
83. Lu X, Zhong Y, Ma Z et al (2019) Structural imaging biomarkers for bipolar disorder: meta-analyses of whole-brain voxel-based morphometry studies. Depress Anxiety 36(4):353–364
84. Redlich R, Almeida JJR, Grotegerd D et al (2014) Brain morphometric biomarkers distinguishing unipolar and bipolar depression. A voxel-based morphometry-pattern classification approach. JAMA Psychiat 71(11):1222–1230

Nosology of Primary Progressive Aphasia and the Neuropathology of Language

M. -Marsel Mesulam, Christina Coventry,
Eileen H. Bigio, Changiz Geula,
Cynthia Thompson, Borna Bonakdarpour,
Tamar Gefen, Emily J. Rogalski,
and Sandra Weintraub

Introduction

Primary progressive aphasia (PPA) is a major syndrome of frontotemporal lobar degeneration (FTLD) and accounts for nearly 25% of all FTLD cases [1]. Approximately 60% of PPA is associated with FTLD and the remaining 40% with the neuropathology of Alzheimer's disease (AD). Information on PPA prevalence is limited. One study from the UK suggests an approximate prevalence of 3–4/100,000, a level comparable to what has been reported for ALS [1]. The one common denominator for all PPA, whether caused by FTLD or AD, is the preferential degeneration of the language network, usually located in the left hemisphere of the brain. Current research on primary progressive aphasia is evolving in multiple directions. For one, the variety of the aphasic disturbances continues to fuel discussion on nomenclature and clinical classification. Second, the selective dissolution of individual language domains is offering new paradigms for exploring the functional anatomy of language, a pursuit that has already prompted modifications of classic models. Third, the multiplicity of the underlying degenerative diseases is generating new insights on the heterogeneity of dementias, the probabilistic relationship of syndrome to

M.-M. Mesulam (✉) · B. Bonakdarpour
Mesulam Center for Cognitive Neurology
and Alzheimer's Disease;
Department of Neurology,
Northwestern University,
Chicago, IL, USA
e-mail: mmesulam@northwestern.edu;
bbk@northwestern.edu

C. Coventry · C. Geula · S. Weintraub
Mesulam Center for Cognitive Neurology
and Alzheimer's Disease,
Northwestern University,
Chicago, IL, USA
e-mail: christina.coventry@northwestern.edu;
c-geula@northwestern.edu;
sweintraub@northwestern.edu

E. H. Bigio
Mesulam Center for Cognitive Neurology
and Alzheimer's Disease; Department of Pathology,
Northwestern University, Chicago, IL, USA
e-mail: e-bigio@northwestern.edu

C. Thompson
Mesulam Center for Cognitive Neurology
and Alzheimer's Disease; Department
of Communication Sciences and Disorders;
Department of Neurology, Northwestern University,
Evanston, IL, USA
e-mail: ckthom@northwestern.edu

T. Gefen · E. J. Rogalski
Mesulam Center for Cognitive Neurology
and Alzheimer's Disease; Department of Psychiatry,
Northwestern University, Chicago, IL, USA
e-mail: tamar.gefen@northwestern.edu;
e-rogalski@northwestern.edu

© Springer Nature Switzerland AG 2021
B. Ghetti et al. (eds.), *Frontotemporal Dementias*, Advances in Experimental Medicine and Biology
1281, https://doi.org/10.1007/978-3-030-51140-1_3

pathology, and the mechanisms of selective vulnerability. Fourth, there is lively interest in formulating personalized interventions aimed not only at the nature of the language disturbance but also at the biology of the underlying disease entity. These are some of the current trends that will be reviewed in this chapter. Given the constraints of space and the vast literature on PPA, the account will be selective and based predominantly on the PPA research programs at Northwestern University where a cohort of 235 PPA patients have been enrolled, 97 of whom have come to brain autopsy.

Diagnosis, Nomenclature, and Subtyping

The existence of progressive language disorders had been known for more than 100 years. Pick, Sérieux, Dejerine, Franceschi, and Rosenfeld were among the first to report such patients during the late nineteenth and early twentieth centuries [2–7]. However, this topic did not attract much, if any, attention during most of the twentieth century. The current resurgence of interest in this condition can be traced to the 1982 report of six patients who experienced a slowly progressive aphasia without other cognitive or behavioral impairments [8]. The syndrome was named "primary progressive aphasia," and diagnostic criteria were formulated [9, 10]. The following decades witnessed a rapidly expanding literature on PPA and on overlapping entities designated progressive nonfluent aphasia (PNFA) and semantic dementia (SD) [11]. For a number of years, research on PNFA and SD developed in parallel to research on PPA. In 2011, an international group of investigators presented classification guidelines that incorporated PNFA and SD under the PPA umbrella [12]. This unitary approach stimulated rapid progress in this field.

Three features define PPA: (1) adult-onset and progressive impairment of language (not just speech), (2) absence of other consequential behavioral or cognitive deficits for approximately the first 2 years, and (3) neurodegenerative disease as the only cause of impairment [10]. These criteria help to filter out patients where progressive aphasias arise in conjunction with equally prominent speech apraxia, behavioral disturbances, loss of memory for recent events, associative agnosias, or visuospatial deficits. In the course of diagnostic evaluation, patients may show subtle impairments in non-language tasks, especially those related to memory and executive function. Such abnormalities of test performance do not by themselves preclude a PPA diagnosis unless they are associated with limitations of daily life in the corresponding non-language domains.

Many neuropsychological tests require verbal responses and verbal instructions. The clinician needs to consider the influence of the aphasia on these aspects of performance. For example, a patient with PPA who cannot name a famous face is not necessarily prosopagnosic, a patient who cannot verbalize the nature of an object does not necessarily lack knowledge of the object, and a patient who cannot learn a word list is not necessarily amnestic. Conversely, patients who cannot produce words because of articulation deficits, those who cannot repeat language because of general working memory limitations, those who misname objects or faces they do not recognize, or those who have impoverished speech because of abulia or impaired executive function are not necessarily aphasic. As in the case of many other syndromes, the diagnosis of PPA relies on the judgment and experience of the clinician. While clear-cut cases do exist, there are also cases where the salience and primacy of the aphasia will generate debate, especially if the patient is examined a few years after symptom onset. In some patients, the aphasia will remain the only salient feature for over a decade [13]. Other patients, however, may first come to a specialty clinic at a time when the disease has progressed to encompass other cognitive domains. The term "PPA plus" (PPA+) can be used to designate such patients, based on the assumption that the disease had started as PPA, but that it had since spread beyond the language network [14].

In contrast to many other dementias, where the patient has little insight into the predicament, patients with PPA are usually the first to notice and report the difficulty. At those stages of the disease, MRI and metabolic positron emission tomography (PET) scans may be negative. The absence of positive neurodiagnostic tests, combined with lack of recognition of these symptoms in general practice, may lead to unwarranted referrals to otolaryngologists or psychiatrists [15]. Patients and families often ask whether the diagnosis is PPA or AD. When AD biomarkers (such as amyloid and phospho-tau in cerebrospinal fluid [CSF] or amyloid PET scans) are positive, the clinician will have to explain that the patient has both PPA and AD, that PPA refers to the symptoms that bring the patient to the clinic, and that AD refers to the abnormal amyloid and tau proteins in the brain that attack the language centers. There was a time when PPA was underdiagnosed. There are now instances where it seems to be overdiagnosed, probably because language impairments can be so prominent during the office evaluation that other equally substantial cognitive and behavioral impairments become overlooked. This issue comes up most commonly in patients with prominent apraxia of speech or executive dysfunction who are also aphasic. We give these patient descriptive diagnoses such as "apraxia of speech with aphasia" or "aphasic frontal syndrome."

Language impairment can encompass word retrieval, object naming, sentence construction, or language comprehension, either singly or in combination. Once the PPA diagnosis is established, the subtyping exercise can be initiated. At the time of writing, the 2011 guidelines dominate this process [12]. They help to classify PPA into nonfluent/agrammatic, logopenic, and semantic variants. Although this system has been immensely influential and is even frequently mandated during the review of manuscripts submitted for publication, it has widely recognized shortcomings [16–18]. For one, a strict adherence to the 2011 guidelines entails arduous assessment of nearly a dozen separate aspects of language. Second, even if the guidelines are strictly applied, approximately one-third of the patients will fail to be classified into any of the three variants. Third, there are certain feature clusters that allow the same patient to simultaneously fit the designation of both nonfluent/agrammatic and logopenic PPA. Yet another challenge is posed by the evolution over time, so that a patient who fits the logopenic subtype initially may fit criteria for one of the other two subtypes as the disease progresses.

The following modifications have helped us address some of these concerns [16]. (1) The relative preservation of both grammar and comprehension is made to be a core feature of the logopenic variant. This prevents the double assignment problem. (2) In contrast to the 2011 guidelines, repetition impairment is not considered an obligatory core feature of the logopenic variant. This practice reduces the number of unclassifiable patients. (3) Patients with combined impairments of grammar and word comprehension even early in the disease, and who would therefore remain unclassifiable by the 2011 guidelines, make up a fourth variant of "mixed" PPA. (4) The semantic variant is diagnosed when poor word comprehension is the principal feature. When additional and equally prominent impairments of object or face recognition (not just naming) are detected, a diagnosis of semantic dementia (SD) is made [11]. This recommendation is at odds with the 2011 guidelines, which would diagnose semantic PPA even in patients with significant face and object recognition impairment (i.e., visual associative agnosia). The justification for the distinction of PPA from SD is summarized in the section on the anatomy of language.

The modifications listed above lead to a classification method based on a template where the Y-axis represents worsening impairment in the grammaticality of sentence construction and the X-axis represents worsening impairment in single word comprehension [15]. Each of the four PPA subtypes will cluster within a different quadrant of this template. The *nonfluent/agrammatic* PPA patients, for example, will cluster in the upper left quadrant (impaired grammar but spared comprehension); the *semantic* PPA patients will cluster in the lower right quadrant (impaired comprehension but spared grammar); the *mixed* PPA patients will cluster in the lower left quadrant (combined impairments of grammar and comprehension); and the *logopenic* PPA patients will cluster in the upper right quadrant (relatively spared grammar and comprehension). The logopenic group would have met the PPA criteria through impairments of word retrieval, naming, and spelling. Specific tests for assessing grammaticality of sentence construction and word comprehension and their normative values have been reported [15]. As patterns of agrammatism vary greatly from language to language, considerable attention is being directed to the adaptation of grammar tests for languages other than English [19].

Some logopenic patients maintain fluency as they circumvent word finding failures through circumlocution; others pause after word retrieval failures and produce halting nonfluent speech that appears similar to what is seen in patients with nonfluent/agrammatic PPA. Word finding impairments and paraphasias may make it impossible to gauge a sentence grammaticality. The delineation of logopenic from agrammatic PPA can thus be quite challenging [17]. Quantitative analyses of speech samples show that the nonfluent/agrammatic patients make word finding pauses that are longer before verbs, whereas logopenic patients make pauses that are longer before nouns [20]. Furthermore, patients with nonfluent/agrammatic PPA display a preferential impairment of verb rather than object naming, whereas the converse may be seen in logopenic PPA [21]. When research objectives necessitate such distinctions, these features may help to establish a quantitative differentiation of nonfluent/agrammatic from logopenic forms of PPA. Subtyping need not become an end onto itself. For purposes of both research and treatment, the emphasis could also be on single parameters, such as grammar or naming, across all subjects and regardless of subtype.

The 2011 guidelines did not prescribe acronyms for the three variants. At present, *non-fluent variant* (nfvPPA), *logopenic variant* (lvPPA), and *semantic variant* (svPPA) are the most popular choices. Alternative acronyms such as naPPA, agPPA, PPA-NFV, LPA, and PPA-SV have also been used, albeit more rarely [22–25]. The "nfv" prefix is particularly problematic because it appears to overlook grammar, which is the single most characteristic impairment of this subtype. The choice of "nfv" was probably based on experience derived from stroke aphasia where low fluency can be used as a proxy for agrammatism. In PPA, grammar and fluency can be dissociated, especially in logopenic patients where long word finding pauses diminish fluency but without grammatical impairment [26]. Based on these considerations and also in order to underscore the primacy of the PPA diagnosis, we have used the alternative acronyms of PPA-G, PPA-L, PPA-S and PPA-M for the nonfluent/agrammatic, logopenic, semantic and mixed variants, respectively. It may take another collective international effort to determine whether the 2011 consensus guidelines should be modified along the lines listed above and whether the acronyms can be harmonized.

Clinical progression patterns vary by subtype and are likely to reflect the differential anatomical trajectories of disease spread. In PPA-S, the spread of atrophy from the anterior temporal lobe to orbitofrontal, insular, or contralateral temporal lobe can lead to the additional face and object recognition impairments of SD, and to the behavioral abnormalities seen in behavioral variant frontotemporal dementia (bvFTD). In PPA-G, spread of atrophy from the inferior frontal gyrus (IFG) to other premotor and frontal cortices can lead to the abnormalities seen in apraxia of speech, corticobasal syndrome, supranuclear ophthalmoplegia, and frontal-type executive dysfunction. In PPA-L, spread of atrophy from the temporoparietal junction (TPJ) to surrounding cortices can lead to additional impairments of explicit memory and constructions. For all subtypes, the spread of atrophy tends to be more pronounced in the left hemisphere, and there are substantial interindividual differences in the speed and trajectory of progression [27].

Contributions to the Anatomy of Language

The classic Wernicke-Lichtheim-Geschwind model of language revolved around two epicenters, namely Broca's area in the inferior frontal gyrus (IFG) and Wernicke's area in the temporoparietal junction (TPJ), a region that can be said to encompass parts of the inferior parietal lobule and the posterior segments of the superior and middle temporal gyri [28] (Fig. 1). The former has been linked to fluency and grammar and the latter to language comprehension. The literature of the past 150 years displays greater agreement on the location and function of Broca's area than of Wernicke's area [28]. These two epicenters are connected through the arcuate fasciculus, which is thought to play a critical role in language repetition [29]. This basic model has undergone major revisions through investigations with functional imaging, event-related potentials, and sophisticated neuropsychological assessments [30–32].

Each of these approaches has advantages and disadvantages. Cerebrovascular lesions cause sudden and irreversible destruction of the core lesion site. However, the damage usually extends into deep white matter. The exact contribution of the damaged cortical region to the ensuing language impairment is therefore difficult to specify. Functional mapping approaches based on MRI and electrical recordings, on the other hand, can reveal activity confined to the cerebral cortex but cannot differentiate areas that are critical for a function from those that have collateral participatory roles.

Investigations based on focal cortical atrophy can circumvent some of these shortcomings. Regions where the magnitude of cortical thinning correlates with the magnitude of impairment can be said to have critical (rather than participatory) roles in maintaining the integrity of that function. Consequently, PPA has offered new tools for investigating the cortical anatomy of the language network without the deep white matter problem of stroke or the collateral activation dilemma of functional brain mapping. Nonetheless, clinicoanatomical correlations in PPA are not without caveats. For one, the slow evolution of the lesion is likely to trigger compensatory plasticity that may complicate the interpretation of correlations. Second, even areas of peak atrophy may contain residual neurons that could sustain some functionality of that region [33]. Third, each neuropathologic entity

may trigger a different pattern of cortical injury. For example, the neurofibrillary tangles of AD have a predilection for deep cortical layers whereas the opposite is the case for Pick's disease.

Despite these potential complications, clinicoanatomical investigations on PPA have generated new insights into the functional anatomy of language. Each PPA variant is associated with a characteristic location of peak atrophy, for instance, Broca's area (IFG) in PPA-G, Wernicke's area (TPJ) in PPA-L, and the anterior half of the temporal lobe (ATL) in PPA-S [34–36]. The anatomical correlate of PPA-G is in keeping with prevailing models of language, which give Broca's area a critical role in the maintenance of fluency and grammar [37]. The relationships in PPA-L and PPA-S, however, are in conflict with classic aphasiology and also with most contemporary models of language. For one, traditional models of language exclude the ATL. For example, an influential review published at the height of twentieth-century aphasiology states that the probability that a lesion would impair comprehension is "very high in or near the first temporal gyrus, and fades out with different gradients (varying among individuals) toward the poles. And by the time it gets to any pole (occipital, temporal, or frontal) the probability is essentially zero" [38]. Research on PPA-S has contradicted this statement by showing that damage to the left ATL, including the temporal pole and anterior fusiform gyrus, causes severe impairments of word comprehension. Based on this finding, a proposal has been made that this region should be considered a core component of the language network [28].

This proposal has generated considerable debate. The disagreement revolves around the alternative characterization of ATL as an amodal hub for all semantic knowledge, verbal and nonverbal. Consequently, ATL damage should cause more than a language impairment (i.e., aphasia) and should give rise to a universal loss of semantic knowledge not only for words but also for faces and objects [39]. Based on this point of view, the syndrome of ATL damage was designated semantic dementia (SD), a syndrome

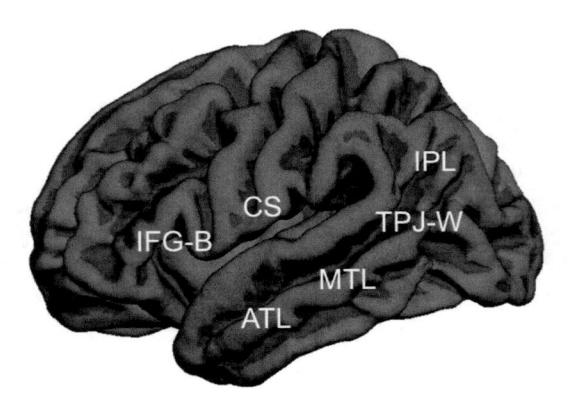

Fig. 1 Major components of the left hemisphere language network – ATL: The acronym ATL will be used to refer to the anterior third of the temporal lobe including the temporal pole; CS (the central sulcus) is shown as a reference point, IFG-B (the inferior frontal gyrus) contains Broca's area, IPL (inferior parietal) lobule, MTL (the middle third of the temporal lobe), TPJ-W (the temporoparietal junction) contains the posterior third of the temporal lobe and the immediately adjacent parts of the inferior parietal lobule. Although the exact site of Wernicke's area remains ambiguous, it is usually considered to be located within the TPJ-W and adjacent parts of the MTL

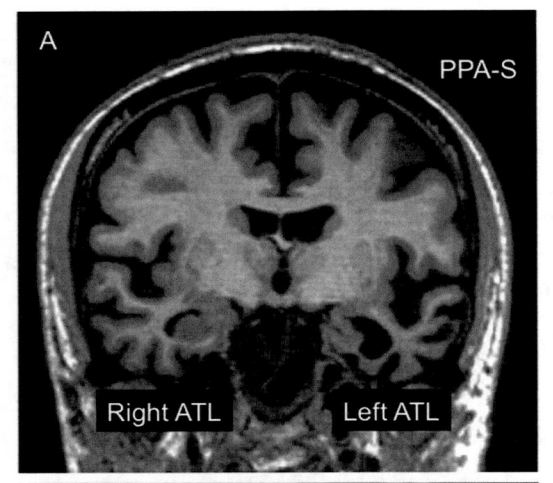

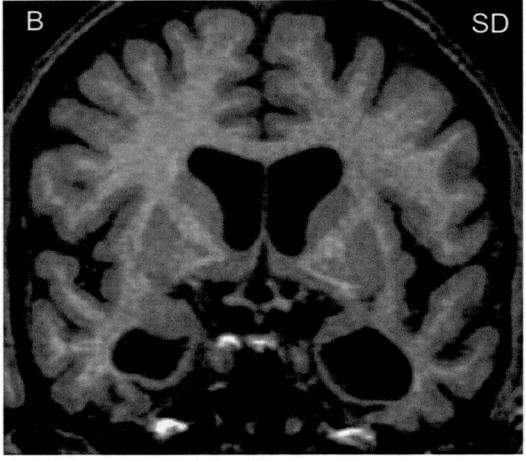

Fig. 2 PPA-S versus SD. Figure 2a shows the MRI scan of a right-handed man with symptom onset at the age of 59. On examination, 7 years later, the clinical pattern was PPA-S and atrophy was much more prominent in the left anterior temporal lobe (ATL). At that time, he had severe word comprehension impairments but no difficulty with non-verbal object recognition either in testing or in everyday life. In comparison, Fig. 2b shows the MRI scan of a right-handed man with symptom onset at the age of 65. Three years later, at his initial visit, ATL atrophy was bilateral. He had prominent word comprehension and object recognition impairments. This combination led to a subsequent diagnosis of semantic dementia (SD)

defined by the combination of semantic aphasia (word comprehension deficit) with visual associative agnosia (loss of face and object recognition) [11, 39]. Such patients would not fit the diagnostic criteria for PPA since the aphasia would no longer constitute the dominant feature.

The disagreement on the nature of the syndrome caused by ATL damage can be resolved by considering the influence of hemispheric specialization [28, 40, 41]. Clinical observations and specially designed experimental tasks show that PPA-S is a selective aphasic syndrome of the left anterior temporal lobe, whereas the SD syndrome reflects a wider deficit with a more bilateral anatomical substrate [42–45]. The patients with left ATL damage may not be able to name objects and faces but are generally cognizant of their identity and nature [46]. It should be pointed out, however, that many PPA-S patients may also have minor atrophy in the right anterior temporal lobe, and that further spread of neurodegeneration within the right hemisphere may lead some, but not all, to eventually develop the additional face and object recognition deficits of SD. It is not surprising, therefore, that some authors have considered PPA-S and SD to be the two sides of the same coin [40, 41]. The question is whether syndromic designations should be based on clinical presentation at disease onset, as we advocate, or based on possible progression trajectories (Fig. 2). When ATL atrophy is predominantly right-sided, the patient may present with one of three syndromes, SD, non-aphasic associative agnosia, or bvFTD [47, 48].

Exactly how the left ATL contributes to word comprehension is a topic of active investigation. Resting state functional imaging experiments show that the left ATL has left-sided asymmetric functional connectivity patterns that support its inclusion within the language network [49]. In our cohort, all right-handed patients with severe word comprehension impairment have also had substantial left ATL atrophy extending all the way into the pole. However, some patients with such a location of atrophy may have severe anomia in the absence of word comprehension impairment. In these patients, the distinctive comprehension impairment of PPA-S emerges as

the atrophy extends posteriorly from the anterior tip of the left temporal lobe into adjacent parts of the middle portion of the temporal lobe (MTL), especially the middle temporal gyrus (MTG) [28]. In keeping with this observation, functional MRI studies in PPA and clinicoanatomical correlations in stroke have shown that the connectivity of the mid-to-posterior parts of the MTG with ATL and other parts of the language network may have important roles in sustaining word comprehension [50, 51]. In our experience, isolated atrophy of the middle parts of the temporal lobe in PPA has not been associated with impairment of this function [28]. Damage to the left ATL may therefore be necessary but not always sufficient for word recognition impairment. Posterior expansion of damage into the middle parts of the temporal lobe may also be required.

Patients with PPA-S have severe naming impairments principally because they do not understand the meaning of the word that denotes the object they are asked to name [46]. The impairment initially undermines the comprehension of a word at its specific level of meaning (does the word denote a strawberry or a cherry) but later generalizes to the generic meaning of the word (does the word denote a fruit or an animal) [52]. Based on these observations in PPA-S, the left ATL can be conceptualized as a transmodal region of cortex where sensory word form information is linked to the multimodal associations that collectively encode the meaning of the word [28, 53]. Word recognition at a specific level of meaning requires more extensive associative elaboration and would therefore be more vulnerable to early stages of neurodegeneration.

Another unexpected outcome of research on PPA was the finding that patients with the logopenic variant have normal single word comprehension despite peak atrophy sites that encompass Wernicke's area as defined above. In fact, regression analyses in 73 PPA patients showed no correlation between atrophy in Wernicke's area and impairment of word comprehension [28, 54]. In addition to clinicoanatomical correlations in PPA-L, which have shown that severe cortical degeneration of Wernicke's area does not impair single word comprehension, investigations on PPA-S have shown that an intact Wernicke's area is not sufficient to sustain word comprehension if the ATL is damaged. The body of work on PPA therefore leads to the conclusion that the cortex of Wernicke's area is neither necessary nor sufficient for word comprehension. This conclusion can be reconciled with classic aphasiology by keeping in mind that nearly all reports linking Wernicke's area to word comprehension are based on cerebrovascular lesions. Such lesions include not only the cortex of Wernicke's area but also deep white matter axons, such as those in the middle longitudinal fasciculus [55], that are likely to carry projections of otherwise intact distal posterior and contralateral cortices. The resultant additional cortical disconnections may explain why stroke in Wernicke's *region* impairs word comprehension while neurodegeneration in Wernicke's *cortex* does not [54].

The large-scale network model posits that each network node mediates critical (or essential) as well as ancillary (or sustaining) functions related to its principal cognitive domain [56, 57]. While damage to a given node may not cause fixed impairments of its ancillary functionalities, the overall computational flexibility of the network for mediating that task may be compromised. These principles apply to the role of Wernicke's area in language comprehension. For example, agrammatic and logopenic PPA patients whose atrophy encompasses Wernicke's area but not the ATL, and who have normal word comprehension in standard tests and daily life, display abnormally prolonged semantic interference effects and loss of the N400 semantic incongruence potential [52, 58]. Furthermore, functional magnetic resonance imaging (fMRI) investigations using synonym identification tasks revealed activations not only in the anterior temporal lobe but also in regions overlapping Wernicke's area [59, 60]. The cerebral cortex within Wernicke's area therefore serves an ancillary role in word comprehension. Multiple lines of evidence show that Wernicke's area plays a critical role in language repetition, a finding that is in keeping with observations in stroke aphasia [54]. This area is important for language repetition presumably because it links phonologic word form codes to their articulatory sequences [61–63].

An additional contribution of PPA to the anatomy of language comes through the discovery of the aslant tract, a pathway that connects the core language network with dorsal premotor cortex and appears to play a major role in sustaining fluency [64]. Patients with PPA may also show patterns of aphasia that have not been observed in other settings. For example, some patients may show a preferential inability to name objects orally but not in writing and fail to understand words they hear but not those they read [65]. These patients do not fit the pattern seen in pure word deafness because they are anomic and they do not fit the pattern of auditory agnosia because they can match objects to their characteristic sounds. Investigations on this small group of patients have helped to explore the functionality of a putative "auditory word form area" that sits at the confluence of modality-specific pathways for word comprehension and language repetition.

The totality of these investigations on PPA depicts a large-scale language network built upon the interactive functionalities of dorsal and ventral (rather than anterior and posterior) streams of processing [31]. The dorsal route mediates phonological encoding, repetition, articulatory programming, fluency, word retrieval and also the sequencing of morphemes and words into grammatically correct sentences. The ventral route mediates the lexicosemantic processes of object naming and word comprehension. Word finding in speech is a joint function of both routes and therefore the most common presenting complaint in PPA.

Asymmetry of Neuropathology and Genetics

In our group of 97 consecutive autopsies, the primary neuropathology was FTLD with tauopathy (FTLD-tau) in 29%, FTLD with transactive response DNA-binding protein 43 (FTLD-TDP) in 25%, and AD in 44%. All three major neuropathologic forms of FTLD-tau (Pick's disease,

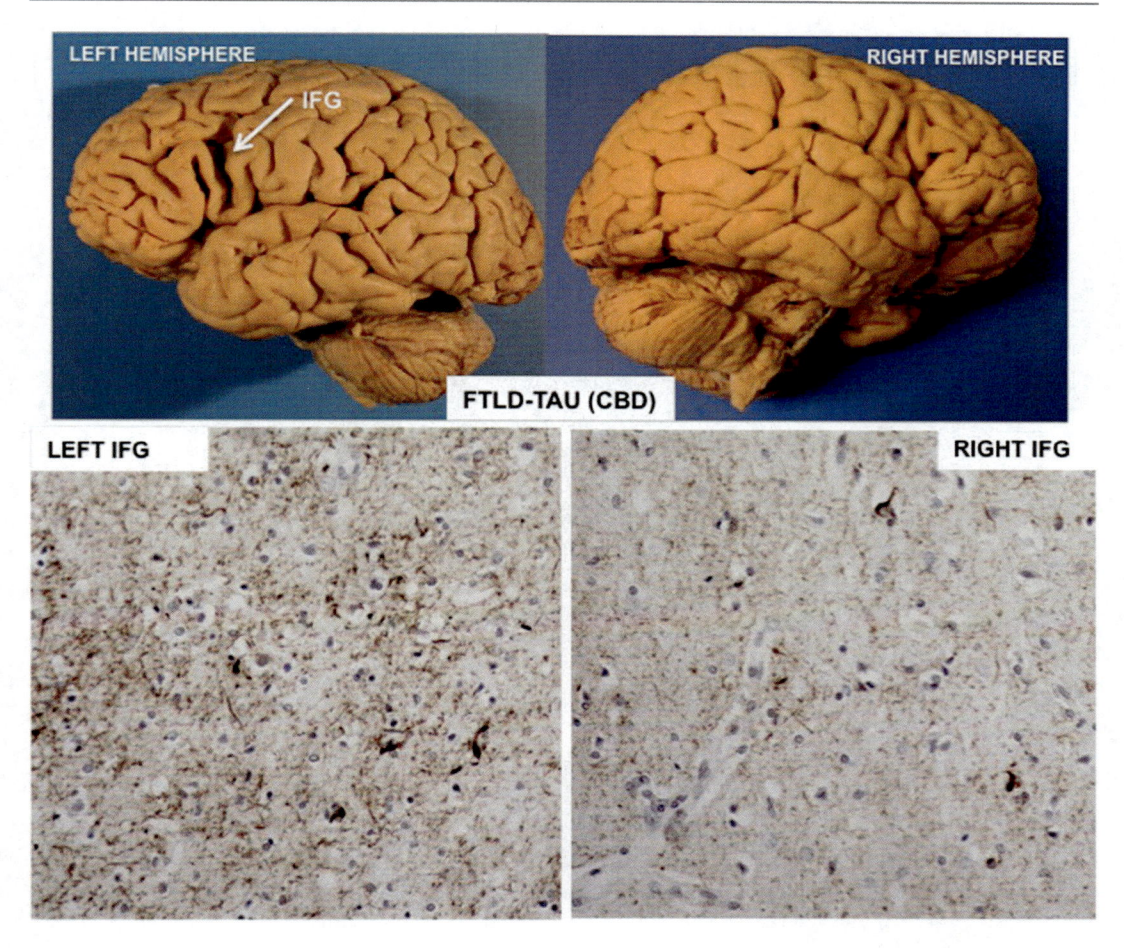

Fig. 3 Asymmetry of neurodegeneration. Postmortem examination of a right-handed woman with symptom onset at the age of 72 and findings of agrammatic PPA with prominent word finding impairments. Death occurred 6 years later. The primary neuropathology was found to be FTLD-tau of the CBD type. The top figures show the profound asymmetry of atrophy. There is an almost cystic area of atrophy around the left inferior frontal gyrus (IFG) but no comparable atrophy of the right. The photomicrographs at the bottom, based on phosphotau immunostaining in the same patient, show the tauopathy to be more intense in the left IFG than in the right

corticobasal degeneration [CBD], progressive supranuclear palsy [PSP]), and all three major forms of FTLD-TDP (types A, B and C) were represented. There were some disease-specific preferential patterns of atrophy. For example, AD almost always led to peak atrophy that included the temporoparietal junction; TDP-C almost always led to severe anterior temporal atrophy; Pick's disease routinely caused combined atrophy of anterior temporal and prefrontal cortex; and PSP and CBD tended to be associated with surprisingly modest cortical atrophy, usually in dorsal premotor or inferior frontal cortex. The

one common denominator of nearly all cases is the leftward asymmetry of the atrophy (Figs. 3 and 4). What is surprising is that the asymmetry is almost always maintained up to the time of death. The initial predilection of the language-dominant left hemisphere is therefore not a random event at disease onset but a core biological feature of the syndrome.

There was nearly equal representation of males and females in our autopsy cohort. Age of onset varied from 41 to 80 with a mean of 61 ± 8 years. Survival from symptom onset to death varied from 2 to 23 years with a mean of

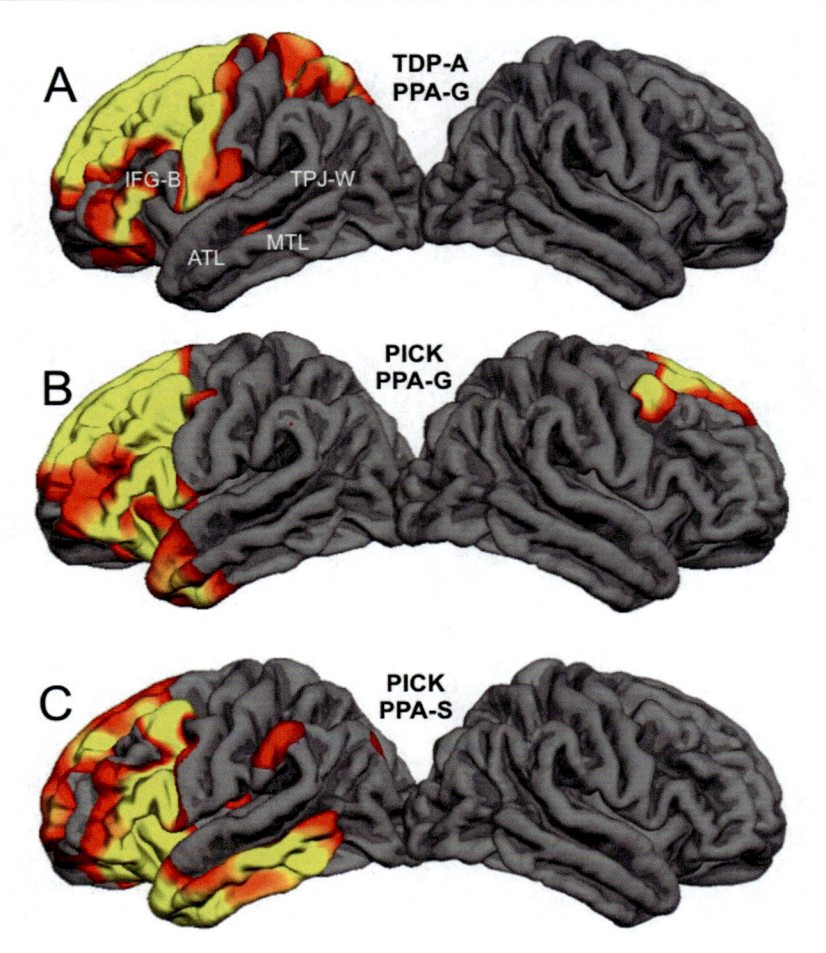

Fig. 4 Correspondences of pathology, atrophy, and syndrome. Quantitative MRI morphometry in three right-handed patients who had come to postmortem brain autopsy. Areas of significant cortical thinning compared to controls are shown in red and yellow. (**a**) Onset of PPA-G was at the age of 65. The scan was obtained 2 years after onset. At postmortem, the primary pathology was FTLD-TDP type A. (**b**) Onset of PP-G was at the age of 57. The scan was obtained 5 years after onset. At postmortem, the primary pathology was Pick's disease. (**c**) Onset of PPA-S was at the age of 62. The scan was obtained 5 years after onset. At postmortem, the primary pathology was Pick's disease. Despite the differences in neuropathology and clinical syndrome, the one common denominator is the profound leftward asymmetry of atrophy. Abbreviations: ATL anterior third of the temporal lobe, IFG-B inferior frontal gyrus where Broca's area is located, MTL middle third of the temporal lobe, TPJ-W temporoparietal junction where Wernicke's area is located

9.69 ± 3.93. Survival tended to be the longest for those with AD (10.8 ± 4.4) and FTLD-TDP type C (12.4 ± 2.6) and shortest for those with FTLD-TDP types A and B (5.8 ± 2.2). In keeping with these different rates of progression, FTLD-TDP aggregates extracted from subjects with type A pathology were shown to be more cytotoxic than aggregates from subjects with type C pathology [66].

The relationship of PPA variants to the underlying neuropathologic entity is probabilistic rather than absolute [67]. Autopsy data show that the vast majority of PPA-S cases have had TDP-C pathology but approximately 20% have had Pick's disease; the majority of PPA-G cases have had FTLD-tau (all types) but approximately 30% have had FTLD-TDP or AD; the majority of PPA-L cases have had AD but 30% have shown

FTLD-tau or FTLD-TDP. Figure 4 illustrates the clinicopathologic heterogeneity of PPA, namely that the same neuropathologic entity can cause more than one aphasic variant and that the same PPA variant may be caused by more than one neuropathologic entity. As shown in Fig. 4a and b, FTLD-TDP type A and Pick's disease cause nearly identical peak atrophy patterns that extend into the frontal components of the language network known to underlie grammar and fluency, giving rise to the concordant syndrome of PPA-G. Figure 4b and c raise challenging questions. They show atrophy patterns in two different patients with Pick's disease at autopsy, one with PPA-G (Fig. 4b), the other with PPA-S (Fig. 4c). As explained in the section on the anatomy of language, the semantic aphasia associated with Fig. 4c could be attributed to the posterior expansion of atrophy from ATL into more middle sections of the temporal lobe. However, it is difficult to understand why the patient in Fig. 4c was not also agrammatic since the frontal atrophy is nearly as extensive as in the other two cases with PPA-G. Perhaps this discrepancy can be blamed on vagaries of cortical morphometry performed on single subjects or, alternatively, on individual variations in the functional anatomy of the language network.

During life, cortical thinning (i.e., atrophy) and hypometabolism are the two most conspicuous markers of asymmetric neurodegeneration. Considerable progress has been made in exploring the potential cellular substrates of the asymmetrical atrophy (Fig. 3). For example, neurofibrillary tangles (NFT) (but not the amyloid plaques) of AD, tauopathy of CBD/PSP, Pick bodies, abnormal TDP-43 deposits of FTLD-TDP, activated microglia, and the extent of neuronal atrophy/loss tend to be more prominent in the left hemisphere than in the right hemisphere and also more prominent in language-related than other cortical areas of the left hemisphere [68–73]. In one left-handed PPA patient with documented right hemisphere language dominance and FTLD-TDP neuropathology, cortical atrophy and neurodegeneration markers were more prominent in the right hemisphere [74]. In at least some PPA patients with AD neuropathology, NFT may be more numerous in the language-related cortices of the left hemisphere than in the medial temporal areas, a distribution that deviates from the Braak and Braak pattern of neuropathology and underlies the atypical preservation of episodic memory in these patients [71, 73].

Quantitative investigations have also looked into the concordance of PPA subtypes with regional variations of neurodegeneration markers. A study of four right-handed PPA patients with FTLD-TDP type A neuropathology showed that the two patients with PPA-G displayed the highest density of TDP-43 precipitates in the frontal components of the language network, whereas the two with PPA-L displayed the highest density of precipitates in the temporoparietal components of the language network [69]. The cellular pathology in PPA can therefore asymmetrically target parts of the language-dominant hemisphere in a way that also mirrors the anatomical predilection patterns of the specific PPA variant. In the future, it would be useful to conduct similar analyses based on synaptic density. Some patients, especially those with PPA-G and FTLD-tau, may have no detectable cortical atrophy in the initial years of disease. These patients display abnormalities of functional connectivity, suggesting that physiological perturbations of the language network may precede atrophy [75]. In this group of patients, the neurodegeneration may be particularly prominent in subcortical white matter [76]. It is important to keep in mind that the identity of the disease marker that shows the best correlation with clinical dysfunction can change over time. Inclusions are likely to reflect leading indicators and would be expected to show the best correlation with clinical patterns in early disease stages, whereas neuronal death is likely to represent a trailing indicator more closely aligned with clinical patterns late in the disease.

In our autopsy cohort of 97 cases, a third of TDP-A cases had granulin (*GRN*) mutations. No other disease-causing mutations were encountered. Other studies have also shown that mutations in the *GRN* gene constitute the most common genetic correlate of familial PPA [77]. In such *GRN* families, some members may have PPA and others bvFTD [78, 79]. Rarely, all affected members of a *GRN* family will have PPA [80]. Even then, the type of aphasia may differ from one sibling to another and there is considerable heterogeneity of PPA subtypes associated with *GRN* mutations [81, 82]. The literature also contains rare associations of PPA with mutations in the presenilin (*PSEN1*), tau (*MAPT*), and *C9orf72* genes [83–85]. The most common clinical variants associated with dominantly inherited diseases are PPA-G and PPA-L, but rare cases of PPA-S have been reported [82]. The cellular neuropathology is FTLD-TDP type A in *GRN* mutations, FTLD-TDP type B in *C9orf72* mutations, and any one of the major FTLD-tau types in *MAPT* mutations. FTLD-TDP type C is very rarely, if ever, associated with known disease-causing mutations [86–88].

The heterogeneity of phenotypes encountered within *GRN* families shows that molecular underpinnings alone are not sufficient to account for the patterns of selective vulnerability and their clinical manifestation. The biological mechanisms underlying the selective and asymmetric involvement of the language-dominant hemisphere in PPA remain to be elucidated. One line of investigation has focused on the significantly higher frequency of learning disabilities, including dyslexia, in PPA patients and their first-degree relatives compared to control populations and patients with other dementias [89–91]. Follow-up research has replicated this association and raised the possibility that it may be peculiar to PPA-L [92]. Some families of PPA probands have strikingly high prevalence of developmental dyslexia in siblings or children [89]. We saw one family where seven of nine siblings of a PPA patient had findings indicative of developmental dyslexia [93]. As a group, the dyslexic siblings in this family had decreased functional connectivity within the language network although none had any findings of PPA. These observations led to the speculation that at least some cases of PPA could be arising on a developmentally or genetically based vulnerability of the left hemisphere language network. In some family members, this vulnerability would interfere with the acquisition of language and lead to dyslexia, while in others, it would make the language

network a locus of least resistance for the effects of an independently arising neurodegenerative process, leading to PPA [33]. So far, linkage studies addressing this hypothesis have not detected an association between PPA and known dyslexia genes [77]. Given the polygenic nature of dyslexia, negative results may reflect an insufficient number of cases.

Therapeutic Interventions

The heterogeneity of PPA highlights the need to individualize therapeutic approaches. Interventions in individual patients should target the underlying disease as well as the symptom complex. The former step requires the use of in vivo biomarkers. There are excellent CSF and PET biomarkers for detecting PPA patients with AD neuropathology and blood-based biomarkers may be on the horizon. However, current tau ligands for PET do not yet offer reliable identification of non-AD tauopathies associated with CBD, PSP, and Pick's disease [94]. When such biomarkers become available, they will enable the identification of PPA patients with FTLD-tau and, by exclusion, those with FTLD-TDP. The goal of these diagnostic investigations is to prescribe approved medications (e.g., cholinesterase inhibitors if AD) and to channel the patient to relevant disease-specific clinical trials. Although clinical examination is rarely sufficient to specify the underlying disease entity, we have found that prominent single word comprehension deficits that arise as the most salient feature of PPA are never associated with AD. The presence of this feature may therefore be used to forego AD biomarker testing.

The nonpharmacologic interventions aimed at the language impairment include speech therapy and brain stimulation modalities such as transcranial magnetic stimulation (TMS) or transcranial direct current stimulation (tDCS) [95]. Promising effects have been reported following left hemisphere tDTS in PPA-S [96]. If confirmed, this may well be the first time that brain stimulation will be shown to have therapeutic effects in an FTLD syndrome. Evidence for the effectiveness of speech-language therapy in PPA is emerging [97–99]. Utilization of this intervention modality is low in part due to the misconception that speech-language therapy is not appropriate for neurodegenerative syndromes where worsening is inevitable [100, 101]. An additional barrier is the lack of familiarity of speech-language pathologists with neurodegenerative conditions. Speech-language therapy in PPA requires personalization to fit the pattern of impairment and its evolution over time. For example, there are patients with modality-selective impairments of naming and word comprehension who could benefit from treatments emphasizing the relatively spared channels of language processing [65]. Additional questions to be resolved in the course of speech-language therapy include the relative usefulness of multicomponent, impairment-based, or compensatory approaches and the comparative benefits of group, dyadic, or patient-only approaches [102]. In each case, ecologically meaningful and statistically robust outcome measures will need to be devised.

Recent developments in telemedicine raise the possibility of delivering speech-language therapy in the home of the individual living with PPA [103, 104]. Communication Bridge, for example, is a two-arm, randomized control trial of speech-language intervention delivered through video chat for individuals with PPA [104]. The experimental arm uses a client-informed, dyadic approach for individuals with PPA and their communication partner. Impairment-based exercises using personalized stimuli and compensatory strategies are utilized to address real-world communication difficulties. The trial includes an individually tailored web application with native practice exercises and education materials that participants rehearse between treatment sessions. To evaluate whether treatment gains are relevant to the daily functions of the participant, outcomes are measured using a communication confidence rating scale and goal attainment scores. This method allows the targeting of individualized goals of high relevance to participants. In the future, transcranial stimulation could be combined with speech-language therapy to attain even more effective benefits [95].

Conclusions

Despite its relative rarity, PPA has led to conceptual advances in understanding the heterogeneity of dementia, the principles of selective brain vulnerability, and the neuroanatomy of the language network. PPA was arguably the first entity to show that there is more to dementia than memory loss, that the same clinical syndrome can be caused by multiple neuropathologies, that the same neuropathology can cause multiple syndromes, and that the relationship of syndrome to neuropathology is probabilistic rather than deterministic. Future work on PPA is likely to shed new light on the anatomical tropisms of neurodegenerative diseases and on the internal architecture of the language network.

Acknowledgments SUPPORT: R01 DCOO8552 and K23 DC014303 from the National Institute of Deafness and Communication Disorders; P30 AG013854 and R01 AG056258 from the National Institute on Aging, R01 NS085770 from the National Institute of Neurological Disorders and Stroke, the Davee Foundation and the Jeanine Jones Fund.

References

1. Coyle-Gilchrist TS, Dick KM, Patterson K, Rodriquez PV, Wehmann E, Wilcox A et al (2016) Prevalence, characteristics, and survival of frontotemporal lobar degeneration syndromes. Neurology 86:1736–1743
2. Rosenfeld M (1909) Die partielle Grosshirnatrophie. J Psychol Neurol 14:115–130
3. Pick A (1892) Ueber die Beziehungen der senilen Hirnatrophie zur Aphasie. Prager Medizinische Wochenschrift 17:165–167
4. Pick A (1904) Zur Symptomatologie der linksseitigen Schlaffenlappenatrophie. Monatsschr Psychiatr Neurol 16:378–388
5. Franceschi F (1908) Gliosi perivasculare in un caso de demenza afasica. Ann Neurol 26:281–290
6. Sérieux P (1893) Sur un cas de surdité verbale pure. Revue de Medecine 13:733–750
7. Dejerine J, Sérieux P (1897) Un cas de surdité verbale pure terminée par aphasie sensorielle, suivie d'autopsie. Comptes Rendues des Séances de la Société de Biologie 49:1074–1077
8. Mesulam MM (1982) Slowly progressive aphasia without generalized dementia. Ann Neurol 11(6):592–598

9. Mesulam MM (1987) Primary progressive aphasia – differentiation from Alzheimer's disease [editorial]. Ann Neurol 22(4):533–534

10. Mesulam M-M (2001) Primary progressive aphasia. Ann Neurol 49:425–432

11. Neary D, Snowden JS, Gustafson L, Passant U, Stuss D, Black S et al (1998) Frontotemporal lobar degeneration. A consensus on clinical diagnostic criteria. Neurology 51:1546–1554

12. Gorno-Tempini ML, Hillis A, Weintraub S, Kertesz A, Mendez MF, Cappa SF et al (2011) Classification of primary progressive aphasia and its variants. Neurology 76:1006–1014

13. Weintraub S, Rubin NP, Mesulam MM (1990) Primary progressive aphasia. Longitudinal course, neuropsychological profile, and language features. Arch Neurol 47(12):1329–1335

14. Mesulam M-M, Weintraub S (2008) Primary progressive aphasia and kindred disorders. In: Duyckaerts C, Litvan I (eds) Handbook of clinical neurology. Elsevier, New York, pp 573–587

15. Mesulam M-M, Wieneke C, Thompson C, Rogalski E, Weintraub S (2012) Quantitative classification of primary progressive aphasia at early and mild impairment stages. Brain 135:1537–1553

16. Mesulam M, Weintraub S (2014) Is it time to revisit the classification of primary progressive aphasia? Neurology 82:1108–1109

17. Sajjadi SA, Patterson K, Arnold RJ, Watson PC, Nestor PJ (2012) Primary progressive aphasia: a tale of two syndromes and the rest. Neurology 78:1670–1677

18. Wicklund MR, Duffy JR, Strand EA, Machulda MM, Whitwell JL, Josephs KA (2014) Quantitative application of the primary progressive aphasia consensus criteria. Neurology 82(13):1119–1126

19. Canu E, Agosta F, Imperiale F, Ferraro PM, Fontana A, Magnani G et al (2019) Northwestern anagram test-Italian (NAT-I) for primary progressive aphasia. Cortex 119:497–510

20. Mack JE, Chandler SD, Meltzer-Asscher A, Rogalski E, Weintraub S, Mesulam M-M et al (2015) What do pauses in narrative production reveal about the nature of word retrieval deficits in PPA. Neuropsychologia 77:211–222

21. Hillis AE, Tuffiash E, Caramazza A (2002) Modality-specific deterioration in naming verbs in nonfluent primary progressive aphasia. J Cog Neurosci 14:1099–1108

22. Mendez MF, Sabadash V (2015) Clinical amyloid imaging in logopenic progressive aphasia. Alzheimer Dis Assoc Disord 29:94–96

23. Tree J, Kay J (2014) Longitudinal assessment of short-term memory deterioration in logopenic variant primary progressive aphasia with post-mortem confirmed Alzheimer's disease pathology. J Neuropsychol 9:184–202

24. Josephs KA, Duffy J, Strand EA, Machulda MM, Senjem ML, Lowe VJ et al (2013) Syndromes domi-nated by apraxia of speech show distinct characteristics from agrammatic PPA. Neurology 81:337–345

25. Grossman M (2012) The non-fluent/agrammatic variant of primary progressive aphasia. Lancet Neurol 11:545–555

26. Thompson CK, Cho S, Hsu C-J, Wieneke C, Rademaker A, Weitner BB et al (2012) Dissociations between fluency and agrammatism in primary progressive aphasia. Aphasiology 26:20–43

27. Rogalski E, Cobia D, Martersteck AC, Rademaker A, Wieneke CA, Weintraub S et al (2014) Asymmetry of cortical decline in subtypes of primary progressive aphasia. Neurology 83:1184–1191

28. Mesulam M-M, Thompson CK, Weintraub S, Rogalski EJ (2015) The Wernicke conundrum and the anatomy of language comprehension in primary progressive aphasia. Brain 138:2423–2437

29. Catani M, Mesulam M (2008) The arcuate fasciculus and the disconnection theme in language and aphasia: history and current state. Cortex 44:953–961

30. Friederici AD (2011) The brain basis of language processing: from structure to function. Physiol Rev 91:1357–1392

31. Hickok G, Poeppel D (2007) The cortical organization of speech processing. Nat Rev Neurosci 8:293–402

32. Hagoort P (2013) MUC (memory, unification, control) and beyond. Front Psychol 4:1–13

33. Mesulam M-M, Rogalski E, Wieneke C, Hurley RS, Geula C, Bigio E et al (2014) Primary progressive aphasia and the evolving neurology of the language network. Nat Rev Neurol 10:554–569

34. Hodges JR, Patterson K, Oxbury S, Funnell E (1992) Semantic dementia. Progressive fluent aphasia with temporal lobe atrophy. Brain 115:1783–1806

35. Gorno-Tempini ML, Dronkers NF, Rankin KP, Ogar JM, Phengrasamy L, Rosen HJ et al (2004) Cognition and anatomy in three variants of primary progressive aphasia. Ann Neurol 55:335–346

36. Rogalski E, Cobia D, Harrison TM, Wieneke C, Weintraub S, Mesulam M-M (2011) Progression of language impairments and cortical atrophy in subtypes of primary progressive aphasia. Neurology 76:1804–1810

37. Hagoort P (2014) Nodes and networks in the neural architecture for language: Broca's region and beyond. Curr Opin Neurobiol 28:136–141

38. Bogen JE, Bogen GM (1976) Wernicke's region-where is it? Ann N Y Acad Sci 280:834–843

39. Patterson K, Nestor P, Rogers TT (2007) Where do you know what you know? The representation of semantic knowledge in the human brain. Nat Rev Neurosci 8:976–988

40. Adlam A-LR, Patterson K, Rogers TT, Nestor PJ, Salmond CH, Acosta-Cabronero J et al (2006) Semantic dementia and fluent primary progressive aphasia: two sides of the same coin? Brain 129:3066–3080

41. Bright P, Moss ME, Stamatakis EA, Tyler LK (2008) Longitudinal studies of semantic dementia: the rela-

tionship between structural and functional changes over time. Neuropsychologia 46:2177–2188

42. Lambon Ralph MA, Cipolotti L, Manes F, Patterson K (2010) Taking both sides: do unilateral anterior temporal lobe lesions disrupt semantic memory? Brain 133:3243–3255

43. Mesulam M-M, Rogalski E, Wieneke C, Cobia D, Rademaker A, Thompson C et al (2009) Neurology of anomia in the semantic subtype of primary progressive aphasia. Brain 132:2553–2565

44. Gefen T, Wieneke C, Martersteck AC, Whitney K, Weintraub S, Mesulam M-M et al (2013) Naming vs knowing faces in primary progressive aphasia. A tale of two hemispheres. Neurology 81:658–664

45. Hurley RS, Mesulam M-M, Sridhar J, Rogalski E, Thompson CK (2018) A nonverbal route to conceptual knowledge involving the right anterior temporal lobe. Neuropsychologia 117:92–101

46. Mesulam M-M, Wieneke C, Hurley RS, Rademaker A, Thompson CK, Weintraub S et al (2013) Words and objects at the tip of the left temporal lobe in primary progressive aphasia. Brain 136:601–618

47. Nakachi R, Muramatsu T, Kato M, Akiyama T, Saito F, Yoshino F et al (2007) Progressive prosopagnosia at a very early stage of frontotemporal lobar degeneration. Psychogeriatrics 7:155–162

48. Snowden J, Harris JM, Thompson JC, Kobylecki C, Jones M, Richardson AMT et al (2018) Semantic dementia and the left and right temporal lobes. Cortex 107:188–203

49. Hurley RS, Bonakdarpour B, Wang X, Mesulam M-M (2015) Asymmetric connectivity between the anterior temporal lobe and the language network. J Cog Neurosci 27:464–473

50. Bonakdarpour B, Hurley RS, Wang A, Fereira HR, Basu A, Chatrathi A et al (2019) Perturbations of language network connectivity in primary progressive aphasia. Cortex 121:468–480

51. Turken AU, Dronkers NF (2011) The neural architecture of the language comprehension network: converging evidence from lesion and connectivity analysis. Front Syst Neurosci 5:1–20

52. Hurley RS, Paller K, Rogalski E, Mesulam M-M (2012) Neural mechanisms of object naming and word comprehension in primary progressive aphasia. J Neurosci 32:4848–4855

53. Seckin M, Mesulam MM, Voss JL, Huang W, Rogalski EJ, Hurley RS (2016) Am I looking at a cat or a dog? Gaze in semantic variant of primary progressive aphasia is subject to excessive taxonomic capture. J Neurolinguistics 37:68–81

54. Mesulam M-M, Rader B, Sridhar J, Nelson MJ, Hyun J, Rademaker A et al (2019) Word comprehension in temporal cortex and Wernicke area: a PPA perspective. Neurology 92:e224–e233

55. Luo C, Makaretz S, Stepanivic M, Papadimitrou G, Quimby M, Palanivelu S et al (2020) Middle longitudinal fasciculus is associated with semantic processing deficits in primary progressive aphasia. NeuroImage 25:1–7

56. Mesulam M-M (1990) Large-scale neurocognitive networks and distributed processing for attention, language, and memory. Ann Neurol 28(5):597–613

57. Mesulam M-M (1998) From sensation to cognition. Brain 121:1013–1052

58. Thompson C, Cho S, Rogalski E, Wieneke C, Weintraub S, Mesulam M-M (2012) Semantic interference during object naming in agrammatic and logopenic primary progressive aphasia (PPA). Brain Lang 120:237–250

59. Gitelman DR, Nobre AC, Sonty S, Parrish TB, Mesulam M-M (2005) Language network specializations: an analysis with parallel task design and functional magnetic resonance imaging. NeuroImage 26:975–985

60. Sonty SP, Mesulam M-M, Weintraub S, Johnson NA, Parrish TP, Gitelman DR (2007) Altered effective connectivity within the language network in primary progressive aphasia. J Neurosci 27:1334–1345

61. Gorno-Tempini ML, Brambati SM, Ginex V, Ogar J, Dronkers NF, Marcone A et al (2008) The logopenic/phonological variant of primary progressive aphasia. Neurology 71:1227–1234

62. Binder JR (2015) The Wernicke area. Neurology 85:1–6

63. Binder J (2017) Current controversies on Wernicke's area and its role in language. Curr Neurol Neurosci Rep 17:1–10

64. Catani M, Mesulam M-M, Jacobsen E, Malik F, Martersteck A, Wieneke C et al (2013) A novel frontal pathway underlies verbal fluency in primary progressive aphasia. Brain 136:2619–2628

65. Mesulam M-M, Nelson MJ, Hyun J, Rader B, Hurley RS, Rademakers R et al (2019) Preferential disruption of auditory word representations in primary progressive aphasia with the neuropathology of FTLD-TDP type A. Cogn Behav Neurol 32(1):46–53

66. Laferriére F, Maniecka Z, Pérez-Berlanga M, Hruska-Plochan M, Gilhrspy L, Hock E-M et al (2019) TDP-43 extracted from frontotemporal lobar degeneration subject brains displays distinct aggregate assemblies and neurotoxic effects reflecting disease progression rates. Nat Neurosci 22:65–77

67. Mesulam M-M, Weintraub S, Rogalski EJ, Wieneke C, Geula C, Bigio EH (2014) Asymmetry and heterogeneity of Alzheimer and frontotemporal pathology in primary progressive aphasia. Brain 137:1176–1192

68. Gliebus G, Bigio E, Gasho K, Mishra M, Caplan D, Mesulam M-M et al (2010) Asymmetric TDP-43 distribution in primary progressive aphasia with progranulin mutation. Neurology 74:1607–1610

69. Kim G, Ahmadian SS, Peterson M, Parton Z, Memon R, Weintraub S et al (2016) Asymmetric pathology in primary progressive aphasia with progranulin mutations and TDP inclusions. Neurology 86:627–636

70. Kim G, Bolbolan K, Gefen T, Weintraub S, Bigio E, Rogalski E et al (2018) Atrophy and microglial dis-

tribution in primary progressive aphasia with transactive response DNA-binding protein-43 kDa. Ann Neurol 83(6):1096–1104

71. Gefen T, Gasho K, Rademaker A, Lalehzari M, Weintraub S, Rogalski E et al (2012) Clinically concordant variations of Alzheimer pathology in aphasic versus amnestic dementia. Brain 135:1554–1565

72. Giannini LAA, Xie SX, McMillan CT, Liang M, Williams A, Jester C et al (2019) Divergent patterns of TDP-43 and tau pathologies in primary progressive aphasia. Ann Neurol 85:630–643

73. Ohm DT, Fought AJ, Rademaker A, Kim G, Sridhar J, Coventry C et al (2019) Neuropathologic basis of in vivo cortical atrophy in the aphasic variant of Alzheimer's disease. Brain Pathol 30:332–344

74. Kim G, Vahedi S, Gefen T, Weintraub S, Bigio E, Mesulam M-M et al (2018) Asymmetric TDP pathology in primary progressive aphasia with right hemisphere language dominance. Neurology 90:e396–e403

75. Bonakdarpour B, Rogalski E, Wang A, Sridhar J, Mesulam M-M, Hurley RS (2017) Functional connectivity is reduced in early-stage primary progressive aphasia when atrophy is not prominent. Alzheimer Dis Assoc Disord 31:101–106

76. Caso F, Mandelli ML, Henry ML, Gesierich B, Bettcher BM, Ogar J et al (2014) In vivo signatures of nonfluent/agrammatic primary progressive aphasia caused by FTLD pathology. Neurology 82:239–247

77. Ramos EM, Dokuru ER, Van Berlo V, Wojta K, Wang Q, Huang AY et al (2019) Genetic screen in a large series of patients with primary progressive aphasia. Alzheimers Dement 15:553–560

78. Baker M, Mackenzie IR, Pickering-Brown SM, Gass J, Rademakers R, Lindholm C et al (2006) Mutations in progranulin cause tau-negative frontotemporal dementia linked to chromosome 17. Nature 442:916–919

79. Gass J, Cannon A, Mackenzie I, Boeve B, Baker M, Adamson J et al (2006) Mutations in *progranulin* are a major cause of ubiquitin-positive frontotemporal lobar degeneration. Hum Mol Genet 15(20):2988–3001

80. Mesulam M, Johnson N, Krefft TA, Gass JM, Cannon AD, Adamson JL et al (2007) Progranulin mutations in primary progressive aphasia. Arch Neurol 64:43–47

81. Coppola C, Oliva M, Saracino D, Pappata S, Zampella E, Cimini S et al (2019) One novel GRN null mutation, two different aphasia phenotypes. Neurobiol Age 87:e9–e14

82. LeBer I, Camuzat A, Hannequin D, Pasquier F, Guedj E, Rovelet-Lecrux A et al (2008) Phenotype variability in progranulin mutation carriers: a clinical, neuropsychological, imaging and genetic study. Brain 131:732–746

83. Simón-Sánchez J, Dopper EGP, Cohn-Hokke PE, Hukema RK, Nicolau N, Seelar H et al (2012) The clinical and pathological phenotype of

C9ORF72 hexanucleotide repeat expansions. Brain 135:723–735

84. Munoz DG, Ros R, Fatas M, Bermejo F, de Yebenes J (2007) Progressive nonfluent aphasia associated with a new mutation V363I in tau gene. Am J Alzheimers Dis Other Dement 22:294–299

85. Godbolt AK, Beck JA, Collinge J, Garrard P, Warren JD, Fox NC et al (2004) A presenilin 1 R278I mutation presenting with language impairment. Neurology 63:1702–1704

86. Josephs K, Whitwell JL, Murray ME, Parisi JE, Graff-Radford N, Knopman D et al (2013) Corticospinal tract degeneration associated with TDP-43 type C pathology and semantic dementia. Brain 136:455–470

87. Lee EB, Porta S, Michael Baer G, Xu Y, Suh E, Kwong LK et al (2017) Expansion of the classification of FTLD-TDP: distinct pathology associated with rapidly progressive frontotemporal degeneration. Acta Neuropathol 134:65–78

88. Rohrer JD, Lashley T, Schott JM, Warren JE, Mead S, Isaacs AM et al (2011) Clinical and neuroanatomical signatures of tissue pathology in frontotemporal lobal degeneration. Brain 134:2565–2581

89. Rogalski E, Johnson N, Weintraub S, Mesulam M-M (2008) Increased frequency of learning disability in patients with primary progressive aphasia and their first degree relatives. Arch Neurol 65:244–248

90. Mesulam M-M, Weintraub S (1992) Primary progressive aphasia: sharpening the focus on a clinical syndrome. In: Boller F, Forette F, Khachaturian Z, Poncet M, Christen Y (eds) Heterogeneity of Alzheimer's disease. Springer-Verlag, Berlin, pp 43–66

91. Rogalski EJ, Rademaker A, Wieneke C, Bigio EH, Weintraub S, Mesulam M-M (2014) Association between the prevalance of learning disabilities and primary progressive aphasia. JAMA Neurol 71:1576–1577

92. Miller ZA, Mandelli MA, Rankin KP, Henry ML, Babiak MC, Frazier DT et al (2013) Handedness and language learning disability differentially distribute in progressive aphasia variants. Brain 136(Pt 11):3461–3473

93. Weintraub S, Rader B, Coventry C, Sridhar J, Wood J, Guillaume K et al (2020) Familial langue network vulnerability in primary progressive aphasia. Neurology 22:1–9

94. Marquié M, Normandin MD, Vanderburg CR, Costantino IM, Bien EA, Rycyna LG et al (2015) Validating novel tau positron emission tomography tracer [F-18]-AV-1451 (T807) on postmortem brain tissue. Ann Neurol 78:787–800

95. Cotelli M, Manenti R, Ferrari C, Gobbi E, Macis A, Cappa SF (2020) Effectiveness of language training and non-invasive brain stimulation on oral and written naming performance in primary progressive aphasia: a meta-analysis and systematic review. Neurosci Biobehav Rev 108:498–525

96. Teichmann M, Lesoil C, Godard J, Vernet M, Bertrand A, Levy R et al (2016) Direct current stimulation over the anterior temporal areas boosts semantic processing in primary progressive aphasia. Ann Neurol 80:693–707

97. Carthery-Goulart MT, da Silveira AC, Machado TH, Mansur LL, Parente MM, Senaha MLH et al (2013) Interventions for cognitive impairments following primary progressive aphasia. Dement Neuropsychol 7:121–131

98. Volkmer A, Spector A, Meitanis V, Warren JD, Beeke S (2019) Effects of functional communication interventions for people with primary progressive aphasia and their caregivers: a systematic review. Aging Ment Health 28:1–13

99. Henry ML, Hubbard HI, Grasso SM, Mandelli ML, Wilson SM, Sathishkumar MT et al (2018) Retraining speech production and fluency in nonfluent/agrammatic primary progressive aphasia. Brain 141:1799–1814

100. Riedl L, Last D, Danek A, Diehl-Schmid J (2014) Long-term follow-up in primary progressive aphasia: clinical course and health care utilization. Aphasiology 28:981–992

101. Taylor C, Kingma RM, Croot K, Nickels L (2009) Speech pathology services for primary progressive aphasia: exploring an emerging area of preactice. Aphasiology 23:161–174

102. Jokel R, Meltzer J (2017) Group intervention for individuals with primary progressive aphasia and their spouses: who comes first? J Commun Disord 66:51–64

103. Meyer AM, Getz HR, Brennan DM, Hu TM, Friedman RB (2016) Telerehabilitation of anomia in primary progressive aphasia. Aphasiology 30:483–507

104. Rogalski E, Saxon M, McKenna H, Wieneke C, Rademaker A, Corden M et al (2016) Communication bridge: a pilot feasibility study of internet-based speech-language therapy for individuals with primary progressive aphasia. Alzheimers Dement (N Y) 2:213–221

Measuring Behavior and Social Cognition in FTLD

Katherine P. Rankin

Introduction

Among neurodegenerative disorders, the frontotemporal lobar degeneration (FTLD) syndromes have a uniquely focal, and in some cases devastating, impact on socioemotional behavior. Because of this, research investigating the clinical neuropsychology of non-Alzheimer's dementias over the past 20 years has needed to expand beyond traditional cognitive domains like memory in order to accurately represent what are often the primary deficits in patients with FTLD. This requirement has occasioned many significant advances in the measurement of socioemotional behavior and cognition in patients with progressive cognitive deficits, while also revealing many unforeseen challenges.

Any investigation into behavior in the FTLD syndromes must start with an understanding of the neuroanatomic circuits affected by these diseases, and their contribution to healthy social and emotional behavior. While the FTLD syndromes have diverse and even somewhat individualized patterns of initial neuronal damage and spread, it has become clear that specific intrinsically connected networks (ICNs) [1] show distinct patterns of selective vulnerability in the different major neurodegenerative syndromes [2]. Predictably, the ICN initially impacted in typical Alzheimer's disease (AD) syndrome is the brain's network for performing memory operations (i.e., the default mode network, or DMN) [3, 4]. However, in behavioral variant frontotemporal dementia (bvFTD), it is an ICN underpinning salience-driven attention (the salience network, SN) [5], and in semantic variant primary progressive aphasia it is an ICN involved in both general and socioemotional semantic knowledge (the semantic appraisal network, SAN) [2, 6]. The realization that damage to the SN is both necessary and sufficient to create a catastrophic behavior syndrome in bvFTD patients has had a widespread impact over the past decade, not only on the study of the FTLD syndromes, but on the way social affective neuroscientists have understood normal salience-driven attention [7]. Similarly, as the FTLD community has consistently confirmed the existence of an overlapping but qualitatively different set of socioemotional impairments associated with svPPA syndrome, particularly when right frontotemporal circuits involved in the evaluation of semantic information become damaged [8–10], it has highlighted the central importance of the SAN for key socioemotional functions such as visceral emotional

K. P. Rankin (✉)
Department of Neurology, The Memory and Aging Center, University of California San Francisco, San Francisco, CA, USA
e-mail: kate.rankin@ucsf.edu

© Springer Nature Switzerland AG 2021
B. Ghetti et al. (eds.), *Frontotemporal Dementias*, Advances in Experimental Medicine and Biology 1281, https://doi.org/10.1007/978-3-030-51140-1_4

experience and expression, evaluating hedonic signals, and decoding social and emotional cues [11, 12].

Using Behavior to Evaluate Key FTLD Brain Circuits

Because the primary utility of neuropsychological testing in the FTLD syndromes is to identify and diagnose patients, and to mark the degree of disease progression, the best neuropsychological tests are those that reflect the functional integrity of these circuits that are specific to FTLD. Measurement is complicated by the cognitive deficits, loss of insight, and failure to cooperate fully with testing procedures that are typical of patients with these syndromes; thus, the best tests ideally show some degree of robustness and domain specificity in reflecting the intended circuits despite these obstacles. FTLD researchers have examined many such tests over the past two decades; while the majority of tests purporting to show specific brain–behavior relationships have been validated using structural MRI data reflecting neurodegenerative atrophy as a measure of brain circuit damage [13–15], investigators are increasingly showing correspondence of such tests to functional connectivity in these networks [5, 16, 17]. This is a particularly welcome advance, not only because such tests will be more sensitive in patients in the earliest stages of neurodegeneration before frank atrophy can be discerned, but also because there is substantial evidence that bvFTD-type behavior deficits may emerge as a result of a "disconnection syndrome" affecting these FTLD-specific circuits, both in patients with a more "subcortical" variation of bvFTD with little cortical involvement [18], and also in patients with other FTLD syndromes such as Progressive Supranuclear Palsy (PSP) [19].

Socioemotional deficits in the FTLD syndromes must be conceptualized as encompassing two distinct targets that are evaluated through very different measurement approaches: (1) socioemotional behavior or reactivity, and (2) socioemotional cognition or information processing. The first category, socioemotional behavior, includes all the physiological and behavioral *responses produced by the patient* in a socioemotional context or simply when presented with socioemotional stimuli. Typically, FTLD researchers investigating altered reactivity in patients with the FTLD syndromes have relied on precision laboratory approaches such as standard psychophysiological measurement and detailed observation and behavioral coding [20–23], including facial and vocal emotion coding [24, 25]. Increasingly, task-based fMRI studies are also being used to directly quantify altered patterns of neural response in FTLD patients. These measurement approaches have the benefit of scientific rigor and reproducibility, but also require special equipment and sophisticated user training for data collection and analysis, thus are suited only to research investigations and cannot easily translate into neuropsychological assessment approaches for patient identification and classification in broad clinical or even clinical trial settings.

A more holistic but imprecise measurement of FTLD patient behavioral responsiveness is also performed via observational methods such as home visit-based ethnographic coding and real-world challenge paradigms [21], which again require sophisticated training and cannot scale up for clinical use, but may be less equipment -heavy. Clinician quantification of spontaneous behavior during patient visits [26, 27] is a less precise but more scalable quantitative option. A fourth approach to documenting patients' holistic socioemotional responsiveness that has been widely used with FTLD patients is interview- or questionnaire-based informant reports on the patient's typical behavior, attitudes, and personality [10, 28–31]. These require little to no specialized training for data collection and have published normative reference sets available for interpretation, thus are accessible options for clinical and clinical trial use, though of course they lack the precision afforded by laboratory-based observational measures.

The second main category of socioemotional measurement in the FTLD syndromes is the more traditional measurement of social cognition, or more specifically, whether or not the patient is able to *identify and discriminate socioemotional stimuli and make correct interpretations of social scenarios*. Based in the tradition of neuropsychological assessment of cognition, this approach emphasizes direct face-to-face testing of the patient's abilities, evaluating whether the patient is able to achieve a test score at a normal threshold. The most obvious and widely used example of this in the FTLD field is testing whether a patient can accurately discriminate or name emotional faces from static pictures or videos, though investigators have developed or adapted many tests evaluating socioemotional cue detection and interpretation of social scenarios [15, 32, 33]. This category of assessment is often idealized because theoretically it represents the most practically useful balance between precise but equipment-heavy laboratory approaches and more holistic but nebulous informant-based observational measures, in part because a precedent has already been set for performing neuropsychological evaluations with Alzheimer's disease patients both in clinical trials and at clinic, thus direct socioemotional testing has the potential to be rigorous yet still scale up for general use. However, the direct-testing approach for evaluation of social cognition in the FTLD syndromes has met with a number of important pitfalls that have limited success in this area, despite substantial effort being expended to develop and validate such tests.

The most obvious caveat to the value of these face-to-face tests of socioemotional functioning is that they are limited to the cooperativeness and cognitive ability of the patient, and thus become invalid measurement tools once a patient becomes behaviorally disordered enough that they refuse to participate or fail to engage appropriately with the testing situation, a reaction that often occurs only a year or two after initial presentation in the bvFTD patients for whom this testing is most important, that is, at a merely intermediate stage of disease progression. A core deficit in bvFTD, corresponding to its primary selectively vulnerable ICN, the salience network, is an early and progressive loss of the ability to care about meeting expectations in social situations, of which test-taking is a clear example. Thus, paradoxically, the most relevant socioemotional brain network to test in these patients is the one for which moderate dysfunction creates invalid test performance, logically limiting the utility of any face-to-face test for measuring SN progression. Furthermore, the SN is involved in attending and responding to stimuli that are personally and emotionally salient to the individual, thus early deficits are likely to appear as a highly personalized and focal failure to notice or care about a specific event or person in the course of daily life. Thus, at earlier phases of disease progression, these deficits can be difficult to evoke or observe during a homogenous, standardized testing situation, limiting the value of face-to-face testing for early detection of SN involvement. Finally, investigation of the range of normal intrinsic functional connectivity in healthy individuals over the past decade suggests that there is substantial normal inter-individual variability in SN function among healthy individuals [16, 17], corresponding to an equally wide range of normal socioemotional functioning. This means that, unlike a typical "achievement" test of memory or language functioning, where healthy individuals demonstrate a fairly narrow range of premorbid ability and thus cognitive deficits are easy to ascertain, it will not be initially apparent whether an individual's score on a test of SN function represents a relative deficit (i.e., a decline from premorbid functioning) for that individual, or is simply a reflection of lifelong weakness in the socioemotional domain.

While "achievement" in relation to the SN is exceedingly difficult to test in a valid face-to-face manner, direct patient testing of socioemotional abilities related to SAN function has been marginally more successful. In particular, a number of tests measuring comprehension of social and emotional semantics have been developed and validated for use with FTLD patients, and have been particularly effective at identifying the subset of bvFTD and svPPA patients who have right frontotemporal dysfunction [15, 34–37]. Some more difficult tests seem to be sensitive to early neurodegenerative dysfunction [38], though patients with semantic loss quickly hit the psychometric floor of such tests. As with neuropsychological tests measuring any cognitive domain, however, a patient's performance can be confounded by cognitive deficits in other domains or by dysfunction in correlated ICNs. For example, the impact of generalized, non-social semantic deficits must be accounted for with any test purporting to measure comprehension of specifically socioemotional semantics. Emotion naming tests that provide labels and ask the patient to select among them rather than asking them to spontaneously label the emotion may be more accurate with FTLD patients.

Finally, a last challenge to socioemotional testing in the FTLD syndromes has been the need for tests to be valid for use in multicultural contexts across international boundaries. While cultural and linguistic influences are important when translating tests in traditional neuropsychological domains like memory and executive functioning, the interpretation of whether a socioemotional behavior is normal or abnormal is wholly dependent on the cultural context of the patient, making many such measures completely culturally invalid despite correct linguistic translation. While formal tests of patients' ability to recognize basic socioemotional cues such as facial emotions are marginally more cross-culturally robust, tests of higher order comprehension of social stimuli, and the expressive social behavior of the patients themselves, are often subject to very different rules governing social context and expectations. Thus far, the most effective response of the worldwide FTLD community has been for clinical researchers to design and validate sets of socioemotional measures within their own cultural context. This results in customized, local, and thus more accurate, socioemotional testing; however, this approach is extremely time intensive and inefficient, and leaves out countries and cultures without investigators focused on developing such tasks. While independent suites of effective socioemotional tests have been developed and validated for use with FTLD syndrome patients in North American, European, and Australian/New Zealander English-speaking contexts, as well as for South American Spanish-speaking patients [39–42], work is only beginning to develop such culturally valid batteries in Chinese- or Hindi-speaking patients. Individual groups in Europe have also developed and used socioemotional tests in non-English-speaking FTLD patients [43], and the GENFI study in Europe is currently developing and validating a set of socioemotional tests for use across its more than 10 linguistically distinct countries. Furthermore, even within linguistically similar groups, important cultural differences in social norms and expectations can substantially influence testing, further confounding the question of whether certain behaviors or test results reflect clinically abnormal socioemotional behavior. Using a patient as their own control over time to detect declines from baseline could partly mitigate these issues from a research design perspective; however, it is clear that development of cross-cultural evaluation methods for socioemotional functioning is a critical, ongoing need in the FTLD community.

Practical Socioemotional Testing in FTLD

The following is a discussion of a number of practical tests of socioemotional functions validated for use in FTLD patients and which have no major equipment or training requirements, thus have the potential for broader adoption either in clinical trials or in clinic. Rather than attempting to provide a comprehensive review of all such tests, the following section takes this opportunity to provide a more in-depth discussion of both published and unpublished data on a number of tests developed and validated by our group, with which we are of course most familiar. For some of the socioemotional functions discussed, there are additional valid alternatives developed by other investigators that are currently being used with FTLD patients, which can be found in the literature should the reader be interested in further study of these measures.

Measures Reflecting Salience Network Dysfunction in FTLD

As described earlier, the network most central to the bvFTD syndrome is the cingulo-insular-subcortical SN [44–46], a network that integrates sensory stimuli with interoceptive, hedonic, affective, and motivational information via the anterior insula (AI), and which works to adjust attention and emotional arousal on the basis of the relevance of these signals. Subcortical SN nodes providing interoceptive signals include the dorsomedial thalamus, hypothalamus, amygdala, and midbrain periaqueductal gray (PAG) [5], and a node in the anterior cingulate provides motivational and top-down regulation in the SN [47, 48]. Because the SN is the hub of selective vulnerability in bvFTD, tests that are sensitive and specific to SN dysfunction are the most ideal measures for use in detecting and monitoring progression of the bvFTD syndrome. For the reasons explained earlier, face-to-face patient tests of the SN have been elusive, but a number of observer-based behavioral measures have been successfully validated as both sensitive and specific to SN dysfunction.

Revised Self-Monitoring Scale (RSMS). One of the measures most extensively validated for use in FTLD patients, and for which there is strong support for its correspondence with salience network structure and function, is the RSMS [49]. With neurodegenerative disease patients, this 13-item questionnaire has primarily been used as an informant-reported observational measure of the patient's typical spontaneous behavior in real-life social settings. The RSMS has been thoroughly validated for use in other non-neurodegenerative populations, and has good psychometric characteristics, including strong internal consistency and test–retest reliability [50, 51] as well as appropriate construct validity to predict related traits such as social anxiety and sociability [52]. It measures sensitivity and responsiveness to subtle emotional expressions during face-to-face interactions. Sample items include "In conversations, the patient is sensitive to even the slightest change in the facial expression of the other person they are conversing with," and "In social situations, the patient has the ability to alter their behavior if they feel that something else is called for."

The RSMS has been used in a number of studies with neurodegenerative disease patients, and seems to be particularly sensitive to the core social deficits inherent to bvFTD syndrome. Multiple studies have shown that not only do bvFTD patients score abnormally low, but they also are rated as having worse social sensitivity than patients with other syndromes such as svPPA, PSP, or AD [13, 17, 53] (Fig. 1). Importantly, there is strong evidence for the correspondence of RSMS score to structural integrity of the SN. In one study, Shdo et al. (2017) [13] examined 275 individuals with bvFTD, svPPA, nfvPPA, PSP, and AD, as well as healthy older controls, and performed a voxel-based morphometry whole-brain analysis to discover linear relationships between RSMS score and structural gray matter volume regardless of syndromes. They found that RSMS score predicted volume in medial and lateral temporal as well as inferior frontal structures, and found that subcortical structures including the amygdala, thalamus, caudate, putamen, and globus pallidus corresponded with RSMS. RSMS score has also been correlated with white matter integrity measured via DTI analysis. Examining 145 participants, including 105 patients with bvFTD, svPPA, and nfvPPA as well as 40 healthy controls, Toller et al. (2020) [54] used TBSS to perform a voxel-wise analysis of whole-brain white matter tracts to determine how white matter FA was predicted by RSMS score. Higher RSMS score was significantly associated with higher FA values in the right uncinate fasciculus (UF), a white matter structure that connects the anterior temporal lobe with inferior frontal regions. This effect was not only found in the entire sample (patients plus controls), but was also found to be significant in the subset of 40 healthy controls alone, suggesting the RSMS is sensitive not only to disease-related social deficits, but also mild normal variations in white matter structural integrity in the right UF. Patients with bvFTD and svPPA both had significantly lower FA in the right (M ± SD; bvFTD: 0.35 ± 0.01; svPPA: 0.36 ± 0.01; NC: 0.41 ± 0.01) and left (bvFTD: 0.34 ± 0.01; svPPA: 0.33 ± 0.01; NC: 0.39 ± 0.01) UF compared to NCs, though neither right nor left UF integrity was abnormal in patients with nfvPPA. This study also found an interesting dissociation between svPPA and bvFTD patients in terms of contribution of right frontotemporal gray matter volume versus white matter integrity

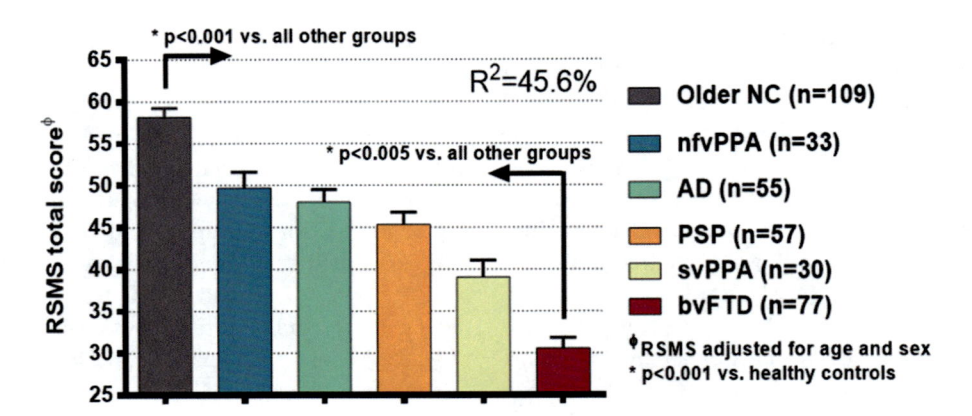

Fig. 1 The RSMS (Revised Self-Monitoring Scale, Lennox and Wolfe [49]) shows high accuracy differentiating bvFTD patients from healthy older controls and from all other FTLD syndromes and AD

of the UF to RSMS score. Though FA in the UF did not significantly predict RSMS score in the bvFTD group alone, lower gray matter volume in the right medial OFC ROI did. Thus, though right UF integrity alone was able to predict socioemotional sensitivity in both healthy controls and svPPA syndrome patients, in patients with bvFTD gray matter volume in the right medial OFC cortex adjacent to the UF tract predicted socioemotional behavior than UF integrity.

Perhaps the strongest evidence for the value of using informant-reported RSMS to reflect SN integrity comes from studies directly linking RSMS score with functional connectivity in the SN. In a study of 168 participants, including patients with bvFTD, svPPA, nfvPPA, PSP, and AD syndromes, and healthy controls, Toller et al. (2018) [17] found that higher functional connectivity in the SN significantly predicted higher RSMS score, even controlling for atrophy and for diagnostic group membership. Region-of-interest analysis of connectivity within the SN showed that RSMS score could be predicted by connectivity among cortical structures (bilateral AI and ACC), as well as between the right AI and subcortical structures. Not only did this result occur across the whole sample, but, in a second analysis of a subsample of 98 healthy controls across the age spectrum (age range 19–87), RSMS score showed this same significant linear relationship, again suggesting that RSMS score not only reflects disease-related social insensitiv-

ity caused by damage and dysfunction in the SN, but it actually reflects normally occurring individual differences in socioemotional sensitivity in a manner specific enough to reflect normal SN connectivity.

Finally, the longitudinal sensitivity of the RSMS to disease progression in bvFTD patients has also been established, using a large multi-site cohort of 475 participants who had behavioral Mild Cognitive Impairment, bvFTD, or were asymptomatic controls (Toller 2020) [53]. This study showed a main effect of disease severity (measured by CDR® plus NACC FTLD score) in which RSMS decreased significantly at every disease stage as CDR worsened. Linear mixed effects models showed a significant main effect of disease duration in which RSMS decreases linearly in patients at a rate of 5 points per year (average RSMS slope per year: -2.13 ± 1.29) in bvFTD. An additional voxelwise analysis of structural brain volume showed that more rapid declines on the RSMS were associated with faster progression of gray matter atrophy in regions of the SN and SAN, including the right AI, dorsal ACC, and OFC. Sub-regional analysis by disease progression showed some evidence that worsening score on the RSMS tracks with loss of volume in the thalamus, primarily in very mild and mild disease stages, but to a lesser degree later in the disease. This study also examined whether the RSMS was able to differentiate between mutation carriers and non-carriers, though no differences were found. This study also used the Zarit Burden measure to show that worse RSMS score predicts greater self-reported burden for bvFTD caregivers, which provides evidence that the loss of socioemotional sensitivity measured by the RSMS reflects a clinically meaningful symptom in FTLD patients, an important consideration for its potential inclusion as a clinical trial outcome measure.

Interpersonal Adjectives Scales – Warmth Subscale (IAS-Warmth). Another measure for which there is solid evidence that it reflects SN function is the Warmth subscale of the IAS [55]. The IAS has been used with FTLD patients as an informant-reported personality questionnaire designed to measure trait-level expression of interpersonal characteristics, including dimensions of dominance and affiliation. Informants rate on an 8-point Likert scale the degree to which patients can be accurately described using a list of adjectives descriptive of an interpersonal behavior (e.g., "self-assured"; "shy"; "iron-hearted"). The IAS as a whole produces ratings of 8 traits: dominance, arrogance, coldness, introversion, submissiveness, ingenuousness, warmth, and extraversion. Numerous studies of its characteristics in FTLD patients have been published, which include evidence that depending on their syndrome, patients show characteristic changes in personality [56], that these changes correspond with neuropsychological features [57], and that patients with the most significant personality changes, i.e., those with bvFTD and svPPA, are least likely to be aware of those personality changes [58]. Studies have also demonstrated the unique patterns of atrophy corresponding with different IAS facets [59].

While a number of IAS facets appear to change with FTLD, one dimension in particular seems to correspond with SN structure and function, the Warmth–Coldness axis. One study of the structural gray matter correlates of the IAS included 239 individuals, comprised of patients diagnosed with bvFTD, svPPA, nfvPPA, corticobasal syndrome, PSP, and AD syndrome, as well as healthy controls [59]. Warmth scores were significantly lower in the bvFTD and svPPA groups, and the scores for their opposite trait, Coldness, were significantly higher than in NCs, though this effect was not seen in any of the other neurodegenerative disease syndromes. Warmth score correlated with primarily right-sided structures reflecting SN and SAN regions, specifically the gray matter volume in predominantly right frontal and anterior temporal lobe structures, including the right posterior caudal orbitofrontal cortex, the right anterior and medial insula, the subgenual cingulate region, the anterior medial prefrontal cortex, the right caudate head, the anterior parahippocampus and hippocampus, amygdala, and superior temporal pole. This apparent correspondence of IAS-Warmth with both SN and SAN, however, was further clarified in another study directly examining the relationship between IAS-Warmth and functional connectivity. Toller et al. (2019) [16] studied 132 participants, including healthy controls and patients with bvFTD, svPPA, nfvPPA, and AD. Their analysis showed that while all patient groups had significantly lower IAS-Warmth scores than NCs, only the bvFTD group scored outside of the normal range (-2 SD below average), while the other patient groups averaged less than 1 SD below average (T-score $\pm$ SD; bvFTD: 31.1 $\pm$ 2.7, AD: 46.65 $\pm$ 2.3, svPPA: 42.3 $\pm$ 2.8, nfvPPA: 47.73 $\pm$ 2.6; NC 56.3 $\pm$ 2.5). They found a significant interaction between diagnostic group and time (premorbid versus current IAS-Warmth) showing that patients with bvFTD ($p < 0.013$) and svPPA ($p < 0.013$) had significant declines from premorbid to current warmth compared to the NC group. When they investigated whether current functional connectivity in the SN, SAN, or DMN predicted current warmth score across the entire sample (controls and patients), only higher connectivity in the SN predicted higher current IAS-Warmth score after atrophy correction, and SAN connectivity dropped out of the model, suggesting that SN connectivity was the primary driver of IAS-Warmth score. The study furthermore documented the divergence across different patients within the bvFTD and svPPA groups in the degree of change they experienced in both warmth and SN connectivity from estimated premorbid to current levels. SN connectivity did not predict IAS-Dominance, thus connectivity in this network appears to be specific to warmth. Overall, these studies suggest that interpersonal warmth is a trait characteristic that decreases in many patients with the FTLD syndrome, and that it acts as an index of the degree of decrease in SN connectivity in those patients.

Measures Illustrating the Gating Mechanism of the SN in Socioemotional Behavior

Evidence has accrued from numerous sources that the SN plays a role in activating certain downstream networks [60], and this influences higher order social cognitive processes like moral reasoning [61] and theory of mind [62] that predominantly rely on those downstream ICNs [63], particularly the DMN and the frontoparietal adaptive task network (FPN) [64]. Studies using these kinds of complex socioemotional tasks with FTLD patients suggest that bvFTD-related SN dysfunction directly impacts their decision-making, likely through the mechanism of altering patients' ability to notice salient cues while processing complex, often multimodal information from the realistic scenarios these tasks often employ. Patients with most neurodegenerative syndromes, including AD, show nonspecific impairments on these types of difficult face-to-face tests because of their complexity and reliance on multiple cognitive functions [33, 62], thus they are not useful for differential diagnosis or for isolating SN dysfunction; however, in FTLD patients, scores often do correlate with SN structure and function, thus they may have some utility for early detection.

Chiong and colleagues [61] performed a moral reasoning task during task-based fMRI with healthy older controls and 13 early bvFTD patients, and found not only that bvFTD patients tended to respond to scenarios in a more utilitarian manner but also that this tendency was directly explained by differences in the way bvFTD patients activated the underlying networks. When healthy controls deliberated about moral scenarios where personal relationships

often supersede practical logic, SN activation led to activation of the DMN; however, in the early bvFTD patients, the SN failed to exert this downstream influence on the DMN, and instead the FPN was more likely to activate, resulting in decisions that relied on logical rather than personal considerations.

This same relationship has been found when FTLD patients are asked to perform complex social reasoning and make "theory of mind" inferences from realistic social scenarios. One direct face-to-face task that has been used in a number of studies of FTLD patients to test this ability [15, 33] is The Awareness of Social Inference Test (TASIT)–Social Inference Enriched subtest (SI-E) [65]. This test has shown differential diagnostic utility in discriminating bvFTD-specific socioemotional deficits in comparison to patients with the aphasia syndromes or AD, and has even shown sensitivity to social reasoning deficits in PSP [33]. This subtest of the TASIT consists of 16 short video vignettes in which either a visual or verbal enrichment is given to provide unambiguous cues about the social situation, the state of knowledge of each character, and the characters' social intentions. After watching each video, four questions related to what the characters in the video do, think, say, or feel are used to assess the patient's understanding of the social interaction they just viewed. To correctly interpret the videos, realistic contextual and paralinguistic cues have to be selectively attended to and integrated, which makes the TASIT-SIE an appropriate tool to measure the patient's ability to make ToM social inferences from complex dynamic multimodal information in real life. While the ecological validity of this test, and by extension its ability to reflect real-life impairments, is its strength, the drawback is that, to correctly interpret a scenario, patients must successfully perform many complex social and non-social cognitive operations, thus task failure may be due to problems with memory, executive function, language, or visuospatial functioning, not necessarily due to deficits in socioemotional processing per se. This makes the test, and all other complex social cognition tests like it, nonspecific for differential diagnosis among the

FTLD syndromes because patients from many groups underperform or fail [33].

However, the TASIT-SIE has been used to model how SN damage in FTLD patients can create a powerful cascade effect in which patients are unable to make incorrect social inferences even when other brain networks required for theory of mind are intact. In that study, Rijpma and colleagues [62] performed the TASIT-SIE with a total of 179 participants, including patients with bvFTD, PSP, and AD syndromes. They examined how gray matter volume in three ICNs, the DMN, FPN, and SN, influenced patients' ability to infer others' intentions using the TASIT-SIE, and found that while lower volume in all three networks appeared to predict poorer social reasoning, when all three networks were included in the same model, task performance was entirely accounted for by SN volume and not by DMN and FPN volume. While numerous other studies have found that theory of mind reasoning is typically performed by the DMN and FPN, this study used the realistic video vignettes of the TASIT-SIE to further confirm the SN gating hypothesis, showing that if a patient is unable to recognize and selectively attend to key socioemotional cues while viewing a complex scene due to SN damage, then they cannot successfully engage the DMN and FPN to perform downstream social cognitive operations.

Measures Reflecting Semantic Appraisal Network Dysfunction in FTLD

While the SN is of central importance in understanding the socioemotional deficits in the FTLD syndromes, a second intrinsically connected network is also responsible for many of the severe symptoms we associate with FTLD. As described earlier, the SAN appears to be selectively vulnerable not only in the svPPA syndrome, but also in a large proportion of patients with bvFTD [18]. A seed placed at the medial boundary of either temporal pole reveals a normally occurring ICN that includes both medial anterior temporal and subgenual cingulate cortex, as well as the head of the caudate, the nucleus accumbens, and the amygdala [2, 35, 66, 67]. The functions of this network are less well understood, but FTLD patients with early and focal damage to this network have a disproportionate number of deficits reading emotions and other social cues, even compared to other bvFTD patients, and are more likely to be described as having social disinhibition (i.e., rudeness) during the first year of their disease [18, 37]. Recent work suggests that these socioemotional symptoms may reflect disruption of the inferior orbitofrontal/basal ganglia structures that facilitate hedonic evaluation, and their links to anterior temporal areas involved in semantic knowledge [35]. Tests that reflect the ability to make judgments about socioemotional semantics, including emotion and social cue identification, seem to provide the best reflection of SAN function, though a more thorough investigation of this connection is still needed in the FTLD field.

Tests of Emotion Reading: Numerous studies of emotion reading ability in patients with the FTLD syndromes have been published, using a variety of testing modalities and stimuli. While these measures are typically direct, face-to-face patient tests, they have included measures of facial, vocal, and bodily expression of emotion, both static (picture) and dynamic (video clip) stimuli, single modality versus multimodal, as well as full expression versus degraded or morphing gradations of emotion. A number of commonalities have arisen out of these studies that can provide guidance for using these measures in patients with the FTLD syndromes. First, the identification of emotion reading impairments in patients is confounded by the fact that there is a wide range of emotion reading ability among healthy individuals, thus a "low average" score on a test could be a normal, unchanged performance for one patient whose premorbid capacity for emotion reading was always low, while it could represent a substantial deficit for another patient whose premorbid ability was high. Emotion tests that are particularly difficult, such as those requiring fine-grained distinctions of facial affect, or those requiring affect reading with complex or low signal-to-noise stimuli, are particularly problematic to interpret for these reasons, as dementia patients rapidly approach an impaired threshold on these tests, often hitting floor levels early in the disease process [68]. Tests placing a high demand on non-emotional systems, such as auditory processing for prosody tests, visuospatial processing for facial emotion tests, or semantic processing for tests where fine distinctions among emotions must be spontaneously labeled, can result in test failure despite intact emotion systems. For these reasons, even with unambiguous stimuli designed to realistically reflect real-life emotion reading ability, patients without emotion system dysfunction, such as patients with AD syndrome, may perform just as poorly as patients with bvFTD or svPPA, making these tests imprecise for differential diagnosis in FTLD.

However, when understood as a window into regional neural dysfunction, emotion reading tests do have some clinical utility. One test that has been used in studies with FTLD patients is another subtest of the TASIT, called the Emotion Evaluation Test (EET), which consists of 28 videos of about ~20-second duration in which actors express emotions through congruent facial, vocal, and gestural modalities using realistic but semantically neutral scripts. Patients are asked to select the correct label for the video from among the six basic emotions (happy, sad, disgusted, surprised, angry, frightened, plus neutral). One benefit of these stimuli is that they are realistically dynamic (i.e., video based) and multimodal (with multiple, congruent cues of the emotion), thus are undemanding enough that patients with non-social cognitive deficits are less likely to fail them due to difficulty processing the stimuli, yet they still detect true emotion reading impairments. Patients with bvFTD and svPPA both perform more poorly than AD patients on the TASIT-EET [15, 18, 68]. Studies examining the gray matter correlates of the test show broad correspondence with bilateral temporal lobe structures, as well as inferior frontal cortex [15, 69].

Another similar but freely available test that is increasingly used with FTLD patients is the Dynamic Affect Recognition Test (DART). This shorter measure is a video-based test comprised of 12 20-second vignettes of an actor expressing one of the six basic emotions (happy, surprised, sad, angry, fearful, disgusted) via ecologically realistic and congruent facial, vocal, and postural cues, and with semantically neutral scripts. Each vignette involves only one actor, whose facial emotions were coded via the Facial Action Coding System (FACS) [70] to ensure valid and reliable emotional expression. One study [71] compared FTLD patients' performance on the DART, the TASIT-EET, and a third static facial emotion test, the Affect Matching subtest of the Comprehensive Affect Testing System (CATS-AM) [72]. The study examined emotion identification performance on the three tasks with 153 participants, including patients with bvFTD, svPPA, nfvPPA, PSP, and AD, along with older healthy controls. ROC modeling comparing all three tests showed they had similar sensitivity discriminating bvFTD patients from healthy controls (AUC: DART = 0.94; TASIT-EET = 0.91; CATS = 0.81). A VBM analysis of gray matter showed that DART score had a linear relationship with volume in the left superior medial temporal pole, left medial temporal pole, left inferior temporal pole, left hippocampus, left caudate head nucleus accumbens, right caudate head/nucleus accumbens, right dorsal anterior insula, and the right anterior inferior temporal gyrus. When the DART was analyzed controlling for patients' semantic loss, in order to model emotion reading distinct from any language deficits that might interfere with their ability to correctly label the videos, the resulting VBM revealed primarily right-sided structures, retaining insula and caudate/accumbens regions, while correlations with ventrolateral temporal regions did not appear. Overall, the TASIT-EET and the DART appear to function similarly in patients with the FTLD syndromes, and reveal very similar structural anatomic substrates in ventral frontal and temporal regions.

Another recent study [69] examined the functional correlates of the TASIT-EET, and demonstrated more conclusively the correspondence of these emotion reading tests with SAN function rather than SN or other brain networks. In this study, a total of 185 individuals were studied, including patients with bvFTD, svPPA, nfvPPA, PSPs, and AD, along with older healthy controls. As expected, they found that patients with bvFTD and svPPA had significantly lower TASIT-EET scores than controls. However, they also modeled TASIT-EET score against functional connectivity in the SN and SAN ICN, and found that when SAN and SN were modeled together, mean connectivity in the SAN independently predicted TASIT-EET scores but the SN did not, when SAN was accounted for, and this strong relationship with SAN connectivity remained after atrophy correction and error checking for the confounding effects of diagnostic group membership. ROI-level analysis showed that connectivity between the right anterior temporal pole and other parts of the SAN, including regions of the subgenual cingulate involved in making hedonic evaluations, was primarily responsible for differences in patients' performance on the TASIT-EET. Overall, these results suggest that though these video-based emotion labeling tests seem to correspond with general frontotemporal anatomy in FTLD syndrome patients, the functional anchors for test performance are the right temporal pole and the medial orbitofrontal cortex, regions located in the SAN and which are selectively vulnerable in a subset of patients with FTLD.

Tests of Socioemotional Semantics. Another aspect of socioemotional cognition that deteriorates in patients with some variants of FTLD is the knowledge of social rules, expectations, categories, and concepts. This symptom often becomes specifically associated with a "right temporal" syndrome of bvFTD or svPPA [34, 36, 73], though it is actually an element of many bvFTD or svPPA patients' socioemotional behavior deficits. Many FTLD clinics administer informal tests of "famous faces" to determine if a patient's semantic knowledge about individuals they should know is intact, though these tests are notoriously difficult to standardize because they rely on culturally specific knowledge of famous individuals from different historic epochs.

Another approach to evaluating socioemotional semantics is to directly test patients'

knowledge of social information. One test that has been designed and validated for FTLD patients is the Social Norms Questionnaire (SNQ) [74, 75], a set of 22 yes–no questions asking patients whether specific social behaviors are appropriate. The measure contains two subscales, one control scale (*Overadhere*) with items describing appropriate behaviors (e.g., "Is it socially acceptable to tell someone your age?"), and the test scale (*Break*) with items describing inappropriate behaviors (e.g., "Is it socially acceptable to spit on the floor?"). The test was initially validated using data from 200 well-educated neurologically healthy predominantly Caucasian controls aged 45–90 to confirm response agreement was over 90% for all items. A study [75] of the differential diagnostic utility of the test with 283 patients, including those with bvFTD, svPPA, nfvPPA, PSP, CBS, and AD, showed that only bvFTD and svPPA patients had significantly higher Break norms errors than controls, even though additional patient groups (PSP, nfvPPA, and bvFTD) made significantly more Overadhere control task errors than the healthy group. VBM analysis showed that SNQ score had a strong linear relationship with gray matter volume in right>left anterior temporal lobes as well as inferior frontal regions and the head of the caudate. Thus, the Break subtest of the SNQ appears to be specific to the socioemotional semantic loss of bvFTD and svPPA patients, and corresponds with structural anatomy found in the SAN.

Another newer test of socioemotional semantics, designed for use with FTLD patients but modeled on non-social tests of semantic knowledge like the Peabody Picture Vocabulary Test (PPVT) and the Pyramids and Palm Trees test (PPT), is the Social Interaction Vocabulary Test (SIVT). This 18-item test examines patients' ability to understand and label interpersonal dynamics. It is designed as a multiple-choice picture/word socioemotional vocabulary matching task. Patients are given a word describing a specific socioemotional interaction (e.g., "consoling"), and then are asked to choose from among 4 pictures depicting two interacting individuals in order to identify the image that best represents the meaning of the word. Pictures are carefully matched for visual complexity, body posture, and gestures of the actors, and are arranged with 3 subscales corresponding to easy, moderate, and difficult vocabulary words. The test was normed with 52 neurologically healthy, predominantly Caucasian and well-educated individuals aged 21–87, who performed at or near ceiling. When SIVT performance was examined in 213 individuals, including patients with bvFTD, svPPA, nfvPPA, PSP, CBS, AD, and healthy controls, only patients with bvFTD and svPPA had significantly lower total SIVT scores than healthy controls. A VBM analysis showed a linear relationship between SIVT score and both frontotemporal and subcortical structures known to be in the SAN, including right>left subgenual cingulate, temporal pole, and the nucleus accumbens and rostral caudate structures (Fig. 2). These results suggest that the ability to make these socioemotional associations is not only mediated by temporal structures known to convey semantic knowledge, but also dependent on ventromedial frontal-subcortical circuits involved in hedonic evaluation.

Conclusions

Numerous tests have been used to evaluate socioemotional behavior in FTLD patients, and most if not all of them are capable of revealing the impairments typical of bvFTD; however, the most useful tests are those that are sensitive and specific enough to reveal the dysfunction of the two key brain networks known to mediate the majority of socioemotional deficits in the FTLD syndromes: the salience and semantic appraisal networks. While SN integrity has proven difficult to test in face-to-face clinical encounters, questionnaire-based accounts of patient behavior from informants who are in a position to observe them well yield a surprisingly accurate reflection of individual differences in SN function. Socioemotional functions associated with the SAN are easier to evaluate through traditional patient-facing cognitive testing, and may include tests of emotion reading, and assessment of the vocabulary for social interactions and personal traits. Further progress needs to be made by the field toward refining and fully validating the various potential tests to precisely

Fig. 2 SAN network gray matter correlates of the Social Interaction Vocabulary Task (SIVT). VBM analysis of regions of the brain showing a linear relationship between socioemotional semantic loss and volume loss in patients with neurodegenerative disease

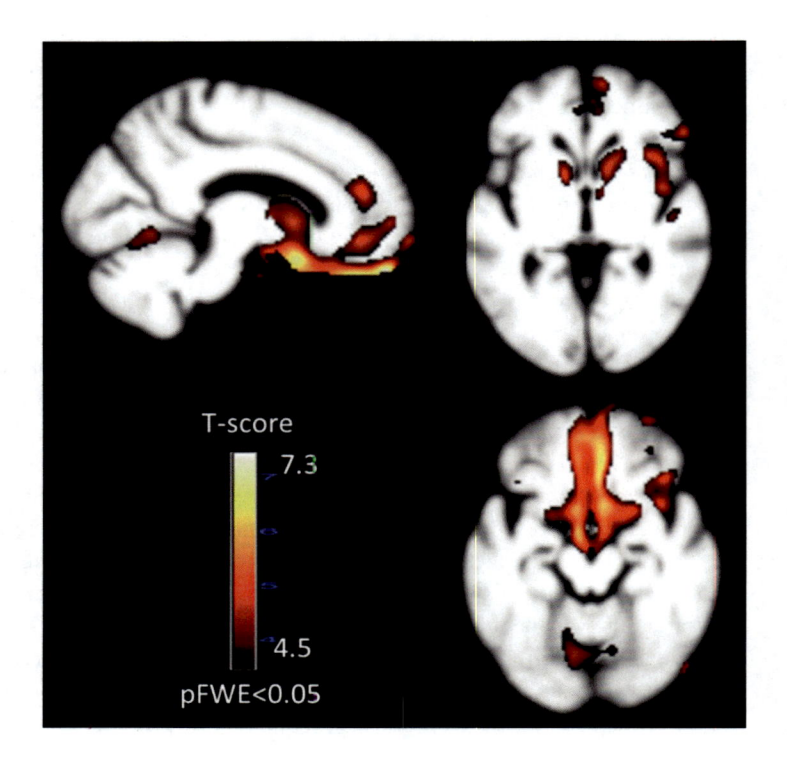

evaluate socioemotional functioning in patients with the FTLD syndromes; however, it is clear that this set of disorders has provided the impetus for much more neuroscientifically rigorous evaluation of this cognitive domain than has previously been available in the field of clinical neuropsychology.

References

1. Yeo BT, Krienen FM, Eickhoff SB, Yaakub SN, Fox PT, Buckner RL et al (2016) Functional specialization and flexibility in human association cortex. Cereb Cortex 26(1):465–465
2. Seeley WW, Crawford RK, Zhou J, Miller BL, Greicius MD (2009) Neurodegenerative diseases target large-scale human brain networks. Neuron 62(1):42–52
3. Zhou J, Greicius MD, Gennatas ED, Growdon ME, Jang JY, Rabinovici GD et al (2010) Divergent network connectivity changes in behavioural variant frontotemporal dementia and Alzheimer's disease. Brain J Neurol 133(Pt 5):1352–1367. PMCID: PMC2912696
4. Greicius MD, Srivastava G, Reiss AL, Menon V (2004) Default-mode network activity distinguishes Alzheimer's disease from healthy aging: evidence from functional MRI. Proc Natl Acad Sci U S A 101(13):4637–4642
5. Seeley WW, Menon V, Schatzberg AF, Keller J, Glover GH, Kenna H et al (2007) Dissociable intrinsic connectivity networks for salience processing and executive control. J Neurosci 27(9):2349–2356
6. Seeley WW, Zhou J, Kim EJ (2012) Frontotemporal dementia: what can the behavioral variant teach us about human brain organization? The neuroscientist: a review journal bringing neurobiology. Neuroscientist 18(4):373–385
7. Bolt T, Nomi JS, Arens R, Vij SG, Riedel M, Salo T et al (2020) Ontological dimensions of cognitive-neural mappings. Neuroinformatics 18(3):451–463.
8. Coon EA, Whitwell JL, Parisi JE, Dickson DW, Josephs KA (2012) Right temporal variant frontotemporal dementia with motor neuron disease. J Clin Neurosci 19(1):85–91. PMCID: PMC3248959
9. Seeley WW, Bauer AM, Miller BL, Gorno-Tempini M, Kramer JH, Weiner M et al (2005) The natural history of temporal variant frontotemporal dementia. Neurology 64(8):1384–1390
10. Rankin KP, Gorno-Tempini M, Allison SC, Stanley CM, Glenn S, Weiner MW et al (2006) Structural anatomy of empathy in neurodegenerative disease. Brain 129:2945–2956
11. Stein JL, Wiedholz LM, Bassett DS, Weinberger DR, Zink CF, Mattay VS et al (2007) A validated network of effective amygdala connectivity. NeuroImage 36(3):736–745
12. Patterson K, Nestor PJ, Rogers TT (2007) Where do you know what you know? The representation of semantic knowledge in the human brain. Nat Rev Neurosci 8(12):976–987
13. Shdo SM, Ranasinghe KG, Gola KA, Mielke CJ, Sukhanov PV, Miller BL et al (2018) Deconstructing empathy: neuroanatomical dissociations between affect sharing and prosocial motivation using a patient lesion model. Neuropsychologia 116(Pt A): 126–135.
14. Goodkind MS, Sturm VE, Ascher EA, Shdo SM, Miller BL, Rankin KP et al (2015) Emotion recognition in frontotemporal dementia and Alzheimer's disease: a new film-based assessment. Emotion 15(4):416–427
15. Kipps CM, Nestor PJ, Acosta-Cabronero J, Arnold R, Hodges JR (2009) Understanding social dysfunction in the behavioral variant of frontotemporal dementia: the role of emotion and sarcasm processing. Brain 132(Pt 3):592–603.
16. Toller G, Yang WFZ, Brown J, Ranasinghe K, Shdo S, Kramer JH et al (2019) Divergent patterns of loss of interpersonal warmth in frontotemporal dementia syndromes are predicted by altered intrinsic network connectivity. Neuroimage Clin 22:101729. (ePub)
17. Toller G, Brown J, Sollberger M, Shdo SM, Bouvet L, Sukhanov P et al (2018) Individual differences in socioemotional sensitivity are an index of salience network function. Cortex 103:211–223
18. Ranasinghe KG, Rankin KP, Pressman PS, Perry DC, Lobach IV, Seeley WW et al (2016) Distinct subtypes of behavioral variant frontotemporal dementia based on patterns of network degeneration. JAMA Neurol 73(9):1078–1088
19. Gardner RC, Boxer AL, Trujillo A, Mirsky JB, Guo CC, Gennatas ED et al (2013) Intrinsic connectivity network disruption in progressive supranuclear palsy. Ann Neurol 73(5):603–616
20. Sturm VE, Brown JA, Hua AY, Lwi SJ, Zhou J, Kurth F et al (2018) Network architecture underlying basal autonomic outflow: evidence from frontotemporal dementia. J Neurosci Off J Soc Neurosci 38(42):8943–8955
21. Sturm VE, Sible IJ, Datta S, Hua AY, Perry DC, Kramer JH et al (2018) Resting parasympathetic dysfunction predicts prosocial helping deficits in behavioral variant frontotemporal dementia. Cortex 109:141–155
22. Guo CC, Sturm VE, Zhou J, Gennatas ED, Trujillo AJ, Hua AY et al (2016) Dominant hemisphere lateralization of cortical parasympathetic control as revealed by frontotemporal dementia. Proc Natl Acad Sci U S A 113(17):E2430–E2439
23. Pressman PS, Shdo S, Simpson M, Chen KH, Mielke C, Miller BL et al (2018) Neuroanatomy of shared conversational laughter in neurodegenerative disease. Front Neurol 9:464

24. Nevler N, Ash S, Jester C, Irwin DJ, Liberman M, Grossman M (2017) Automatic measurement of prosody in behavioral variant FTD. Neurology 89(7):650–656

25. Kumfor F, Hazelton JL, Rushby JA, Hodges JR, Piguet O (2019) Facial expressiveness and physiological arousal in frontotemporal dementia: phenotypic clinical profiles and neural correlates. Cogn Affect Behav Neurosci 19(1):197–210

26. Rankin KP, Santos-Modesitt W, Kramer JH, Pavlic D, Beckman V, Miller BL (2008) Spontaneous social behaviors discriminate behavioral dementias from psychiatric disorders and other dementias. J Clin Psychiatry 69(1):60–73

27. Toller G, Cobigo Y, Ljubenkov PA, Appleby BS, Dickerson BC, Domoto-Reilly K et al (2020) Social Behavior Observer Checklist measures early frontotemporal progression. Annals of Neurology. Under review

28. Mendez MF, Yerstein O, Jimenez EE (2020) Vicarious embarrassment or "fremdscham": overendorsement in frontotemporal dementia. J Neuropsychiatry Clin Neurosci 32(3):274–279

29. Sturm VE, Yokoyama JS, Seeley WW, Kramer JH, Miller BL, Rankin KP (2013) Heightened emotional contagion in mild cognitive impairment and Alzheimer's disease is associated with temporal lobe degeneration. Proc Natl Acad Sci U S A 110(24):9944–9949

30. Sollberger M, Stanley CM, Wilson SM, Gyurak A, Beckman V, Growdon M et al (2009) Neural basis of interpersonal traits in neurodegenerative diseases. Neuropsychologia 47(13):2812–2827

31. Rosen HJ, Allison SC, Schauer GF, Gorno-Tempini M, Weiner MW, Miller BL (2005) Neuroanatomical correlates of behavioural disorders in dementia. Brain 128:2612–2625

32. Synn A, Mothakunnel A, Kumfor F, Chen Y, Piguet O, Hodges JR et al (2018) Mental states in moving shapes: distinct cortical and subcortical contributions to theory of mind impairments in dementia. J Alzheimers Dis 61(2):521–535

33. Shany-Ur T, Poorzand P, Grossman SN, Growdon ME, Jang JY, Ketelle RS et al (2012) Comprehension of insincere communication in neurodegenerative disease: lies, sarcasm, and theory of mind. Cortex 48(10):1329–1341

34. Luzzi S, Baldinelli S, Ranaldi V, Fabi K, Cafazzo V, Fringuelli F et al (2017) Famous faces and voices: differential profiles in early right and left semantic dementia and in Alzheimer's disease. Neuropsychologia 94:118–128

35. Yang WF, Toller G, Shdo S, Kotz SA, Brown J, Seeley WW et al (2020) Resting functional connectivity in the semantic appraisal network predicts accuracy of emotion identification. Unpublished manuscript

36. Binney RJ, Henry ML, Babiak M, Pressman PS, Santos-Santos MA, Narvid J et al (2016) Reading words and other people: a comparison of exception word, familiar face and affect processing in the left and right temporal variants of primary progressive aphasia. Cortex 82:147–163

37. Rankin KP, Salazar A, Gorno-Tempini ML, Sollberger M, Wilson SM, Pavlic D et al (2009) Detecting sarcasm from paralinguistic cues: anatomic and cognitive correlates in neurodegenerative disease. NeuroImage 47(4):2005–2015

38. Vogel A, Jorgensen K, Larsen IU (2020) Normative data for emotion hexagon test and frequency of impairment in behavioral variant frontotemporal dementia, Alzheimer's disease and Huntington's disease. Appl Neuropsychol Adult 14:1-6

39. Baez S, Pinasco C, Roca M, Ferrari J, Couto B, Garcia-Cordero I et al (2019) Brain structural correlates of executive and social cognition profiles in behavioral variant frontotemporal dementia and elderly bipolar disorder. Neuropsychologia 126:159–169

40. García-Cordero I, Sedeño L, de la Fuente L, Slachevsky A, Forno G, Klein F et al (2016) Feeling, learning from and being aware of inner states: interoceptive dimensions in neurodegeneration and stroke. Philos Trans R Soc Lond B Biol Sci 371(1708):1–10

41. Baez S, Manes F, Huepe D, Torralva T, Fiorentino N, Richter F et al (2014) Primary empathy deficits in frontotemporal dementia. Front Aging Neurosci 6:262

42. Couto B, Manes F, Montanes P, Matallana D, Reyes P, Velasquez M et al (2013) Structural neuroimaging of social cognition in progressive non-fluent aphasia and behavioral variant of frontotemporal dementia. Front Hum Neurosci 7:467

43. Dodich A, Cerami C, Canessa N, Crespi C, Iannaccone S, Marcone A et al (2015) A novel task assessing intention and emotion attribution: Italian standardization and normative data of the story-based empathy task. Neurol Sci 36(10):1907–1912

44. Dopper EG, Rombouts SA, Jiskoot LC, Heijer T, de Graaf JR, Koning I et al (2013) Structural and functional brain connectivity in presymptomatic familial frontotemporal dementia. Neurology 80(9):814–823

45. Seeley WW, Crawford R, Rascovsky K, Kramer JH, Weiner M, Miller BL et al (2008) Frontal paralimbic network atrophy in very mild behavioral variant frontotemporal dementia. Arch Neurol 65(2):249–255

46. Whitwell JL, Jack CR, Parisi JE, Knopman DS, Boeve BF, Petersen RC et al (2011) Imaging signatures of molecular pathology in behavioral variant frontotemporal dementia. J Mol Neurosci 45(3):372

47. Craig AD, Craig A (2009) How do you feel – now? The anterior insula and human awareness. Nat Rev Neurosci 10(1):59–70

48. Craig AD (2003) Interoception: the sense of the physiological condition of the body. Curr Opin Neurobiol 13(4):500–505

49. Lennox RD, Wolfe RN (1984) Revision of the self-monitoring scale. J Pers Soc Psychol 46(6):1349–1364

50. Anderson LR (1991) Test-retest reliability of the revised self-monitoring scale over a two-year period. Psychol Rep 68(3):1057–1058

51. O'Cass A (2000) A psychometric evaluation of a revised version of the Lennox and wolfe revised self-monitoring scale. Psychol Mark 17(5):397–419

52. Wolfe RN, Lennox RD, Cutler BL (1986) Getting along and getting ahead: empirical support for a theory of protective and acquisitive self-presentation. J Pers Soc Psychol 50(2):356

53. Toller G, Ranasinghe K, Cobigo Y, Staffaroni A, Appleby B, Brushaber D et al (2020) Revised self-monitoring scale: a potential endpoint for frontotemporal dementia clinical trials. Neurology 94(22):e2384–e2395

54. Toller G, Mandelli ML, Cobigo Y, Rosen HJ, Kramer JH, Miller BL et al (2020) Right uncinate fasciculus supports socioemotional sensitivity in health and neurodegenerative disease. Brain Imaging Behav. Under review

55. Jerry S. Wiggins, IAS. and Adjective scales. (1995). Interpersonal adjective scales: professional manual. 1st Edition. Psychological Assessment Resources. Lutz, FL

56. Rankin KP, Kramer JH, Mychack P, Miller BL (2003) Double dissociation of social functioning in frontotemporal dementia. Neurology 60(2):266–271

57. Sollberger M, Stanley CM, Ketelle R, Beckman V, Growdon M, Jang J et al (2012) Neuropsychological correlates of dominance, warmth, and extraversion in neurodegenerative disease. Cortex 48(6):674–682

58. Rankin KP, Baldwin E, Pace-Savitsky C, Kramer JH, Miller BL (2005) Self-awareness and personality change in dementia. J Neurol Neurosurg Psychiatry 75(5):632–639

59. Sollberger M, Rosen HJ, Shany-Ur T, Ullah J, Stanley CM, Laluz V et al (2014) Neural substrates of socioemotional self-awareness in neurodegenerative disease. Brain Behav 4(2):201–214

60. Menon V, Uddin LQ (2010) Saliency, switching, attention and control: a network model of insula function. Brain Struct Funct 214(5–6):655–667

61. Chiong W, Wilson SM, D'Esposito M, Kayser AS, Grossman SN, Poorzand P et al (2013) The salience network causally influences default mode network activity during moral reasoning. Brain 136(6):1929–1941

62. Rijpma MG, Shdo SM, Toller G, Shany-Ur T, Kramer JH, Miller BL et al (2020) Salience-driven attention is pivotal to understanding others' intentions. Cognitive Neuropsychology. In press

63. Greene JD, Nystrom LE, Engell AD, Darley JM, Cohen JD (2004) The neural bases of cognitive conflict and control in moral judgment. Neuron 44(2):389–400

64. Dosenbach NU, Fair DA, Miezin FM, Cohen AL, Wenger KK, Dosenbach RA et al (2007) Distinct brain networks for adaptive and stable task control in humans. Proc Natl Acad Sci U S A 104(26):11073–11078

65. McDonald S, Bornhofen C, Shum D, Long E, Saunders C, Neulinger K (2006) Reliability and validity of the awareness of social inference test (TASIT): a clinical test of social perception. Disabil Rehabil 28(24):1529–1542

66. Yeo BT, Krienen FM, Sepulcre J, Sabuncu MR, Lashkari D, Hollinshead M et al (2011) The organization of the human cerebral cortex estimated by intrinsic functional connectivity. J Neurophysiol 106(3):1125–1165

67. Guo CC, Gorno-Tempini ML, Gesierich B, Henry M, Trujillo A, Shany-Ur T et al (2013) Anterior temporal lobe degeneration produces widespread network-driven dysfunction. Brain J Neurol 136(Pt 10):2979–2991

68. Ranasinghe KG, Rankin KP, Lobach IV, Kramer JH, Sturm VE, Bettcher BM et al (2016) Cognition and neuropsychiatry in behavioral variant frontotemporal dementia by disease stage. Neurology 86(7):600–610

69. Shany-Ur T, Lee A, Harley A, Poorzand P, Grossman S, Nguyen L et al (2020) Neuroanatomy of identifying dynamic emotions in patients with neurodegenerative diseases. Unpublished Manuscript

70. Ekman P, Friesen WV (1978) The facial affect coding system. Consulting Psychologists Press, Palo Alto

71. Radke A. (2020) The Dynamic Affect Recognition Test: Behavior and neuroanatomy in neurodegenerative disease. Unpublished manuscript

72. Froming K, Levy M, Schaffer S, Ekman P. (2006) The comprehensive affect testing system. Psychology Software, Inc. Available online at: http://www.psychologysoftware.com/CATS.htm

73. Suarez-Gonzalez A, Crutch SJ (2016) Relearning knowledge for people in a case of right variant frontotemporal dementia. Neurocase 22(2):130–134

74. Kramer JH, Mungas D, Possin KL, Rankin KP, Boxer AL, Rosen HJ et al (2014) NIH EXAMINER: conceptualization and development of an executive function battery. J Int Neuropsychol Soc 20(1):11–19

75. Rankin KP, Adhimoolam B, Nguyen LA, Toofanian P, Zhong A, Pishori A et al (2020) The Social Norms Questionnaire: Psychometrics and neural correlates. Unpublished manuscript

Clinical Update on *C9orf72*: Frontotemporal Dementia, Amyotrophic Lateral Sclerosis, and Beyond

Dario Saracino and Isabelle Le Ber

Introduction

The last decades have marked a turning point in the knowledge of frontotemporal dementia (FTD) and amyotrophic lateral sclerosis (ALS), two diseases forming a clinical continuum and sharing common pathogenic mechanisms and genetic etiologies. In particular, the identification of the pathogenic GGGGCC repeat expansion in the first intron of the *C9orf72* gene, responsible for familial FTD and ALS, in 2011, represented a break-

through discovery in these domains. Most healthy individuals in control populations harbor less than 24 GGGGCC repeats, and most often only two to eight repeats [1]. Although the exact pathogenic threshold remains uncertain, expansions above 30 repeats are usually considered pathogenic in most studies. However, the vast majority of patients carry much larger expansions ranging from several hundred to thousand repeats [1, 2].

The discovery of the *C9orf72* expansion led to strong scientific emulation and important advances, but the molecular and biological mechanisms by which the expansion might produce neurodegeneration are not completely elucidated. Three pathogenic mechanisms, not mutually exclusive, are proposed: (i) loss of function caused by reduced C9orf72 protein levels in brain, possibly mediated by methylation of a CpG island upstream to the expansion repeat [3], (ii) toxicity of mutant RNA that aggregates into nuclear foci, and (iii) accumulation of dipeptide-repeat (DPR) proteins generated by non-ATG dependent translation of the expanded repeat [4]. At the pathological level, the mutation is mostly associated with FTLD-TDP type A or type B subtypes, or with a combination of both [5, 6], together with widespread p62-positive inclusions, as well as intranuclear RNA foci and cytoplasmic DPR inclusions, the lesional signatures of *C9orf72* disease [7, 8].

The relevance of *C9orf72* expansion within the FTD/ALS spectrum is now well established [9]. In

D. Saracino
Hôpitaux de Paris, Hôpital Pitié-Salpêtrière, Paris, France

Inserm U1127, CNRS UMR 7225, Institut du Cerveau et la Moelle épinière (ICM), Hôpital Pitié-Salpêtrière, Sorbonne Université, Paris, France

IMMA, Centre de référence des démences rares ou précoces, Hôpitaux de Paris, Hôpital Pitié-Salpêtrière, Paris, France

I. Le Ber (✉)
Hôpitaux de Paris, Hôpital Pitié-Salpêtrière, Paris, France

Inserm U1127, CNRS UMR 7225, Institut du Cerveau et la Moelle épinière (ICM), Hôpital Pitié-Salpêtrière, Sorbonne Université, Paris, France

IMMA, Centre de référence des démences rares ou précoces, Hôpitaux de Paris, Hôpital Pitié-Salpêtrière, Paris, France

ICM-UMRS_1127, ICM, Hôpital de la Salpêtrière, Paris Cedex 13, France
e-mail: isabelle.leber@upmc.fr

B. Ghetti et al. (eds.), *Frontotemporal Dementias*, Advances in Experimental Medicine and Biology 1281, https://doi.org/10.1007/978-3-030-51140-1_5

most countries, *C9orf72* represents the most frequent genetic cause of familial FTD (8–25%) and ALS (27–47%) and is the gene most commonly found in families with a combination of both disorders (65–80%) [1, 2, 9–12], although mutation frequencies vary according to specific demographic and geographic contexts. Notably, the prevalence of the expansion is significantly lower in Asian populations than in those of European and North-American ancestry [12, 13], both being different from intermediate frequencies in Indian populations [14]. It is also the most frequent genetic cause of "apparently sporadic" FTD or ALS implicated in 4–20% of patients without any family history of neurodegenerative diseases [1, 2, 9, 15, 16], suggesting that genetic testing should be proposed in apparently sporadic cases as well.

A decade after the discovery of *C9orf72*, several genetic and clinical questions remain unsolved. For example, the occurrence of an anticipation – at the clinical and molecular level – in *C9orf72* kindreds is still debated. There have been an increasing number of reports suggesting that the spectrum of *C9orf72*-related phenotypes is broader than the behavioral variant of FTD (bvFTD) and ALS, encompassing psychiatric disorders and, possibly, parkinsonian syndromes; however, the factors driving this variability are still unknown. The reliable cutoff for the pathogenic repeat number and the implication of intermediate alleles in FTD, ALS, or in other neurological phenotypes are still uncertain as well. All these questions have a significant impact not only in clinical practice for diagnosis and genetic counseling but also in a research context for the initiation of therapeutic trials. In this chapter, we will address all those issues and summarize the recent updates about clinical aspects of *C9orf72* disease, focusing on both the common and the less typical phenotypes.

Characteristics of *C9orf72*-related Frontotemporal Dementia

The behavioral variant of FTD is, by far, one of the most common phenotypes in *C9orf72* carriers. Patients carrying the expansion fulfill the possible and probable bvFTD criteria rather well, but the sensitivity is lower compared to neuropathologically confirmed bvFTD cohorts [17]. Overall, no specific cognitive and behavioral profile distinguishes *C9orf72* patients from other genetic or sporadic FTD cases [18]. However, specific cognitive domains might be altered early in the disease course since mild deficits in cognitive inhibition [19] and semantic access [20] have been evidenced in presymptomatic *C9orf72* carriers. As part of the clinical phenotype, a proportion of patients with bvFTD or FTD/ALS phenotype may develop parkinsonism occurring during disease course whose severity is milder than in other genetic or non-genetic FTD [18, 21].

The patterns of neuroimaging changes are distinctive according to FTD genotypes, and *C9orf72* patients have relatively symmetrical and diffuse volume loss [22] and variable rates of atrophy, with some patients progressing rapidly and others very slowly [22]. Multiple studies have shown early involvement of subcortical structures, such as basal ganglia, and hippocampus, amygdala [22–24], and thalamus as well [25]. Involvement of the cerebellum is seen particularly in patients with *C9orf72* mutations [26].

Overall, the mean disease duration is shorter in *C9orf72* (6.4 ± 4.9 years) than in other FTD genotypes, in particular *GRN* (7.1 ± 3.9) or *MAPT* (9.3 ± 6.4) [12]. This is partly explained by the deleterious impact of ALS, which shortens the survival of a proportion of *C9orf72* carriers. Besides, a particular subset of *C9orf72* expansion carriers have a remarkably different and slow progression corresponding to the "FTD phenocopy" syndrome. These patients exhibit behavioral and cognitive changes indistinguishable from those of typical bvFTD, but with remarkably slow progression [27–30]. They present only mild executive deficits and slow brain imaging changes, as illustrated in Fig. 1. Clinical and cognitive deficits remain relatively stable over time and the survival time may by longer than 20 years. Even if this repre-

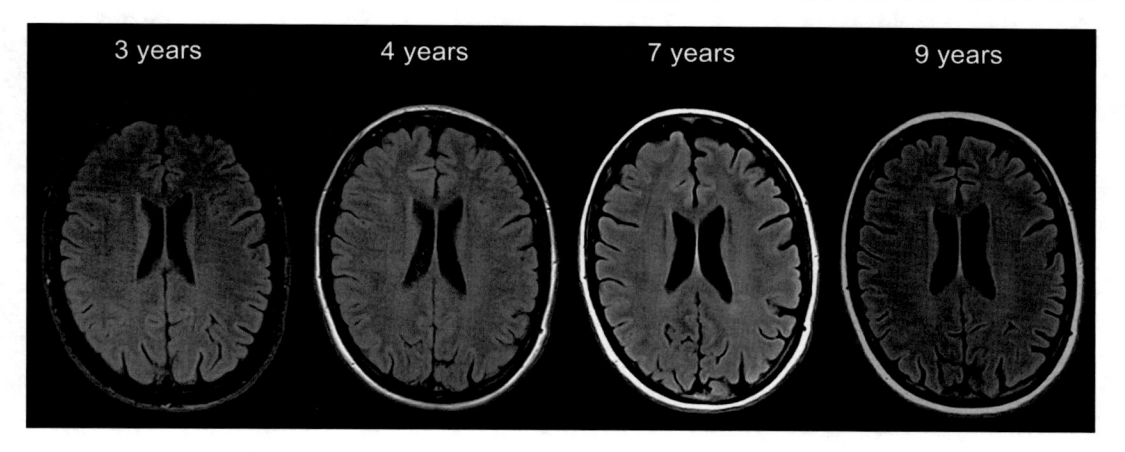

Fig. 1 Neuroimaging characteristics (brain MRI, axial sections) of a *C9orf72* patient with a slow progression evocative of FTD phenocopy. Follow-up over a 9-year disease duration

sents a possible phenotype in some *C9orf72* carriers, a recent review assesses that *C9orf72* expansion explains only a minority (less than 2%) of "FTD phenocopies" [31]. The predictors and the genetic/environmental factors contributing to the slow progression of neurodegeneration in these cases are not determined, but their identification will undoubtedly improve our knowledge on the mechanisms and disease modifiers implicated in *C9orf72* progression.

Characteristics of Amyotrophic Lateral Sclerosis Phenotypes Associated with *C9orf72* Expansion

In *C9orf72* carriers presenting with ALS, the disease has no particular distinguishing features; nevertheless, bulbar onset and co-occurring cognitive deterioration (56% of the patients) are overall more frequent with respect to non-mutated cases [11, 32, 33]. The *C9orf72* ALS population presented relatively more homogeneous clinical features than the non-*C9orf72* ALS population [34]. Disease duration is shorter in *C9orf72* than in other genetic forms of ALS [9, 32, 34, 35], and the *C9orf72* group included a significantly smaller fraction of slow-progressing individuals [34].

Psychiatric Symptoms in *C9orf72* Carriers: A Continuum with Frontotemporal Dementia and Amyotrophic Lateral Sclerosis

The high prevalence of psychotic symptoms in *C9orf72* families has highlighted the potential links and shared predispositions between FTD, ALS, and psychiatric disorders [10, 36–41]. Psychiatric symptoms or syndromes in *C9orf72* carriers variably include hallucinations (mostly auditory and visual), delusions (especially persecutory, jealousy, and grandiose delusions), and other psychotic (schizophrenia, bipolar disorder/hypomania) or obsessive-compulsive disorders. They are present in 10–50% of *C9orf72* patients according to the populations and may either occur concomitantly with FTD/ALS symptoms, or precede their onset by several years or decades. A large study in *C9orf72* kindreds underlined that family members of FTD or ALS patients also had increased risk of autism and other major psychiatric disorders, but the absence of genetic analyses in family members with psychiatric diseases was a limit of this study [40].

While there is now strong evidence supporting the association of psychotic symptoms with *C9orf72* expansion, the neuroanatomical substrate of those neuropsychiatric syndromes remains elusive. A functional deficit of the thalamus, detected at the early stage of *C9orf72* disease, does not appear to be specifically associated with psychotic symptoms [42]. Psychiatric symptoms in *C9orf72* rather correlated with a cortical and subcortical network implicated in schizophrenia including frontal, temporal, and occipital cortices as well as thalamus, striatum, and cerebellum [40, 43]. Interestingly, presymptomatic carriers showed abnormally low gyrification in left frontal and parieto-occipital regions, which was detected years before symptom onset [44]. Together, these findings provide the first clues to understand the neuroanatomical bases and networks implicated in *C9orf72*, but further investigations are warranted to fully elucidate their mechanisms, and further studies will clarify whether these abnormalities are part of a neurodevelopmental phenomenon.

Finally, the discovery of *C9orf72* gene has enlarged the clinical continuum linking FTD and ALS to psychiatric diseases. It led to reconsider the frontier between frontal behavioral and psychiatric disorders, as both are linked by similar molecular mechanisms and common functional network alterations in *C9orf72* carriers. Besides defining new exciting links between neuro-psychiatric disorders, this also raises new important questions that should be considered in clinical practice. First, the recommendations of genetic analysis in patients from *C9orf72* families presenting with psychiatric symptoms need to be clarified. More broadly, it opens the questions of the indications, criteria, and limitations of genetic analyses that could be proposed to patients suffering typical or atypical psychiatric diseases, outside of a familial context of *C9orf72* mutation [45]. These relevant questions must be addressed in the context of international consortia of experts in FTD, involving psychiatrists and geneticists too, in order to provide a framework and guidelines defining the indications and limitations of genetic testing in these patients.

Primary Progressive Aphasias Are Rare Phenotypes in *C9orf72* Patients

Unlike bvFTD, primary progressive aphasias (PPAs) are rarely associated with *C9orf72* expansion, as shown by the low frequency (1%) of *C9orf72* mutation carriers in a large North-American population of 403 PPA patients [46] and in other studies [16, 21, 36, 47]. Only few cases of *C9orf72*-related PPA have been described in detail in the literature [46, 48, 49]. Their phenotypes were mostly consistent with non-fluent and semantic PPA variants [12, 46]. Furthermore, a large European-Canadian study evidenced an important discrepancy between *C9orf72* and *GRN* genotypes, as only 3% of 1433 *C9orf72* included in this study have initially presented PPA, compared to 14% of *GRN* carriers [12]. This suggests that gene-specific effects lead to selective vulnerability of brain structures that differentially affect language networks. Further studies should clarify the biological determinants of selective lesion tropism for the language networks in genetic patients displaying PPA.

Implication of *C9orf72* in Dementia Syndromes Other than Frontotemporal Dementia

An association of *C9orf72* with Alzheimer's disease (AD) has also been examined. Episodic memory disorders at onset, mimicking Alzheimer's disease, are rarely associated with *C9orf72* expansion [16, 50–53]. Several studies have detected expansions in clinically diagnosed or pathologically confirmed AD patients, with a low prevalence accounting for less than 1% of AD populations [54, 55]. Although rare, this association has suggested a possible interrelationship between transactive response DNA-binding protein 43 (TDP-43) and amyloid pathological processes. In some cases, however, TDP-43 lesions were detected in absence of Alzheimer pathology [53, 54]. Amnestic symptoms misleading to a clinical diagnosis of AD might be related to hippocampal sclerosis detected in a proportion of *C9orf72* patients [53, 55].

Relation Between *C9orf72* Expansion and Non-amyotrophic Lateral Sclerosis Motor Diseases

Following the discovery of *C9orf72*, the clinical significance of its expansion in Parkinson's disease (PD), atypical parkinsonian syndromes, and other movement disorders has been debated.

Some studies initially suggested an excess of parkinsonian symptoms, PD, or atypical parkinsonism in FTD/ALS families who carry the *C9orf72* expansion [10, 21, 56–58]. However, the connections between the *C9orf72* expansion and PD, parkinsonism and other movement disorders have not been further clarified in the literature and its role beyond the FTD/ALS spectrum is still uncertain [59]. A prevalence of *C9orf72* expansion close to 1% was found in the first two investigations of large European and North American PD populations [47, 60]. However, all other cohort studies failed to replicate any association [61–65], and the frequency of *C9orf72* expansion appeared to be similar between PD patients and controls from two large meta-analyses [66, 67].

Rare *C9orf72* cases presenting with atypical parkinsonism have been described also suggesting a shared predisposition. However, further investigations in large populations of patients with multiple system atrophy (MSA) syndrome [65, 68–70], progressive supranuclear palsy (PSP) [16, 70], corticobasal syndrome (CBS) [60, 68, 70], Lewy body dementia [60], and essential tremor [61] showed that *C9orf72* has little or no contribution to the abovementioned movement disorders. Together, all these findings evidence that large repeat expansions do not play a major role in the pathogenesis of PD and related disorders, and that *C9orf72* testing should not be widely offered to these patients, but only to those who have overt symptoms and/or family history of FTD/ALS.

The implication of *C9orf72* in other movement disorders has also been investigated. *C9orf72* expansion appears to be responsible for 1–2% of patients with chorea or with Huntington-like phenocopies that are negative for *IT15* expansion [71–73]. Anecdotal cases of patients with isolated or complex cerebellar ataxia carrying *C9orf72* expansion were described [68, 74–76]. However, no association was evidenced in cerebellar ataxia study cohorts, thus suggesting that this association might be exceptional or even coincidental [77, 78].

Intermediate Alleles in Clinical Practice: A Genetic Risk Factor for Neurodegenerative Diseases?

Besides the pathogenic effect of large repeat expansion (> 100 GGGGCC), the role of intermediate alleles (20–30 GGGGCC) in neurodegenerative diseases is uncertain and their significance is still debated. In ALS, a meta-analysis provided evidence of an association of 24–30 repeat alleles with the disease, suggesting that these alleles should be considered as pathogenic [79].

Many cases with PD or parkinsonian phenotypes carrying intermediate repeats were reported [60, 66], therefore, also questioning the role of these alleles as a susceptibility factor for PD. However, associations were not systematically found in autopsy-confirmed cohorts, and meta-analyses detected a potential but low effect of intermediate repeats (10 repeats or larger) in PD susceptibility [47]. Similarly, a lack or a low association of intermediate alleles with MSA, PSP, and essential tremor likely supports that variations in *C9orf72* gene do not play a major role in the susceptibility to these diseases [63, 70]. Intermediate expansions in *C9orf72* also seem to weakly contribute to AD and dementia with Lewy bodies [47, 52, 80]. Results in CBS were more debatable, with controversial conclusions [70].

Anticipation Phenomenon: A Still Debated Question

The clinical heterogeneity of *C9orf72* is additionally characterized by an important variability in the age at onset (AO), ranging from the third decade of life to a nearly incomplete penetrance in elderly carriers [12]. As for other expansion diseases, the question of an association between expansion size and AO is raised. A clinical anticipation in *C9orf72* carriers, characterized by an earlier and more severe phenotype linked to increasing repeat number over successive generations, is also an important issue in clinical practice. There is no clear evidence for intergenerational anticipation in *C9orf72* families so far. Studies evaluating the correlation between AO and expansion size have provided conflicting results [81, 82]. Several reasons explain the difficulty in identifying these factors and precluding definite conclusions. Most studies have analyzed a correlation between AO and *C9orf72* expansion sizes in blood, but age at sampling appears as a major confounding factor [83, 84]. Additionally, the number of GGGGCC repeats in peripheral lymphocytes appears to unpredictably vary over time in subjects with multiple blood samples, as well as through generations in parents-offspring pairs [83, 84]. Finally, as in other expansion diseases, *C9orf72* expansions are unstable across tissues, producing somatic mosaicism [81]. The size variations among different tissues and the level of imprecision of Southern blot in determining the expansion size are also strong limitations to translate the observations from blood to brain tissue. Other factors such as *TMEM106B* gene and *C6orf10* locus may influence AO and survival time in *C9orf72* patients but need replications as well [85–88].

The identification of factors explaining the emergence of FTD or ALS phenotypes in *C9orf72* carriers is also a major issue. The repeat size, detected in blood, frontal cortex, cerebellum, or the spinal cord, does not appear to be associated to the clinical FTD or ALS phenotypes in any case [81]. The size of the expansion itself therefore does not seem to contribute significantly to the phenotypic variability. Intermediate alleles of the *ataxin-2 gene* (*ATXN2*) constitute a known risk factor for ALS in non-genetic populations, and several studies have evidenced that they could also drive the phenotype toward ALS in *C9orf72* patients [86, 87]. So far, our knowledge in this domain is largely incomplete, and the identification of the modifiers driving the phenotype to FTD or ALS remains a major research challenge.

Conclusions and Perspectives

Knowledge on genetic diseases, their associated phenotypes, and parameters such as penetrance and expressivity is indispensable to offer appropriate genetic diagnosis to the patients and genetic counseling to their families. In the last years, a major breakthrough has been achieved in the understanding of the genetics and molecular biology of FTD and ALS with the fast development of next generation sequencing (NGS). It is nowadays possible to analyze most genes by NGS, except *C9orf72* gene, whose expansion should be looked for separately with repeat primed polymerase chain reaction (PCR) or Southern blot. However, the interpretation of the huge load of data and the considerable number of variants of uncertain significance (VUS) generated by NGS now represent new challenges for geneticists and clinicians. Caution must be taken when interpreting uncertain results, and a good expertise in the genotypes underlying clinical phenotypes is needed.

A higher level of complexity comes from the identification of double mutations in rare patients [89]. Notably, in rare *C9orf72* carriers, an additional pathogenic mutation in another *FTD/ALS* gene have been identified [55, 90, 91]. This questions about the frequency of the coincidental occurrence of two mutations in FTD/ALS patients. It also suggests that a second mutational hit could contribute to the disease penetrance or influence the phenotypic presentation. Therefore, an extensive analysis of all known *FTD/ALS* genes may be recommended in *C9orf72* carriers, as in patients with other *FTD/ALS* gene mutations.

Although important advances have been made in *C9orf72* disease, many unsolved questions also remain a source of difficulties for family counseling. For example, the pathogenic threshold conferring a risk for FTD or ALS is not firmly established, the factors contributing to age and phenotype variability have not been clearly identified, and there is still a lack of precise information about age-related penetrance.

Important advances in our knowledge of *C9orf72*-mediated disease and its underlying pathogenesis have paved the way to new therapeutic perspectives. It is noteworthy that we are now reaching a turning point as regards the development of potentially preventive therapies. The study of presymptomatic stage in mutation carriers is currently of outmost importance, as it represents the optimal time window to test therapeutic molecules. There is a need to better clarify the definition of the presymptomatic and prodromal stages of the disease and to establish firm criteria for phenoconversion. These are new clinical challenges that, once completed, will hopefully expand the scope of potentially modifying therapies targeting the earliest disease stage in genetic FTD.

Conflict of Interest The authors declare no conflicts of interest.

References

1. DeJesus-Hernandez M, Mackenzie IR, Boeve BF, Boxer AL, Baker M, Rutherford NJ et al (2011) Expanded GGGGCC Hexanucleotide repeat in noncoding region of C9ORF72 causes chromosome 9p-linked FTD and ALS. Neuron 72(2):245–256
2. Renton AE, Majounie E, Waite A, Simón-Sánchez J, Rollinson S, Gibbs JR et al (2011) A Hexanucleotide repeat expansion in C9ORF72 is the cause of chromosome 9p21-linked ALS-FTD. Neuron 72(2):257–268
3. Xi Z, Rainero I, Rubino E, Pinessi L, Bruni AC, Maletta RG et al (2014) Hypermethylation of the CpG-island near the C9orf72 G4C2-repeat expansion in FTLD patients. Hum Mol Genet 23(21):5630–5637
4. Cruts M, Gijselinck I, Van Langenhove T, van der Zee J, Van Broeckhoven C (2013) Current insights into the C9orf72 repeat expansion diseases of the FTLD/ALS spectrum. Trends Neurosci 36(8):450–459
5. Mackenzie IRA, Neumann M, Baborie A, Sampathu DM, Du Plessis D, Jaros E et al (2011) A harmonized classification system for FTLD-TDP pathology. Acta Neuropathol 122(1):111–113
6. Mackenzie IR, Neumann M (2020) Subcortical TDP-43 pathology patterns validate cortical FTLD-TDP subtypes and demonstrate unique aspects of C9orf72 mutation cases. Acta Neuropathol 139(1):83–98
7. Troakes C, Maekawa S, Wijesekera L, Rogelj B, Siklós L, Bell C et al (2012) An MND/ALS phenotype associated with C9orf72 repeat expansion: abundant p62-positive, TDP-43-negative inclusions in cerebral cortex, hippocampus and cerebellum but without associated cognitive decline: p 62 proteinopathy. Neuropathology 32(5):505–514
8. Mackenzie IRA, Frick P, Neumann M (2014) The neuropathology associated with repeat expansions in the C9ORF72 gene. Acta Neuropathol 127(3):347–357
9. Majounie E, Renton AE, Mok K, Dopper EG, Waite A, Rollinson S et al (2012) Frequency of the C9orf72 hexanucleotide repeat expansion in patients with amyotrophic lateral sclerosis and frontotemporal dementia: a cross-sectional study. Lancet Neurol 11(4):323–330
10. Mahoney CJ, Beck J, Rohrer JD, Lashley T, Mok K, Shakespeare T et al (2012) Frontotemporal dementia with the C9ORF72 hexanucleotide repeat expansion: clinical, neuroanatomical and neuropathological features. Brain 135(3):736–750
11. Millecamps S, Boillée S, Le Ber I, Seilhean D, Teyssou E, Giraudeau M et al (2012) Phenotype difference between ALS patients with expanded repeats in C9ORF72 and patients with mutations in other ALS-related genes. J Med Genet 49(4):258–263
12. Moore KM, Nicholas J, Grossman M, McMillan CT, Irwin DJ, Massimo L et al (2019) Age at symptom onset and death and disease duration in genetic frontotemporal dementia: an international retrospective cohort study. Lancet Neurol 19:145–156. S1474442219303941
13. Liu X, He J, Gao F-B, Gitler AD, Fan D (2018 Aug) The epidemiology and genetics of amyotrophic lateral sclerosis in China. Brain Res 1693:121–126
14. Shamim U, Ambawat S, Singh J, Thomas A, Pradeep-Chandra-Reddy C, Suroliya V et al (2020) C9orf72 hexanucleotide repeat expansion in Indian patients with ALS: a common founder and its geographical predilection. Neurobiol Aging 88:156.e1–156.e9. S0197458019304531
15. Byrne S, Elamin M, Bede P, Shatunov A, Walsh C, Corr B et al (2012) Cognitive and clinical characteristics of patients with amyotrophic lateral sclerosis carrying a C9orf72 repeat expansion: a population-based cohort study. Lancet Neurol 11(3):232–240
16. Le Ber I, Guillot-Noel L, Hannequin D, Lacomblez L, Golfier V, Puel M et al (2013) C9ORF72 repeat expansions in the frontotemporal dementias Spectrum of diseases: a flow-chart for genetic testing. JADA 34(2):485–499
17. Solje E, Aaltokallio H, Koivumaa-Honkanen H, Suhonen NM, Moilanen V, Kiviharju A et al (2015) The phenotype of the C9ORF72 expansion carriers

according to revised criteria for bvFTD. Dermaut B, editor. PLoS One 10(7):e0131817

18. Heuer HW, Wang P, Rascovsky K, Wolf A, Appleby B, Bove J et al (2020 Jan) Comparison of sporadic and familial behavioral variant frontotemporal dementia (FTD) in a north American cohort. Alzheimers Dement 16(1):60–70

19. Montembeault M, Sayah S, Rinaldi D, Le Toullec B, Bertrand A, Funkiewiez A et al (2020) Cognitive inhibition impairments in presymptomatic C9orf72 carriers. J Neurol Neurosurg Psychiatry 91:366–372. jnnp-2019-322242

20. Moore K, Convery R, Bocchetta M, Neason M, Cash DM, Greaves C et al (2020) A modified camel and Cactus test detects presymptomatic semantic impairment in genetic frontotemporal dementia within the GENFI cohort. Appl Neuropsychol Adult Feb 5:1–38. Online ahead of print

21. Boeve BF, Boylan KB, Graff-Radford NR, DeJesus-Hernandez M, Knopman DS, Pedraza O et al (2012) Characterization of frontotemporal dementia and/or amyotrophic lateral sclerosis associated with the GGGGCC repeat expansion in C9ORF72. Brain 135(3):765–783

22. Convery R, Mead S, Rohrer JD (2019) Review: clinical, genetic and neuroimaging features of frontotemporal dementia. Neuropathol Appl Neurobiol 45(1):6–18

23. Bocchetta M, Gordon E, Cardoso MJ, Modat M, Ourselin S, Warren JD et al (2018) Thalamic atrophy in frontotemporal dementia—not just a C9orf72 problem. NeuroImage: Clinical 18:675–681

24. Olney NT, Ong E, Goh S-YM, Bajorek L, Dever R, Staffaroni AM et al (2020) Clinical and volumetric changes with increasing functional impairment in familial frontotemporal lobar degeneration. Alzheimers Dement 16(1):49–59

25. Bocchetta M, Iglesias JE, Neason M, Cash DM, Warren JD, Rohrer JD (2020) Thalamic nuclei in frontotemporal dementia: Mediodorsal nucleus involvement is universal but pulvinar atrophy is unique to C9orf72. Hum Brain Mapp 41(4):1006–1016

26. Cash DM, Bocchetta M, Thomas DL, Dick KM, van Swieten JC, Borroni B et al (2018) Patterns of gray matter atrophy in genetic frontotemporal dementia: results from the GENFI study. Neurobiol Aging 62:191–196

27. Khan BK, Yokoyama JS, Takada LT, Sha SJ, Rutherford NJ, Fong JC et al (2012) Atypical, slowly progressive behavioural variant frontotemporal dementia associated with C9ORF72 hexanucleotide expansion. J Neurol Neurosurg Psychiatry 83(4):358–364

28. Gómez-Tortosa E, Serrano S, de Toledo M, Pérez-Pérez J, Sainz MJ (2014) Familial benign frontotemporal deterioration with C9ORF72 hexanucleotide expansion. Alzheimers Dement 10:S284–S289

29. Suhonen N-M, Kaivorinne A-L, Moilanen V, Bode M, Takalo R, Hänninen T et al (2015) Slowly progressive frontotemporal lobar degeneration caused by

the C9ORF72 repeat expansion: a 20-year follow-up study. Neurocase 21(1):85–89

30. Llamas-Velasco S, García-Redondo A, Herrero-San Martín A, Puertas Martín V, González- Sánchez M, Pérez-Martínez DA et al (2018 Jan 2) Slowly progressive behavioral frontotemporal dementia with C9orf72 mutation. Case report and review of the literature. Neurocase 24(1):68–71

31. Valente ES, Caramelli P, Gambogi LB, Mariano LI, Guimarães HC, Teixeira AL et al (2019) Phenocopy syndrome of behavioral variant frontotemporal dementia: a systematic review. Alz Res Therapy 11(1):30

32. Glasmacher SA, Wong C, Pearson IE, Pal S. (2019) Survival and prognostic factors in C9orf72 repeat expansion carriers: a systematic review and meta-analysis. JAMA Neurol. [cited 2020 Feb 27]

33. Chiò A, Moglia C, Canosa A, Manera U, D'Ovidio F, Vasta R et al (2020) ALS phenotype is influenced by age, sex, and genetics: a population-based study. Neurology 94(8):e802–e810

34. Cammack AJ, Atassi N, Hyman T, van den Berg LH, Harms M, Baloh RH et al (2019) Prospective natural history study of C9orf72 ALS clinical characteristics and biomarkers. Neurology

35. Chio A, Calvo A, Mazzini L, Cantello R, Mora G, Moglia C et al (2012) Extensive genetics of ALS: a population-based study in Italy. Neurology 79(19):1983–1989

36. Snowden JS, Rollinson S, Thompson JC, Harris JM, Stopford CL, Richardson AMT et al (2012) Distinct clinical and pathological characteristics of frontotemporal dementia associated with C9ORF72 mutations. Brain 135(3):693–708

37. Kertesz A, Ang LC, Jesso S, MacKinley J, Baker M, Brown P et al (2013) Psychosis and hallucinations in frontotemporal dementia with the C9ORF72 mutation: a detailed clinical cohort. Cogn Behav Neurol 26(3):146–154

38. Galimberti D, Reif A, Dell'Osso B, Kittel-Schneider S, Leonhard C, Herr A et al (2014) The C9ORF72 hexanucleotide repeat expansion is a rare cause of schizophrenia. Neurobiol Aging 35(5):1214.e7–1214.e10

39. Shinagawa S, Naasan G, Karydas AM, Coppola G, Pribadi M, Seeley WW et al (2015) Clinicopathological study of patients with C9ORF72-associated frontotemporal dementia presenting with delusions. J Geriatr Psychiatry Neurol 28(2):99–107

40. Devenney EM, Ahmed RM, Halliday G, Piguet O, Kiernan MC, Hodges JR (2018) Psychiatric disorders in C9orf72 kindreds: study of 1,414 family members. Neurology 91:e1498–e1507

41. Sellami L, St-Onge F, Poulin S, Laforce R (2019) Schizophrenia phenotype preceding behavioral variant frontotemporal dementia related to C9orf72 repeat expansion. Cogn Behav Neurol 32(2):120–123

42. Diehl-Schmid J, Licata A, Goldhardt O, Förstl H, Yakushew I, Otto M et al (2019) FDG-PET underscores the key role of the thalamus in frontotemporal

lobar degeneration caused by C9ORF72 mutations. Transl Psychiatry 9(1) [cited 2019 Feb 14]

43. Sellami L, Bocchetta M, Masellis M, Cash DM, Dick KM, van Swieten J et al (2018) Distinct neuroanatomical correlates of neuropsychiatric symptoms in the three Main forms of genetic frontotemporal dementia in the GENFI cohort. JAD 13:1–16

44. Caverzasi E, Battistella G, Chu SA, Rosen H, Zanto TP, Karydas A et al (2019) Gyrification abnormalities in presymptomatic c9orf72 expansion carriers. J Neurol Neurosurg Psychiatry 90(9):1005–1010

45. Watson A, Pribadi M, Chowdari K, Clifton S, Wood J, Miller BL et al (2016) C9orf72 repeat expansions that cause frontotemporal dementia are detectable among patients with psychosis. Psychiatry Res 235:200–202

46. Ramos EM, Dokuru DR, Van Berlo V, Wojta K, Wang Q, Huang AY et al (2019) Genetic screen in a large series of patients with primary progressive aphasia. Alzheimers Dement 15(4):553–560

47. Xi Z, Zinman L, Grinberg Y, Moreno D, Sato C, Bilbao JM et al (2012) Investigation of C9orf72 in 4 neurodegenerative disorders. Arch Neurol 69(12):1583

48. Saint-Aubert L, Sagot C, Wallon D, Hannequin D, Payoux P, Nemmi F et al (2014) A case of Logopenic primary progressive aphasia with C9ORF72 expansion and cortical Florbetapir binding. JAD 42(2):413–420

49. Flanagan EP, Baker MC, Perkerson RB, Duffy JR, Strand EA, Whitwell JL et al (2015) Dominant frontotemporal dementia mutations in 140 cases of primary progressive aphasia and speech apraxia. Dement Geriatr Cogn Disord 39(5–6):281–286

50. Wallon D, Rovelet-Lecrux A, Deramecourt V, Pariente J, Auriacombe S, Le Ber I et al (2012) Definite behavioral variant of frontotemporal dementia with C9ORF72 expansions despite positive Alzheimer's disease cerebrospinal fluid biomarkers. JAD 32(1):19–22

51. Cacace R, Van Cauwenberghe C, Bettens K, Gijselinck I, van der Zee J, Engelborghs S et al (2013) C9orf72 G4C2 repeat expansions in Alzheimer's disease and mild cognitive impairment. Neurobiol Aging 34(6):1712.e1–1712.e7

52. Harms M, Benitez BA, Cairns N, Cooper B, Cooper P, Mayo K et al (2013) C9orf72 Hexanucleotide repeat expansions in clinical Alzheimer disease. JAMA Neurol 70(6):736

53. Murray ME, Bieniek KF, Banks Greenberg M, DeJesus-Hernandez M, Rutherford NJ, van Blitterswijk M et al (2013) Progressive amnestic dementia, hippocampal sclerosis, and mutation in C9ORF72. Acta Neuropathol 126(4):545–554

54. Majounie E, Abramzon Y, Renton AE, Perry R, Bassett SS, Pletnikova O et al (2012) Repeat expansion in C9ORF72 in Alzheimer's disease. N Engl J Med 366(3):283–284

55. Ramos EM, Dokuru DR, Van Berlo V, Wojta K, Wang Q, Huang AY et al (2020) Genetic screening of a large series of north American sporadic and familial frontotemporal dementia cases. Alzheimers Dement 16(1):118–130

56. Simón-Sánchez J, Dopper EGP, Cohn-Hokke PE, Hukema RK, Nicolaou N, Seelaar H et al (2012) The clinical and pathological phenotype of C9ORF72 hexanucleotide repeat expansions. Brain 135(3):723–735

57. Annan M, Beaufils É, Viola U-C, Vourc'h P, Hommet C, Mondon K (2013) Idiopathic Parkinson's disease phenotype related to C9ORF72 repeat expansions: contribution of the neuropsychological assessment. BMC Res Notes 6(1):343

58. Cooper-Knock J, Frolov A, Highley JR, Charlesworth G, Kirby J, Milano A et al (2013) C9ORF72 expansions, parkinsonism, and Parkinson disease: a clinicopathologic study. Neurology 81(9):808–811

59. Bourinaris T, Houlden H (2018) C9orf72 and its relevance in parkinsonism and movement disorders: a comprehensive review of the literature. Mov Disord Clin Pract 5(6):575–585

60. Lesage S, Le Ber I, Condroyer C, Broussolle E, Gabelle A, Thobois S et al (2013) C9orf72 repeat expansions are a rare genetic cause of parkinsonism. Brain 136(2):385–391

61. DeJesus-Hernandez M, Rayaprolu S, Soto-Ortolaza AI, Rutherford NJ, Heckman MG, Traynor S et al (2013) Analysis of the C9orf72 repeat in Parkinson's disease, essential tremor and restless legs syndrome. Parkinsonism Relat Disord 19(2):198–201

62. Daoud H, Noreau A, Rochefort D, Paquin-Lanthier G, Gauthier MT, Provencher P et al (2013) Investigation of C9orf72 repeat expansions in Parkinson's disease. Neurobiol Aging 34(6):1710.e7–1710.e9

63. Nuytemans K, Inchausti V, Beecham GW, Wang L, Dickson DW, Trojanowski JQ et al (2014) Absence of C9ORF72 expanded or intermediate repeats in autopsy-confirmed Parkinson's disease. Mov Disord 29(6):827–830

64. Cannas A, Solla P, Borghero G, Floris GL, Chio A, Mascia MM et al (2015) C9ORF72 intermediate repeat expansion in patients affected by atypical parkinsonian syndromes or Parkinson's disease complicated by psychosis or dementia in a Sardinian population. J Neurol 262(11):2498–2503

65. Chen X, Chen Y, Wei Q, Ou R, Cao B, Zhao B et al (2016) C9ORF72 repeat expansions in Chinese patients with Parkinson's disease and multiple system atrophy. J Neural Transm 123(11):1341–1345

66. Nuytemans K, Bademci G, Kohli MM, Beecham GW, Wang L, Young JI et al (2013) C9ORF72 intermediate repeat copies are a significant risk factor for Parkinson disease. Ann Hum Genet 77(5):351–363

67. Theuns J, Verstraeten A, Sleegers K, Wauters E, Gijselinck I, Smolders S et al (2014) Global investigation and meta-analysis of the C9orf72 (G4C2) n repeat in Parkinson disease. Neurology 83(21):1906–1913

68. Lindquist S, Duno M, Batbayli M, Puschmann A, Braendgaard H, Mardosiene S et al (2013) Corticobasal and ataxia syndromes widen the spectrum of C9ORF72 hexanucleotide expansion disease. Clin Genet 83(3):279–283

69. Goldman JS, Quinzii C, Dunning-Broadbent J, Waters C, Mitsumoto H, Brannagan TH et al (2014) Multiple system atrophy and amyotrophic lateral sclerosis in a family with Hexanucleotide repeat expansions in C9orf72. JAMA Neurol 71(6):771

70. Schottlaender LV, Polke JM, Ling H, MacDoanld ND, Tucci A, Nanji T et al (2015) The analysis of C9orf72 repeat expansions in a large series of clinically and pathologically diagnosed cases with atypical parkinsonism. Neurobiol Aging 36(2):1221.e1–1221.e6

71. Hensman Moss DJ, Poulter M, Beck J, Hehir J, Polke JM, Campbell T et al (2014) C9orf72 expansions are the most common genetic cause of Huntington disease phenocopies. Neurology 82(4):292–299

72. Kostić VS, Dobričić V, Stanković I, Ralić V, Stefanova E (2014) C9orf72 expansion as a possible genetic cause of Huntington disease phenocopy syndrome. J Neurol 261(10):1917–1921

73. Ida CM, Butz ML, Lundquist PA, Dawson DB (2018) C9orf72 repeat expansion frequency among patients with Huntington disease genetic testing. Neurodegener Dis 18(5–6):239–253

74. Corcia P, Vourc'h P, Guennoc A-M, Del Mar Amador M, Blasco H, Andres C et al (2016) Pure cerebellar ataxia linked to large C9orf72 repeat expansion. Amyotroph Lat Scl Fr 17(3–4):301–303

75. Zhang M, Xi Z, Misquitta K, Sato C, Moreno D, Liang Y et al (2017) C9orf72 and ATXN2 repeat expansions coexist in a family with ataxia, dementia, and parkinsonism. Mov Disord 32(1):158–162

76. Meloni M, Farris R, Solla P, Mascia MM, Marrosu F, Cannas A (2017) C9ORF72 intermediate repeat expansion in a patient with psychiatric disorders and progressive cerebellar Ataxia. Neurologist 22(6):245–246

77. Fogel BL, Pribadi M, Pi S, Perlman SL, Geschwind DH, Coppola G (2012) C9ORF72 expansion is not a significant cause of sporadic spinocerebellar ataxia. Mov Disord 27(14):1835–1836

78. Hsiao C-T, Tsai P-C, Liao Y-C, Lee Y-C, Soong B-W (2014) C9ORF72 repeat expansion is not a significant cause of late onset cerebellar ataxia syndrome. J Neurol Sci 347(1–2):322–324

79. Iacoangeli A, Al Khleifat A, Jones AR, Sproviero W, Shatunov A, Opie-Martin S et al (2019) C9orf72 intermediate expansions of 24–30 repeats are associated with ALS. Acta Neuropathol Commun 7(1):115

80. Orme T, Guerreiro R, Bras J (2018) The genetics of dementia with Lewy bodies: current understanding and future directions. Curr Neurol Neurosci Rep 18(10):67

81. van Blitterswijk M, DeJesus-Hernandez M, Niemantsverdriet E, Murray ME, Heckman MG, Diehl NN et al (2013) Association between repeat sizes and clinical and pathological characteristics in carriers of C9ORF72 repeat expansions (Xpansize-72): a cross-sectional cohort study. Lancet Neurol 12(10):978–988

82. Van Mossevelde S, van der Zee J, Cruts M, Van Broeckhoven C (2017) Relationship between C9orf72 repeat size and clinical phenotype. Curr Opin Genet Dev 44:117–124

83. Fournier C, Barbier M, Camuzat A, Anquetil V, Lattante S, Clot F et al (2019) Relations between C9orf72 expansion size in blood, age at onset, age at collection and transmission across generations in patients and presymptomatic carriers. Neurobiol Aging 74:234.e1–234.e8

84. Jackson JL, Finch NA, Baker MC, Kachergus JM, DeJesus-Hernandez M, Pereira K et al (2020) Elevated methylation levels, reduced expression levels, and frequent contractions in a clinical cohort of C9orf72 expansion carriers. Mol Neurodegener 15(1):7

85. Gallagher MD, Suh E, Grossman M, Elman L, McCluskey L, Van Swieten JC et al (2014) TMEM106B is a genetic modifier of frontotemporal lobar degeneration with C9orf72 hexanucleotide repeat expansions. Acta Neuropathol 127(3):407–418

86. Lattante S, Le Ber I, Galimberti D, Serpente M, Rivaud-Péchoux S, Camuzat A et al (2014) Defining the association of TMEM106B variants among frontotemporal lobar degeneration patients with GRN mutations and C9orf72 repeat expansions. Neurobiol Aging 35(11):2658.e1–2658.e5

87. van Blitterswijk M, Mullen B, Wojtas A, Heckman MG, Diehl NN, Baker MC et al (2014) Genetic modifiers in carriers of repeat expansions in the C9ORF72 gene. Mol Neurodegener 9(1):38

88. Zhang M, Ferrari R, Tartaglia MC, Keith J, Surace EI, Wolf U et al (2018) A C6orf10/LOC101929163 locus is associated with age of onset in C9orf72 carriers. Brain 141(10):2895–2907

89. Pottier C, Bieniek KF, Finch N, van de Vorst M, Baker M, Perkersen R et al (2015) Whole-genome sequencing reveals important role for TBK1 and OPTN mutations in frontotemporal lobar degeneration without motor neuron disease. Acta Neuropathol 130(1):77–92

90. van Blitterswijk M, Baker MC, DeJesus-Hernandez M, Ghidoni R, Benussi L, Finger E et al (2013) C9ORF72 repeat expansions in cases with previously identified pathogenic mutations. Neurology 81(15):1332–1341

91. Testi S, Tamburin S, Zanette G, Fabrizi GM (2015) Co-occurrence of the C9ORF72 expansion and a novel GRN mutation in a family with alternative expression of frontotemporal dementia and amyotrophic lateral sclerosis. JAD 44(1):49–56

Clinical and Neuroimaging Aspects of Familial Frontotemporal Lobar Degeneration Associated with *MAPT* and *GRN* Mutations

Bradley F. Boeve and Howard Rosen

Abbreviations

ALS	amyotrophic lateral sclerosis
bvFTD	behavioral variant frontotemporal dementia
c9FTD/ALS	frontotemporal dementia and/or amyotrophic lateral sclerosis linked to chromosome 9
C9orf72	gene encoding *chromosome 9 open reading frame 72*
CBS	corticobasal syndrome
CNS	central nervous system
FDG-PET	fluorodeoxyglucose positron emission tomography
FTD	frontotemporal dementia
FTD/ALS	frontotemporal dementia with amyotrophic lateral sclerosis
FTLD	frontotemporal lobar degeneration
GRN	gene encoding progranulin (or granulin)
MAPT	gene encoding microtubule-associated protein tau
MRI	magnetic resonance imaging
PPA	primary progressive aphasia

B. F. Boeve (✉)
Department of Neurology, Mayo Clinic, Rochester, MN, USA
e-mail: bboeve@mayo.edu

H. Rosen
Department of Neurology, University of California San Francisco, San Francisco, CA, USA

Introduction

Frontotemporal lobar degeneration (FTLD) is an overarching term for a group of neurodegenerative disorders that are typically associated with progressive degeneration in the frontal and anterior temporal lobes [1]. FTLD can involve many structures in the central nervous system (CNS), and therefore can present with a wide variety of symptoms, including the behavioral variant of frontotemporal dementia (bvFTD), the nonfluent and semantic variants of primary progressive aphasia (nfvPPA and svPPA), progressive supranuclear palsy syndrome (PSP), corticobasal syndrome (CBS), and amyotrophic lateral sclerosis, with or without other features of frontotemporal disease (ALS and FTD/ALS) [2]. In about 20% of people, FTLD is caused by genetic mutations, and therefore affects multiple members within the same family (familial frontotemporal lobar degeneration [f-FTLD]). F-FTLD has been associated with mutations in a number of genes, including the genes encoding microtubule-associated protein tau *(MAPT)*, progranulin *(GRN)*, chromosome 9 open reading frame 72 gene *(C9orf72)*, and less commonly valosin-containing protein *(VCP)*, TAR DNA-binding protein *(TARDBP)*, and fused in sarcoma *(FUS)*, among others. The clinical and imaging findings associated with the mutation in *C9orf72* are described in a separate chapter, and this chapter will focus on those findings in *MAPT* and *GRN*.

© Springer Nature Switzerland AG 2021
B. Ghetti et al. (eds.), *Frontotemporal Dementias*, Advances in Experimental Medicine and Biology 1281, https://doi.org/10.1007/978-3-030-51140-1_6

Historical Perspectives

The search for causative genes in f-FTLD was initiated with the identification of linkage between the symptoms of FTLD and a locus on the q21–22 region of chromosome 17 [3]. This finding prompted the first conference devoted to this topic in Ann Arbor, Michigan, in 1996, which focused on the clinical, pathological, and genetic features of frontotemporal dementia with parkinsonism (FTDP) linked to chromosome 17, which had been identified in a number of families up until then [4]. Researchers who characterized these early families noted that, despite common themes of disinhibition and parkinsonism across these kindreds, there was still considerable clinical heterogeneity within and across families [3, 4]. Beyond their implications for our understanding of the causes and pathogenesis of FTLD, these descriptions highlighted an important aspect of FTLD, which is that the same pathological mechanism can be associated with a variety of clinical manifestations. This observation has continued to be reinforced as a core clinical feature as knowledge about f-FTLD has expanded. As discussed below, this represents a challenge for diagnosis and clinical trials in FTLD and presents a fascinating biological mystery regarding the mechanisms by which the same molecular mechanism can be manifest by dysfunction in different neurological systems across people. In addition, the variation in clinical features across families prompted speculation that there might be another locus on chromosome 17 that accounts for disease in some families.

Soon after the link to chromosome 17 was discovered, mutations in MAPT were identified [5]. Even after the discovery that FTLD could be caused by mutations in MAPT, it was noted that a significant minority of patients with FTDP linked to chromosome 17 had no identifiable mutations in MAPT, nor did they have any tau-positive inclusions at autopsy [6]. This mystery was solved when it was discovered that mutations in GRN, which is also on chromosome 17, could also be associated with f-FTLD [7, 8], accounting for FTLD in all of these remaining chromosome 17-linked families. Subsequently, many other kindreds carrying mutations in GRN have been reported, thereby solving a decade-long conundrum and providing a remarkable insight into genetic diversity. The GRN gene is only 1.7 Mb centromeric to MAPT on chromosome 17, demonstrating that two apparently different genes reside very close to each other on the same chromosome and cause a similar phenotype.

This chapter summarizes the demographic/inheritance characteristics, histopathology, pathophysiology, key clinical aspects, and neuroimaging findings of disease due to MAPT and GRN mutations using data and findings from prior reports that included large numbers of informative cases and kindreds, in particular some recent publications that have provided extensive data [9–12]. While the features of FTLD due to GRN and MAPT mutations show many similarities, there are also unique features associated with mutations in each of these genes. These features are discussed in more detail below, and some of them are summarized in Table 1.

Demographic and Inheritance Characteristics

There are currently 67 known mutations in MAPT and 130 in GRN [9]. Additionally, there are over 790 affected individuals among at least 250 kindreds with mutations in MAPT, and over 1170 affected individuals among at least 480 kindreds with mutations in GRN [9]. The vast majority of what is known about f-FTLD comes from kindreds identified in the United States and Europe [9, 13], and thus, inferences about the relative prevalence of FTLD mutations across races and ethnicities is largely unknown. That said, studies in China, Korea, and Japan have identified mutations in the MAPT and GRN genes, as well as C9orf72 and other genes [14–19]. Case reports describing MAPT and GRN mutations in people of African descent [20, 21] and GRN mutations in Latinos [22] have also been published.

The male-to-female ratio for mutations in both genes is close to 1:1, although recent studies have suggested a slight predilection for females in GRN [9]. The mean age of onset is around

Table 1 Comparisons between key characteristics associated with mutations in microtubule-associated protein tau (*MAPT*) and progranulin (*GRN*)

	MAPT	GRN
Known number of distinct mutations[a]	67	130
Known number of affected individuals[a]	>790	>1170
Known number of kindreds[a]	250	480
Mode of inheritance	Autosomal dominant	Autosomal dominant
Penetrance	>95%	90% by age 70
Sex[a]	F 51%/M 49%	F 58%/M 42%
Onset age – mean (years)[a]	50	61
Onset age – range (years)[a]	17–82	25–90
Duration of illness – mean (years)[a]	9	7
Duration of illness – range (years)[a]	0–45	0–27
Cognitive features		
Executive dysfunction	++++	++++
Language impairment	+++	++++
Memory impairment	++	++
Visuospatial impairment	+	++
Behavioral features		
Personality/behavior changes	++++	++++
Psychotic features	+	+
Motor features		
Limb apraxia	+	+ – +++
Parkinsonism	++	+++
Motor neuron disease	+	+
Clinical syndromes		
Behavioral variant FTD	++++	++++
Nonfluent variant PPA	+	+++
Semantic variant PPA	+	+
Logopenic variant PPA	0	+
Amnestic mild cognitive impairment	+	+

(continued)

Table 1 (continued)

	MAPT	GRN
Alzheimer's type dementia	+	++
Corticobasal syndrome	+	++
Posterior cortical atrophy	0	+
Parkinson's disease	+	+
Progressive supranuclear palsy	+	0
Parkinson's disease + dementia	+	+
Dementia with Lewy bodies	+	+
Amyotrophic lateral sclerosis	+	+
MRI and FDG-PET findings		
Frontal abnormalities	+++	++++
Temporal abnormalities	++++	+++
Parietal abnormalities	+	++
Occipital abnormalities	0	0
Parenchymal signal changes on MRI	+	++
Symmetry vs. asymmetry	Usually symmetric	Often asymmetric

++++ = very frequently reported, +++ = frequently reported, ++ infrequently reported, + = rare, 0 = no definite cases reported to date

AD autosomal dominant, *F* female, *FDG-PET* fluorodeoxyglucose positron emission tomography, *f-FTLD* familial frontotemporal lobar degeneration, *FTD* frontotemporal dementia, *M* male, *MRI* magnetic resonance imaging, *PPA* primary progressive aphasia

[a]data from Moore et al. Lancet Neurol 2019

50 years for *MAPT* and around 61 years for *GRN*, with a wide range in age of onset in both mutations (17–82 years for *MAPT* and 25–90 years for *GRN*) [9]. Duration of clinical symptoms also varies widely, with *MAPT* being 0–45 years and *GRN* being 0–27 years [9].

F-FTLD associated with mutations in both *MAPT* and *GRN* is inherited in an autosomal dominant fashion. Penetrance is high but not complete, with penetrance appearing to be greater in *MAPT* than *GRN* [9, 11]. Seemingly "sporadic" cases have been identified in both genetic cohorts [9, 11]; some of these appear to be de novo mutations, while some have been found to reflect more convincing germline mosaics [23]. Anticipation has not been strongly suggested, but there is evidence that a slightly lower age of onset can occur in successive generations in both genes [9]. This may be explained by presentation bias, such that symptoms may be recognized earlier in individuals in successive generations due to their family history.

In contrast to familial Alzheimer's disease, where estimating the age of onset among asymptomatic mutation carriers based on parental and other relatives' age of onset is relatively reliable [24, 25], the predictability in *MAPT* and particularly *GRN* kindreds is more challenging [9]. Ongoing natural history studies such as the ARTFL LEFFTDS Longitudinal Frontotemporal Lobar Degeneration (ALLFTD; https://www. allftd.org/) consortium in North America [26], the Genetic Frontotemporal Dementia Initiative (GENFI; http://genfi.org.uk/) consortium in Europe, the Dominantly Inherited Non-Alzheimer's Disease (DINAD) study in Australia, and recently formed Frontotemporal Dementia Prevention Initiative (FPI; http://genfi.org.uk/fpi. html) consortium involving global consortia are seeking to improve this predictability by combining clinical and biomarker data to generate predictive models to facilitate current and future clinical trials involving potential disease-modifying therapies [12, 27].

Histopathology and Pathophysiology

Frontotemporal lobar degeneration is thought to be caused most commonly by the accumulation of one of two proteins in CNS tissue: microtubule-associated protein tau, or transactive response DNA binding protein molecular weight 43 (abbreviated TDP), each of which is thought to account for roughly half of FTLD cases. *MAPT* and *GRN* mutations represent genetic causes for each of these two common FTLD-associated protein disorders. In the case of *MAPT*, autopsy studies demonstrate aggregation of microtubule-

associated tau protein in paired helical filaments and neurofibrillary tangles in CNS tissue [28]. The tau protein plays an important role in the stabilization of microtubules and therefore in maintenance of the neuronal cytoskeleton that maintains neuronal structure and intraneuronal transport, among other functions [29, 30]. Although the precise mechanisms by which *MAPT* mutations cause cellular injury are not known, the mutations clearly have effects on tau function, with the specific effect depending on the precise mutation, and these differing effects can be associated with specific features of tau protein aggregation. Most *MAPT* mutations affect the region on the gene that encodes the microtubule-binding portion of the protein. They can be classified as missense mutations, which alter the sequence of the protein and usually influence the affinity of tau for microtubules, or splicing mutations, which influence the isoform of tau that is produced, favoring a form that contains either three or four repetitions of the carboxy-terminal sequence that is encoded by exon 10 of the gene, called 3-repeat (3R) or 4-repeat (4R) tau. In the absence of *MAPT* mutations, most cells contain the 3R and 4R isoforms of tau in roughly equal proportions, and this is also true in many *MAPT* mutations [29, 30]. Depending on how the mutation affects the protein and the post-translational modifications, tau protein aggregates can be found predominantly in the neurons, or in both glia and neurons. Furthermore, the features of the tau inclusions can vary and take on a wide variety of characteristics, including aggregation in inclusions similar to Pick bodies, as can be seen with the G389R mutations, or neurofibrillary tangles that are very similar to those seen in Alzheimer's disease, as can be seen with the V337M and R406W mutations [31]. Recent work has also highlighted the fact that abnormal tau proteins can induce other tau proteins to take on a pathological confirmation in a prion-like fashion [29, 30], and this has been proposed as a mechanism by which tau-mediated neurodegenerative disease spreads through neural systems [32].

In contrast to *MAPT*, *GRN* mutations are invariably associated with accumulation of TDP pathology [33], specifically in the form of TDP type A pathology, which is characterized by lentiform neuronal aggregates of TDP [34]. The mechanisms by which *GRN* mutations cause accumulation of TDP and neurodegeneration are incompletely understood. The primary effect of all *GRN* mutations is reduction in the production of the progranulin (PGRN) protein [35]. PGRN is normally broken into cleavage products called granulins, and PGRN and various granulin proteins have been shown to regulate survival and morphology of neurons, with PGRN promoting neuronal growth and various granulin proteins having different effects on neurons [35]. Beyond its effects on neuronal growth, PGRN also has important roles in the inflammatory response and in lysosomal function [35–37]. Thus, there are at least three major biological pathways through which PGRN deficiency may lead to disease, but which of these is most important, or whether disease results from a combination of these effects, is not yet known. The available knowledge on PGRN biology and its potential relationship to FTLD is extensively reviewed in another chapter in this book.

Cognitive and Behavioral Features

Mutations that cause FTLD are associated with a wide variety of clinical symptoms, and patients often present with one of the syndromes that are seen in sporadic FTLD, but some types of symptoms are seen more commonly in f-FTLD compared with sporadic FTLD. The bvFTD syndrome, characterized by early socioemotional changes that include disinhibition, apathy, loss of sympathy and empathy, stereotyped or compulsive/ritualistic behavior, and dietary changes, with or without early executive dysfunction, and usually sparing memory and visuospatial function [38–40], is the most common presentation with both mutations, occurring in about 40% of carriers [9]. These clinical changes are accompanied by atrophy and hypometabolism in orbitofrontal cortex and anterior cingulate cortex, along with dorsolateral prefrontal cortex – a pattern typical of bvFTD. Both the clinical and imaging

abnormalities in bvFTD due to mutations are very similar to the features that characterize sporadic bvFTD [41–46].

Language impairment is also relatively common in both mutation groups, but the qualitative aspects tend to be different. In *MAPT* mutation carriers, the topography of degeneration usually begins in the medial and anterior temporal lobes, leading to loss of semantic knowledge that begins with word and proper name retrieval problems and progresses to more generalized loss of knowledge about words and objects (svPPA; [47]). Loss of semantic knowledge can also affect knowledge about people (e.g., family members) and previously familiar locations [48, 43–46, 49]. Affected patients or their family members may state that they have "memory problems" when in fact they are describing problems recalling the names of people or objects rather than problems remembering recent events (episodic memory). Over time, verbal and semantic fluency become impaired as more frontal and posterior temporal language networks become affected.

In *GRN* mutation carriers, the topography of degeneration tends to be far more focal and asymmetric, often with disproportionate involvement of either the left or the right hemisphere [45, 46, 50]. If the anterior temporal lobe of the dominant hemisphere is the initial focus of degeneration, a semantic naming problem is prominent, similar to what is described above for *MAPT*; if the anterior temporal lobe of the nondominant hemisphere is affected, early semantic aphasia as well as object agnosia/prosopagnosia are the major manifestations. A nonfluent/agrammatic aphasia syndrome can occur in *GRN* mutation carriers (nfvPPA), with degeneration being prominent around Broca's area and the adjacent supplementary motor area of the dominant hemisphere [50, 51].

While the most common cognitive and behavioral features in both *MAPT* and *GRN* mutations are similar to those seen in sporadic FTLD, several other clinical syndromes can be seen with these mutations that are not typically associated with FTLD. A prominent amnestic syndrome (i.e., amnestic mild cognitive impairment, or memory-predominant dementia syndrome) can occur in both genetic groups, which has led to many such individuals being diagnosed with clinically suspected Alzheimer's disease (AD) [52].

Focal/asymmetric cortical degeneration syndromes similar to those seen in AD can also occur and are far more common in *GRN* than *MAPT* mutations. For instance, while *GRN* mutations can be associated with nfvPPA and svPPA, both of which are typical of FTLD, degeneration in the posterior perirolandic region of the dominant hemisphere can also occur with *GRN* mutations, leading to a logopenic primary progressive aphasia syndrome (lvPPA), which is characterized by word retrieval problems but without semantic loss, and progresses to empty, poorly directed speech. When this syndrome occurs outside of the setting of mutations, it is usually due to AD pathology [47]. Posterior degeneration in the right-greater-than-left hemisphere can also occur with *GRN* mutations, resulting in prominent visuospatial impairment, optic ataxia, ocular apraxia, simultanagnosia, dressing apraxia, spatial disorientation, and/or hemineglect: these are the typical elements of the posterior cortical atrophy syndrome (PCA; [51, 53]) which is also often due to AD pathology when it occurs outside of the setting of mutations [54, 55]. Visuospatial presentations are rare with *MAPT* mutations. Corticobasal syndrome (CBS) is a syndrome of unilateral or asymmetric limb apraxia and limb rigidity, which can also include other features such alien limb phenomenon, myoclonus, dystonia, etc. [56]. CBS can occur with either *MAPT* or *GRN* mutations [50, 51]. While this syndrome is typically associated with FTLD, it is also commonly caused by AD [57], and so a patient presenting with this syndrome without a known genetic mutation or strong family history of FTLD may be mistaken as a case of AD.

Lastly, more bizarre behavioral manifestations that are not core features of sporadic FTLD, including bizarre, schizophrenia-like delusions, visual or auditory hallucinations, and manic symptomatology, can occur with FTLD mutations. These features appear to be most common in *C9orf72* mutation carriers [58, 59], but they can occur in *MAPT* and *GRN* carriers as well, leading to erroneous diagnoses of schizophrenia, bipolar disorder, or other psychiatric illnesses [3, 22, 60, 61].

Motor Features

While the documentation of other clinical features has varied across reports, many cases develop parkinsonism as the course progresses in both *MAPT* and *GRN* carriers [51, 62, 63]. Among *MAPT* mutation carriers, bradykinesia, rigidity, and postural instability are most common, with a few patients having a tremor-predominant syndrome [64, 65]. All of these features can lead to patients being diagnosed with Parkinson's disease or dementia with Lewy bodies. Many develop a PSP/Richardson's syndrome-like phenotype in the more advanced stage of the disorder [65]. Limb apraxia is very rarely documented in *MAPT* [66], but can occur in *GRN* mutation carriers, particularly when the parietal lobe is affected [50, 51]. Other features such as alien limb syndrome, myoclonus, dystonia, etc. also occur in *GRN* mutation carriers [50, 51]. Upper and/or lower motor neuron dysfunction suggestive of ALS is rare in both groups.

Variation Within and Across Families and Mutation-Specific Syndromes

One of the most striking features of FTLD due to genetic mutations is the clinical heterogeneity. As reviewed above, patients can present with any of the FTLD syndromes, in addition to the typical and atypical syndromes associated with AD pathology. While there is some association between the type of mutation, the family history, and the clinical presentation, the ability to predict the clinical syndrome that will emerge in an individual based on these factors is limited. *MAPT* mutations tend to have more similarity within families compared with *GRN* mutations. The clinical phenotype among affected relatives in the same *MAPT* kindred tend to be relatively similar (with some notable exceptions). In addition, recent work has shown that age of symptom onset for an individual within a *MAPT* family is significantly correlated with the age of onset in their parents and other relatives [9]. There is evidence that some of the clinical symptomatology in *MAPT* mutations can be accounted for by the specific type of mutations (see Table 2). For example, mutations that are associated with accumulation of AD-like tau inclusions (e.g., V337M and R406W) are more likely to present with amnestic symptoms similar to AD, and some mutations, such as the N279K and IVS10 + 16 > T, have a particularly strong association with parkinsonism. Age of onset is also somewhat predicted by mutation, with some mutations such as the L266V, G335S, and G335V being associated with ages of onset lower than 30. A recent analysis confirmed that mutation type explained a significant amount of variation in age of onset, in addition to familial age of onset [9]. That said, ages of onset and clinical phenotypes across individuals can still be quite variable within kindreds, and families with individual ages of onset as disparate as 60 years apart have been reported. Some of the most common mutations, including the P301L and N279K, are associated with significant variability in clinical phenotype and age of onset. The reason for the significant genetic variability even within families is not known. Genetic background is certainly one possible mechanism, but no genetic modifiers that influence phenotype or age of onset have been identified in *MAPT*-associated disease. Such an analysis would have a low power to detect modifier effects because of the relatively small number of known symptomatic carriers.

Phenotypic variability within *GRN* families is very common [10, 50, 51]. Some reports have linked specific *GRN* mutations to specific phenotypes (Table 2). For instance, the IVS1 + 5G > C mutation is commonly associated with nfvPPA, while the T52fs and T272fs mutations often present with an AD-like phenotype [51]. The Q300X and IVS7-2A > G mutations have been associated with an ALS phenotype [9]. Despite these associations, the clinical presentations within *GRN* mutation families are quite variable, and a given family can certainly include one person

Table 2 Notable features associated with specific mutations in genes encoding microtubule-associated protein tau (*MAPT*) and progranulin (*GRN*)

MAPT	
L266V	Many have a very early age of onset (age < 30)
N279K	Parkinsonism early in the course; phenotype can be classic PD syndrome with rest tremor at least partially responsive to levodopa therapy. One of the more common mutations in *MAPT*
P301L	Most common mutation in *MAPT* (>230 known individuals). A minority have a very early age of onset (age < 30)
IVS10 + 3G > A	Highly variable age of onset; minority have a very early age of onset (age < 30)
IVS10 + 16C > T	One of the more common mutations in *MAPT* (>140 known individuals); a PD-predominant syndrome occurs in a minority
G335S, G335V	Most have very young onset (age < 30), can be rapidly progressive
Q336H	While this mutation is very rare, the FTD/ALS phenotype has occurred in multiple affected individuals
V337M	Often prominent amnestic component early in the course, and an AD-like phenotype can occur; can be very slow rate of progression
G389R	Very young onset, rapidly progressive. Pathology shows Pick bodies
R406W	Often prominent amnestic component early in the course, and an AD-like phenotype is common; can be very slow rate of progression
GRN	
IVS1 + 5G > C	Nonfluent variant PPA is the predominant phenotype
T52fs	An AD-like phenotype is common
T272fs	The most common mutation in *GRN* (>200 known individuals); an AD-like phenotype is common

(continued)

Table 2 (continued)

R493X	One of the more common mutations in *GRN* (>55 known individuals); an AD-like phenotype is common
Q300X	One of the few *GRN* mutations associated with ALS phenotype
IVS7-2A > G	One of the few *GRN* mutations associated with ALS phenotype
A472fs	While this mutation is very rare, the semantic variant PPA phenotype has occurred in multiple affected individuals

AD Alzheimer's disease, *ALS* amyotrophic lateral sclerosis, *FTD/ALS* combined frontotemporal dementia and amyotrophic lateral sclerosis, *PD* Parkinson's disease, *PPA* primary progressive aphasia

with bvFTD, another with a PPA, and a third with an AD-like presentation. In addition, individual age of onset varies considerably in *GRN* families and is much less predictable based on parental and familial age of onset compared with *MAPT* families [9]. Similarly, the specific type of *GRN* mutation does not strongly predict age of onset or disease duration [9]. Because of this variability and the stronger association of *GRN* mutations with AD-like clinical presentations, family histories in these families can potentially be interpreted as suggesting familial AD. Alternatively, the family history can be characterized by different individuals having apparently different diseases, so that the family and the clinician may not initially suspect that a single mutation is affecting the family. The variation across individuals carrying *GRN* mutations is particularly remarkable, given that the vast majority of mutations have the same primary effect on biology, which is a roughly 50% reduction in production of the PGRN protein. Again, genetic background has been considered as a potential explanation, and variation in the *TMEM106B* and *GFRA2* genes has been identified as conferring some protection, but no clear genetic modifiers that associate with age of onset or phenotypic presentation have been identified [67, 68].

The variation in age of onset and phenotype is a significant impediment in developing treatments for disease due to *MAPT* and *GRN* mutations. One reason is that studies focusing on symptomatic carriers will face obstacles in developing outcome measures suitable for participants with changes in social-emotional function as well as patients with language dysfunction and amnesia who would all be in the same study [27]. In addition, one goal for intervention studies would be to begin treatment before the onset of symptoms in order to demonstrate delay or prevention of symptoms. Such studies would have to focus on individuals who are most likely to develop symptoms within 1 to 2 years, the duration of a typical study. The significant variability in age of onset, along with the variation in early symptoms, makes it very difficult to identify such patients. The large studies referred to above are all seeking to develop better methods for studying participants with FTLD-causing mutations in clinical trials [12].

Neuroimaging and Other Biological Markers

The neuroimaging features in *MAPT* versus *GRN* mutation carriers are quite different, and the patterns of regional topography can provide clues to the presence of a mutation when combined with the clinical phenotype and family history.

As noted above, the evolution of degeneration in *MAPT* mutation carriers tends to begin with symmetric involvement of the bilateral medial and anterior temporal lobes with eventual involvement of the bilateral frontotemporal neural networks [10, 12, 45, 46, 69]. This topography underlies the semantic loss, object agnosia, and other cognitive and behavioral manifestations. Hippocampal atrophy is the most common feature in *MAPT* mutation carriers [10, 70], thereby explaining prominent memory impairment when this is present. Since the parietal and occipital lobes are relatively spared, these regions appear preserved on MRI and FDG-PET scans.

While clinical syndromes associated with symmetric changes on MRI and FDG-PET scans can occur in *GRN* mutation carriers, focal, asymmetric atrophy and hypometabolism early in the course are far more common, with the focality or asymmetry persisting over the course of the disorder [45, 46, 69, 71, 72]. As one would expect, the topography of atrophy or hypometabolism correlates with the clinical syndrome (e.g., left hemisphere disease being associated with language disorders, right hemisphere disease with visuospatial dysfunction). Furthermore, the tendency for imaging abnormalities to "respect the midline" in *GRN* mutation carriers is striking, with no atrophy/hypometabolism in one cerebral hemisphere despite profound abnormalities in the affected hemisphere. Other neuroimaging modalities, such as MRI-based diffusion-weighting imaging, magnetic resonance spectroscopy, etc., and longitudinal studies provide additional insights on f-FTLD [12, 73–79]. Representative MRI and FDG-PET scans from affected individuals are shown in Fig. 1.

Recent work has sought to identify additional biological markers in these disorders. The development of PET ligands that bind tau protein in AD [80] led to studies examining the utility of these agents in FTLD. While these studies have indicated increased uptake in patients with *MAPT* mutations that cause AD-like inclusions, other types of *MAPT* mutations unfortunately do not show increased binding [81, 82]. Furthermore, uptake can also be seen in *GRN* and *C9orf72* mutations, indicating that non-specific binding occurs with currently available tau PET ligands, severely limiting their utility.

Fluid biomarkers are also a major focus for development. In *GRN* carriers, genetic haploinsufficiency results in reduced production of PGRN messenger RNA and protein, which can be quantified in the CSF or blood [83, 84]. These reduced levels are present throughout life in carriers and have no relationship with any clinical features and therefore have no utility for tracking the natural history of disease or prognosis [85]. However, increases in PGRN levels in the setting of drug trials may indicate significant biological effects of a potential treatment and may predict clinical response. No fluid biomarkers that track tau concentrations in *MAPT* carriers have yet been developed.

Neurofilament light chain (NfL) is a neuronal cytoskeletal protein that is elevated in symptomatic FTLD and other neurodegenerative diseases

[86]. While not specific to any particular mutation or form of FTLD, higher levels of CSF and blood NfL are associated with greater degrees of atrophy and clinical impairment, both cross-sectionally and longitudinally, indicating that NfL is an indicator of the intensity of neurodegeneration [87, 88]. Recent work has shown that NfL levels predict increased rates of decline and shorter survival in f-FTLD mutation carriers, including *MAPT* and *GRN* mutations [89]. Furthermore, higher levels of NfL and rises in NfL over time are associated with development of symptoms in asymptomatic carriers [90].

A recent study also showed that plasma glial fibrillary acidic protein (GFAP) is elevated in symptomatic *GRN*, but not *MAPT* carriers, and that increased GFAP levels are associated with lower cognitive scores and brain volumes in asymptomatic *GRN* mutation carriers [91]. Measures of immune activation, including sCD163, CCL18, LBP, sCD14, IL-18, and CRP, correlate with severity of clinical symptoms and brain atrophy in *GRN*-associated disease [92]. All of these markers may provide useful indicators of emerging disease and also allow tracking of therapeutic effects in drug trials.

Prediction of Symptom Onset in Asymptomatic Carriers

Reliable approaches for predicting the age when symptoms will develop are important for individuals carrying these mutations to help with life planning. In addition, as noted above, prediction tools are important for selection of participants in clinical trials. Although studies have shown that abnormal performance on cognitive testing can be seen prior to frank development of symptoms in f-FTLD [43], the specific tests that show impairment vary considerably across individuals [10], even when carrying the same mutation, so that monitoring of a single or just a few tests is unlikely to provide adequate sensitivity and specificity. Several studies have indicated that reductions in brain volume precede onset of symptoms by up to 10 years in f-FTLD mutation carriers [43], and that quantification of the degree and

A

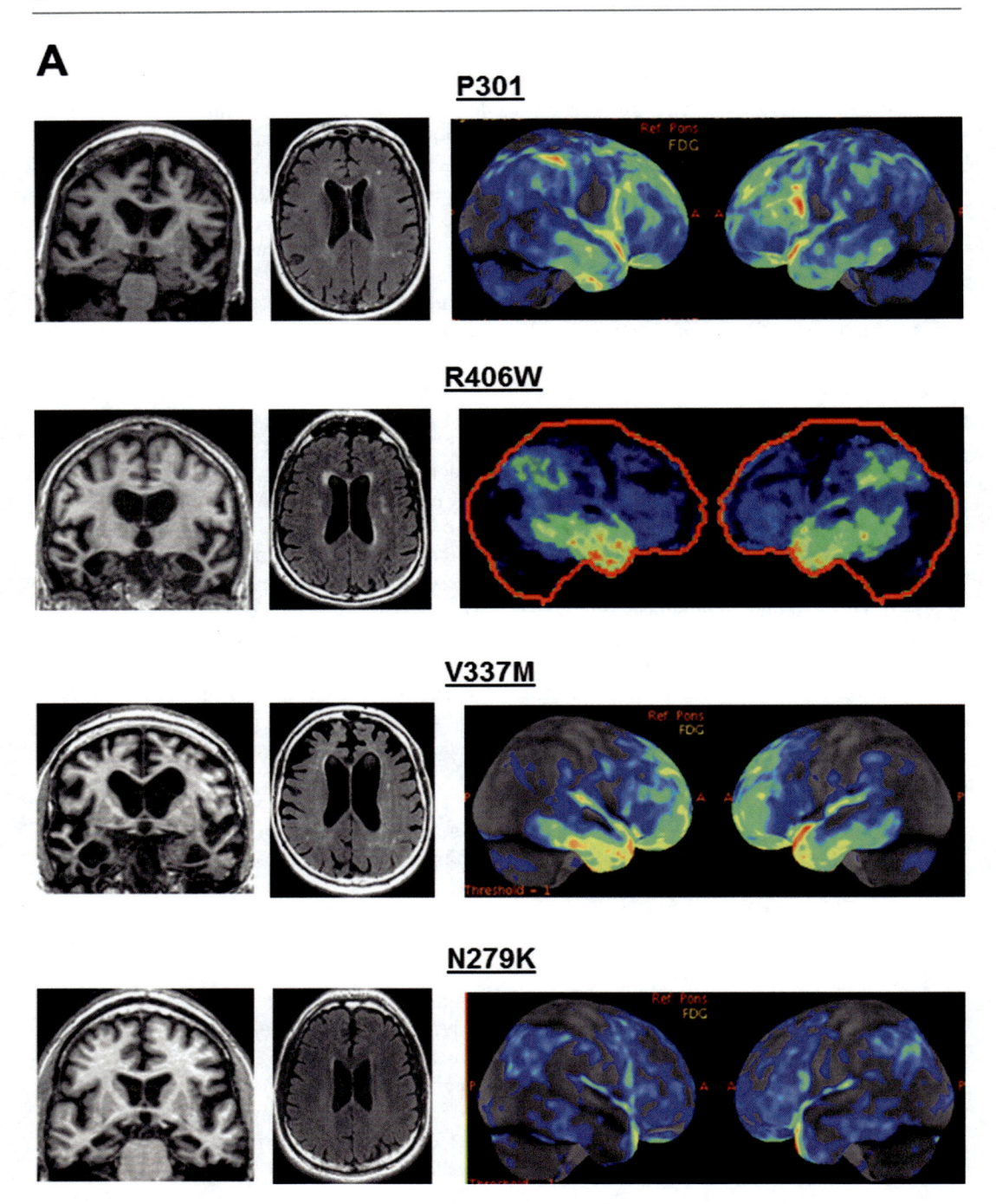

Fig. 1 Representative scans of individuals with *MAPT* (**a**) and *GRN* (**b**) mutations, showing coronal T1-weighted magnetic resonance images (far left), axial fluid attenuation inversion recovery magnetic resonance images (middle), and right and left lateral statistical map images from FDG-PET scans (right). Increasing degrees of hypometabolism on FDG-PET are represented by the following color scheme: black/gray (normal) – blue – green – yellow – orange – red (maximally abnormal). The scans in (**a**) are from symptomatic *MAPT* mutation carriers: P301L (age 53 with bvFTD features for 3 years), R406W (age 68 with early amnestic features followed by more typical yet slowly progressive bvFTD features for >20 years), V337M (age 67 with slowly progressive bvFTD features for >20 years), and N279K (age 50 with early amnestic and parkinsonian features followed by a mixed bvFTD/PSP phenotype for 3 years). Note the hippocampal atrophy in the R406W, V337, and N279K mutation cases, and frontal

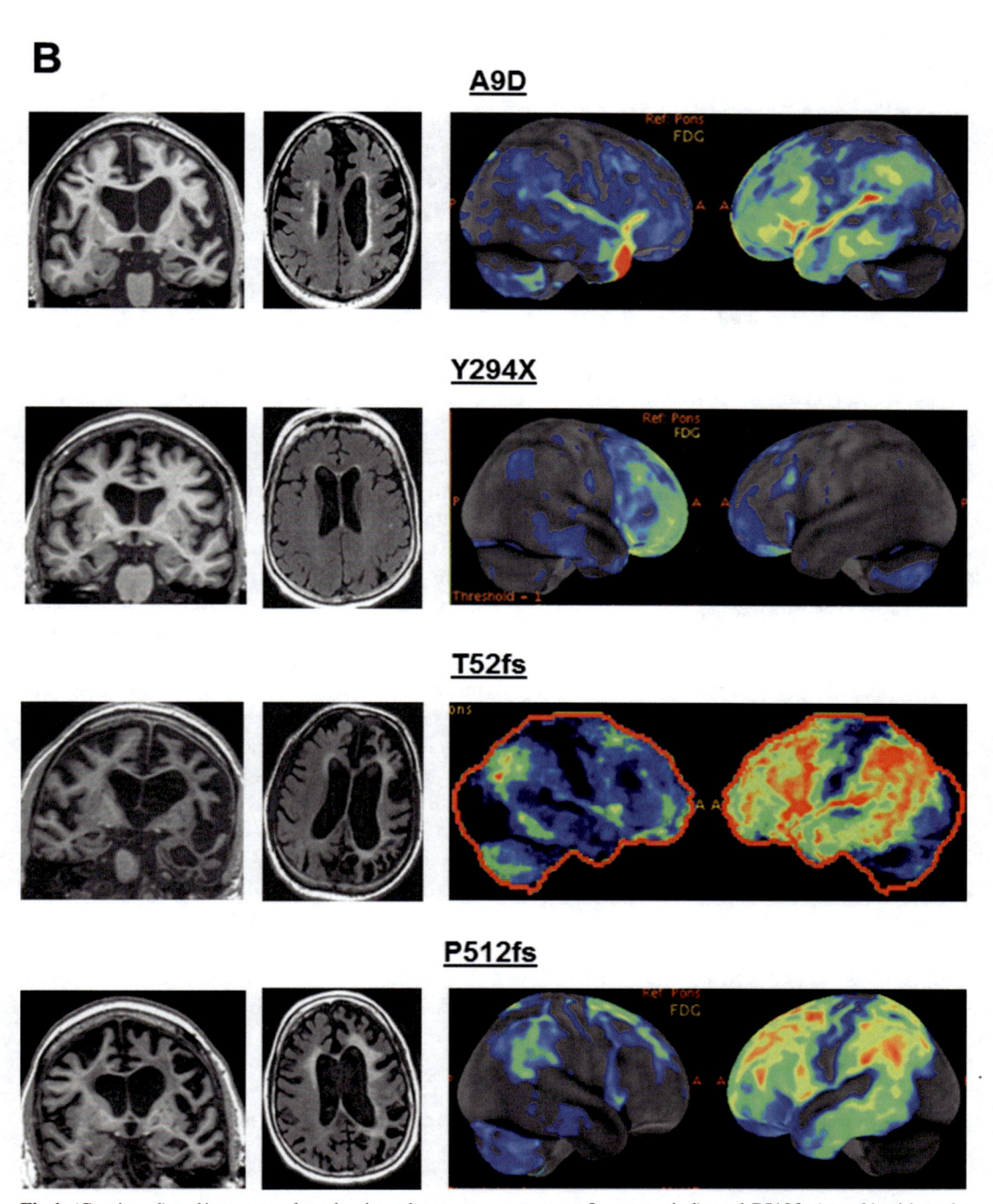

Fig 1 (Continued) and/or temporal predominant hypometabolism on the FDG-PET scans. Also note that the topographic distribution of atrophy and hypometabolism is relatively symmetric, which is typical of affected *MAPT* mutation carriers. The scans in (**b**) are from symptomatic *GRN* mutation carriers: A9D (age 72 with mixed PPA/bvFTD features for 3 years), Y294X (age 70 with bvFTD features for 3 years), T52fs (age 68 with early amnestic features followed by PPA and then CBS features evolving over an >8 year period), and P512fs (age 64 with early PPA features followed by CBS features for 2 years). Note the striking degree of asymmetry in all cases, which is typical of affected *GRN* mutation carriers. Abbreviations: bvFTD behavioral variant frontotemporal dementia, CBS corticobasal syndrome, FDG-PET fluorodeoxyglucose positron emission tomography, PPA primary progressive aphasia, PSP progressive supranuclear palsy

Table 3 Clues to suspect a mutation in microtubule-associated protein tau (*MAPT*) or progranulin (*GRN*)

MAPT
bvFTD +/− parkinsonism phenotype in the setting of a positive family history of dementia and/or parkinsonism
Very early age of onset and/or rapidly progressive course regardless of the clinical phenotype and presence/absence of a family history of dementia, parkinsonism, or ALS
Prominent memory impairment in the context of otherwise classic bvFTD clinical features
Relatively symmetric temporal and/or frontal abnormalities on MRI or FDG-PET

GRN
Any neurodegenerative syndrome in the context of a positive family history of dementia or parkinsonism – particularly if:
The patient's clinical findings and/or imaging abnormalities have focal or asymmetric features
One or more affected relatives have features or diagnoses (e.g., PPA, CBS) that suggest focal or asymmetric abnormalities

ALS amyotrophic lateral sclerosis, *bvFTD* behavioral variant frontotemporal dementia, *CBS* corticobasal syndrome, *FDG-PET* fluorodeoxyglucose positron emission tomography, *MRI* magnetic resonance imaging, *PPA* primary progressive aphasia

pattern of brain atrophy in individuals can significantly improve prediction of symptom development compared with age alone [93]. Furthermore, it has been shown that acceleration of the rates of brain volume loss and white matter degradation occurs within the few years prior to development of symptoms in *MAPT* carriers [12, 74–77]. Abnormalities in other types of imaging, such as MR spectroscopy and FLAIR, can also be seen in asymptomatic *MAPT* carriers prior to development of symptoms [78, 94]. Lastly, as noted above, rises in levels of NfL appear to predict development of symptoms. Most of these findings have been identified in relatively small groups of individuals who have been observed to convert from asymptomatic to symptomatic. Additional studies will need to be done to verify the utility of these measures, to quantify their predictive value, and to develop models based on combinations of these predictors to identify individuals close to this conversion with high sensitivity and specificity.

Clues for the Clinician to Suspect an *MAPT* or *GRN* Mutation

An important consideration for any clinician evaluating a patient for changes in cognition, behavior, or motor functioning is when to be suspicious of a mutation in *MAPT* or *GRN*. Family history is one of the most important features, and a thorough family history should be taken in any individual with an FTLD syndrome, a dementia syndrome with an age of onset younger than 65, or a rapidly progressive dementia syndrome. Any evidence of neurodegenerative disease in multiple family members, even if the syndromes differ across individuals, should raise concerns, and a mixture of motor, cognitive, and behavioral symptoms across family members is typical. It is worth noting that mutations are a much more common cause of f-FTLD than a cause of familial AD, which is only due to a mutation in a very small proportion of AD patients [95].

Another scenario that should raise concerns about a mutation would be the presence of a neurodegenerative syndrome with features of FTLD that are not entirely typical of the syndrome usually seen in sporadic cases. For instance, a patient with relatively symmetric atrophy of the anterior and medial temporal lobes and mild semantic loss mixed with amnesia of the type seen in typical AD has some features of svPPA but would be atypical because of the symmetry of temporal lobe changes and episodic memory loss. Such a patient might be an *MAPT* mutation carrier. Marked asymmetry on brain imaging is a finding that should prompt consideration of a *GRN* mutation. These clinical clues are summarized in Table 3.

Conclusions and Future Directions

F-FTLD syndromes due to mutations in the *MAPT* and *GRN* genes are important entities, because they offer many opportunities to understand the pathophysiology that leads to aggregation of tau and transactive response DNA-binding protein 43 (TDP-43), which are the two most common proteins associated with FTLD. The variability in clinical presentation across individuals with these mutations also offers an opportunity to understand how a single biological change can be manifest in many different effects in the CNS. Carriers of these mutations are also important for clinical trials of FTLD treatments because such studies could be assured that all individuals enrolling in the study are affected by the targeted biological mechanism if they recruit mutation carriers. The natural history studies described above (i.e., ALLFTD, GENFI, DINAD, FPI, others) will expand the characterization of kindreds with f-FTLD, as academic and industry investigators continue efforts to develop therapies that may slow the rate of progression, delay the onset of symptoms, and ultimately prevent the development of symptoms among those with mutations in *MAPT* or *GRN* [12, 27].

Acknowledgments This work was supported by NIH grants AG045390, NS092089, AG062677, AG063911, AG016574, AG019724, AG062422, AG056749, and AG045333.

We thank our many collaborators across the ALLFTD, GENFI, DINAD, and FPI consortia. The images in the Figure are courtesy of Drs. Val Lowe, Clifford Jack Jr., and Kejal Kantarci. We particularly thank the patients and their families for participating in aging and neurodegenerative disease research.

References

1. Mackenzie IR, Neumann M, Bigio EH, Cairns NJ, Alafuzoff I, Kril J et al (2009) Nomenclature for neuropathologic subtypes of frontotemporal lobar degeneration: consensus recommendations. Acta Neuropathol 117(1):15–18
2. Olney NT, Spina S, Miller BL (2017) Frontotemporal Dementia. Neurol Clin 35(2):339–374
3. Wilhelmsen KC, Lynch T, Pavlou E, Higgins M, Nygaard TG (1994) Localization of disinhibition-dementia-parkinsonism-Amyotrophy complex to 17q21–22. Am J Hum Genet 55:1159–1165
4. Foster NL, Wilhelmsen K, Sima AA, Jones MZ, D'Amato CJ, Gilman S (1997) Frontotemporal dementia and parkinsonism linked to chromosome 17: a consensus conference. Ann Neurol 41(6):706–715
5. Hutton M, Lendon CL, Rizzu P, Baker M, Froelich S, Houlden H et al (1998) Association of missense and 5′-splice-site mutations in tau with the inherited dementia FTDP-17. Nature 393(6686):702–705
6. Rademakers R, Cruts M, Dermaut B, Sleegers K, Rosso SM, Van den Broeck M et al (2002) Tau negative frontal lobe dementia at 17q21: significant fine-mapping of the candidate region to a 4.8 cM interval. Mol Psychiatry 7(10):1064–1074
7. Baker M, Mackenzie IR, Pickering-Brown SM, Gass J, Rademakers R, Lindholm C et al (2006) Mutations in progranulin cause tau-negative frontotemporal dementia linked to chromosome 17. Nature 442(7105):916–919
8. Cruts M, Gijselinck I, van der Zee J, Engelborghs S, Wils H, Pirici D et al (2006) Null mutations in progranulin cause ubiquitin-positive frontotemporal dementia linked to chromosome 17q21. Nature 442(7105):920–924
9. Moore KM, Nicholas J, Grossman M, McMillan CT, Irwin DJ, Massimo L et al (2020) Age at symptom onset and death and disease duration in genetic frontotemporal dementia: an international retrospective cohort study. Lancet Neurol 19(2):145–156
10. Olney NT, Ong E, Goh SM, Bajorek L, Dever R, Staffaroni AM et al (2020) Clinical and volumetric changes with increasing functional impairment in familial frontotemporal lobar degeneration. Alzheimers Dement 16(1):49–59
11. Ramos EM, Dokuru DR, Van Berlo V, Wojta K, Wang Q, Huang AY et al (2020) Genetic screening of a large series of North American sporadic and familial frontotemporal dementia cases. Alzheimers Dement 16(1):118–130
12. Rosen HJ, Boeve BF, Boxer AL (2020) Tracking disease progression in familial and sporadic frontotemporal lobar degeneration: recent findings from ARTFL and LEFFTDS. Alzheimers Dement 16(1):71–78
13. Sirkis DW, Geier EG, Bonham LW, Karch CM, Yokoyama JS (2019) Recent advances in the genetics of frontotemporal dementia. Curr Genet Med Rep 7(1):41–52
14. Che XQ, Zhao QH, Huang Y, Li X, Ren RJ, Chen SD et al (2017) Genetic features of MAPT, GRN, C9orf72 and CHCHD10 gene mutations in Chinese patients with frontotemporal dementia. Curr Alzheimer Res 14(10):1102–1108
15. Ikeuchi T, Kaneko H, Miyashita A, Nozaki H, Kasuga K, Tsukie T et al (2008) Mutational analysis in early-onset familial dementia in the Japanese population. The role of PSEN1 and MAPT R406W mutations. Dement Geriatr Cogn Disord 26(1):43–49
16. Kim EJ, Kwon JC, Park KH, Park KW, Lee JH, Choi SH et al (2014) Clinical and genetic analysis of MAPT, GRN, and C9orf72 genes in Korean patients with frontotemporal dementia. Neurobiol Aging 35(5):1213 e13–1213 e17
17. Kim HJ, Oh KW, Kwon MJ, Oh SI, Park JS, Kim YE et al (2016) Identification of mutations in Korean patients with amyotrophic lateral sclerosis using multigene panel testing. Neurobiol Aging 37:209 e9–209e16
18. Ogaki K, Li Y, Takanashi M, Ishikawa K, Kobayashi T, Nonaka T et al (2013) Analyses of the MAPT,

PGRN, and C9orf72 mutations in Japanese patients with FTLD, PSP, and CBS. Parkinsonism Relat Disord 19(1):15–20

19. Wei Q, Chen X, Chen Y, Ou R, Cao B, Hou Y et al (2019) Unique characteristics of the genetics epidemiology of amyotrophic lateral sclerosis in China. Sci China Life Sci 62(4):517–525

20. Perry DC, Lehmann M, Yokoyama JS, Karydas A, Lee JJ, Coppola G et al (2013) Progranulin mutations as risk factors for Alzheimer disease. JAMA Neurol 70(6):774–778

21. Van Deerlin VM, Forman MS, Farmer JM, Grossman M, Joyce S, Crowe A et al (2007) Biochemical and pathological characterization of frontotemporal dementia due to a Leu266Val mutation in microtubule-associated protein tau in an African American individual. Acta Neuropathol 113(4):471–479

22. Momeni P, DeTucci K, Straub RE, Weinberger DR, Davies P, Grafman J et al (2010) Progranulin (GRN) in two siblings of a Latino family and in other patients with schizophrenia. Neurocase 16(3):273–279

23. Boeve BF, Tremont-Lukats IW, Waclawik AJ, Murrell JR, Hermann B, Jack CR Jr et al (2005) Longitudinal characterization of two siblings with frontotemporal dementia and parkinsonism linked to chromosome 17 associated with the S305N tau mutation. Brain 128(Pt 4):752–772

24. Bateman RJ, Xiong C, Benzinger TL, Fagan AM, Goate A, Fox NC et al (2012) Clinical and biomarker changes in dominantly inherited Alzheimer's disease. N Engl J Med 367(9):795–804

25. Ryman DC, Acosta-Baena N, Aisen PS, Bird T, Danek A, Fox NC et al (2014) Symptom onset in autosomal dominant Alzheimer disease: a systematic review and meta-analysis. Neurology 83(3):253–260

26. Boeve B, Bove J, Brannelly P, Brushaber D, Coppola G, Dever R et al (2020) The longitudinal evaluation of familial frontotemporal dementia subjects protocol: framework and methodology. Alzheimers Dement 16(1):22–36

27. Boxer AL, Gold M, Feldman H, Boeve BF, Dickinson SL, Fillit H et al (2020) New directions in clinical trials for frontotemporal lobar degeneration: methods and outcome measures. Alzheimers Dement 16(1):131–143

28. Irwin DJ (2016) Tauopathies as clinicopathological entities. Parkinsonism Relat Disord 22(Suppl 1):S29–S33

29. Bodea LG, Eckert A, Ittner LM, Piguet O, Gotz J (2016) Tau physiology and pathomechanisms in frontotemporal lobar degeneration. J Neurochem 138(Suppl 1):71–94

30. Wang Y, Mandelkow E (2016) Tau in physiology and pathology. Nat Rev Neurosci 17(1):5–21

31. Karch CM, Kao AW, Karydas A, Onanuga K, Martinez R, Argouarch A et al (2019) A comprehensive resource for induced pluripotent stem cells from patients with primary Tauopathies. Stem Cell Reports 13(5):939–955

32. Vaquer-Alicea J, Diamond MI (2019) Propagation of protein aggregation in neurodegenerative diseases. Annu Rev Biochem 88:785–810

33. Mackenzie IR, Neumann M (2016) Molecular neuropathology of frontotemporal dementia: insights into disease mechanisms from postmortem studies. J Neurochem 138(Suppl 1):54–70

34. Cairns NJ, Bigio EH, Mackenzie IR, Neumann M, Lee VM, Hatanpaa KJ et al (2007) Neuropathologic diagnostic and nosologic criteria for frontotemporal lobar degeneration: consensus of the consortium for frontotemporal lobar degeneration. Acta Neuropathol 114(1):5–22

35. Galimberti D, Fenoglio C, Scarpini E (2018) Progranulin as a therapeutic target for dementia. Expert Opin Ther Targets 22(7):579–585

36. Bateman A, Cheung ST, Bennett HPJ (2018) A brief overview of progranulin in health and disease. Methods Mol Biol 1806:3–15

37. Kao AW, McKay A, Singh PP, Brunet A, Huang EJ (2017) Progranulin, lysosomal regulation and neurodegenerative disease. Nat Rev Neurosci 18(6):325–333

38. Neary D, Snowden JS, Gustafson L, Passant U, Stuss D, Black S et al (1998) Frontotemporal lobar degeneration: a consensus on clinical diagnostic criteria. Neurology 51(6):1546–1554

39. Rascovsky K, Hodges JR, Kipps CM, Johnson JK, Seeley WW, Mendez MF et al (2007) Diagnostic criteria for the behavioral variant of frontotemporal dementia (bvFTD): current limitations and future directions. Alzheimer Dis Assoc Disord 21(4):S14–S18

40. Rascovsky K, Hodges JR, Knopman D, Mendez MF, Kramer JH, Neuhaus J et al (2011) Sensitivity of revised diagnostic criteria for the behavioural variant of frontotemporal dementia. Brain 134(Pt 9):2456–2477

41. Rosen HJ, Gorno-Tempini ML, Goldman WP, Perry RJ, Schuff N, Weiner M et al (2002) Patterns of brain atrophy in frontotemporal dementia and semantic dementia. Neurology 58(2):198–208

42. Heuer HW, Wang P, Rascovsky K, Wolf A, Appleby B, Bove J et al (2020) Comparison of sporadic and familial behavioral variant frontotemporal dementia (FTD) in a North American cohort. Alzheimers Dement 16(1):60–70

43. Rohrer JD, Nicholas JM, Cash DM, van Swieten J, Dopper E, Jiskoot L et al (2015) Presymptomatic cognitive and neuroanatomical changes in genetic frontotemporal dementia in the genetic frontotemporal dementia initiative (GENFI) study: a cross-sectional analysis. Lancet Neurol 14(3):253–262

44. Whitwell JL, Jack CR Jr, Parisi JE, Knopman DS, Boeve BF, Petersen RC et al (2011) Imaging signatures of molecular pathology in behavioral variant frontotemporal dementia. J Mol Neurosci 45(3):372–378

45. Whitwell JL, Weigand SD, Boeve BF, Senjem ML, Gunter JL, DeJesus-Hernandez M et al (2012) Neuroimaging signatures of frontotemporal dementia

genetics: C9ORF72, tau, progranulin and sporadics. Brain 135(Pt 3):794–806

46. Whitwell JL, Boeve BF, Weigand SD, Senjem ML, Gunter JL, Baker MC et al (2015) Brain atrophy over time in genetic and sporadic frontotemporal dementia: a study of 198 serial magnetic resonance images. Eur J Neurol 22(5):745–752

47. Gorno-Tempini ML, Hillis AE, Weintraub S, Kertesz A, Mendez M, Cappa SF et al (2011) Classification of primary progressive aphasia and its variants. Neurology 76(11):1006–1014

48. Snowden JS, Pickering-Brown SM, Mackenzie IR, Richardson AM, Varma A, Neary D et al (2006) Progranulin gene mutations associated with frontotemporal dementia and progressive non-fluent aphasia. Brain 129:3091–3102

49. Whitwell JL, Jack CR Jr, Boeve BF, Senjem ML, Baker M, Ivnik RJ et al (2009) Atrophy patterns in IVS10+16, IVS10+3, N279K, S305N, P301L, and V337M MAPT mutations. Neurology 73(13):1058–1065

50. Beck J, Rohrer JD, Campbell T, Isaacs A, Morrison KE, Goodall EF et al (2008) A distinct clinical, neuropsychological and radiological phenotype is associated with progranulin gene mutations in a large UK series. Brain 131(Pt 3):706–720

51. Kelley BJ, Haidar W, Boeve BF, Baker M, Graff-Radford NR, Krefft T et al (2009) Prominent phenotypic variability associated with mutations in progranulin. Neurobiol Aging 30(5):739–751

52. Ygland E, van Westen D, Englund E, Rademakers R, Wszolek ZK, Nilsson K et al (2018) Slowly progressive dementia caused by MAPT R406W mutations: longitudinal report on a new kindred and systematic review. Alzheimers Res Ther 10(1):2

53. Caroppo P, Belin C, Grabli D, Maillet D, De Septenville A, Migliaccio R et al (2015) Posterior cortical atrophy as an extreme phenotype of GRN mutations. JAMA Neurol 72(2):224–228

54. Crutch SJ, Schott JM, Rabinovici GD, Boeve BF, Cappa SF, Dickerson BC et al (2013) Shining a light on posterior cortical atrophy. Alzheimers Dement 9(4):463–465

55. Wolk DA, Price JC, Madeira C, Saxton JA, Snitz BE, Lopez OL et al (2012) Amyloid imaging in dementias with atypical presentation. Alzheimers Dement 8(5):389–398

56. Boeve BF, Lang AE, Litvan I (2003) Corticobasal degeneration and its relationship to progressive supranuclear palsy and frontotemporal dementia. Ann Neurol 54(Suppl 5):S15–S19

57. Lee SE, Rabinovici GD, Mayo MC, Wilson SM, Seeley WW, DeArmond SJ et al (2011) Clinicopathological correlations in corticobasal degeneration. Ann Neurol 70(2):327–340

58. Boeve BF, Graff-Radford NR (2012) Cognitive and behavioral features of c9FTD/ALS. Alzheimers Res Ther 4(4):29

59. Snowden JS, Rollinson S, Thompson JC, Harris JM, Stopford CL, Richardson AM et al (2012) Distinct

clinical and pathological characteristics of frontotemporal dementia associated with C9ORF72 mutations. Brain 135(Pt 3):693–708

60. Khan BK, Woolley JD, Chao S, See T, Karydas AM, Miller BL et al (2012) Schizophrenia or neurodegenerative disease prodrome? Outcome of a first psychotic episode in a 35-year-old woman. Psychosomatics 53(3):280–284

61. Block NR, Sha SJ, Karydas AM, Fong JC, De May MG, Miller BL et al (2016) Frontotemporal dementia and psychiatric illness: emerging clinical and biological links in gene carriers. Am J Geriatr Psychiatry 24(2):107–116

62. Warren JD, Rohrer JD, Rossor MN (2013) Clinical review. Frontotemporal dementia. BMJ 347:f4827

63. Bang J, Spina S, Miller BL (2015) Frontotemporal dementia. Lancet 386(10004):1672–1682

64. Wszolek ZK, Pfeiffer RF (1992) Genetic considerations in movement disorders. Curr Opin Neurol Neurosurg 5(3):324–330

65. Wszolek ZK, Kardon RH, Wolters EC, Pfeiffer RF (2001) Frontotemporal dementia and parkinsonism linked to chromosome 17 (FTDP-17): PPND family. A longitudinal videotape demonstration. Mov Disord 16(4):756–760

66. Rossi G, Marelli C, Farina L, Laura M, Maria Basile A, Ciano C et al (2008) The G389R mutation in the MAPT gene presenting as sporadic corticobasal syndrome. Mov Disord 23(6):892–895

67. Nicholson AM, Rademakers R (2016) What we know about TMEM106B in neurodegeneration. Acta Neuropathol 132(5):639–651

68. Pottier C, Zhou X, Perkerson RB 3rd, Baker M, Jenkins GD, Serie DJ et al (2018) Potential genetic modifiers of disease risk and age at onset in patients with frontotemporal lobar degeneration and GRN mutations: a genome-wide association study. Lancet Neurol 17(6):548–558

69. Whitwell JL, Weigand SD, Gunter JL, Boeve BF, Rademakers R, Baker M et al (2011) Trajectories of brain and hippocampal atrophy in FTD with mutations in MAPT or GRN. Neurology 77(4):393–398

70. Frank AR, Wszolek ZK, Jack CR Jr, Boeve BF (2007) Distinctive MRI findings in pallidopontonigral degeneration (PPND). Neurology 68(8):620–621

71. Jacova C, Hsiung GY, Tawankanjanachot I, Dinelle K, McCormick S, Gonzalez M et al (2013) Anterior brain glucose hypometabolism predates dementia in progranulin mutation carriers. Neurology 81(15):1322–1331

72. McDade E, Boeve BF, Burrus TM, Boot BP, Kantarci K, Fields J et al (2012) Similar clinical and neuroimaging features in monozygotic twin pair with mutation in progranulin. Neurology 78(16):1245–1249

73. Chen Q, Boeve BF, Tosakulwong N, Lesnick T, Brushaber D, Dheel C et al (2019) Brain MR spectroscopy changes precede frontotemporal lobar degeneration Phenoconversion in Mapt mutation carriers. J Neuroimaging 29(5):624–629

74. Chen Q, Boeve BF, Senjem M, Tosakulwong N, Lesnick TG, Brushaber D et al (2019) Rates of lobar atrophy in asymptomatic MAPT mutation carriers. Alzheimers Dement (N Y) 5:338–346

75. Chen Q, Boeve BF, Senjem M, Tosakulwong N, Lesnick T, Brushaber D et al (2019) Trajectory of lobar atrophy in asymptomatic and symptomatic GRN mutation carriers: a longitudinal MRI study. Neurobiol Aging

76. Chen Q, Boeve BF, Schwarz CG, Reid R, Tosakulwong N, Lesnick TG et al (2019) Tracking white matter degeneration in asymptomatic and symptomatic MAPT mutation carriers. Neurobiol Aging 83:54–62

77. Jiskoot LC, Panman JL, Meeter LH, Dopper EGP, Donker Kaat L, Franzen S et al (2019) Longitudinal multimodal MRI as prognostic and diagnostic biomarker in presymptomatic familial frontotemporal dementia. Brain 142(1):193–208

78. Sudre CH, Bocchetta M, Cash D, Thomas DL, Woollacott I, Dick KM et al (2017) White matter hyperintensities are seen only in GRN mutation carriers in the GENFI cohort. NeuroImage: Clinical 15:171–180

79. Tavares TP, Mitchell DGV, Coleman K, Shoesmith C, Bartha R, Cash DM et al (2019) Ventricular volume expansion in presymptomatic genetic frontotemporal dementia. Neurology 93(18):e1699–ee706

80. Choi Y, Ha S, Lee YS, Kim YK, Lee DS, Kim DJ (2018) Development of tau PET imaging ligands and their utility in preclinical and clinical studies. Nucl Med Mol Imaging 52(1):24–30

81. Ikeda A, Shimada H, Nishioka K, Takanashi M, Hayashida A, Li Y et al (2019) Clinical heterogeneity of frontotemporal dementia and parkinsonism linked to chromosome 17 caused by MAPT N279K mutation in relation to tau positron emission tomography features. Mov Disord 34(4):568–574

82. Tsai RM, Bejanin A, Lesman-Segev O, LaJoie R, Visani A, Bourakova V et al (2019) (18)F-flortaucipir (AV-1451) tau PET in frontotemporal dementia syndromes. Alzheimers Res Ther 11(1):13

83. Schofield EC, Halliday GM, Kwok J, Loy C, Double KL, Hodges JR (2010) Low serum progranulin predicts the presence of mutations: a prospective study. J Alzheimers Dis 22(3):981–984

84. Guven G, Bilgic B, Tufekcioglu Z, Erginel Unaltuna N, Hanagasi H, Gurvit H et al (2019) Peripheral GRN mRNA and serum progranulin levels as a potential indicator for both the presence of splice site mutations and individuals at risk for frontotemporal dementia. J Alzheimers Dis 67(1):159–167

85. Galimberti D, Fumagalli GG, Fenoglio C, Cioffi SMG, Arighi A, Serpente M et al (2018) Progranulin plasma levels predict the presence of GRN mutations in asymptomatic subjects and do not correlate with brain atrophy: results from the GENFI study. Neurobiol Aging 62:245 e9–245e12

86. Zhao Y, Xin Y, Meng S, He Z, Hu W (2019) Neurofilament light chain protein in neurodegenerative dementia: a systematic review and network meta-analysis. Neurosci Biobehav Rev 102:123–138

87. Rohrer JD, Woollacott IO, Dick KM, Brotherhood E, Gordon E, Fellows A et al (2016) Serum neurofilament light chain protein is a measure of disease intensity in frontotemporal dementia. Neurology 87(13):1329–1336

88. Ljubenkov PA, Staffaroni AM, Rojas JC, Allen IE, Wang P, Heuer H et al (2018) Cerebrospinal fluid biomarkers predict frontotemporal dementia trajectory. Ann Clin Transl Neurol 5(10):1250–1263

89. Meeter LH, Dopper EG, Jiskoot LC, Sanchez-Valle R, Graff C, Benussi L et al (2016) Neurofilament light chain: a biomarker for genetic frontotemporal dementia. Ann Clin Transl Neurol 3(8):623–636

90. van der Ende EL, Meeter LH, Poos JM, Panman JL, Jiskoot LC, Dopper EGP et al (2019) Serum neurofilament light chain in genetic frontotemporal dementia: a longitudinal, multicentre cohort study. Lancet Neurol 18(12):1103–1111

91. Heller C, Foiani MS, Moore K, Convery R, Bocchetta M, Neason M et al (2020) Plasma glial fibrillary acidic protein is raised in progranulin-associated frontotemporal dementia. J Neurol Neurosurg Psychiatry 91(3):263–270

92. Ljubenkov PA, Miller Z, Mumford P, Zhang J, Allen IE, Mitic L et al (2019) Peripheral innate immune activation correlates with disease severity in GRN Haploinsufficiency. Front Neurol 10:1004

93. Staffaroni AM, Cobigo Y, Goh SM, Kornak J, Bajorek L, Chiang K et al (2020) Individualized atrophy scores predict dementia onset in familial frontotemporal lobar degeneration. Alzheimers Dement 16(1):37–48

94. Chen Q, Boeve BF, Tosakulwong N, Lesnick T, Brushaber D, Dheel C et al (2019) Frontal lobe (1)H MR spectroscopy in asymptomatic and symptomatic MAPT mutation carriers. Neurology 93(8):e758–e765

95. Cohn-Hokke PE, Elting MW, Pijnenburg YA, van Swieten JC (2012) Genetics of dementia: update and guidelines for the clinician. Am J Med Genet B Neuropsychiatr Genet 159B(6):628–643

Neuroimaging in Frontotemporal Lobar Degeneration: Research and Clinical Utility

Sheena I. Dev, Bradford C. Dickerson, and Alexandra Touroutoglou

Introduction

The clinical and pathological heterogeneity in frontotemporal lobar degeneration (FTLD) presents a variety of challenges to clinicians and researchers, including accurate and timely diagnosis, prognostication, monitoring, and identification of appropriate endpoints in clinical trials. Neuroimaging offers a powerful set of tools to visualize structural, functional, and pathological changes associated with FTLD. This is particularly true of the three most common clinical presentations of sporadic FTLD—behavioral variant frontotemporal dementia (bvFTD), semantic variant primary progressive aphasia (svPPA), and nonfluent/agrammatic variant PPA (nfvPPA)—as the neuroanatomical and hypometabolic signatures of these three syndromes are generally well defined.

In this chapter, we selectively review evidence supporting the utility of neuroimaging biomarkers in bvFTD, svPPA, and nfvPPA, with an emphasis on current and future clinical applications. We begin by discussing patterns of abnormalities and diagnostic utility among those neuroimaging methods most commonly used in clinical settings, including structural T1 and fluorodeoxyglucose positron emission tomography (FDG-PET). This is followed by a review of imaging methods used in research settings that show a promising role in clinical settings or as endpoints in clinical trials.

Imaging Modalities Commonly Used in Clinical Settings

Consensus diagnostic criteria for bvFTD, svPPA, and nfvPPA include neuroimaging as the major biomarker to aid in confident clinical diagnosis with an emphasis on structural magnetic resonance imaging (MRI) and 18F-fluorodeoxyglucose positron emission tomography (FDG-PET) [1, 2].

Structural Magnetic Resonance Imaging

Structural MRI is widely used in clinical practice to visualize regional brain atrophy, with quantitative methods employed in research [3–7] and in some clinical practice settings [8]. In bvFTD, structural imaging studies consistently

S. I. Dev
Department of Psychiatry, Massachusetts General Hospital/Harvard Medical School, Charlestown, MA, USA
e-mail: sidev@mgh.harvard.edu

B. C. Dickerson (✉) · A. Touroutoglou
Department of Neurology, Massachusetts General Hospital/Harvard Medical School, Charlestown, MA, USA
e-mail: brad.dickerson@mgh.harvard.edu;
atouroutoglou@mgh.harvard.edu

© Springer Nature Switzerland AG 2021
B. Ghetti et al. (eds.), *Frontotemporal Dementias*, Advances in Experimental Medicine and Biology
1281, https://doi.org/10.1007/978-3-030-51140-1_7

demonstrate atrophy patterns that implicate frontal (orbitofrontal, middle frontal gyrus, rostromedial prefrontal cortex, pre-supplementary motor cortex), insula (dorsal and ventral anterior insula extending to posterior insula at late stages), anterior/mid cingulate cortex (ACC/MCC), anterior temporal lobes and subcortical structures (basal ganglia, thalamus, hippocampus), as well as cerebellum (see Fig. 1) [9–11]. For review, see reference [6], and for meta-analyses, see references [12–14]. Despite a relatively predictable atrophy pattern, there remains considerable heterogeneity across patients. One study employed cluster analyses and identified four distinct neuroanatomical subtypes: frontal dominant, temporal dominant, frontotemporal, and distributed temporofrontoparietal [15]. A subcortical dominant subtype has also been described [16, 17]. Brain atrophy in fronto-insula-cingulate regions is present at the earliest stages of the disease, although less pronounced [9, 18]. The right temporal variant of FTLD has been variably characterized as semantic demen-

tia or bvFTD but is generally associated with prominent behavioral symptoms, largely typical of bvFTD, often accompanied by prosopagnosia and semantic memory loss [19–22]. With time, atrophy progresses to include more distributed frontal, temporal, and insular cortices, as well as parietal regions and ventricular expansion (see Fig. 2) [9, 23–25]. Automated longitudinal MRI volumetry has demonstrated that structural MRI is sensitive to frontal atrophy progression in a period as short as 6 months after baseline [26].

Peak atrophy at baseline in svPPA has been reported in the anterior temporal pole (left > right hemisphere), extending to include frontoinsula, subgenual ACC, left middle and inferior temporal gyri, fusiform gyri, amygdala and basal forebrain (see Fig. 1) [12, 27–33]. In this regard, svPPA presents with similar atrophy distribution as the temporal dominant subtype of bvFTD with more atrophy in left relative to right hemisphere [15]. However, in contrast to bvFTD, svPPA has greater atrophy in the fusiform gyrus and relatively spared frontal as well as dorsal anterior

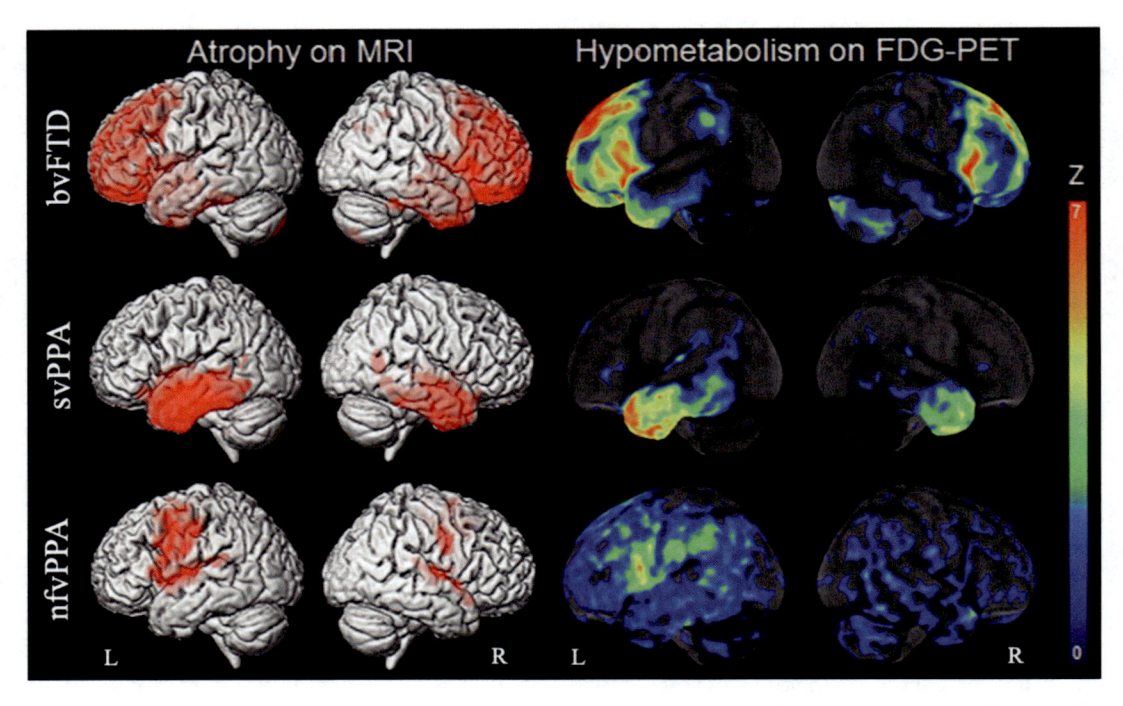

Fig. 1 Group-level patterns of atrophy and hypometabolism associated with FTLD clinical syndromes demonstrate similar patterns to single-subject scans (see Fig. 3). svPPA, semantic variant primary progressive aphasia; bvFTD, behavioral variant frontotemporal dementia; L, left; R, right. (Image adapted from Whitwell [17])

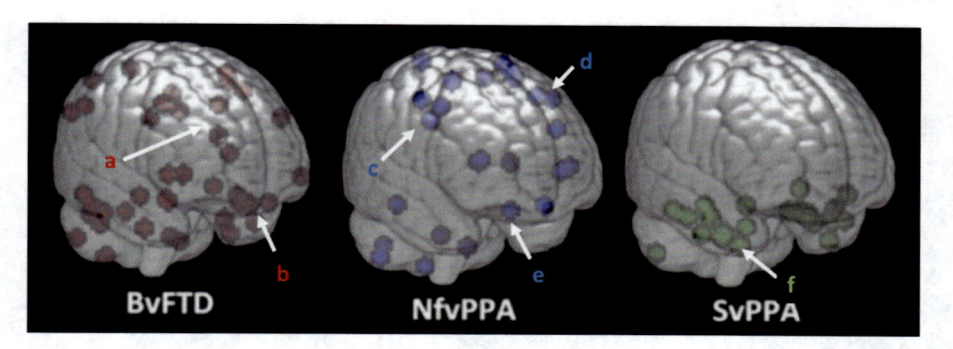

Fig. 2 Peak regions that display the highest rate of gray matter atrophy over a one-year follow-up. All three variants demonstrate differential patterns of longitudinal loss in gray matter volume, with highly clustered regions of change in svPPA (anterior temporal lobe) and more distributed changes in frontotemporal regions among bvFTD and nfvPPA variants. (**a**) dorsolateral prefrontal cortex; (**b**) orbitofrontal cortex; (**c**) premotor cortex; (**d**) superior frontal cortex; (**e**) inferior frontal gyrus (pars orbitalis); (**f**) anterior temporal lobe. (Image adapted from Binney, Pankov [53])

cingulate cortex [23, 34]. With disease progression, atrophy in svPPA becomes more distributed to include the middle and inferior frontal gyri, posterior temporal gyrus, and inferior parietal lobule [7, 28, 34–36]. Atrophy usually continues to be left-lateralized but spreads to homologous regions of the right hemisphere (see Fig. 2) [36, 37]. The rate of temporal gray matter loss can be 3–4% per six months [26], which is a higher rate of temporal atrophy than any other FTD variant [38].

In nfvPPA, peak atrophy is found in the inferior frontal gyrus, dorsolateral prefrontal cortex, and supramarginal gyrus of the left hemisphere (see Fig. 1) [32, 36]. Compared to both svPPA and bvFTD, nfvPPA demonstrates greater progressive atrophy in the parietal lobes [23] but relatively spared bilateral temporal pole, parahippocampal, entorhinal, fusiform, inferior temporal, middle, and left superior temporal gyrus [35]. Over time, atrophy in nfvPPA progresses to include more widespread left superior and middle frontal gyri, anterior insula, superior temporal gyri, transverse temporal gyrus as well as premotor areas, and caudate (see Fig. 2) [30, 31, 35, 36, 39]. In general, as atrophy progresses in both PPA variants, the pattern becomes less specific (though remains generally left-lateralized) and merges with each other to include major regions involved in language processing [40].

Within clinical settings, T1-weighted MRI is used nearly universally to increase confidence in the likely pathological changes in patients with these clinical syndromes [41]. In most clinical practice settings, planar images are inspected visually (see Fig. 3), which has been shown to be predictive of likely neuropathologic changes postmortem [42]. Although not yet routinely used in clinical practice, a variety of fully automated, observer-independent atrophy quantification methods have been developed, including voxel-based morphometry, single-subject whole-cortex general linear models, and machine learning–based individual subject classification models [26, 32, 42–45]. A few of these methods are beginning to be employed in clinical practice settings [46, 47].

A number of studies have examined the sensitivity and specificity of quantitative analysis of MRI volumetrics for diagnosis. In a recent multicenter structural MRI study, Meyer and Mueller [48] applied pattern recognition algorithms to regional brain atrophy and predicted diagnosis of bvFTD (vs. healthy controls) with high accuracy of up to 84.6%. In another study, gray matter density-based machine learning classification of bvFTD versus Alzheimer's disease (AD) outperformed the classification that was based on neuropsychological test results [49]. Similarly, diagnostic criteria for PPA sup-

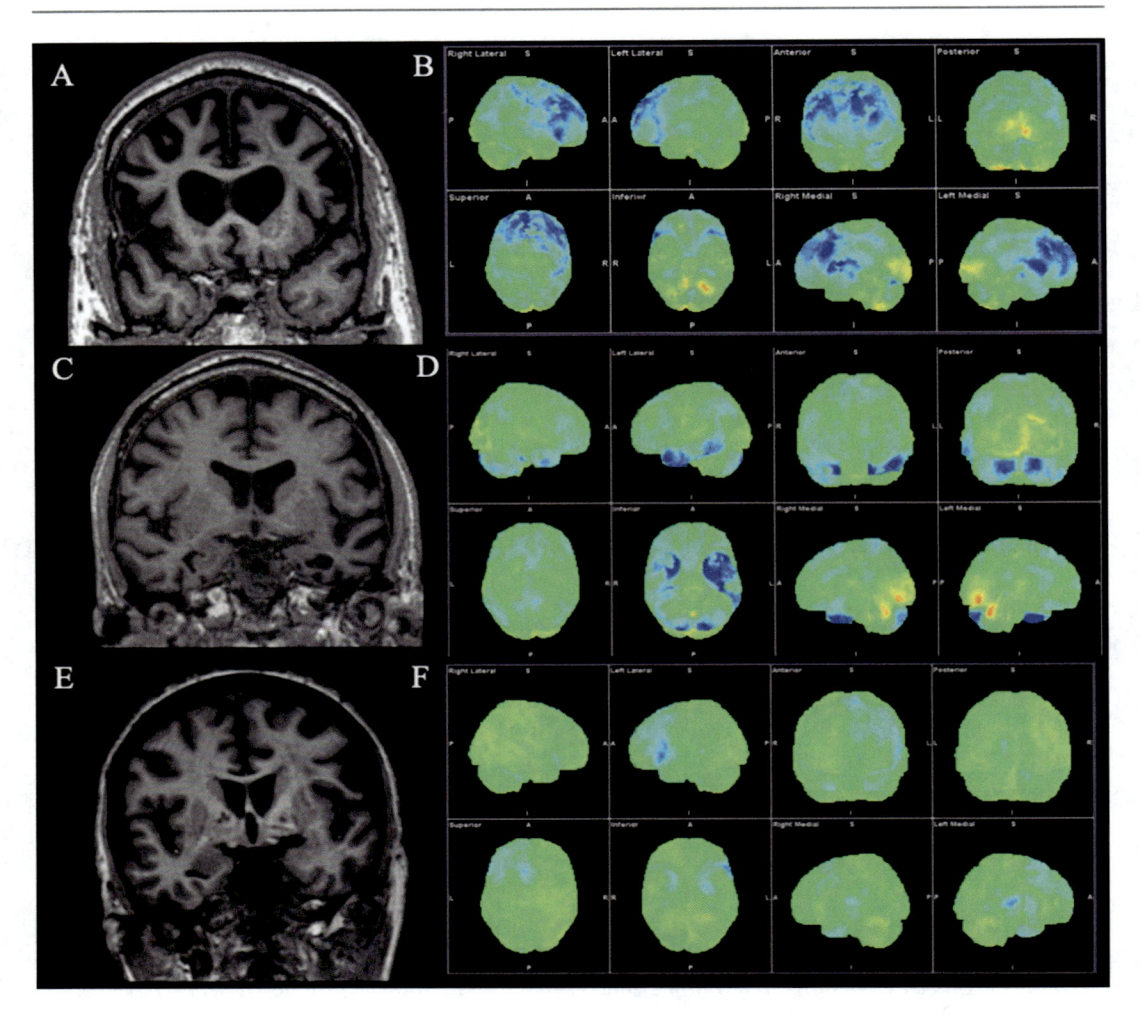

Fig. 3 T1-weighted MRI images and FDG-PET surface projections obtained in clinical practice in individual patients with typical FTLD clinical syndromes. In bvFTD, atrophy (**a**) and hypometabolism (**b**) are observed in bilateral dorsolateral and dorsomedial prefrontal, anterior and mid-cingulate, and insular cortices. BvFTD is heterogeneous, and patients often present with orbitofrontal or right anterior temporal atrophy. In svPPA, atrophy (**c**) and hypometabolism (**d**) are most notable in anterior temporal lobe (L > R) regions and extend to left lateral temporal cortices. In nfvPPA, atrophy (**e**) and hypometabolism (**f**) are seen in left insular cortex, inferior and middle frontal gyri, and the amygdala

port patterns of brain atrophy specific to the regions outlined earlier. Recently, MRI-based cortical thickness was shown to classify a single patient as belonging to one subtype of FTD with high accuracy: 86% healthy controls versus dementia (FTD and AD), 90.8% of AD versus FTLD, 86.9% bvFTD versus PPA, and 92.1% svPPA versus nfvPPA [44]. Atrophy in right frontotemporal regions successfully discriminates bvFTD from PPA, atrophy in the left frontal lobe discriminates nfvPPA from svPPA, and atrophy in the bilateral anterior temporal cortex discriminates svPPA from bvFTD and nfvPPA [44, 50]. Another study reported a similar high accuracy (78%) of discrimination between svPPA and nfvPPA [43]. Despite the high predictive power of structural MRI in distinguishing FTLD subtypes from each other and from other neurodegenerative diseases, more research is needed to validate findings in unselected cases in clinical practice settings [51, 52].

Studies are underway to identify the best MRI measures that could serve as clinical trial endpoints in FTLD. Findings are variable without clear convergence on optimal regions of interest across FTLD variants [24, 53]. Even within variants, differences in peak atrophy across patients impact effect sizes. Binney and Pankov [53] estimated a sample size of 103 bvFTD, 31 nfvPPA, and 10 svPPA to be able to detect a 40% reduction in annual rate of regional atrophy. In that study, the most sensitive regions of interest were the medial and lateral frontal gyri; the insula, striatum, and temporoparietal junction bilaterally for bvFTD; the superior and ventral anterior temporal, mid-to-posterior lateral temporal, and medial frontal cortices for svPPA; and the dorsomedial and lateral frontal cortices with predominant involvement of the precentral and perisylvian regions for nfvPPA. A study of PPA [40] with all three major subtypes found that atrophy in the left perisylvian temporal cortex, including insula and surrounding temporal regions, may be a highly sensitive measure of disease progression and a promising endpoint for clinical trials. As small as ten participants per arm could have 80% power to detect 40% slowing of atrophy [40]. A follow-up proof-of-concept study of optimal MRI endpoints in clinical trials in PPA further found that a composite with weighted averages of regional volumes within the left perisylvian temporal cortex can reduce sample size relative to total region of interest (ROI) by 38% [54]. Most recently, Staffaroni and Ljubenkov [55] reported that gray matter volume in frontal and temporal lobes predicted longitudinal change across all FTLD subtypes. Further, sample size predictions to detect a 40% reduction in decline following a therapeutic intervention (54 bvFTD, 34 svPPA, and 29 nfvPPA) were better than or comparable to estimates for clinical measures alone (e.g., functional assessment questionnaire). Notably, bvFTD yielded the largest confidence intervals

across all measures and metrics of white matter integrity (discussed later) yielded the smallest predicted sample sizes.

Fluorodeoxyglucose Positron Emission Tomography

PET with 18F-fluorodeoxyglucose tracer enables the quantification of cerebral glucose metabolism as a proxy measure of neural activity. This biomarker has been found to have greater sensitivity and specificity to neurodegeneration relative to measures of perfusion using single-photon emission computed tomography (SPECT; [56]), although SPECT is more widely available. Patterns of hypometabolism in FTLD generally precede and correlate with the spread of atrophy (see Fig. 1) [57]. In samples with bvFTD, hypometabolism has been identified in the orbitofrontal, dorsolateral and medial prefrontal, insular, and cingulate cortices [58, 59]. Subcortical structures, particularly the caudate nucleus, are also affected [60–62]. Cluster analyses have delineated two bvFTD subgroups with frontal (dorsolateral, medial, and ventromedial prefrontal cortices) and temporal-limbic (temporal poles, hippocampal formation, lateral temporal cortex, amygdala, thalamus) hypometabolic signatures. The frontal subgroup was associated with greater executive dysfunction and a faster rate of clinical decline, suggesting that differential patterns of hypometabolism can predict clinical outcomes [63, 64]. However, less is known about metabolic changes with disease progression. Some evidence suggests that over a 1–2-year follow-up period, worsening hypometabolism is observed in regions implicated at baseline, accompanied by a progression of hypometabolic activity into inferior frontal, parietal, and temporal regions [58, 65].

Among PPA variants, reduced metabolism is primarily left lateralized, particularly earlier in the disease. Hypometabolism in svPPA has been most consistently reported in the temporal poles, middle and inferior temporal gyri, and insula [63, 66, 67], with some studies also demonstrating thalamic [68], medial temporal (hippocampus and amygdala), fusiform, and superior temporal involvement. nfvPPA is associated with a heterogenous metabolic pattern [69], with evidence of hypometabolism in superior and inferior (particularly the pars opercularis and pars triangularis) frontal, dorsolateral prefrontal, anterior cingulate, and insular regions [67, 70]. Reduced metabolic activity in the thalamus and temporal cortices has also been reported (See Fig. 1) [68]. Hypometabolism spreads posteriorly toward the precentral gyrus in those with nfvPPA who progress to develop parkinsonism and toward the anterior temporal lobe in those who developed motor neuron disease [67].

FDG-PET has been approved for reimbursement by the US Center for Medicare Services, but, unfortunately, many private insurance companies still do not reimburse for its use in the US. Consensus groups have identified FDG-PET as an effective and recommended diagnostic tool to identify FTLD (see Fig. 3) [71] and differentiate FTLD from AD or Lewy body disease pathology [72]. Utilizing FDG-PET to identify FTLD in mild cognitive impairment (MCI) stages is also recommended, though this is understudied and requires additional formal investigation [73]. Visual assessments of FDG-PET scans in a clinical setting are generally accurate, with 89.6–92% accuracy, 81–86% sensitivity, and 94–98% specificity, in the differential diagnosis between AD and FTLD [74, 75]. The distinguishing pattern of hypometabolism that informs this differential follows a dissociation between anterior and posterior cortical areas, with reduced metabolic activity in frontal, but not posterior, regions predicting FTLD pathology [76]. Statistical regions of interest and parametric mapping analyses improve diagnostic accuracy and further underscore the utility of this biomarker in clinical settings [74, 77–80]. Only a small handful of studies have investigated the diagnostic utility of FDG-PET in PPA variants, but these have provided compelling evidence supporting its use in clinical settings [81]. One study found that visual ratings resulted in 87.8% sensitivity and 90% specificity to differentiate between PPA and cognitively normal patients;

statistical analyses improved these numbers to 95.70–96.9% and 90%, respectively [82]. The diagnostic accuracy of using FDG-PET to differentiate between PPA variants also yields sensitivity, specificity, and accuracy values above 90% [82]. A number of case studies have also documented the utility of FDG-PET in diagnosing PPA [83, 84].

Amyloid Positron Emission Tomography

PET imaging with amyloid tracers measures insoluble fibrillar amyloid, largely reflecting neuritic plaques, one of two core neuropathologic changes seen in AD. Amyloid PET has little clinical utility in the evaluation of a patient with a typical presentation of nfvPPA, svPPA, or bvFTD. Its primary value is in patients presenting with a dysexecutive-behavioral or complex language syndrome that is not typical of FTLD or AD, but where AD is a possibility [85, 86]. Most experts would recommend that FDG-PET be performed first, and if the case is still ambiguous, amyloid PET could be considered. The important challenge, though, is that even though amyloid PET is approved by the US Food and Drug Administration and by the European Medicines Agency, reimbursement is generally not available outside the context of research. An amyloid PET scan showing elevated signal in a patient with a typical FTLD clinical syndrome may be an indicator that AD is a coexisting pathologic change along with FTLD [87, 88]. Comorbid AD and FTLD pathologies are not infrequent [89], especially in older age. Analyses of elevated signal on amyloid PET in svPPA and nfvPPA have generally concluded that the frequency of "positive" amyloid PET scans increase with age at the same rate as in cognitively normal adults [90–92]. Thus, amyloid PET imaging may be useful in clinical settings with complicated cases in which AD remains on the differential diagnosis after comprehensive workup, but it is not considered a routine element of the workup of a patient with a typical FTLD clinical syndrome.

Multimodal Imaging in Frontotemporal Lobar Dementia in Clinical Practice

A multimodal approach to differential diagnosis appears to improve classification accuracy, particularly when integrating structural and metabolic imaging. The combination of T1-weighted MRI and FDG-PET imaging (see Fig. 3) distinguishes between other neurodegenerative disorders and bvFTD with 82.5% accuracy, svPPA with 97.5% accuracy, and nfvPPA with 87.5% accuracy [93]. Combined T1-weighted MRI and FDG-PET also distinguish between bvFTD and psychiatric disease with 96% sensitivity and 73% specificity, suggesting that this approach may reduce the number of bvFTD patients that are misdiagnosed with psychiatric illness [49]. Ultimately, sensitive and specific molecular biomarkers are badly needed for these conditions.

Imaging Modalities Utilized in Research Contexts Only

Although not yet used clinically, there are a number of additional informative neuroimaging modalities that can elucidate the pathophysiology underlying FTLD, inform differential diagnosis, and, in some cases, have the potential to be valuable in clinical trials. These include diffusion tensor imaging (DTI), resting-state functional MRI (rs-fMRI), arterial spin labeling (ASL) MRI, and tau PET.

Diffusion Tensor Imaging

DTI characterizes white matter microstructural integrity by measuring directionality (fractional anisotropy [FA]) and diffusivity (mean diffusivity [MD]) of water molecules along white matter tracts. Decreased FA and increased MD suggest degeneration of white matter fibers and compromised structural connectivity within the brain. There is a robust literature demonstrating widespread alterations in white matter tracts in FTLD (see Fig. 4). When compared to both AD and con-

trol groups, bvFTD is associated with bilateral white matter alterations in tracts underlying frontal and temporal lobes [94]. Studies have demonstrated either reduced FA or increased MD in the superior longitudinal fasciculus, anterior cingulum, corpus callosum, and uncinate fasciculus [95–98]. The inferior longitudinal fasciculus and inferior fronto-occipital fasciculus, along with fronto-striatal and fronto-thalamic pathways, have also been implicated [94, 99–102]. Longitudinal studies have demonstrated progression of abnormalities in these tracts, particularly in the uncinate fasciculus, corpus callosum, and paracossal cingulum [103, 104], and suggest that further progression to parietal and occipital white matter can be expected [105]. White matter alterations in bvFTD are associated with greater behavioral symptom severity [96, 100] and reduced integrity in the corpus callosum over a 2-year follow-up is associated with a decline in executive functioning [104], further underscoring the value of this modality in tracking disease progression in bvFTD.

In contrast, the PPA subtypes demonstrate more focal white matter alterations in tracts that originate from and terminate in brain regions important for language. svPPA is associated with relatively circumscribed alterations to white matter in ventral pathways projecting to the left temporal lobe early in the disease, with particular emphasis on uncinate and inferior longitudinal fasciculi [54, 70, 94, 106]. Reduced FA in the external capsule and cingulum bundle has also been reported [107]. With disease progression, further degeneration extends bilaterally to the nondominant frontotemporal, uncinate, and anterior inferior longitudinal fasciculi [54, 108, 109]. Cross-sectional alterations in nfvPPA relative to control samples are most notable in left frontotemporal and frontoparietal projections of the superior longitudinal and uncinate fasciculi [54, 94]. Alterations in the frontal aslant tract and in white matter projections from the basal ganglia to premotor and motor cortical areas have also been reported [70, 106, 110, 111]. Over time, white matter degeneration in both PPA variants spread from left to right hemisphere, though posterior tracts remain relatively spared [54].

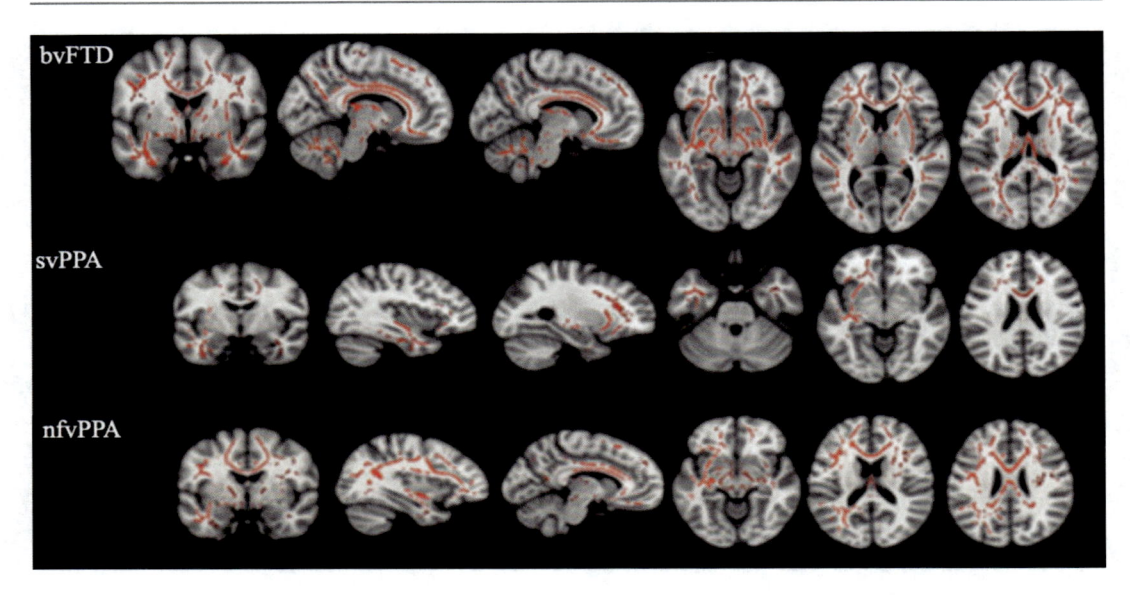

Fig. 4 Diffusion tensor imaging demonstrates white matter abnormalities in tracts projecting to and from regions atrophied in each variant. bvFTD is associated with widespread bilateral white matter changes in orbitofrontal and anterior temporal tracts (i.e., corpus callosum, inferior and superior longitudinal fasciculi, anterior thalamic radiation, and uncinate fasciculus). White matter changes in svPPA are seen primarily in the temporal lobe (i.e., anterior portions of the inferior longitudinal fasciculus and the uncinate fasciculus, L > R). In nfvPPA, white matter changes are observed in frontotemporal tracts (i.e., superior and inferior longitudinal fasciculi and corpus callosum). (Image adapted from Lam, Halliday [105])

Although not used clinically, DTI has proven to be a promising biomarker to inform differential diagnosis and monitor disease progression in FTLD. FA in the corpus callosum and uncinate fasciculus is particularly helpful in differentiating between FTLD and controls [112], AD patients [113], and among FTLD variants [95]. Some studies have shown that DTI outperforms structural gray matter volumetric and FDG-PET in differentiating bvFTD from controls [114] and from other FTLD variants at the group level [115]. However, another study found that DTI remains less sensitive than FDG-PET at the single subject level [116]. Further studies are needed to evaluate the potential added value of combining DTI with structural MRI or FDG-PET. Specifically, FA in the corpus callosum and gray matter volume in the precuneus and posterior cingulate provided optimal classification between a combined FTLD group and AD. A similar white and gray matter solution was found to optimally distinguish between FTLD variants, yielding classification accuracies of 90% in bvFTD, 80% for svPPA, and 100% for nfvPPA [5, 117]. A combination of gray (left temporal pole and pars opercularis) and white matter (left uncinate and inferior longitudinal fasciculi) structural integrity also appears to maximally distinguish between PPA variants with 89% accuracy, 92% sensitivity, and 85% specificity [118]. Finally, with regard to clinical trials, DTI measures of corpus callosum integrity are potential clinical trial endpoints that requires the smallest sample sizes to detect clinical change [55, 112]. Thus, while it is unlikely that DTI will take the place of traditional neuroimaging biomarkers currently used in clinical settings, it may offer additive value that is worth further investigation in the context of clinical trials.

Resting-State Functional Magnetic Resonance Imaging

rs-fMRI measures intrinsic functional connectivity between brain regions to detect synchronous patterns of low-frequency fluctuations in blood oxygen level–dependent signals. From a network perspective, this modality has been critical in identifying groups of brain regions in healthy populations that are functionally related yet spatially distinct [119–122]. In general, rs-fMRI studies in FTLD demonstrate altered connectivity in established networks that closely follow distributed atrophy patterns identified in FTLD (see Fig. 4).

The most robust finding in bvFTD is reduced functional connectivity within the salience network [123–125], with primary nodes in the ACC/MCC, frontoinsula, middle frontal gyrus, and subcortical regions in the striatum and amygdala [120, 121]. Abnormalities in this network are associated with compromise in behavioral and socio-emotional functioning that are a hallmark of bvFTD [121, 126]. Reduced connectivity between frontal and limbic structures within the salience network has also been observed [125, 127, 128]. Paralleling other reports of neuroanatomical and functional subgroups within bvFTD, Ranasinghe and Rankin [16] reported four distinct patterns of network dysfunction encompassing the frontotemporal dominant salience network, frontal dominant salience network, a subcortical network, and a semantic appraisal network (temporal pole, ventral striatum, subgenual cingulate, and basolateral amygdala). Graph theory models have documented a decline in major network nodes in frontotemporal regions with relative sparing of posterior cortical areas [129, 130]. Indeed, several studies have demonstrated either equivalent or even increased connectivity within the default mode network relative to controls and AD, suggesting the possibility of a compensatory response in the context of alterations in frontotemporal networks [124, 131]. Over time, however, network disruption does progress posteriorly to involve frontoparietal and default mode networks [132].

Alterations in rs-fMRI also closely follow atrophy patterns among PPA variants. Consistent with the posterior-medial anterior-temporal framework proposed to dissociate regions underlying semantic and episodic memory [133], networks anchored in anterior temporal regions important for semantic memory appear to be most vulnerable to disruption in svPPA. Specifically, there is reduced connectivity between the anterior temporal lobe and distributed cortical and subcortical areas, including modality-specific cortical regions, which support the role of the anterior temporal lobe as a critical transmodal hub within the semantic network (see Fig. 5) [33, 134, 135]. Others have documented decreased connectivity and reduced network hubs in ventral regions of (i.e., middle temporal gyrus and angular gyrus), while increased connectivity in more dorsal regions were observed (i.e., inferior frontal gyrus and superior portion of angular gyrus) [136–138]. Only a small handful of studies have investigated functional connectivity in nfvPPA; these have documented reduced connectivity within the speech and language network (SLN), which encompasses left inferior frontal, dorsal insular, supplementary motor, and inferior parietal regions, that are responsible for speech and language production [119]. Relative to control samples, nfvPPA patients demonstrate reduced connectivity within this network, but not among regions belonging to the default mode network [139]. Research designs employing graph theory to characterize nodes within the SLN report reduced efficiency and number of nodes, particularly in left parietal regions [140, 141].

There is also a growing body of literature demonstrating the utility of rs-fMRI in predicting future atrophy in FTLD. Network disruption in svPPA and nfvPPA among key nodes of language networks identified in control samples predict longitudinal gray matter thinning in those regions in respective patient groups [33, 140], suggesting that neurodegeneration may propagate along functional pathways in large-scale networks (see Fig. 5). Building on this work, Brown and Deng [142] demonstrated that individualized "epicenters" of atrophy at baseline (i.e., atrophied regions whose functional connectivity guides disease spread within a network) in bvFTD (anterior cingulate and fronto-insular cortex), and svPPA (anterior temporal lobe) predicted longitudinal gray matter loss for patients in mild-to-moderate clinical stages.

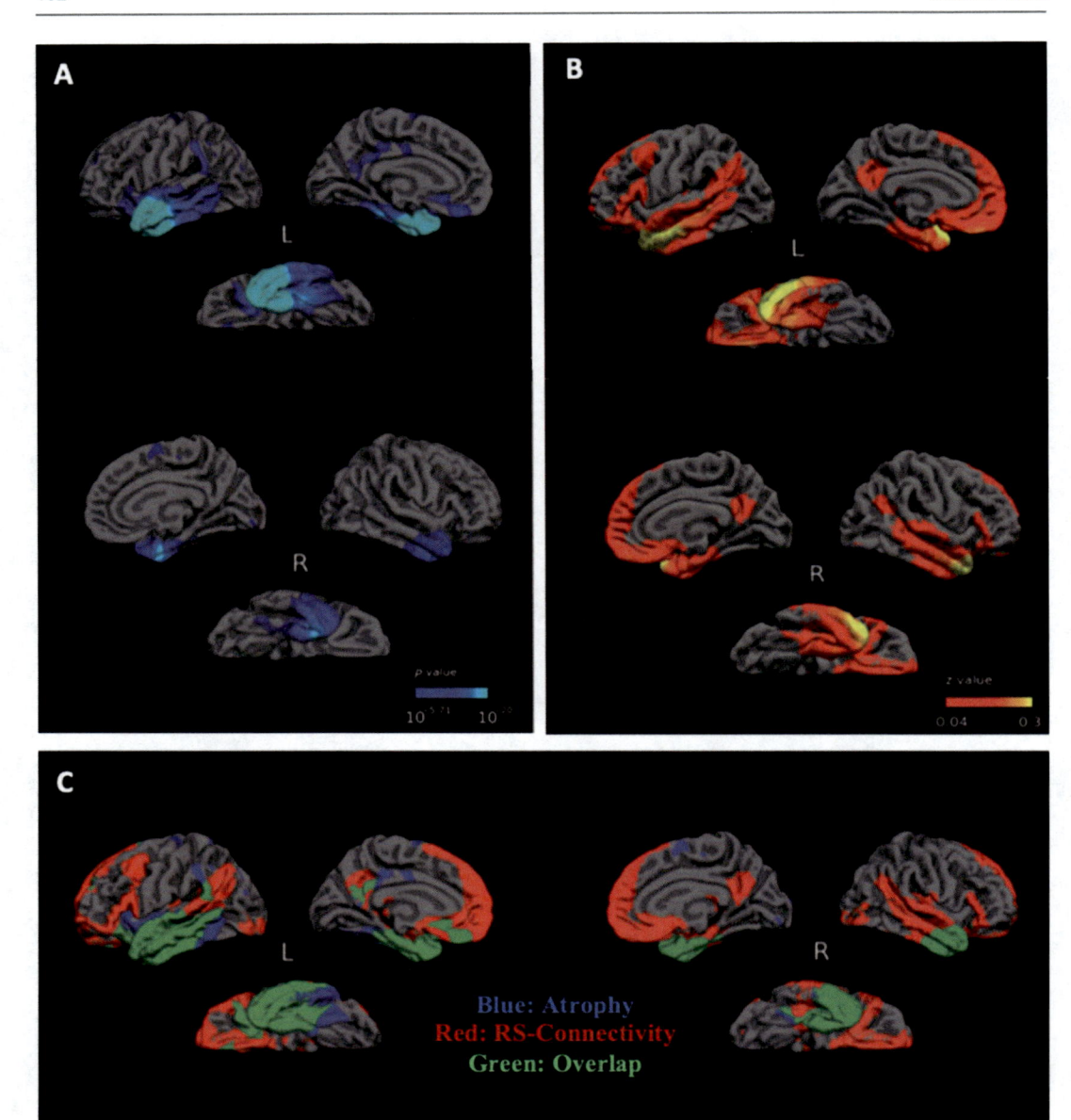

Fig. 5 Network-specific degeneration in svPPA. The temporal pole area of greatest atrophy in svPPA (**a**) anchors a large-scale intrinsic functional connectivity network important for semantic cognition (shown in healthy adults, **b**), which overlaps with the pattern of cortical atrophy in svPPA (**c**, overlap shown in green). (Image created from data published in Collins, Montal [33])

Brain atrophy tends to spread from the epicenter to its neighboring and adjacent functionally connected regions that also exhibit some intermediate atrophy at the baseline MRI. These findings support the possibility for rs-fMRI to be utilized

as a marker to make individualized predictions of future gray matter loss in FTLD, but this has not yet been fully investigated.

The few studies that describe the utility of rs-fMRI to inform differential diagnosis in clinical settings have focused on bvFTD and AD samples. bvFTD is associated with network disruption more specific to frontal and temporal nodes, while studies in AD demonstrate more widespread global alterations of network connectivity with preferential involvement of posterior regions [143]. This pattern has been demonstrated to distinguish between bvFTD and AD with 92% accuracy [124]. Introducing rs-fMRI to a multimodal solution appears to yield mixed results. One study documented only a small improvement in distinguishing between bvFTD and control participant using a multimodal solution that incorporated morphometric features (i.e., temporal and frontal gray matter volume) and network connectivity between and within frontal, temporal, and parietal nodes [144]. In contrast, the combination of rs-fMRI, DTI, and structural MRI was found to provide the strongest diagnostic classification between AD and bvFTD [145].

Arterial Spin Labeling

ASL is an MRI sequence that magnetically labels arterial water as an endogenous tracer to quantify cerebral blood flow (CBF). This method offers several advantages over traditional PET/SPECT perfusion methods, as it is less expensive, has shorter acquisition times, and does not require intravenous contrast agents. Despite its promise, only a small handful of studies have utilized ASL techniques to assess CBF in sporadic FTLD. These studies have documented hypoperfusion in frontal lobes and the anterior cingulate cortex in bvFTD (see Fig. 6) [62, 146, 147]. svPPA is associated with hypoperfusion in the left temporal lobe and insula, and hyperperfusion in the right superior temporal, inferior parietal, and orbitofrontal cortices. Alterations in CBF adjacent to regions that were atrophied at baseline predict subsequent gray matter loss at follow-up [148], suggesting a role for this modality in predicting future regional cortical degeneration.

Evidence to date suggests that ASL may be useful in differentiating between FTLD and AD patients. Similar to other modalities discussed in this chapter, perfusion between these two patient populations appears to follow an anterior-posterior dissociation; FTLD cases demonstrate frontotemporal hypoperfusion, while AD patients exhibit hypoperfusion in parietal regions [146]. One study reported that whole-brain ASL accurately classified bvFTD and AD groups with 83% sensitivity and 93% specificity [149], while another reported 77% sensitivity and 76% specificity in the precuneus [147]. Two studies have suggested that ASL and FDG-PET have equivalent diagnostic utility [62, 149, 150], while another found reduced classification accuracy with ASL relative to FDG-PET [151]. Thus, ASL imaging in FTLD is relatively understudied but merits further investigation to better document its clinical value relative to other modalities.

Tau Positron Emission Tomography

In 2013, when Brad Dickerson was putting the final touches on editing Hodges' Frontotemporal Dementia [152], we were so enthusiastic about the potential of new radioligands for measuring tau in the living human brain that we put our first FTD patient's scan on the cover of the book. This was a patient with *MAPT* P301L-related mild-stage FTD who had clearly elevated signal in all of the right places. Similar enthusiasm was generated when we saw our first nfvPPA and our first progressive supranuclear palsy (PSP) case. Unfortunately, when we presented a summary of our first series of cases at the Human Amyloid Imaging meeting in Miami in 2014 [153], we also had to reveal that we saw substantially elevated signal in a *GRN*-related FTD patient and in a svPPA patient, both of whom eventually were confirmed by autopsy not to have FTLD tau pathology, but rather the expected FTLD TDP43 pathology. Looking back at the accrued knowledge from the perspective of the Tau 2020 meeting in Washington D.C., where Gil Rabinovici presented a masterful summary of tau PET in the non-Alzheimer dementia spectrum [154], we have learned a number of specific lessons. First, the current generation of tau PET tracers works

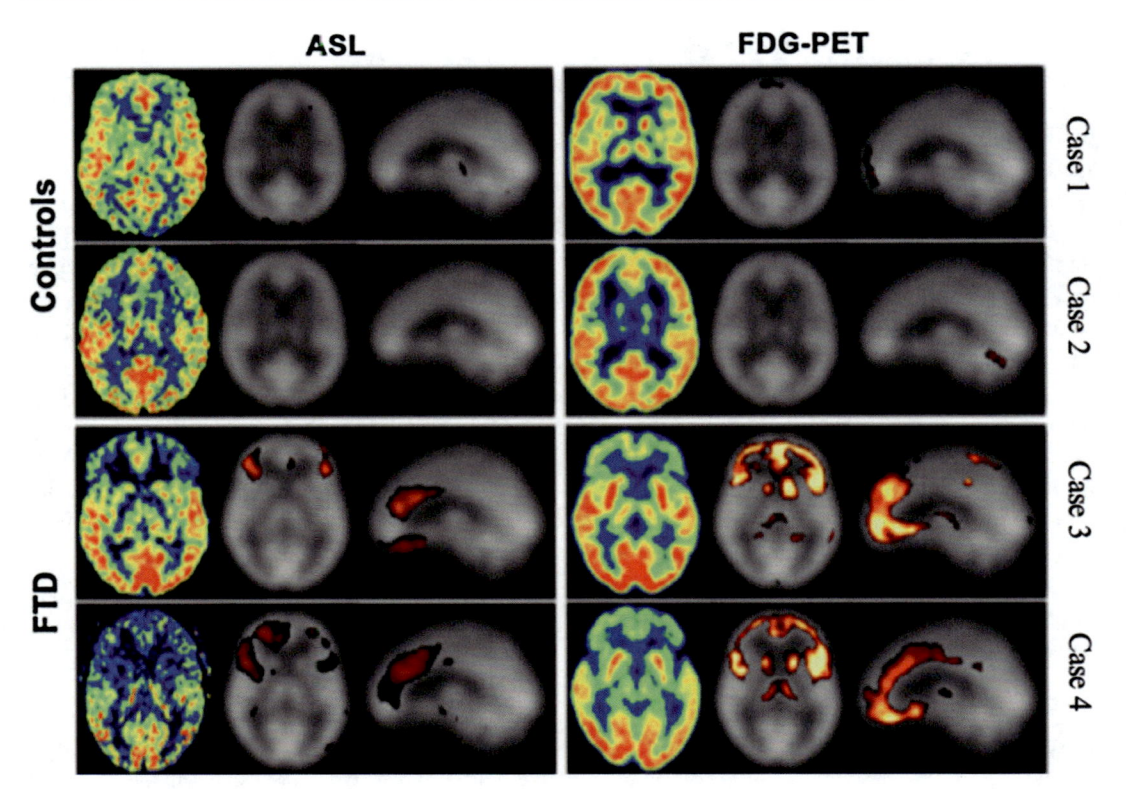

Fig. 6 Single-subject images showing ASL and FDG-PET in control (cases 1 and 2) and bvFTD (cases 3 and 4) participants. For each modality, the two right columns show statistical comparisons to controls. Correlations between hypoperfusion in ASL and hypometabolism in FDG-PET have led some investigators to suggest that ASL, collected in an MRI session with other sequences, could potentially serve as a surrogate for FDG-PET. (Image adapted from Fällmar, Haller [150])

generally very well for measuring AD-related paired helical filament tau pathology [155], and there appears to be weak, but topographically appropriate, signal in PSP [156] and CBS likely due to CBD pathology [157]. And in *MAPT* mutation carriers, signal is more elevated in those with mutations associated with tau aggregation that has conformational shapes more similar to those of AD (e.g., the R406W mutation) [158, 159]. But there is also consistently elevated signal in svPPA [160–163] and variably elevated signal in *GRN*-related or *C9orf72*-related FTLD, calling into serious question the specificity of binding of the first tracers to tau pathology. Furthermore, autoradiographic studies show little or no binding of as ^{18}F-Flortaucipir or as ^{18}F-MK6240 to FTLD tau pathology [164–168].

Some of this can be understood with our advancing knowledge of the 3D shape of tau inclusions and other work on the fundamental biology of tau by pioneers, including Michel Goedert and Maria Grazia Spillantini and Bernardino Ghetti and colleagues [169–172].

Thus, while advances in tau PET imaging over the past 7 years is tremendous and is having a prominent impact on the AD field, its value in FTLD is not yet clear and will require substantial further work which is ongoing [173, 174]. There also remains an urgent need to develop imaging biomarkers of FTLD TDP-43, and work is ongoing to try to measure glial cell responses that may contribute to FTLD-related neurodegeneration [175].

Conclusions

Recent advances in neuroimaging have enabled a more complete understanding of the pathophysiology of FTLD and have offered improved diagnosis of sporadic clinical syndromes associated with FTLD. Overwhelmingly across all modalities, abnormalities in brain structure and function have been identified predominantly in frontal, temporal, and subcortical areas, with eventual progression posteriorly. At present, gray matter morphometry and glucose metabolism, captured by structural MRI and FDG-PET, respectively, appear to have the most robust evidence supporting differential diagnosis in clinical settings. However, multimodal imaging protocols are gaining traction and may serve to improve diagnostic accuracy and longitudinal monitoring, particularly when including DTI and/or rs-fMRI.

Substantial multicenter efforts are underway to identify ideal biomarkers that are sensitive to preclinical stages, track disease progression, and predict underlying pathology, particularly the Genetic Frontotemporal Dementia Initiative (GENFI) [176, 177] and the Advancing Research and Treatment for Frontotemporal Lobar Degeneration (ARTFL)/Longitudinal Evaluation of Familial Frontotemporal Lobar Dementia Subjects (LEFFTDS)/Longitudinal Frontotemporal Lobal Degeneration (ALLFTD) [178] initiatives. While these and many other smaller studies are ongoing, evidence to date supports the value of a variety of imaging biomarkers in clinical trials, aiming to develop novel therapeutics for these devastating diseases.

References

1. Rascovsky K, Hodges JR, Knopman D, Mendez MF, Kramer JH, Neuhaus J et al (2011) Sensitivity of revised diagnostic criteria for the behavioural variant of frontotemporal dementia. Brain 134(Pt 9):2456–2477
2. Gorno-Tempini ML, Hillis AE, Weintraub S, Kertesz A, Mendez M, Cappa SF et al (2011) Classification of primary progressive aphasia and its variants. Neurology 76(11):1006–1014
3. Lee JS, Jung NY, Jang YK, Kim HJ, Seo SW, Lee J et al (2017) Prognosis of patients with behavioral variant frontotemporal dementia who have focal versus diffuse frontal atrophy. J Clin Neurol 13(3):234–242
4. Borroni B, Benussi A, Premi E, Alberici A, Marcello E, Gardoni F et al (2018) Biological, neuroimaging, and neurophysiological markers in frontotemporal dementia: three faces of the same coin. J Alzheimers Dis 62(3):1113–1123
5. McMillan CT, Avants BB, Cook P, Ungar L, Trojanowski JQ, Grossman M (2014) The power of neuroimaging biomarkers for screening frontotemporal dementia. Hum Brain Mapp 35(9):4827–4840
6. Meeter LH, Kaat LD, Rohrer JD, van Swieten JC (2017) Imaging and fluid biomarkers in frontotemporal dementia. Nat Rev Neurol 13(7):406–419
7. Gordon E, Rohrer JD, Fox NC (2016) Advances in neuroimaging in frontotemporal dementia. J Neurochem 138(Suppl 1):193–210
8. Brewer JB (2009) Fully-automated volumetric MRI with normative ranges: translation to clinical practice. Behav Neurol 21(1):21–28
9. Seeley WW, Crawford R, Rascovsky K, Kramer JH, Weiner M, Miller BL et al (2008) Frontal paralimbic network atrophy in very mild behavioral variant frontotemporal dementia. Arch Neurol 65(2):249–255
10. Macfarlane MD, Jakabek D, Walterfang M, Vestberg S, Velakoulis D, Wilkes FA et al (2015) Striatal atrophy in the behavioural variant of frontotemporal dementia: correlation with diagnosis, negative symptoms and disease severity. PLoS One 10(6):e0129692
11. Chen Y, Kumfor F, Landin-Romero R, Irish M, Hodges JR, Piguet O (2018) Cerebellar atrophy and its contribution to cognition in frontotemporal dementias. Ann Neurol 84(1):98–109
12. Schroeter ML, Raczka K, Neumann J, Yves von Cramon D (2007) Towards a nosology for frontotemporal lobar degenerations-a meta-analysis involving 267 subjects. NeuroImage 36(3):497–510
13. Pan PL, Song W, Yang J, Huang R, Chen K, Gong QY et al (2012) Gray matter atrophy in behavioral variant frontotemporal dementia: a meta-analysis of voxel-based morphometry studies. Dement Geriatr Cogn Disord 33(2–3):141–148
14. Perry DC, Datta S, Miller ZA, Rankin KP, Gorno-Tempini ML, Kramer JH et al (2019) Factors that predict diagnostic stability in neurodegenerative dementia. J Neurol 266(8):1998–2009
15. Whitwell JL, Przybelski SA, Weigand SD, Ivnik RJ, Vemuri P, Gunter JL et al (2009) Distinct anatomical subtypes of the behavioural variant of frontotemporal dementia: a cluster analysis study. Brain 132(Pt 11):2932–2946
16. Ranasinghe KG, Rankin KP, Pressman PS, Perry DC, Lobach IV, Seeley WW et al (2016) Distinct subtypes of behavioral variant frontotemporal dementia based on patterns of network degeneration. JAMA Neurol 73(9):1078–1088

17. Whitwell JL (2019) FTD spectrum: Neuroimaging across the FTD spectrum. Prog Mol Biol Transl Sci 165:187–223

18. Borroni B, Cosseddu M, Pilotto A, Premi E, Archetti S, Gasparotti R et al (2015) Early stage of behavioral variant frontotemporal dementia: clinical and neuroimaging correlates. Neurobiol Aging 36(11):3108–3115

19. Chan D, Anderson V, Pijnenburg Y, Whitwell J, Barnes J, Scahill R et al (2009) The clinical profile of right temporal lobe atrophy. Brain 132(Pt 5):1287–1298

20. Irish M, Kumfor F, Hodges JR, Piguet O (2013) A tale of two hemispheres: contrasting socioemotional dysfunction in right- versus left-lateralised semantic dementia. Dement and Neuropsychol 7(1):88–95

21. Veronelli L, Makaretz SJ, Quimby M, Dickerson BC, Collins JA (2017) Geschwind syndrome in frontotemporal lobar degeneration: neuroanatomical and neuropsychological features over 9 years. Cortex 94:27–38

22. Henry ML, Wilson SM, Ogar JM, Sidhu MS, Rankin KP, Cattaruzza T et al (2014) Neuropsychological, behavioral, and anatomical evolution in right temporal variant frontotemporal dementia: a longitudinal and post-mortem single case analysis. Neurocase 20(1):100–109

23. Lu PH, Mendez MF, Lee GJ, Leow AD, Lee HW, Shapira J et al (2013) Patterns of brain atrophy in clinical variants of frontotemporal lobar degeneration. Dement Geriatr Cogn Disord 35(1–2):34–50

24. Manera AL, Dadar M, Collins DL, Ducharme S (2019) Frontotemporal lobar degeneration neuroimaging I. deformation based morphometry study of longitudinal MRI changes in behavioral variant frontotemporal dementia. Neuroimage Clin 24:102079

25. Gordon E, Rohrer JD, Kim LG, Omar R, Rossor MN, Fox NC et al (2010) Measuring disease progression in frontotemporal lobar degeneration: a clinical and MRI study. Neurology 74(8):666–673

26. Frings L, Mader I, Landwehrmeyer BG, Weiller C, Hüll M, Huppertz H-J (2012) Quantifying change in individual subjects affected by frontotemporal lobar degeneration using automated longitudinal MRI volumetry. Hum Brain Mapp 33(7):1526–1535

27. Landin-Romero R, Tan R, Hodges JR, Kumfor F (2016) An update on semantic dementia: genetics, imaging, and pathology. Alzheimers Res Ther 8(1):52

28. Chen K, Ding J, Lin B, Huang L, Tang L, Bi Y et al (2018) The neuropsychological profiles and semantic-critical regions of right semantic dementia. Neuroimage Clin 19:767–774

29. Teipel S, Raiser T, Riedl L, Riederer I, Schroeter ML, Bisenius S et al (2016) Atrophy and structural covariance of the cholinergic basal forebrain in primary progressive aphasia. Cortex 83:124–135

30. Botha H, Duffy JR, Whitwell JL, Strand EA, Machulda MM, Schwarz CG et al (2015) Classification and clinicoradiologic features of primary progressive aphasia (PPA) and apraxia of speech. Cortex 69:220–236

31. Gorno-Tempini ML, Dronkers NF, Rankin KP, Ogar JM, Phengrasamy L, Rosen HJ et al (2004) Cognition and anatomy in three variants of primary progressive aphasia. Ann Neurol 55(3):335–346

32. Sapolsky D, Bakkour A, Negreira A, Nalipinski P, Weintraub S, Mesulam MM et al (2010) Cortical neuroanatomic correlates of symptom severity in primary progressive aphasia. Neurology 75(4):358–366

33. Collins JA, Montal V, Hochberg D, Quimby M, Mandelli ML, Makris N et al (2017) Focal temporal pole atrophy and network degeneration in semantic variant primary progressive aphasia. Brain 140(2):457–471

34. Tan RH, Wong S, Kril JJ, Piguet O, Hornberger M, Hodges JR et al (2014) Beyond the temporal pole: limbic memory circuit in the semantic variant of primary progressive aphasia. Brain 137(Pt 7):2065–2076

35. Rohrer JD, Warren JD, Modat M, Ridgway GR, Douiri A, Rossor MN et al (2009) Patterns of cortical thinning in the language variants of frontotemporal lobar degeneration. Neurology 72(18):1562–1569

36. Rogalski E, Cobia D, Harrison TM, Wieneke C, Weintraub S, Mesulam MM (2011) Progression of language decline and cortical atrophy in subtypes of primary progressive aphasia. Neurology 76(21):1804–1810

37. Bocchetta M, Iglesias JE, Russell LL, Greaves CV, Marshall CR, Scelsi MA et al (2019) Segmentation of medial temporal subregions reveals early right-sided involvement in semantic variant PPA. Alzheimers Res Ther 11(1):41

38. Krueger CE, Dean DL, Rosen HJ, Halabi C, Weiner M, Miller BL et al (2010) Longitudinal rates of lobar atrophy in frontotemporal dementia, semantic dementia, and Alzheimer's disease. Alzheimer Dis Assoc Disord 24(1):43–48

39. Botha H, Josephs KA (2019) Primary progressive aphasias and apraxia of speech. Continuum 25(1):101–127

40. Rogalski E, Cobia D, Marsteller A, Rademaker A, Wieneke C, Weintraub S et al (2014) Asymmetry of cortical decline in subtypes of primary progressive aphasia. Neurology 83(13):1184–1191

41. Dickerson B (2014) Neuroimaging, cerebrospinal fluid markers, and genetic testing in dementia. In: Dickerson BC, Atri A (eds) Dementia: comprehensive principles and practice. Oxford University Press, New York, pp 530–564

42. Harper L, Fumagalli GG, Barkhof F, Scheltens P, O'Brien JT, Bouwman F et al (2016) MRI visual rating scales in the diagnosis of dementia: evaluation in 184 post-mortem confirmed cases. Brain 139(Pt 4):1211–1225

43. Bisenius S, Mueller K, Diehl-Schmid J, Fassbender K, Grimmer T, Jessen F et al (2017) Predicting primary progressive aphasias with support vector machine approaches in structural MRI data. Neuroimage Clin 14:334–343

44. Kim JP, Kim J, Park YH, Park SB, Lee JS, Yoo S et al (2019) Machine learning based hierarchical classification of frontotemporal dementia and Alzheimer's disease. Neuroimage Clin 23:101811

45. Dickerson BC, Wolk DA (2013) Alzheimer's disease neuroimaging I. biomarker-based prediction of progression in MCI: comparison of AD signature and hippocampal volume with spinal fluid amyloid-β and tau. Front Aging Neurosci 5:55

46. Goodkin O, Pemberton H, Vos SB, Prados F, Sudre CH, Moggridge J et al (2019) The quantitative neuroradiology initiative framework: application to dementia. Br J Radiol 92(1101):20190365

47. Ferrari BL, GdCC N, Nucci MP, Mamani JB, Lacerda SS, Felício AC et al (2019) The accuracy of hippocampal volumetry and glucose metabolism for the diagnosis of patients with suspected Alzheimer's disease, using automatic quantitative clinical tools. Medicine 98(45):e17824

48. Meyer S, Mueller K, Stuke K, Bisenius S, Diehl-Schmid J, Jessen F et al (2017) Predicting behavioral variant frontotemporal dementia with pattern classification in multi-center structural MRI data. Neuroimage Clin 14:656–662

49. Vijverberg EGB, Wattjes MP, Dols A, Krudop WA, Möller C, Peters A et al (2016) Diagnostic accuracy of MRI and additional [18F]FDG-PET for behavioral variant frontotemporal dementia in patients with late onset behavioral changes. J Alzheimers Dis 53(4):1287–1297

50. Bruun M, Koikkalainen J, Rhodius-Meester HFM, Baroni M, Gjerum L, van Gils M et al (2019) Detecting frontotemporal dementia syndromes using MRI biomarkers. Neuroimage Clin 22:101711

51. Canu E, Agosta F, Imperiale F, Fontana A, Caso F, Spinelli EG et al (2019) Added value of multimodal MRI to the clinical diagnosis of primary progressive aphasia variants. Cortex 113:58–66

52. Dewer B, Rogers P, Ricketts J, Mukonoweshuro W, Zeman A (2016) The radiological diagnosis of frontotemporal dementia in everyday practice: an audit of reports, review of diagnostic criteria, and proposal for service improvement. Clin Radiol 71(1):40–47

53. Binney RJ, Pankov A, Marx G, He X, McKenna F, Staffaroni AM et al (2017) Data-driven regions of interest for longitudinal change in three variants of frontotemporal lobar degeneration. Brain Behav 7(4):e00675

54. Edland SD, Ard MC, Sridhar J, Cobia D, Martersteck A, Mesulam MM et al (2016) Proof of concept demonstration of optimal composite MRI endpoints for clinical trials. Alzheimers Dement: Translat Res and Clin Interv 2(3):177–181

55. Staffaroni AM, Ljubenkov PA, Kornak J, Cobigo Y, Datta S, Marx G et al (2019) Longitudinal multimodal imaging and clinical endpoints for frontotemporal dementia clinical trials. Brain 142(2):443–459

56. Davison CM, O'Brien JT (2014) A comparison of FDG-PET and blood flow SPECT in the diagnosis of

neurodegenerative dementias: a systematic review. Int J Geriatr Psychiatry 29(6):551–561

57. Buhour MS, Doidy F, Laisney M, Pitel AL, de La Sayette V, Viader F et al (2017) Pathophysiology of the behavioral variant of frontotemporal lobar degeneration: a study combining MRI and FDG-PET. Brain Imaging Behav 11(1):240–252

58. Grimmer T, Diehl J, Drzezga A, Förstl H, Kurz A (2004) Region-specific decline of cerebral glucose metabolism in patients with frontotemporal dementia: a prospective 18F-FDG-PET study. Dement Geriatr Cogn Disord 18(1):32–36

59. Ishii K (2014) PET approaches for diagnosis of dementia. AJNR Am J Neuroradiol 35(11):2030–2038

60. Mahapatra A, Sood M, Bhad R, Tripathi M (2016) Behavioural variant frontotemporal dementia with bilateral insular hypometabolism: a case report. J Clin Diagn Res 10(4):VD01–VVD2

61. Shi Z, Liu S, Wang Y, Liu S, Han T, Cai L et al (2017) Correlations between clinical characteristics and neuroimaging in Chinese patients with subtypes of frontotemporal lobe degeneration. Medicine 96(37):e7948

62. Verfaillie SCJ, Adriaanse SM, Binnewijzend MAA, Benedictus MR, Ossenkoppele R, Wattjes MP et al (2015) Cerebral perfusion and glucose metabolism in Alzheimer's disease and frontotemporal dementia: two sides of the same coin? Eur Radiol 25(10):3050–3059

63. Cerami C, Dodich A, Greco L, Iannaccone S, Magnani G, Marcone A et al (2017) The role of single-subject brain metabolic patterns in the early differential diagnosis of primary progressive aphasias and in prediction of progression to dementia. J Alzheimers Dis 55(1):183–197

64. Malpetti M, Carli G, Sala A, Cerami C, Marcone A, Iannaccone S et al (2019) Variant-specific vulnerability in metabolic connectivity and resting-state networks in behavioural variant of frontotemporal dementia. Cortex 120:483–497

65. Diehl-Schmid J, Grimmer T, Drzezga A, Bornschein S, Riemenschneider M, Förstl H et al (2007) Decline of cerebral glucose metabolism in frontotemporal dementia: a longitudinal 18F-FDG-PET-study. Neurobiol Aging 28(1):42–50

66. Iaccarino L, Crespi C, Della Rosa PA, Catricalà E, Guidi L, Marcone A et al (2015) The semantic variant of primary progressive aphasia: clinical and neuroimaging evidence in single subjects. PLoS One 10(3):e0120197

67. Matias-Guiu JA, Cabrera-Martín MN, Moreno-Ramos T, García-Ramos R, Porta-Etessam J, Carreras JL et al (2015) Clinical course of primary progressive aphasia: clinical and FDG-PET patterns. J Neurol 262(3):570–577

68. Bisenius S, Neumann J, Schroeter ML (2016) Validating new diagnostic imaging criteria for primary progressive aphasia via anatomical like-

lihood estimation meta-analyses. Eur J Neurol 23(4):704–712

69. Matias-Guiu JA, Díaz-Álvarez J, Ayala JL, Risco-Martín JL, Moreno-Ramos T, Pytel V et al (2018) Clustering analysis of FDG-PET imaging in primary progressive aphasia. Front Aging Neurosci 10:230

70. Routier A, Habert M-O, Bertrand A, Kas A, Sundqvist M, Mertz J et al (2018) Structural, microstructural, and metabolic alterations in primary progressive aphasia variants. Front Neurol 9:766

71. Nobili F, Arbizu J, Bouwman F, Drzezga A, Agosta F, Nestor P et al (2018) European Association of Nuclear Medicine and European Academy of neurology recommendations for the use of brain (18) F-fluorodeoxyglucose positron emission tomography in neurodegenerative cognitive impairment and dementia: Delphi consensus. Eur J Neurol 25(10):1201–1217

72. Nestor PJ, Altomare D, Festari C, Drzezga A, Rivolta J, Walker Z et al (2018) Clinical utility of FDG-PET for the differential diagnosis among the main forms of dementia. Eur J Nucl Med Mol Imaging 45(9):1509–1525

73. Arbizu J, Festari C, Altomare D, Walker Z, Bouwman F, Rivolta J et al (2018) Clinical utility of FDG-PET for the clinical diagnosis in MCI. Eur J Nucl Med Mol Imaging 45(9):1497–1508

74. Dukart J, Mueller K, Horstmann A, Barthel H, Möller HE, Villringer A et al (2011) Combined evaluation of FDG-PET and MRI improves detection and differentiation of dementia. PLoS One 6(3):e18111

75. Foster NL, Heidebrink JL, Clark CM, Jagust WJ, Arnold SE, Barbas NR et al (2007) FDG-PET improves accuracy in distinguishing frontotemporal dementia and Alzheimer's disease. Brain 130(Pt 10):2616–2635

76. Nazem A, Tang CC, Spetsieris P, Dresel C, Gordon ML, Diehl-Schmid J et al (2018) A multivariate metabolic imaging marker for behavioral variant frontotemporal dementia. Alzheimers Dement: Diagn, Assess & Dis Monit 10:583–594

77. Poljansky S, Ibach B, Hirschberger B, Männer P, Klünemann H, Hajak G et al (2011) A visual [18F] FDG-PET rating scale for the differential diagnosis of frontotemporal lobar degeneration. Eur Arch Psychiatry Clin Neurosci 261(6):433–446

78. Caminiti SP, Ballarini T, Sala A, Cerami C, Presotto L, Santangelo R et al (2018) FDG-PET and CSF biomarker accuracy in prediction of conversion to different dementias in a large multicentre MCI cohort. Neuroimage Clin 18:167–177

79. Perani D, Della Rosa PA, Cerami C, Gallivanone F, Fallanca F, Vanoli EG et al (2014) Validation of an optimized SPM procedure for FDG-PET in dementia diagnosis in a clinical setting. Neuroimage Clin 6:445–454

80. Mosconi L, Tsui WH, Herholz K, Pupi A, Drzezga A, Lucignani G et al (2008) Multicenter standardized 18F-FDG PET diagnosis of mild cognitive

impairment, Alzheimer's disease, and other dementias. J Nucl Med 49(3):390–398

81. Bouwman F, Orini S, Gandolfo F, Altomare D, Festari C, Agosta F et al (2018) Diagnostic utility of FDG-PET in the differential diagnosis between different forms of primary progressive aphasia. Eur J Nucl Med Mol Imaging 45(9):1526–1533

82. Matías-Guiu JA, Cabrera-Martín MN, Pérez-Castejón MJ, Moreno-Ramos T, Rodríguez-Rey C, García-Ramos R et al (2015) Visual and statistical analysis of 18F-FDG PET in primary progressive aphasia. Eur J Nucl Med Mol Imaging 42(6):916–927

83. Ince S, Eroglu E, Karacalioglu AO, Emer O, Alagoz E (2015) 18F-FDG PET/CT findings in a case of a semantic variant of primary progressive aphasia. Hell J Nucl Med 18(2):163–165

84. Jeong YJ, Park KW, Kang D-Y (2018) Role of positron emission tomography as a biologic marker in the diagnosis of primary progressive aphasia: two case reports. Nucl Med Mol Imaging 52(5):384–388

85. Johnson KA, Minoshima S, Bohnen NI, Donohoe KJ, Foster NL, Herscovitch P et al (2013) Appropriate use criteria for amyloid PET: a report of the amyloid imaging task force, the society of nuclear medicine and molecular imaging, and the Alzheimer's association. Alzheimers Dement 9(1):e-16

86. Rabinovici GD, Rosen HJ, Alkalay A, Kornak J, Furst AJ, Agarwal N et al (2011) Amyloid vs FDG-PET in the differential diagnosis of AD and FTLD. Neurology 77(23):2034–2042

87. Mesulam MM, Dickerson BC, Sherman JC, Hochberg D, Gonzalez RG, Johnson KA et al (2017) Case 1-2017. A 70-year-old woman with gradually progressive loss of language. N Engl J Med 376(2):158–167

88. Caso F, Gesierich B, Henry M, Sidhu M, LaMarre A, Babiak M et al (2013) Nonfluent/agrammatic PPA with in-vivo cortical amyloidosis and Pick's disease pathology. Behav Neurol 26(1–2):95–106

89. Serrano GE, Sabbagh MN, Sue LI, Hidalgo JA, Schneider JA, Bedell BJ et al (2014) Positive florbetapir PET amyloid imaging in a subject with frequent cortical neuritic plaques and frontotemporal lobar degeneration with TDP43-positive inclusions. J Alzheimers Dis 42(3):813–821

90. Bergeron D, Gorno-Tempini ML, Rabinovici GD, Santos-Santos MA, Seeley W, Miller BL et al (2018) Prevalence of amyloid-β pathology in distinct variants of primary progressive aphasia. Ann Neurol 84(5):729–740

91. Jansen WJ, Ossenkoppele R, Knol DL, Tijms BM, Scheltens P, Verhey FRJ et al (2015) Prevalence of cerebral amyloid pathology in persons without dementia: a meta-analysis. JAMA 313(19):1924–1938

92. Villarejo-Galende A, Llamas-Velasco S, Gómez-Grande A, Puertas-Martín V, Contador I, Sarandeses P et al (2017) Amyloid pet in primary progressive

aphasia: case series and systematic review of the literature. J Neurol 264(1):121–130

93. Tahmasian M, Shao J, Meng C, Grimmer T, Diehl-Schmid J, Yousefi BH et al (2016) Based on the network degeneration hypothesis: separating individual patients with different neurodegenerative syndromes in a preliminary hybrid PET/MR study. J Nucl Med 57(3):410–415

94. Whitwell JL, Avula R, Senjem ML, Kantarci K, Weigand SD, Samikoglu A et al (2010) Gray and white matter water diffusion in the syndromic variants of frontotemporal dementia. Neurology 74(16):1279–1287

95. Agosta F, Scola E, Canu E, Marcone A, Magnani G, Sarro L et al (2012) White matter damage in frontotemporal lobar degeneration spectrum. Cereb Cortex 22(12):2705–2714

96. Hornberger M, Geng J, Hodges JR (2011) Convergent grey and white matter evidence of orbitofrontal cortex changes related to disinhibition in behavioural variant frontotemporal dementia. Brain 134(9):2502–2512

97. Mahoney CJ, Ridgway GR, Malone IB, Downey LE, Beck J, Kinnunen KM et al (2014) Profiles of white matter tract pathology in frontotemporal dementia. Hum Brain Mapp 35(8):4163–4179

98. Daianu M, Mendez MF, Baboyan VG, Jin Y, Melrose RJ, Jimenez EE et al (2016) An advanced white matter tract analysis in frontotemporal dementia and early-onset Alzheimer's disease. Brain Imaging Behav 10(4):1038–1053

99. Jakabek D, Power BD, Macfarlane MD, Walterfang M, Velakoulis D, van Westen D et al (2018) Regional structural hypo- and hyperconnectivity of frontal-striatal and frontal-thalamic pathways in behavioral variant frontotemporal dementia. Hum Brain Mapp 39(10):4083–4093

100. Borroni B, Brambati SM, Agosti C, Gipponi S, Bellelli G, Gasparotti R et al (2007) Evidence of white matter changes on diffusion tensor imaging in frontotemporal dementia. Arch Neurol 64(2):246–251

101. Meijboom R, Steketee RME, Ham LS, van der Lugt A, van Swieten JC, Smits M (2017) Differential hemispheric predilection of microstructural white matter and functional connectivity abnormalities between respectively semantic and behavioral variant frontotemporal dementia. J Alzheimers Dis 56(2):789–804

102. Kassubek J, Müller H-P, Del Tredici K, Hornberger M, Schroeter ML, Müller K et al (2018) Longitudinal diffusion tensor imaging resembles patterns of pathology progression in behavioral variant frontotemporal dementia (bvFTD). Front Aging Neurosci 10:47

103. Mahoney CJ, Simpson IJA, Nicholas JM, Fletcher PD, Downey LE, Golden HL et al (2015) Longitudinal diffusion tensor imaging in frontotemporal dementia. Ann Neurol 77(1):33–46

104. Yu J, Lee TMC (2019) Frontotemporal lobar degeneration neuroimaging I. the longitudinal decline of white matter microstructural integrity in behavioral variant frontotemporal dementia and its association with executive function. Neurobiol Aging 76:62–70

105. Lam BYK, Halliday GM, Irish M, Hodges JR, Piguet O (2014) Longitudinal white matter changes in frontotemporal dementia subtypes. Hum Brain Mapp 35(7):3547–3557

106. Reyes PA, Rueda ADP, Uriza F, Matallana DL (2019) Networks disrupted in linguistic variants of frontotemporal dementia. Front Neurol 10:903

107. Bouchard L-O, Wilson MA, Laforce R, Duchesne S (2019) White matter damage in the semantic variant of primary progressive aphasia. Can J Neurol Sci 46(4):373–382

108. Tu S, Leyton CE, Hodges JR, Piguet O, Hornberger M (2016) Divergent longitudinal propagation of white matter degradation in logopenic and semantic variants of primary progressive aphasia. J Alzheimers Dis 49(3):853–861

109. Brambati SM, Amici S, Racine CA, Neuhaus J, Miller Z, Ogar J et al (2015) Longitudinal gray matter contraction in three variants of primary progressive aphasia: a tenser-based morphometry study. Neuroimage Clin 8:345–355

110. Galantucci S, Tartaglia MC, Wilson SM, Henry ML, Filippi M, Agosta F et al (2011) White matter damage in primary progressive aphasias: a diffusion tensor tractography study. Brain 134(Pt 10):3011–3029

111. Catani M, Mesulam MM, Jakobsen E, Malik F, Martersteck A, Wieneke C et al (2013) A novel frontal pathway underlies verbal fluency in primary progressive aphasia. Brain 136(Pt 8):2619–2628

112. Elahi FM, Marx G, Cobigo Y, Staffaroni AM, Kornak J, Tosun D et al (2017) Longitudinal white matter change in frontotemporal dementia subtypes and sporadic late onset Alzheimer's disease. Neuroimage Clin 16:595–603

113. Zhang Y, Schuff N, Du A-T, Rosen HJ, Kramer JH, Gorno-Tempini ML et al (2009) White matter damage in frontotemporal dementia and Alzheimer's disease measured by diffusion MRI. Brain 132(Pt 9):2579–2592

114. Santillo AF, Mårtensson J, Lindberg O, Nilsson M, Manzouri A, Landqvist Waldö M et al (2013) Diffusion tensor tractography versus volumetric imaging in the diagnosis of behavioral variant frontotemporal dementia. PLoS One 8(7):e66932

115. Zhang Y, Tartaglia MC, Schuff N, Chiang GC, Ching C, Rosen HJ et al (2013) MRI signatures of brain macrostructural atrophy and microstructural degradation in frontotemporal lobar degeneration subtypes. J Alzheimers Dis 33(2):431–444

116. Krämer J, Lueg G, Schiffler P, Vrachimis A, Weckesser M, Wenning C et al (2018) Diagnostic value of diffusion tensor imaging and positron emission tomography in early stages of frontotemporal dementia. J Alzheimers Dis 63(1):239–253

117. McMillan CT, Brun C, Siddiqui S, Churgin M, Libon D, Yushkevich P et al (2012) White matter imaging contributes to the multimodal diagnosis of frontotemporal lobar degeneration. Neurology 78(22):1761–1768

118. Agosta F, Galantucci S, Magnani G, Marcone A, Martinelli D, Antonietta Volontè M et al (2015) MRI signatures of the frontotemporal lobar degeneration continuum. Hum Brain Mapp 36(7):2602–2614

119. Seeley WW, Crawford RK, Zhou J, Miller BL, Greicius MD (2009) Neurodegenerative diseases target large-scale human brain networks. Neuron 62(1):42–52

120. Touroutoglou A, Hollenbeck M, Dickerson BC, Feldman Barrett L (2012) Dissociable large-scale networks anchored in the right anterior insula subserve affective experience and attention. NeuroImage 60(4):1947–1958

121. Seeley WW, Allman JM, Carlin DA, Crawford RK, Macedo MN, Greicius MD et al (2007) Divergent social functioning in behavioral variant frontotemporal dementia and Alzheimer disease: reciprocal networks and neuronal evolution. Alzheimer Dis Assoc Disord 21(4):S50–SS7

122. Fox MD, Raichle ME (2007) Spontaneous fluctuations in brain activity observed with functional magnetic resonance imaging. Nat Rev Neurosci 8(9):700–711

123. Filippi M, Agosta F, Scola E, Canu E, Magnani G, Marcone A et al (2013) Functional network connectivity in the behavioral variant of frontotemporal dementia. Cortex 49(9):2389–2401

124. Zhou J, Greicius MD, Gennatas ED, Growdon ME, Jang JY, Rabinovici GD et al (2010) Divergent network connectivity changes in behavioural variant frontotemporal dementia and Alzheimer's disease. Brain 133(Pt 5):1352–1367

125. Farb NAS, Grady CL, Strother S, Tang-Wai DF, Masellis M, Black S et al (2013) Abnormal network connectivity in frontotemporal dementia: evidence for prefrontal isolation. Cortex 49(7):1856–1873

126. Day GS, Farb NAS, Tang-Wai DF, Masellis M, Black SE, Freedman M et al (2013) Salience network resting-state activity: prediction of frontotemporal dementia progression. JAMA Neurol 70(10):1249–1253

127. Reyes P, Ortega-Merchan MP, Rueda A, Uriza F, Santamaria-García H, Rojas-Serrano N et al (2018) Functional connectivity changes in behavioral, semantic, and nonfluent variants of frontotemporal dementia. Behav Neurol 2018:9684129

128. Sedeño L, Couto B, García-Cordero I, Melloni M, Baez S, Morales Sepúlveda JP et al (2016) Brain network organization and social executive performance in frontotemporal dementia. J Int Neuropsychol Soc 22(2):250–262

129. Agosta F, Sala S, Valsasina P, Meani A, Canu E, Magnani G et al (2013) Brain network connectivity assessed using graph theory in frontotemporal dementia. Neurology 81(2):134–143

130. Saba V, Premi E, Cristillo V, Gazzina S, Palluzzi F, Zanetti O et al (2019) Brain connectivity and information-flow breakdown revealed by a minimum spanning tree-based analysis of MRI data in behavioral variant frontotemporal dementia. Front Neurosci 13:211

131. Seeley WW, Zhou J, Kim E-J (2012) Frontotemporal dementia: what can the behavioral variant teach us about human brain organization? Neuroscientist 18(4):373–385

132. Hafkemeijer A, Möller C, Dopper EGP, Jiskoot LC, van den Berg-Huysmans AA, van Swieten JC et al (2017) A longitudinal study on resting state functional connectivity in behavioral variant frontotemporal dementia and Alzheimer's disease. J Alzheimers Dis 55(2):521–537

133. Ranganath C, Ritchey M (2012) Two cortical systems for memory-guided behaviour. Nat Rev Neurosci 13(10):713–726

134. Patterson K, Nestor PJ, Rogers TT (2007) Where do you know what you know? The representation of semantic knowledge in the human brain. Nat Rev Neurosci 8(12):976–987

135. Guo CC, Gorno-Tempini ML, Gesierich B, Henry M, Trujillo A, Shany-Ur T et al (2013) Anterior temporal lobe degeneration produces widespread network-driven dysfunction. Brain 136(Pt 10):2979–2991

136. Montembeault M, Chapleau M, Jarret J, Boukadi M, Laforce R Jr, Wilson MA et al (2019) Differential language network functional connectivity alterations in Alzheimer's disease and the semantic variant of primary progressive aphasia. Cortex 117:284–298

137. Battistella G, Henry M, Gesierich B, Wilson SM, Borghesani V, Shwe W et al (2019) Differential intrinsic functional connectivity changes in semantic variant primary progressive aphasia. Neuroimage Clin 22:101797

138. Agosta F, Galantucci S, Valsasina P, Canu E, Meani A, Marcone A et al (2014) Disrupted brain connectome in semantic variant of primary progressive aphasia. Neurobiol Aging 35(11):2646–2655

139. Bonakdarpour B, Rogalski EJ, Wang A, Sridhar J, Mesulam MM, Hurley RS (2017) Functional connectivity is reduced in early-stage primary progressive aphasia when atrophy is not prominent. Alzheimer Dis Assoc Disord 31(2):101–106

140. Mandelli ML, Vilaplana E, Brown JA, Hubbard HI, Binney RJ, Attygalle S et al (2016) Healthy brain connectivity predicts atrophy progression in non-fluent variant of primary progressive aphasia. Brain 139(Pt 10):2778–2791

141. Mandelli ML, Welch AE, Vilaplana E, Watson C, Battistella G, Brown JA et al (2018) Altered topology of the functional speech production network in non-fluent/agrammatic variant of PPA. Cortex 108:252–264

142. Brown JA, Deng J, Neuhaus J, Sible IJ, Sias AC, Lee SE et al (2019) Patient-tailored, connectivity-based forecasts of spreading brain atrophy. Neuron 104(5):856–68.e5

143. Filippi M, Basaia S, Canu E, Imperiale F, Meani A, Caso F et al (2017) Brain network connectivity differs in early-onset neurodegenerative dementia. Neurology 89(17):1764–1772

144. Donnelly-Kehoe PA, Pascariello GO, García AM, Hodges JR, Miller B, Rosen H et al (2019) Robust automated computational approach for classifying frontotemporal neurodegeneration: multimodal/multicenter neuroimaging. Alzheimers Dement: Diagn, Assess & Dis Monit 11:588–598

145. Bouts MJRJ, Möller C, Hafkemeijer A, van Swieten JC, Dopper E, van der Flier WM et al (2018) Single subject classification of Alzheimer's disease and behavioral variant frontotemporal dementia using anatomical, diffusion tensor, and resting-state functional magnetic resonance imaging. J Alzheimers Dis 62(4):1827–1839

146. Du AT, Jahng GH, Hayasaka S, Kramer JH, Rosen HJ, Gorno-Tempini ML et al (2006) Hypoperfusion in frontotemporal dementia and Alzheimer disease by arterial spin labeling MRI. Neurology 67(7):1215–1220

147. Steketee RME, Bron EE, Meijboom R, Houston GC, Klein S, Mutsaerts HJMM et al (2016) Early-stage differentiation between presenile Alzheimer's disease and frontotemporal dementia using arterial spin labeling MRI. Eur Radiol 26(1):244–253

148. Olm CA, Kandel BM, Avants BB, Detre JA, Gee JC, Grossman M et al (2016) Arterial spin labeling perfusion predicts longitudinal decline in semantic variant primary progressive aphasia. J Neurol 263(10):1927–1938

149. Tosun D, Schuff N, Rabinovici GD, Ayakta N, Miller BL, Jagust W et al (2016) Diagnostic utility of ASL-MRI and FDG-PET in the behavioral variant of FTD and AD. Ann Clin Transl Neurol 3(10):740–751

150. Fällmar D, Haller S, Lilja J, Danfors T, Kilander L, Tolboom N et al (2017) Arterial spin labeling-based Z-maps have high specificity and positive predictive value for neurodegenerative dementia compared to FDG-PET. Eur Radiol 27(10):4237–4246

151. Anazodo UC, Finger E, Kwan BYM, Pavlosky W, Warrington JC, Günther M et al (2017) Using simultaneous PET/MRI to compare the accuracy of diagnosing frontotemporal dementia by arterial spin labelling MRI and FDG-PET. Neuroimage Clin 17:405–414

152. Dickerson BC (2016) Hodges' frontotemporal dementia, 2nd edn. Cambridge University Press, Cambridge

153. Dickerson BC (2014) Imaging Tau pathology in vivo in FTLD: initial experience with [18F] T807 PET. 8th Human Amyloid imaging meeting, Miami

154. Rabinovici G (ed) (2020) Tau imaging in non AD Tauopathies. Tau 2020; 2/12/2020, Washington, DC

155. Ossenkoppele R, Rabinovici GD, Smith R, Cho H, Scholl M, Strandberg O et al (2018) Discriminative accuracy of [18F]flortaucipir positron emission tomography for Alzheimer disease vs other neurodegenerative disorders. JAMA 320(11):1151–1162

156. Schonhaut DR, McMillan CT, Spina S, Dickerson BC, Siderowf A, Devous MD Sr et al (2017) (18) F-flortaucipir tau positron emission tomography distinguishes established progressive supranuclear palsy from controls and Parkinson disease: a multicenter study. Ann Neurol 82(4):622–634

157. Smith R, Scholl M, Widner H, van Westen D, Svenningsson P, Hagerstrom D et al (2017) In vivo retention of (18)F-AV-1451 in corticobasal syndrome. Neurology 89(8):845–853

158. Tsai RM, Bejanin A, Lesman-Segev O, LaJoie R, Visani A, Bourakova V et al (2019) (18)F-flortaucipir (AV-1451) tau PET in frontotemporal dementia syndromes. Alzheimers Res Ther 11(1):13

159. Jones DT, Knopman DS, Graff-Radford J, Syrjanen JA, Senjem ML, Schwarz CG et al (2018) In vivo (18)F-AV-1451 tau PET signal in MAPT mutation carriers varies by expected tau isoforms. Neurology 90(11):e947–ee54

160. Makaretz SJ, Quimby M, Collins J, Makris N, McGinnis S, Schultz A et al (2018) Flortaucipir tau PET imaging in semantic variant primary progressive aphasia. J Neurol Neurosurg Psychiatry 89(10):1024–1031

161. Cho H, Kim HJ, Choi JY, Ryu YH, Lee MS, Na DL et al (2019) (18)F-flortaucipir uptake patterns in clinical subtypes of primary progressive aphasia. Neurobiol Aging 75:187–197

162. Josephs KA, Martin PR, Botha H, Schwarz CG, Duffy JR, Clark HM et al (2018) [(18) F]AV-1451 tau-PET and primary progressive aphasia. Ann Neurol 83(3):599–611

163. Bevan-Jones WR, Cope TE, Jones PS, Passamonti L, Hong YT, Fryer TD et al (2018) [(18F]AV-1451 binding in vivo mirrors the expected distribution of TDP-43 pathology in the semantic variant of primary progressive aphasia. J Neurol Neurosurg Psychiatry 89(10):1032–1037

164. Sander K, Lashley T, Gami P, Gendron T, Lythgoe MF, Rohrer JD et al (2016) Characterization of tau positron emission tomography tracer [(18) F]AV-1451 binding to postmortem tissue in Alzheimer's disease, primary tauopathies, and other dementias. Alzheimers Dement 12(11):1116–1124

165. Marquié M, Normandin MD, Vanderburg CR, Costantino IM, Bien EA, Rycyna LG et al (2015) Validating novel tau positron emission tomography tracer [F-18]-AV-1451 (T807) on postmortem brain tissue. Ann Neurol 78(5):787–800

166. Lowe VJ, Curran G, Fang P, Liesinger AM, Josephs KA, Parisi JE et al (2016) An autoradiographic evaluation of AV-1451 Tau PET in dementia. Acta Neuropathol Commun 4(1):58

167. Marquie M, Normandin MD, Meltzer AC, Siao Tick Chong M, Andrea NV, Anton-Fernandez A et al (2017) Pathological correlations of [F-18]-AV-1451 imaging in non-alzheimer tauopathies. Ann Neurol 81(1):117–128

168. Aguero C, Dhaynaut M, Normandin MD, Amaral AC, Guehl NJ, Neelamegam R et al (2019)

Autoradiography validation of novel tau PET tracer [F-18]-MK-6240 on human postmortem brain tissue. Acta Neuropathol Commun 7(1):37

169. Falcon B, Zhang W, Murzin AG, Murshudov G, Garringer HJ, Vidal R et al (2018) Structures of filaments from Pick's disease reveal a novel tau protein fold. Nature 561(7721):137–140

170. Falcon B, Zhang W, Schweighauser M, Murzin AG, Vidal R, Garringer HJ et al (2018) Tau filaments from multiple cases of sporadic and inherited Alzheimer's disease adopt a common fold. Acta Neuropathol 136(5):699–708

171. Falcon B, Zivanov J, Zhang W, Murzin AG, Garringer HJ, Vidal R et al (2019) Novel tau filament fold in chronic traumatic encephalopathy encloses hydrophobic molecules. Nature 568(7752):420–423

172. Zhang W, Tarutani A, Newell KL, Murzin AG, Matsubara T, Falcon B et al (2020) Novel tau filament fold in corticobasal degeneration. Nature 580(7802):283–287

173. Leuzy A, Chiotis K, Lemoine L, Gillberg PG, Almkvist O, Rodriguez-Vieitez E et al (2019) Tau PET imaging in neurodegenerative tauopathies-still a challenge. Mol Psychiatry 24(8):1112–1134

174. Villemagne VL, Dore V, Bourgeat P, Burnham SC, Laws S, Salvado O et al (2017) Abeta-amyloid and Tau imaging in dementia. Semin Nucl Med 47(1):75–88

175. Kim MJ, McGwier M, Jenko KJ, Snow J, Morse C, Zoghbi SS et al (2019) Neuroinflammation in frontotemporal lobar degeneration revealed by (11) C-PBR28 PET. Ann Clin Transl Neurol 6(7):1327–1331

176. Rohrer JD, Nicholas JM, Cash DM, van Swieten J, Dopper E, Jiskoot L et al (2015) Presymptomatic cognitive and neuroanatomical changes in genetic frontotemporal dementia in the genetic frontotemporal dementia initiative (GENFI) study: a cross-sectional analysis. Lancet Neurol 14(3):253–262

177. Rohrer JD, Warren JD, Fox NC, Rossor MN (2013) Presymptomatic studies in genetic frontotemporal dementia. Rev Neurol 169(10):820–824

178. Rosen HJ, Boeve BF, Boxer AL (2020) Tracking disease progression in familial and sporadic frontotemporal lobar degeneration: recent findings from ARTFL and LEFFTDS. Alzheimers Dement 16(1):71–78

The Frontotemporal Dementia Prevention Initiative: Linking Together Genetic Frontotemporal Dementia Cohort Studies

Jonathan D. Rohrer and Adam L. Boxer

Introduction to Genetic Frontotemporal Dementia

Pathogenic mutations are found in around 25–30% of people diagnosed with frontotemporal dementia (FTD). This percentage is higher in those with the behavioural variant (bvFTD) where it is about 40%, and much lower in those with the language variant (known as primary progressive aphasia, PPA) where it is around 5% [7]. Mutations in three genes (*MAPT*, *GRN* and *C9orf72*) account for the majority of genetic frontotemporal dementia, with all having an autosomal dominant pattern of inheritance. However, mutations have also been found less frequently in a number of other genes (*TBK1, VCP, TARDBP, SQSTM1, FUS, CHMP2B*), with individual or limited reports in further genes (*CHCHD10, UBQLN2, OPTN, CCNF, DCTN, TIA1*).

The prevalence of genetic frontotemporal dementia has been poorly studied. Prior studies of FTD as a whole have mainly focused on specific age groups with an estimated point prevalence between the ages of 45 and 64 of 15–22 per 100,000 [15]. However, a recent study in the UK estimated a prevalence across all ages of 11 per 100,000 [5]. Worldwide, around 40% of genetic FTD cases have mutations in *C9orf72*, 35% in *GRN* and 25% in *MAPT*, with only 1–2% having mutations in the other genes [13]. If 30% of FTD is genetic, this equates to a prevalence of ~1.3 per 100,000 (e.g. ~4745 people in North America) for *C9orf72*-related FTD, ~1.2 per 100,000 (e.g. ~4380 people in North America) for *GRN*-related FTD and ~ 0.8 per 100,000 (e.g. ~2920 people in North America) for *MAPT*-related FTD. Overall, prevalence numbers relate to symptomatic mutation carriers, but as their siblings and children are at 50% risk of developing symptoms, there exists a larger population of living presymptomatic mutation carriers as well.

Multicentre Genetic Frontotemporal Dementia Cohort Studies and the Development of the Frontotemporal Dementia Prevention Initiative

Families with genetic FTD have been studied in case reports and series from individual centres over many years (reviewed in [18]). However, given the rarity of genetic FTD, it became

J. D. Rohrer (✉)
Dementia Research Centre, Department of Neurodegenerative Disease, UCL Institute of Neurology, London, UK
e-mail: j.rohrer@ucl.ac.uk

A. L. Boxer
Memory and Aging Center, Department of Neurology, University of California San Francisco, San Francisco, CA, USA
e-mail: adam.boxer@ucsf.edu

B. Ghetti et al. (eds.), *Frontotemporal Dementias*, Advances in Experimental Medicine and Biology 1281, https://doi.org/10.1007/978-3-030-51140-1_8

clear that centres would need to collaborate more closely, in order to better understand the disease, building a joint methodological platform to create a cohort of genetic FTD mutation carriers, and develop robust biomarkers of disease onset and progression. In 2012, a group of centres specializing in FTD within Europe and Eastern Canada came together to create the Genetic FTD Initiative (GENFI). Following this, in 2015, centres in the United States and Western Canada created the overlapping Advancing Research and Treatment for Frontotemporal Lobar Degeneration (ARTFL) and Longitudinal Evaluation of Familial Frontotemporal Dementia Subjects (LEFFTDS) studies. More recently, genetic FTD cohort studies in Australia (Dominantly Inherited Non-Alzheimer Dementias study, DINAD), New Zealand (NZ FTD Genetic Study, FTDGeNZ) and South America (Research Dementia Latin America, ReDLat) have either got started or will be starting soon.

Recognizing the importance of working together across the world, GENFI and ARTFL-LEFFTDS (now ALLFTD) investigators have come together to create the FTD Prevention Initiative (FPI). The overall goal of the group is to work together to promote clinical trials of new therapies to prevent FTD, with the key aims of:

1. Creating an international database of familial FTD research participants who might be eligible for clinical trials
2. Creating uniform standards for the conduct of clinical trials in familial FTD syndromes

The FPI recognizes the importance of involving families with genetic FTD, with patient advocacy groups and foundations involved in FTD research such as the Association for Frontotemporal Degeneration, Bluefield Project to Cure FTD and the FTD Disorders Registry, also part of the initiative.

In this chapter, we describe, firstly, the initial FPI project, which investigated age at symptom onset and disease duration as well as phenotype in genetic FTD [13]; secondly, the current status of outcome measures for genetic FTD trials

(mainly relating to work from the GENFI and ARTFL-LEFFTDS studies) and, lastly, the ongoing and planned trials in genetic FTD.

Other projects that are ongoing as part of the FPI include:

1. Modelling disease progression in genetic FTD with cognitive, brain imaging and fluid biomarkers
2. Predicting phenoconversion to symptomatic FTD
3. Understanding variability in bioassays for progranulin
4. Surveying participants to understand what family members want from genetic FTD trials

Phenotype, Age at Onset and Disease Duration in Genetic Frontotemporal Dementia

The first FPI study bringing together data from across the world on the three main forms of genetic FTD was recently published [13]. From data on 3403 symptomatic individuals with *C9orf72, GRN* and *MAPT* mutations, the project reported a number of key findings:

1. A total of 130 different *GRN* mutations and 67 different *MAPT* mutations are described in the paper, with the most common mutations being T272fs, R493X, IVS7-1G > A, C31fs, G35fs and A9D in *GRN*, and P301L, IVS10 + 16C > T, R406W and N279K in *MAPT*. An updated list of genetic mutations can be found at www.ftdtalk.org/what-is-ftd/genetics/.
2. Geographical variability exists in the distribution of the main genetic FTD groups, with an increased prevalence of *GRN* mutations in some countries, for example, Italy, Spain and Belgium, mainly due to large founder families (T272fs, IVS7-1G > A and IVS1 + 5G > C, respectively). Large *MAPT* families also exist, for example, IVS10 + 16C > T, originally from the North Wales area of the UK, and the PPND family with the N279K mutation.

3. The most common phenotype in each form of genetic FTD is bvFTD. PPA is a more common diagnosis in *GRN* mutation carriers (20%) with the specific variant usually being non-fluent variant PPA or a mixed PPA syndrome, compared with *MAPT* (6%) or *C9orf72* (4%). Corticobasal syndrome is seen not uncommonly in the *GRN* group (6%), to a lesser extent in the *MAPT* group (3%) and only rarely in the *C9orf72* group. In comparison, a classical PSP syndrome (i.e. Richardson's syndrome) is seen in 6% of *MAPT* mutation carriers, but not in the *GRN* group and only in rare cases in *C9orf72* expansion carriers. Amyotrophic lateral sclerosis (ALS) is only a very rare occurrence in *GRN* (2%) or *MAPT* mutation carriers (1%), whereas around 40% of *C9orf72* expansion carriers have either pure ALS (26%) or an FTD-ALS overlap (15%).

4. A wide range of age at symptom onset exists across all of the genetic groups, with onset between the 20s and the 90s for *GRN* and *C9orf72* groups, and from 17 to the 80s in the *MAPT* group (Fig. 1a). There is little difference in age at onset across the different *GRN* mutations (Fig. 1b), but there are key differences across the common *MAPT* mutations, with those with N279K mutations having a lower age at onset (mean 43.8 years), followed by IVS10 + 16C > T (50.9), then P301L (53.0) and, finally, R406W (55.4), having the oldest mean age at onset (Fig. 1c).

5. In all three genetic groups, being given a diagnosis of 'Alzheimer's disease' was associated with an older age at onset, whilst, in *MAPT* mutations, individuals with atypical parkinsonian syndromes were younger at onset.

6. Disease duration was lowest overall in those with *C9orf72* mutations, followed by those with *GRN* and then *MAPT* mutations. In the *C9orf72* group, a diagnosis of ALS was associated with a shorter disease duration (mean 2.9 years) than FTD-ALS (5.0), PPA (7.5) and bvFTD (7.8).

7. Individual age at onset correlated with mean age at onset within the family in all three groups (*C9orf72*, r = 0.36; *GRN* r = 0.18; *MAPT* r = 0.63), as well as with parental age at onset (*C9orf72*, r = 0.32; *GRN* r = 0.22 *MAPT* r = 0.45) (Fig. 2), with the correlation strongest in *MAPT* mutations.

8. Variability in age at onset was explained largely by family membership and the specific mutation in *MAPT* mutations, but not in the *GRN* or *C9orf72* groups.

Overall, the study provides important data that will be useful for future trials. In particular, it tells us that whilst using the mean familial age at onset as a predictor for the age at onset in *MAPT* mutations provides an adequate estimate, it does not do so for *GRN* and *C9orf72* mutations, and better markers of staging during the presymptomatic period will be required.

Potential Outcome Measures for Genetic Frontotemporal Dementia Trials

Much work has been undertaken to understand the pattern of changes occurring in the natural history of genetic FTD, and most recently, this has been mainly through the observational cohort studies that form part of the FPI [2, 7, 19].

Clinical

Few well-validated clinical scales have been developed in genetic FTD. The most well studied is the Clinical Dementia Rating scale plus National Alzheimer Coordinating Center Frontotemporal Lobar Degeneration Module (NACC FTLD) module (previously known as the 'FTLD CDR'), which is able to capture early symptoms in genetic FTD [12]. Less work has been done on the Frontotemporal dementia Rating Scale (FRS), but this also has potential use in trials.

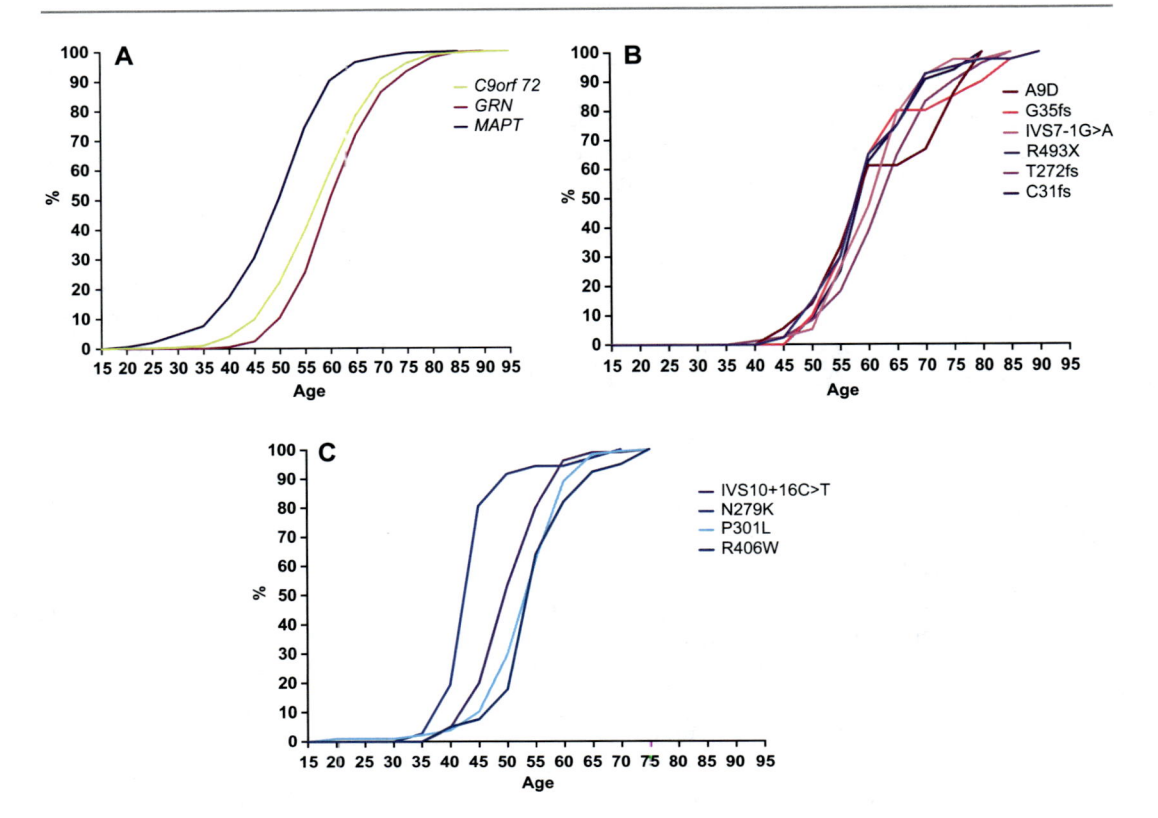

Fig. 1 Range of ages at onset for (**a**) C9orf72, GRN and MAPT mutations overall, (**b**) common GRN mutations and (**c**) common MAPT mutations. (Adapted from Moore et al., Lancet Neurology [13])

Cognitive

Individual neuropsychometric tests have been studied in genetic FTD, with the most sensitive showing change around 5 years before symptom onset [17, 20]. Work is underway to look at cognitive composites, and to look at more novel ways of testing cognition, such as with computerized batteries, wearables and eye tracking.

Magnetic Resonance Imaging

Structural T1 imaging has been the most well studied form of neuroimaging in FTD. Grey matter atrophy occurs at least 10 years before symptom onset in genetic FTD, and probably earlier than this in *C9orf72* mutation carriers [14, 17]. The pattern of atrophy differs across the genetic groups, with early thalamic involvement in

C9orf72 mutation carriers; temporal lobe, particularly medial atrophy early in the disease process in *MAPT* mutation carriers; and insula, frontal and parietal cortical atrophy in *GRN* mutation carriers.

Rates of atrophy are fastest in *GRN* mutation carriers and slowest as a group in *MAPT* mutation carriers, although there is wider variability in the *C9orf72* groups with both fast and slow progressors seen [3, 4].

Use of both whole-brain atrophy rate and specific regions of interest (ROI) atrophy rate (lobar and subcortical structures) are likely to be important in clinical trials, but with more work to be done on identifying the most robust postprocessing methodology that leads to the lowest sample size calculations, and the best ROIs to be used in the different genetic groups.

T2 imaging reveals the presence of white matter hyperintensities in *GRN* mutation carriers. However, a recent GENFI study revealed their

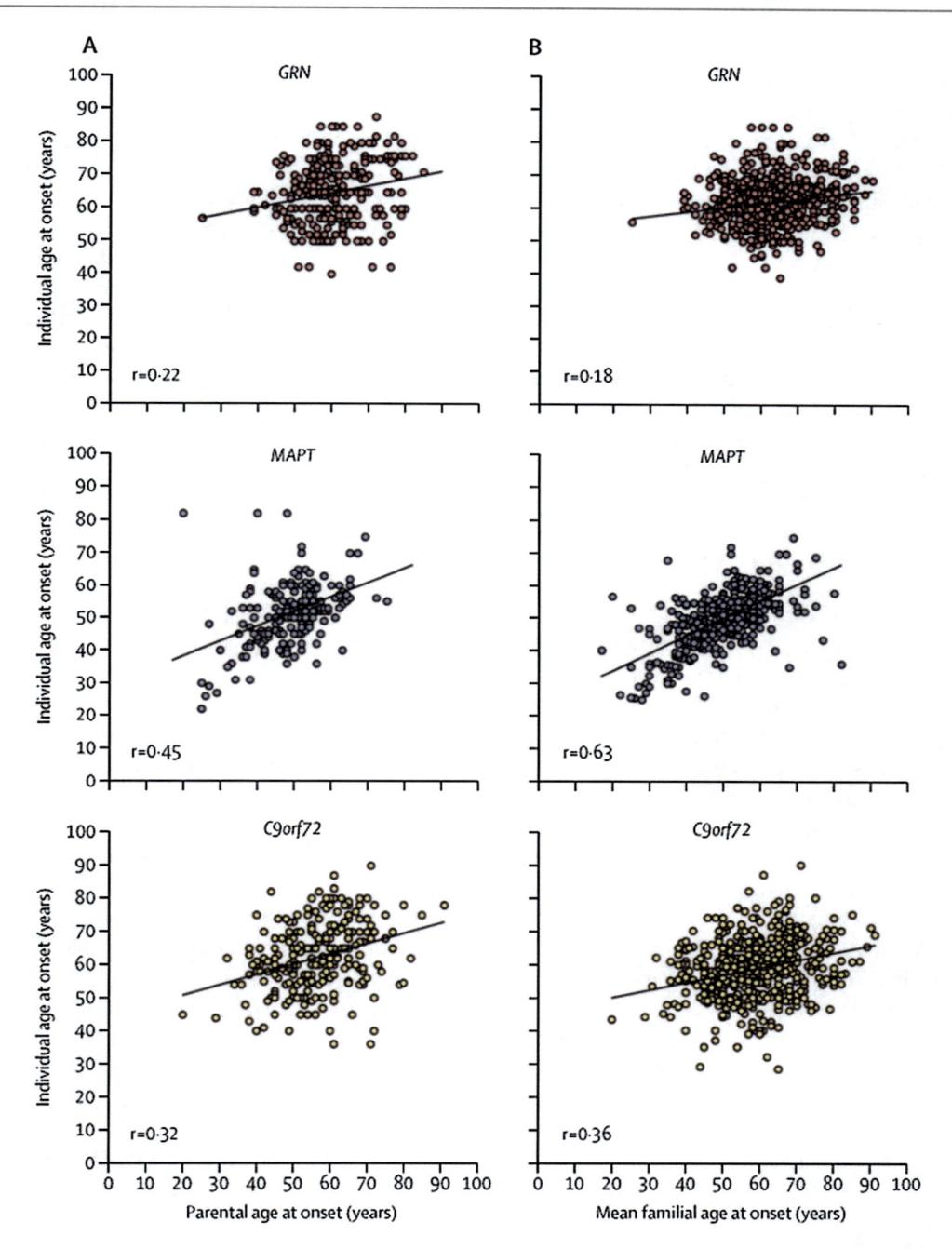

Fig. 2 Correlation of individual age at onset with parental (**a**) and mean familial (**b**) ages at onset. (From Moore et al., Lancet Neurology [13])

presence in only a subset of cases [21], for example, in the symptomatic *GRN* group, 25.0% had none/mild white matter hyperintensity (WMH) load, 37.5% had medium and 37.5% had a severe load. This makes the use of WMH measurement difficult in trials across the entirety of a *GRN* cohort.

Diffusion tensor imaging (DTI) reveals impaired structural connectivity preceding grey matter atrophy [9], and this opens up the opportunity for earlier measurement of change in genetic FTD. However, little work has been performed longitudinally in DTI in genetic FTD, and DTI is more prone to multicentre, cross-scanner issues than T1 imaging, which potentially limits its use in trials.

Other MR imaging modalities such as functional MRI and arterial spin labelling MRI remain poorly studied in genetic FTD, with limited understanding of the variability, extent of longitudinal change and robustness to measurement across multiple scanners within a trial setting.

Positron Emission Tomography Imaging

Relatively less work has been performed in PET imaging than in MRI in FTD. However, hypometabolism using 18F-FDG-PET is also seen up to 10 years prior to symptom onset, although patterns are less clear across the genetic groups than for atrophy using structural imaging.

Whilst the tau PET ligand flortaucipir binds well to the paired helical filament–type tau seen in V337M and R406W *MAPT* mutations, it binds less well to the other forms of tau seen in other *MAPT* mutations [22], and so is not at a stage where it could be adequately used in trials of *MAPT*-related FTD.

Novel tracers are under investigation, including those identifying inflammation, synaptic abnormalities and mitochondrial dysfunction, but these remain some time away from being usable as outcome measures.

Fluid Biomarkers

Two key disease-specific markers are likely to be important outcome measures for genetic FTD trials:

1. Serum, plasma and cerebrospinal fluid (CSF) progranulin have excellent sensitivity and specificity for detecting pathogenic *GRN* mutations [6], with levels low from a young age, and relatively stable over time. Levels are approximately half of 'normal' progranulin levels, and the majority of therapies aimed at *GRN*-related FTD will be aiming to normalize levels by (at least) doubling progranulin measured in biofluids. There is some variability in the different commercially available progranulin assays, and work in the FPI is currently underway to understand that better.

2. Increased CSF poly(GP) levels are seen in both presymptomatic and symptomatic *C9orf72* expansion carriers [11]. Although not felt to be the toxic dipeptide repeat species, poly(GP) levels are currently the best markers available that appear to be a direct surrogate of the pathology seen in *C9orf72*-related FTD, and disease-modifying therapies would be expected to reduce the levels back to 'normal' (essentially zero). However, with current assays, some mutation carriers have very low levels, overlapping with controls. Newer, more sensitive, assays will therefore be required for trials that more clearly separate controls and carriers.

Two other markers have clear potential for use in trials:

1. Neurofilament light chain (NfL, either in CSF or blood) is a measure of disease intensity in FTD, and it predicts progression and survival. Levels increase just prior to symptom onset and appear to continue to increase during the symptomatic period, at least in *GRN* mutations [10, 23].

2. Glial fibrillary acidic protein (GFAP) appears to also increase just before symptom onset in *GRN*-associated FTD [8], although more work is needed to understand longitudinal change in this marker.

More speculatively, markers, such as YKL-40 and chitotriosidase, may index an inflammatory process that occurs in genetic FTD, particularly in *GRN* mutation carriers, although more work is

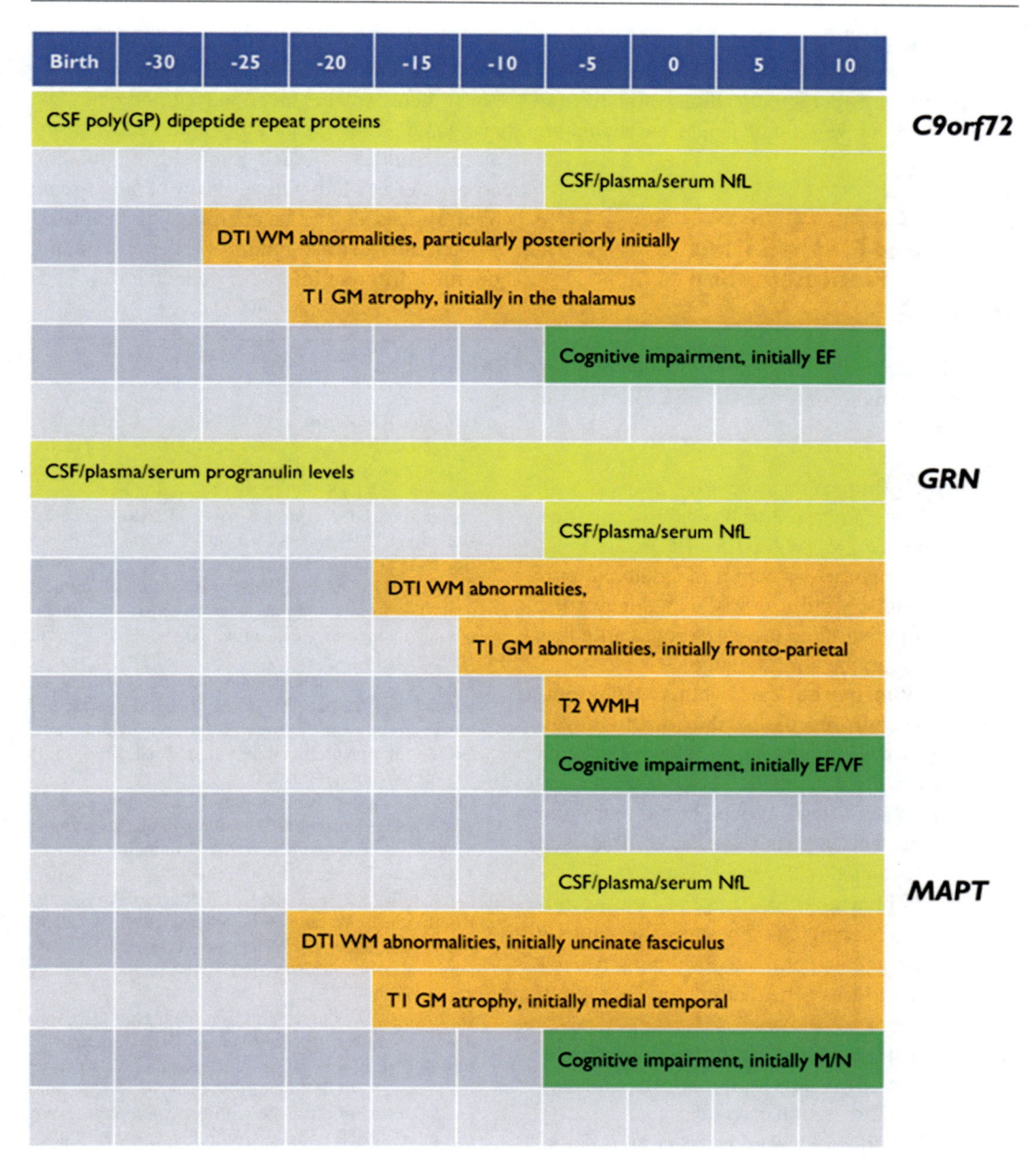

Fig. 3 Schematic of cognitive (green), imaging (orange) and fluid (yellow) biomarker profiles across the lifespan of C9orf72, GRN and MAPT mutation carriers. (From Greaves et al. [7])

needed in this field, as well as in the field of lysosomal and synaptic measures.

A summary of the key biomarker changes that occur through the timeline of the different genetic forms of FTD is shown in Fig. 3. A strategy employed in other rare neurological disorders to improve power to detect treatment effects and better estimate time of disease onset is the construction of Bayesian disease progression models based on multimodal data such as those mentioned earlier [16]. The Dominantly Inherited Alzheimer's Network Treatment Unit (DIAN-TU) has used such a model as the basis for their adaptive clinical trial platform, allowing for streamlined testing of potentially disease-modifying agents for the prevention of genetic Alzheimer's disease [1].

Current and Planned Trials in Genetic Frontotemporal Dementia

The FPI is currently collaborating with a number of pharmaceutical companies on clinical trials for genetic FTD:

1. **Alector** (https://alector.com) is currently undertaking a phase 2 trial for *GRN* mutation carriers at centres within the FPI (and is about to start its phase 3), with a monoclonal antibody against sortilin. Early data suggest the ability of the drug to increase progranulin levels back into the normal range.
2. **Ionis Pharmaceuticals** (https://www.ionispharma.com) in partnership with **Biogen** (https://www.biogen.com) have developed antisense oligonucleotide (ASO) therapies for MAPT and C9orf72, which are currently being tested in Alzheimer's disease and *C9orf72*-related ALS respectively.
3. **Prevail Therapeutics** (https://www.prevailtherapeutics.com) and **Passage Bio** (https://www.passagebio.com) are developing Adeno-Associated Virus gene therapy for *GRN* mutation carriers.
4. **Wave Life Sciences** (https://www.wavelifesciences.com) is developing an ASO therapy for *C9orf72* mutation carriers.
5. **Arkuda** (https://www.arkudatx.com) is developing a therapy for *GRN* mutation carriers.

The Future

We hope to create an international collaborative group of academic FTD research centres, patient advocacy groups and research foundations that are dedicated to finding a cure for genetic FTD. Through this group, we hope to be a voice for genetic FTD family members, working to design and run the best possible clinical trials. Whilst there remains much work to be done, we have come a long way from single-centre observational studies of small numbers of mutation carriers, to a collaborative group of large-scale cohort studies, entering a new era of clinical trials of potentially disease-modifying therapies and a hopefully different future for genetic FTD.

References

1. Bateman RJ, Benzinger TL, Berry S, Clifford DB, Duggan C, Fagan AM et al (2017) DIAN-TU pharma consortium for the dominantly inherited Alzheimer network. The DIAN-TU Next Generation Alzheimer's prevention trial: Adaptive design and disease progression model. Alzheimers Dement 13(1):8–19
2. Boxer AL, Gold M, Feldman H, Boeve BF, Dickinson SL, Fillit H et al (2020) New directions in clinical trials for frontotemporal lobar degeneration: methods and outcome measures. Alzheimers Dement 16(1):131–143
3. Chen Q, Boeve BF, Senjem M, Tosakulwong N, Lesnick T, Brushaber D et al (2020) Trajectory of lobar atrophy in asymptomatic and symptomatic GRN mutation carriers: a longitudinal MRI study Neurobiol Aging 88:42–50
4. Chen Q, Boeve BF, Senjem M, Tosakulwong N, Lesnick TG, Brushaber D et al (2019) Rates of lobar atrophy in asymptomatic *MAPT* mutation carriers. Alzheimers Dement (N Y) 5:338–346
5. Coyle-Gilchrist IT, Dick KM, Patterson K, Vázquez Rodríquez P, Wehmann E, Wilcox A et al (2016) Prevalence, characteristics, and survival of frontotemporal lobar degeneration syndromes. Neurology 86(18):1736–1743
6. Galimberti D, Fumagalli GG, Fenoglio C, Cioffi SMG, Arighi A, Serpente M et al (2018) Genetic FTD Initiative (GENFI) Progranulin plasma levels predict the presence of GRN mutations in asymptomatic subjects and do not correlate with brain atrophy: results from the GENFI study. Neurobiol Aging 62:245.e9–245.e12
7. Greaves CV, Rohrer JD (2019) An update on genetic frontotemporal dementia. J Neurol 266(8):2075–2086
8. Heller C, Foiani MS, Moore K, Convery R, Bocchetta M, Neason M et al (2020 Mar) Plasma glial fibrillary acidic protein is raised in progranulin-associated frontotemporal dementia. J Neurol Neurosurg Psychiatry 91(3):263–270
9. Jiskoot LC, Bocchetta M, Nicholas JM, Cash DM, Thomas D, Modat M et al (2018) Presymptomatic white matter integrity loss in familial frontotemporal dementia in the GENFI cohort: a cross-sectional

diffusion tensor imaging study. Ann Clin Transl Neurol 5(9):1025–1036

10. Meeter LH, Dopper EG, Jiskoot LC, Sanchez-Valle R, Graff C, Benussi L et al (2016) Neurofilament light chain: a biomarker for genetic frontotemporal dementia. Ann Clin Transl Neurol 3(8):623–636

11. Meeter LHH, Gendron TF, Sias AC, Jiskoot LC, Russo SP, Donker Kaat L et al (2018) Poly(GP), neurofilament and grey matter deficits in *C9orf72* expansion carriers. Ann Clin Transl Neurol 5(5):583–597

12. Miyagawa T, Brushaber D, Syrjanen J, Kremers W, Fields J, Forsberg LK et al (2020) Use of the CDR® plus NACC FTLD in mild FTLD: data from the ARTFL/LEFFTDS consortium. Alzheimers Dement 16(1):79–90

13. Moore KM, Nicholas J, Grossman M, McMillan CT, Irwin DJ, Massimo L et al (2020) Age at symptom onset and death and disease duration in genetic frontotemporal dementia: an international retrospective cohort study. Lancet Neurol 19(2):145–156

14. Olney NT, Ong E, Goh SM, Bajorek L, Dever R, Staffaroni AM et al (2020) Clinical and volumetric changes with increasing functional impairment in familial frontotemporal lobar degeneration. Alzheimers Dement 16(1):49–59

15. Onyike CU, Diehl-Schmid J (2013) The epidemiology of frontotemporal dementia. Int Rev Psychiatry 25(2):130–137

16. Quintana M, Shrader J, Slota C, Joe G, McKew JC, Fitzgerald M et al (2019) Bayesian model of disease progression in GNE myopathy. Stat Med 38(8):1459–1474

17. Rohrer JD, Nicholas JM, Cash DM, van Swieten J, Dopper E, Jiskoot L et al (2015) Presymptomatic cognitive and neuroanatomical changes in genetic frontotemporal dementia in the Genetic Frontotemporal dementia Initiative (GENFI) study: a cross-sectional analysis. Lancet Neurol 14(3):253–262

18. Rohrer JD, Warren JD, Fox NC, Rossor MN (2013) Presymptomatic studies in genetic frontotemporal dementia. Rev Neurol (Paris) 169(10):820–824

19. Rosen HJ, Boeve BF, Boxer AL (2020) Tracking disease progression in familial and sporadic frontotemporal lobar degeneration: recent findings from ARTFL and LEFFTDS. Alzheimers Dement 16(1):71–78

20. Staffaroni AM, Bajorek L, Casaletto KB, Cobigo Y, Goh SM, Wolf A et al (2020 Jan) Assessment of executive function declines in presymptomatic and mildly symptomatic familial frontotemporal dementia: NIH-EXAMINER as a potential clinical trial endpoint. Alzheimers Dement 16(1):11–21

21. Sudre CH, Bocchetta M, Heller C, Convery R, Neason M, Moore KM et al (2019) White matter hyperintensities in progranulin-associated frontotemporal dementia: a longitudinal GENFI study. Neuroimage Clin 24:102077

22. Tsai RM, Bejanin A, Lesman-Segev O, LaJoie R, Visani A, Bourakova V et al (2019) $_{18}$F-flortaucipir (AV-1451) tau PET in frontotemporal dementia syndromes. Alzheimers Res Ther 11(1):13

23. van der Ende EL, Meeter LH, Poos JM, Panman JL, Jiskoot LC, Dopper EGP et al (2019 Dec) Serum neurofilament light chain in genetic frontotemporal dementia: a longitudinal, multicentre cohort study. Lancet Neurol 18(12):1103–1111

Fluid Biomarkers of Frontotemporal Lobar Degeneration

Emma L. van der Ende and John C. van Swieten

Introduction

The improved understanding of frontotemporal dementia (FTD) coupled with the emergence of clinical trials has generated much interest in identifying fluid biomarkers that reflect FTD pathophysiology. Generally speaking, a biomarker is a measurable indicator of a normal biological or pathological process. There are currently no FTD-specific fluid biomarkers routinely used in clinical practice. Diagnosing FTD on clinical grounds alone is frequently challenging, especially in the early stages of the disease. A correct and timely diagnosis is needed for appropriate management and support and to exclude treatable causes. At the same time, disease-modifying drugs may be most effective if administered at an early stage, i.e., when neuronal damage is minimal [1]. A biomarker that can identify early disease stages could therefore not only improve clinical management but also have a key position in participant selection for clinical trials. In light of the relative difficulty of quantifying short-term changes in cognitive functioning or atrophy rates, such biomarkers might also be useful as surrogate markers of treatment effect.

Pathologically, FTD is characterized by frontotemporal lobar degeneration (FTLD) with intracellular inclusions, which are most commonly composed of the microtubule-associated protein tau (FTLD-tau) or TAR-DNA-binding protein-43 (TDP-43; FTLD-TDP). Less common forms include FTLD-FUS (inclusions composed of the FUS protein) and FTLD-UPS (ubiquitin-positive inclusions without immunoreactivity for TDP-43 or FUS) [1, 2]. While the underlying neuropathology is known in genetic forms of FTD, with *MAPT* mutations leading to FTLD-tau and *GRN* and *C9orf72* mutations leading to FTLD-TDP, it is not easily predicted in sporadic FTD based on clinical presentation alone [1]. Fluid biomarkers that can help to identify the pathological substrate will be critical to select patients for etiology-directed therapeutic trials.

This chapter explores the current state of fluid biomarkers in sporadic and genetic FTD and discusses challenges in novel fluid biomarker development.

Fluid Biomarker Sources

Cerebrospinal Fluid

Cerebrospinal fluid (CSF) has gathered the most interest as a source of fluid biomarkers in neurodegenerative diseases. Its proximity to the brain and direct connection with the brain inter-

E. L. van der Ende (✉) · J. C. van Swieten (✉)
Erasmus University Medical Center, Department of Neurology and Alzheimer Center,
Rotterdam, The Netherlands
e-mail: e.vanderende@erasmusmc.nl;
j.c.vanswieten@erasmusmc.nl

© Springer Nature Switzerland AG 2021
B. Ghetti et al. (eds.), *Frontotemporal Dementias*, Advances in Experimental Medicine and Biology 1281, https://doi.org/10.1007/978-3-030-51140-1_9

stitial fluid means that it is most likely to contain brain-derived proteins related to neurological disease. CSF can be obtained through a lumbar puncture, a safe procedure with post-lumbar puncture headache being the most significant complication, occurring in approximately 10% of patients [3]. However, lumbar puncture is invasive and inconvenient for monitoring disease progression, and variability in the methods used to collect and store CSF can considerably affect the measurement of certain analytes [4].

Blood

Blood is an attractive alternative to CSF as its collection is minimally invasive and therefore more suitable for repeated collection and disease monitoring. A small fraction of brain proteins that cross the blood–brain barrier can be detected in very low concentrations in the blood [5]. Recent technical developments in the field of ultrasensitive assays and mass spectrometry have greatly improved the detection of these brain-derived proteins [6]. Blood biomarker development poses several challenges, including the possibility that the measured analyte is derived from peripheral tissues instead of the brain, interference with immunoassay platforms by resident blood proteins (albumin, immunoglobulins), and potential degradation or masking of pathological markers through protease or protein carrier activity [5, 7].

Other Biomarker Sources

There is a growing interest in biomarkers in other non-invasively obtained biofluids, including saliva and urine. Several studies have shown that amyloid-β peptides and multiple tau species are detectable in saliva, although results in patients with neurodegenerative diseases are conflicting and require replication [5, 8]. While promising, these biomarkers require considerable work to determine their clinical utility and are not discussed further here.

Amyloid-β and Tau

Background

The CSF biomarkers amyloid-β_{42} (Aβ_{42}), phosphorylated tau$_{181}$ (p-tau$_{181}$), and total tau (t-tau) are increasingly being used in clinical practice to detect Alzheimer's disease (AD) and are thought to directly reflect hallmark pathological changes of AD, namely, cortical amyloid plaques, neurofibrillary tangles, and neuronal loss [9]. Amyloid plaques are extracellular aggregates of Aβ peptides which are formed after sequential cleavage of amyloid precursor protein (APP). While most Aβ peptides are 40 amino acids in length (Aβ_{40}), the larger Aβ_{42} is considered more toxic due to its greater tendency to aggregate and misfold [10, 11]. Neurofibrillary tangles are cytoplasmic aggregates of hyperphosphorylated tau protein and are thought to be neurotoxic [9].

Patients with AD typically have reduced CSF levels of Aβ_{42}, due to cortical amyloid deposition, coupled with increased p-tau$_{181}$ due to tangle formation, and increased total tau, which is attributed to neuronal loss [12–16]. Together, these findings constitute the so-called AD CSF profile, which provides good diagnostic accuracy to identify patients with AD [17] and has recently been incorporated into research diagnostic criteria [12]. Levels of these CSF biomarkers have been shown to correlate with pathological load on post-mortem examination [13, 15, 16, 18]. Of note, these biomarkers can already detect AD pathology in preclinical and prodromal disease stages, and can be used to predict incipient AD in patients with mild cognitive impairment [17, 19].

CSF Amyloid-β and Tau in FTD Diagnosis

In FTD, Aβ_{42} and p-tau$_{181}$ are typically normal and t-tau levels may be normal or elevated, likely due to a release of tau protein following neuronal loss [20–22]. Thus, in the diagnostic workup of FTD, these biomarkers are useful to exclude underlying AD, but cannot confirm or rule out FTD pathology. An elevated ratio of p-tau$_{181}$:Aβ_{42}

or t-tau:$A\beta_{42}$ provides an especially accurate differentiation of FTD from AD (sensitivity 87–89%, specificity 79–80%) [23]. This may be particularly relevant in patients with inconclusive clinical presentations, such as prominent behavioral symptoms, which can be ascribed to behavioral variant FTD or frontal variant AD, or primary progressive aphasia, which can be a feature of either FTD or AD [24]. Correctly identifying patients with underlying AD has become increasingly important with the advent of cholinesterase inhibitors and memantine, which are effective in reducing symptoms in AD but not in FTD and may even worsen FTD symptoms [25, 26].

Remarkably, lower levels of the secreted form of APP (sAPPβ) have been reported in FTD patients compared to both AD patients and cognitively healthy subjects, [27–29] suggesting that APP-derived peptides may be involved in FTD through amyloid-independent mechanisms.

Potential Pitfalls of CSF Amyloid-β and Tau

Importantly, postmortem studies have revealed that FTD patients frequently have some degree of concomitant AD pathology [30]. Especially in patients over the age of 75 years, some degree of AD pathology is common, and up to 30% of cognitively healthy elderly subjects have an AD CSF profile [31, 32]. Therefore, in patients with an AD CSF profile who are clinically suspected of having FTD, the possibility of AD and FTD copathology should be considered.

Furthermore, between-individual variation in overall $A\beta$ production or secretion may cause $A\beta_{42}$ to fall within the normal range despite underlying amyloid pathology; the use of $A\beta_{42/40}$ ratios is thought to provide a more reliable measure [33].

CSF $A\beta$ and tau measurements are sensitive to variations in (pre)analytical conditions. Recommendations for optimal CSF collection include the use of polypropylene collection tubes since $A\beta_{42}$ and other proteins adhere to polystyrene tubes, significantly reducing measured concentrations; similarly, the use of lumbar catheters or manometers should be avoided [4]. Variability also exists between and within commercially available ELISA-based assays, calibration peptides, and platforms, meaning that interlaboratory and interassay consistency is poor, [34] and direct comparisons between laboratories and techniques are not reliable [18]. International efforts are underway to harmonize protocols and assays within and between laboratories [35].

CSF Tau to Differentiate Between Pathological Subtypes of FTD

Two previous studies did not find a difference in CSF p-tau$_{181}$ between FTD patients with or without underlying tau pathology, [36, 37] although one study did reveal an association between the severity of tau pathology and CSF p-tau$_{181}$ levels in FTD patients [38]. The ratio of p-tau$_{181}$ to t-tau is lower in patients with FTLD-TDP than in those with FTLD-tau, [1, 39–42] although this finding may be driven by the presence of concomitant amyotrophic lateral sclerosis (ALS) (leading to more pronounced neuronal loss and thus higher t-tau levels) in some patients with FTLD-TDP. Novel tau fragments to distinguish FTLD-tau from FTLD-TDP pathology have thus far yielded insufficient diagnostic accuracy [18, 43].

Blood Amyloid-β and Tau

There is a growing interest in the measurement of $A\beta$ and tau species in blood as an alternative to CSF. Although previous results were conflicting, recent studies using ultrasensitive analytical assays have demonstrated decreased levels of blood $A\beta_{42}$ in patients with AD. Blood and CSF $A\beta_{42}$ levels are correlated and blood $A\beta_{42}$ appears to reflect AD-associated pathology with a fair diagnostic accuracy [44, 45]. Similarly, plasma tau levels are increased in AD patients compared to controls, although not as clearly as in CSF, hampering its diagnostic use [44, 45]. These results are promising and warrant further research in larger cohorts.

Neurofilament Proteins

Neurofilament proteins (Nfs) are rapidly emerging as the most promising fluid biomarkers for FTD [46]. The discovery that an elevation of neurofilament light chain (NfL), which is thought to reflect neuroaxonal damage, can be measured reliably both in CSF and blood has created much interest in NfL as an easily accessible biomarker across a spectrum of neurological diseases [47, 48].

Background

Nfs are cylindrical heteropolymers located exclusively in the neuronal cytoplasm and are the dominant protein of the axonal cytoskeleton. Nfs consist of three subunits, classified according to molecular weight: neurofilament light chain (NfL), medium chain (NfM), and heavy chain (NfH). Nfs are thought to be critical for stability and radial growth of axons, thereby modulating nerve conduction velocity [49]. Under normal circumstances, Nfs are stable within axons and have a low turnover rate. Upon damage to the axon, Nf molecules are released into the extracellular space, where they traffic to the CSF and, after passing through the blood–brain barrier, enter the bloodstream [49, 50]. The release of Nfs occurs irrespective of the etiology of the neuroaxonal injury; therefore, elevated levels are seen in CSF and blood in patients with various neurological disorders, including dementia, stroke, traumatic brain injury, multiple sclerosis, and Parkinson's disease [47, 48]. Due to its relative abundance and solubility, NfL can be more readily quantified in biofluids than NfH and NfM [49]. Blood and CSF NfL levels are highly correlated [51–53] and advances in ultrasensitive assays (single molecule array, Simoa) have greatly improved the accuracy of blood NfL measurements [54].

Neurofilament Light Chain in FTD

A large number of studies have consistently reported strongly elevated NfL levels in CSF and blood of FTD patients, with a diagnostic accuracy of over 90% to distinguish FTD patients from healthy individuals [55, 56]. These NfL increases occur in all FTD phenotypes, [28, 57–63] with especially high levels in patients with concomitant ALS [52, 60, 64, 65]. Although higher levels have been reported in patients with FTLD-TDP compared to those with FTLD-tau, this difference may be driven by patients with concomitant ALS in the FTLD-TDP group [41, 58, 65, 66].

NfL levels are significantly higher in FTD than in other frequent causes of dementia, including AD, vascular dementia, and dementia with Lewy bodies [55, 62, 64, 67, 68]. The pronounced NfL elevations seen in FTD may be related to the anatomical location of neurodegeneration or due to a higher rate of neuronal death, since especially high NfL levels are also seen in the rapidly progressive Creutzfeldt-Jakob disease [66, 69]. However, considerable overlap in NfL levels exists between dementias, and the discriminatory power of NfL on its own to distinguish FTD from disease mimics is only modest [55]. A promising strategy may be to combine NfL with other fluid biomarkers; for instance, the addition of NfL to core AD CSF biomarkers significantly improves the discrimination between AD and FTD compared to AD CSF biomarkers alone [66, 67].

NfL appears to be a useful diagnostic biomarker to distinguish FTD from non-neurological disorders, including primary psychiatric disorders, in which NfL levels are typically normal [70–73]. Blood NfL may provide an easily accessible, inexpensive screening tool to identify patients who are likely to have an underlying neurological disease and require further investigation.

Importantly, patients with high NfL levels have more brain atrophy, more functional and cognitive impairment, faster disease progression, and shorter overall survival than patients with low NfL levels, [52, 61, 62, 65, 66, 68, 74, 75] demonstrating the value of NfL as a prognostic biomarker. NfL may therefore be a useful tool to inform patients and caregivers about the expected clinical course, and to distinguish patients with clinical hallmarks of FTD who are likely to further decline from those with non-progressing variants (benign or phenocopy FTD syndrome) [76].

Presymptomatic carriers of mutations in *GRN*, *C9orf72*, and *MAPT* typically have low NfL levels, [52, 60, 65] indicating a low rate of axonal turnover, with sharp increases observed at least 1–2 years prior to symptom onset in one longitudinal study [65]. These increases likely reflect early axonal damage in a prodromal disease stage and suggest that NfL could be a valuable selection tool in clinical trials to identify mutation carriers approaching symptom onset. Another promising application of NfL is as a surrogate marker of treatment effect in clinical trials. In multiple sclerosis, NfL decreases have been observed after treatment with anti-inflammatory drugs, [77] and in AD mouse models, NfL decreased after inhibition of amyloid-β production, [78] suggesting that NfL is a dynamic marker of disease activity. As blood NfL levels are generally stable over the course of FTD, [62, 65] a decrease in blood NfL during an FTD trial might reflect a reduced rate of neuroaxonal breakdown in response to the study drug.

CSF and blood NfL levels increase with age among healthy adults, possibly reflecting slow, ongoing axonal turnover as part of physiological aging [47, 48]. This necessitates the development of age-adjusted normal values before NfL can be used in clinical practice. International efforts are underway to harmonize preanalytical and analytical parameters and to develop universal reference values, which will allow a reliable comparison of results from different laboratories [47].

TDP-43

Background

Aggregation of TDP-43 is a hallmark pathological feature of most tau-negative cases of FTD as well as almost all cases of ALS [79–83]. Under normal circumstances, TDP-43 is predominantly localized in the nucleus, [84] where it functions as a transcription factor and regulates important cellular functions such as splicing activity and mRNA stability [85]. In disease, pathological TDP-43 isoforms (phosphorylated and ubiquitinated full-length TDP-43 as well as C-terminal TDP-43 fragments) are redistributed to the cytoplasm, where they form aggregates which are thought to be toxic [79, 82].

Biomarkers of TDP-43 Pathology

Underlying TDP-43 pathology can be predicted in patients with a mutation in *GRN* or *C9orf72*, but not in patients with sporadic FTD [2]. Etiology-specific treatment trials will require biomarkers that can detect TDP-43 pathology during the patients' lifetime to select suitable patients. Thus far, efforts to measure disease-specific forms of TDP-43 in biofluids of ALS and FTD patients have yielded inconsistent results.

TDP-43 antibodies used to date have the ability to detect full-length TDP-43 as well as phosphorylated full-length TDP-43 and longer C-terminal fragments, but not the shorter C-terminal fragments, which are abundant in brain tissue of ALS and FTD patients [86]. Elevated levels of full-length TDP-43 in CSF of patients with ALS or FTD have been reported, [87–89] albeit with considerable overlap between groups, while another study reported decreased CSF TDP-43 in FTD patients [42]. Phosphorylated TDP-43 in CSF was not different in FTD or ALS compared to controls [42, 90]. One small study reported elevated plasma phosphorylated TDP-43 levels in *C9orf72*- or *GRN*-associated FTD; [90] this finding requires replication in larger cohorts.

Accurate quantification of TDP-43 in CSF or blood is challenging for several reasons. TDP-43 is a ubiquitously expressed protein and is abundant under normal circumstances. The majority of CSF TDP-43 appears to be blood-derived and not brain-derived, although it may be possible to enrich for brain-specific fractions of TDP-43 from exosomes in CSF [91]. Monoclonal antibodies which selectively recognize pathological forms of TDP-43, such as short C-terminal TDP-43 fragments, will therefore be critical [86]. Furthermore, quantification of TDP-43 and especially its phosphorylated form appears limited by very low concentrations or low binding affinity of the antibodies in the presence of abundant immunoglobins or albumin, [92, 93] highlighting the need for technical improvements in assays [83].

Markers of Inflammation and Astrogliosis

Background

Increasing clinical, genetic, and cellular evidence suggests that chronic neuroinflammation plays an important role in FTD. Key observations include microglial and astrocytic activation in the frontal and temporal cortices in both postmortem brain tissue and positron-emission tomography (PET), a shared genetic risk between FTD and autoimmune diseases, and a direct link between several FTD-related genes and inflammatory pathways [94–102]. While the exact contribution and timing of neuroinflammation in FTD remains controversial, different disease stages may be characterized by different immune mechanisms, making neuroinflammation an interesting source for potential fluid biomarkers.

Glial Markers

Microglia, the resident macrophages of the central nervous system (CNS), likely play a central role in neuroinflammation. Resting microglia are involved in homeostasis, and can become activated upon exposure to pathogens or inflammatory stimuli to produce a range of signal molecules, including cytokines, chemokines, and complement molecules, which ultimately result in a pro- or an anti-inflammatory CNS microenvironment [102–105]. Similarly, astrocytes are believed to modulate neuroinflammation [102]. The upregulation of microglia and astrocytes in FTD brains has generated interest in biomarkers that can track glial activation in vivo. Candidate glial biomarkers include YKL-40, chitotriosidase-1 (CHIT-1), and glial fibrillary acidic protein (GFAP).

YKL-40, also known as chitinase 3-like protein 1, is produced primarily by reactive astrocytes but also microglia [106]. CSF YKL-40 is elevated in FTD as well as several other dementias, with especially high levels in aggressive and rapidly progressive dementias [27, 28, 107–113]. Although YKL-40 is thought to be a nonspecific biomarker of glial activation, a positive association has been found with tau deposits, suggesting that YKL-40 upregulation may be particularly sensitive to tau aggregation [27].

CHIT-1 is an enzyme which is expressed and secreted by activated microglia. A recent study in 72 FTD patients reported elevated CHIT-1 levels compared to controls, [114] although a previous smaller study did not find these elevations [115]. Importantly, CHIT-1 concentration may be reduced in subjects carrying a *CHIT1* polymorphism common in European populations, complicating its use as a biomarker [115, 116].

GFAP, a cytoskeletal filament protein in astrocytes, is produced and released by astrocytes during neurodegeneration [117]. High levels of GFAP in blood have been found in several neurodegenerative diseases, with remarkably high levels in FTD, [118, 119] suggesting that astrogliosis may be an especially prominent feature of FTD.

The microglial transmembrane receptor TREM2 appears to play a role in microglial homeostatic pathways, [120] and has been investigated as a candidate biomarker for neurodegeneration since genetic variants of *TREM2* are associated with an increased risk of FTD, AD, ALS, and Parkinson's disease [121–126]. Its ectodomain can be released into the extracellular space as a soluble protein (sTREM2), which is measurable in CSF and blood [127]. While small studies reported conflicting results in sTREM2 levels [128–130], a more recent larger study observed no differences between FTD patients versus controls except in a small number of *GRN* mutation carriers [131].

While these glial markers provide further evidence for aberrant microglial and astrocytic activity in FTD, their diagnostic potential is limited, as considerable overlap exists with controls. Similarly elevated levels have been reported in several other neurodegenerative diseases, likely reflecting a shared, nonspecific glial activation. Furthermore, YKL-40, CHIT-1, and TREM2 are expressed by multiple peripheral cell types, which could affect their measurement [107, 120, 132]. GFAP, on the other hand, is brain-specific and therefore may be a more interesting candidate [117].

Cytokine Markers

There is extensive, although somewhat conflicting, evidence for altered pro- and anti-inflammatory cytokine profiles in CSF and blood [102]. For instance, increased levels of blood interleukin-6 (IL-6) were found in *GRN* mutation carriers [133] and sporadic FTD, [134] whereas another study showed no differences in CSF IL-6 in FTD versus controls [135]. Tumor necrosis factor α (TNF-α) was increased in CSF of patients with sporadic FTD, [136] while another study showed a reduction in 10 *GRN* mutation carriers [137]. More consistently elevated levels have been found for monocyte chemoattractant protein 1(MCP-1) [137–139].

These results are mostly derived from small studies and must be interpreted with caution. Peripheral cytokine measurements may be influenced by concurrent infections or other inflammatory conditions outside of the brain. Furthermore, cytokine profiles likely vary depending on the disease stage. Finally, the interpretation of increased or decreased cytokine levels is complex: the original classification of pro-inflammatory and anti-inflammatory cytokines is likely too simplistic, as a given cytokine may behave as either pro- or anti-inflammatory depending on the circumstances [140].

Other Candidate Biomarkers of Neuroinflammation

One study of patients with *GRN*-associated FTD has demonstrated increased CSF levels of the complement proteins C1q and C3b [141]. C1q and C3b are essential components of the classical and alternative complement pathways, which comprise a sequence of protein cleavages and eventually contribute to a pro-inflammatory state. Mouse models have suggested a role for complement proteins in the breakdown of synapses, an early neurodegenerative process, underlining the need for replication of complement protein measurements in CSF and blood [141].

Gene-Specific Biomarkers

Progranulin in GRN Mutation Carriers

Background

Progranulin (PGRN) is a ubiquitously expressed growth factor, which plays important roles in normal tissue development, cell proliferation and regeneration, and inflammation. In the brain, PGRN is expressed in neurons and microglia and promotes neurite outgrowth, neuronal survival, and differentiation, although its exact physiological roles in the nervous system are not fully understood [142]. PGRN also appears to suppress neuroinflammation and modulate neuronal lysosome function, with homozygous mutations in *GRN*, the gene encoding PGRN, leading to the lysosomal storage disease neuronal ceroid lipofuscinosis [143]. PGRN can be cleaved into granulins, which are also biologically active, but often with opposing actions, suggesting that the equilibrium between PGRN and granulins is important for tissue homeostasis [142].

Progranulin in FTD

Heterozygous mutations in *GRN* are among the most common causes of genetic FTD [144–146]. Most pathogenic *GRN* mutations introduce a premature stop codon that triggers nonsense-mediated decay of *GRN* mRNA, leading to a 50% reduction of PGRN protein levels through haploinsufficiency [144, 145]. This reduction in PGRN levels can be detected through immunoassays both in the CSF and blood, enabling accurate recognition of FTD patients due to a *GRN* mutation versus those with sporadic FTD (sensitivity 96%, specificity up to 100%) [147 –154]. PGRN levels are already decreased in the presymptomatic stage, even in the second or third decade of life, indicating that dysregulated PGRN expression is a very early event during the lifespan of mutation carriers. Blood PGRN levels can therefore also distinguish presymptomatic *GRN* mutation carriers from noncarriers with near-perfect sensitivity and specificity [150, 155]. Its concentration does not reflect the extent of neurodegeneration and is therefore not useful as a prognostic or disease staging biomarker [150].

Blood PGRN measurements may be helpful to determine the pathogenicity of novel *GRN* mutations or to detect mutations not found on standard genetic screening, such as large deletions [150]. Since genetic testing is expensive and time-consuming, blood PGRN determination offers a low-cost, minimally invasive screening tool to identify *GRN* mutation carriers, who should then be subjected to further genetic testing. The ability to screen FTD patients for *GRN* mutations on a large scale is particularly important in light of therapeutic trials aimed at increasing PGRN protein levels [150].

Since blood PGRN appears to be stable over time, [150, 155] PGRN levels can be used to monitor whether a trial drug is effective in increasing PGRN levels. It is important to note that CSF and blood PGRN are only moderately correlated, implying a differential regulation of the two [155–157]. Peripheral PGRN levels may not adequately reflect those in the CNS, and the absence of an effect on blood PGRN does not rule out an effect on the brain or PGRN function. Furthermore, the extracellular PGRN levels measured in biofluids might not sufficiently reflect intracellular levels, and it is not clear yet where the loss of PGRN has its main effect [142].

Much variability in PGRN levels exists within individuals, and various genetic and environmental regulators influence PGRN levels. For example, PGRN levels are elevated in inflammation and other clinical conditions including cancer and pregnancy, [158, 159] and certain single nucleotide polymorphisms (SNPs) are associated with increased or decreased PGRN levels, including rs5848 (*GRN*), rs646776 (*SORT1*), and rs1990622 (*TMEM106B*) [155, 156, 160–162]. Further research is needed to elucidate other factors that may confound PGRN measurements. Given the distinct biological properties of granulin peptides, developing antibodies against granulins to study ratios of PGRN to granulins as potential biomarkers could be insightful [149, 142].

Dipeptide Repeat Proteins in C9orf72 Mutation Carriers

Background

The *C9orf72* repeat expansion is the most frequent genetic cause of FTD and ALS [163–168]. While the exact mechanism by which *C9orf72* repeat expansions lead to neurodegeneration is unknown, it has been proposed that toxic dipeptide repeat proteins (DPRs) could play a role [169]. The expanded *C9orf72* repeats are bidirectionally transcribed into repetitive RNA, which forms sense and antisense RNA foci. These RNAs can be translated in every reading frame through repeat-associated non-ATG-initiated translation (RAN translation), generating five DPRs, in order of abundance: poly(GA), poly(GP), poly(GR), poly(PA), and poly(PR) [170]. DPRs are found abundantly in brains of *C9orf72*-FTD/ALS patients, [170–173] mostly in cytoplasmic neuronal inclusions, although DPR burden does not coincide neuropathologically with the degree of neurodegeneration [174–177]. Cell and animal models have shown that poly(GR) and poly(PR), and to a lesser extent poly(GA), are toxic when overexpressed, while poly(PA) and poly(GP) are unlikely to be toxic [169].

Dipeptide Repeat Proteins in CSF

Poly(GP) can be quantitatively detected by ELISAs in CSF, [178] revealing high levels in patients with ALS or FTD due to *C9orf72* repeat expansions. In sporadic cases, on the other hand, poly(GP) levels are generally very low or undetectable, [178, 179] although one study reported high poly(GP) levels in a small number of patients without the repeat expansion [180]. One possible explanation could be somatic mosaicism, where a pathological repeat is present in the CNS but not in peripheral blood, preventing the detection of *C9orf72* repeat expansions in peripheral blood. It has been shown in mice that CSF poly(GP) levels correlate with DPR protein pathology, repeat RNA levels, and RNA foci burden [179].

CSF poly(GP) elevations are already observed in presymptomatic *C9orf72* mutation carriers, [180–182] suggesting that DPR protein production emerges prior to neurodegeneration. This is in agreement with the neuropathological detection of DPRs in young presymptomatic *C9orf72* cases [183–185]. CSF poly(GP) levels do not correlate with the severity of neurodegeneration, disease progression, or other clinical characteristics such as the age of disease onset [179–182], limiting the value of poly(GP) as a disease staging or prognostic biomarker.

Since the RNA transcripts of expanded *C9orf72* repeats are believed to play a key role in *C9orf72* pathogenesis, interventions targeting transcription and translation of the repeat expansion are a promising therapeutic strategy. Poly(GP) levels appear to be stable over time, [179] and therefore measurement of poly(GP) before and during treatment presents a feasible approach to measure target engagement. Antisense oligonucleotides (ASOs) targeting repeat RNAs have been shown in mice to reduce CSF poly(GP) levels [179].

Importantly, poly(GP) can be detected in peripheral blood mononuclear cells (PBMCs); further research is needed to determine its potential as a blood-based biomarker [179].

Most research to date has focused on measuring poly(GP) due to its high abundance and solubility, making it the most likely DPR to be accurately measured [169]. Measurements of poly(GA) and poly(GR) might uncover associations with clinical features not observed for poly(GP). To date, efforts to measure poly(GA) and poly(GR) in CSF have been unsuccessful, possibly because the currently used assays are not sensitive enough to detect very low concentrations of these DPRs [186].

Concluding Remarks and Future Directions

Recent years have seen great advances in identifying both general biomarkers of neurodegeneration, such as NfL, as well as gene-specific biomarkers, including PGRN and DPR proteins.

There remains an unmet need for biomarkers that specifically reflect FTD pathophysiology and, especially with the advent of clinical trials, biomarkers that can predict the underlying neuropathological substrate in sporadic FTD.

The heterogeneity of FTD complicates biomarker development, and the use of a clinical diagnosis as a reference standard is a potential source of heterogeneity given the high false-positive rate of FTD clinical diagnosis [30]. Although novel biomarkers would ideally be validated in postmortem studies, studying genetic forms of FTD, in which the underlying pathological substrate can be accurately predicted during life, provides a valuable alternative. A combination of analytes that reflect different biological processes is likely to yield more information than single biomarkers. Longitudinal studies are needed to determine at what stage during the disease various biomarkers start to become abnormal.

The use of proteomics is a promising strategy to detect differentially regulated proteins in biofluids, although in-depth validation of mass spectrometry results is needed to overcome differences in technical parameters [187]. Candidate proteins that have been identified in multiple independent CSF proteomics studies are likely the most promising and include synaptic proteins, such as neuronal pentraxins and VGF, as well as numerous inflammation-related proteins [187–193].

A crucial step before a biomarker can be implemented in clinical practice is multicenter standardization and harmonization of preanalytical and assay characteristics, as is currently being done for core AD biomarkers [35]. Developing normal reference values and cutpoints is essential and needs to take into account age-related changes in biomarker levels, such as is the case for NfL and several inflammation-related biomarkers [102]. Many of the biomarkers discussed in this chapter are not FTD-specific (e.g., NfL) or brain-specific (e.g., markers of neuroinflammation, TDP-43), and a thorough understanding of potential confounding factors is needed before these biomarkers can be relied upon in a clinical setting.

References

1. Meeter LH, Kaat LD, Rohrer JD, van Swieten JC (2017) Imaging and fluid biomarkers in frontotemporal dementia. Nat Rev Neurol 13:406–419
2. Lashley T, Rohrer JD, Mead S, Revesz T (2015) Review: an update on clinical, genetic and pathological aspects of frontotemporal lobar degenerations. Neuropathol Appl Neurobiol 41:858–881
3. Duits FH, Martinez-Lage P, Paquet C, Engelborghs S, Lleo A, Hausner L et al (2016) Performance and complications of lumbar puncture in memory clinics: results of the multicenter lumbar puncture feasibility study. Alzheimers Dement 12:154–163
4. Ahmed RM, Paterson RW, Warren JD, Zetterberg H, O'Brien JT, Fox NC et al (2014) Biomarkers in dementia: clinical utility and new directions. J Neurol Neurosurg Psychiatry 85:1426–1434
5. Sancesario G, Bernardini S (2019) AD biomarker discovery in CSF and in alternative matrices. Clin Biochem 72:52–57
6. Duffy D (2020) Short keynote paper: single molecule detection of protein biomarkers to define the continuum from health to disease. IEEE J Biomed Health Inform 24:1864–1868
7. Hampel H, O'Bryant SE, Molinuevo JL, Zetterberg H, Masters CL, Lista S et al (2018) Blood-based biomarkers for Alzheimer disease: mapping the road to the clinic. Nat Rev Neurol 14:639–652
8. Ashton NJ, Ide M, Zetterberg H, Blennow K (2019) Salivary biomarkers for Alzheimer's disease and related disorders. Neurol Ther 8:83–94
9. Blennow K, de Leon MJ, Zetterberg H (2006) Alzheimer's disease. Lancet 368:387–403
10. McGowan E, Pickford F, Kim J, Onstead L, Eriksen J, Yu C et al (2005) Abeta42 is essential for parenchymal and vascular amyloid deposition in mice. Neuron 47:191–199
11. Portelius E, Westman-Brinkmalm A, Zetterberg H, Blennow K (2006) Determination of beta-amyloid peptide signatures in cerebrospinal fluid using immunoprecipitation-mass spectrometry. J Proteome Res 5:1010–1016
12. Jack CR Jr, Bennett DA, Blennow K, Carrillo MC, Dunn B, Haeberlein SB et al (2018) NIA-AA research framework: toward a biological definition of Alzheimer's disease. Alzheimers Dement 14:535–562
13. Buerger K, Ewers M, Pirttila T, Zinkowski R, Alafuzoff I, Teipel SJ et al (2006) CSF phosphorylated tau protein correlates with neocortical neurofibrillary pathology in Alzheimer's disease. Brain 129:3035–3041
14. de Souza LC, Chupin M, Lamari F, Jardel C, Leclercq D, Colliot O et al (2012) CSF tau markers are correlated with hippocampal volume in Alzheimer's disease. Neurobiol Aging 33:1253–1257
15. Seppala TT, Nerg O, Koivisto AM, Rummukainen J, Puli L, Zetterberg H et al (2012) CSF biomarkers for Alzheimer disease correlate with cortical brain biopsy findings. Neurology 78:1568–1575
16. Tapiola T, Alafuzoff I, Herukka SK, Parkkinen L, Hartikainen P, Soininen H et al (2009) Cerebrospinal fluid {beta}-amyloid 42 and tau proteins as biomarkers of Alzheimer-type pathologic changes in the brain. Arch Neurol 66:382–389
17. Olsson B, Lautner R, Andreasson U, Ohrfelt A, Portelius E, Bjerke M et al (2016) CSF and blood biomarkers for the diagnosis of Alzheimer's disease: a systematic review and meta-analysis. Lancet Neurol 15:673–684
18. Bjerke M, Engelborghs S (2018) Cerebrospinal fluid biomarkers for early and differential Alzheimer's disease diagnosis. J Alzheimers Dis 62:1199–1209
19. Mattsson N, Zetterberg H, Hansson O, Andreasen N, Parnetti L, Jonsson M et al (2009) CSF biomarkers and incipient Alzheimer disease in patients with mild cognitive impairment. JAMA 302:385–393
20. Riemenschneider M, Wagenpfeil S, Diehl J, Lautenschlager N, Theml T, Heldmann B et al (2002) Tau and Abeta42 protein in CSF of patients with frontotemporal degeneration. Neurology 58:1622–1628
21. Schoonenboom NS, Reesink FE, Verwey NA, Kester MI, Teunissen CE, van de Ven PM et al (2012) Cerebrospinal fluid markers for differential dementia diagnosis in a large memory clinic cohort. Neurology 78:47–54
22. van Harten AC, Kester MI, Visser PJ, Blankenstein MA, Pijnenburg YA, van der Flier WM et al (2011) Tau and p-tau as CSF biomarkers in dementia: a meta-analysis. Clin Chem Lab Med 49:353–366
23. Rivero-Santana A, Ferreira D, Perestelo-Perez L, Westman E, Wahlund LO, Sarria A et al (2017) Cerebrospinal fluid biomarkers for the differential diagnosis between Alzheimer's disease and frontotemporal lobar degeneration: systematic review, HSROC analysis, and confounding factors. J Alzheimers Dis 55:625–644
24. Marshall CR, Hardy CJD, Volkmer A, Russell LL, Bond RL, Fletcher PD et al (2018) Primary progressive aphasia: a clinical approach. J Neurol 265:1474–1490
25. McShane R, Westby MJ, Roberts E, Minakaran N, Schneider L, Farrimond LE et al (2019) Memantine for dementia. Cochrane Database Syst Rev 3:CD003154
26. Noufi P, Khoury R, Jeyakumar S, Grossberg GT (2019) Use of cholinesterase inhibitors in non-Alzheimer's dementias. Drugs Aging 36:719–731
27. Alcolea D, Irwin DJ, Illan-Gala I, Munoz L, Clarimon J, McMillan CT et al (2019) Elevated YKL-40 and low sAPPbeta: YKL-40 ratio in antemortem cerebrospinal fluid of patients with pathologically confirmed FTLD. J Neurol Neurosurg Psychiatry 90:180–186
28. Alcolea D, Vilaplana E, Suarez-Calvet M, Illan-Gala I, Blesa R, Clarimon J et al (2017) CSF sAPPbeta,

YKL-40, and neurofilament light in frontotemporal lobar degeneration. Neurology 89:178–188

29. Illan-Gala I, Pegueroles J, Montal V, Alcolea D, Vilaplana E, Bejanin A et al (2019) APP-derived peptides reflect neurodegeneration in frontotemporal dementia. Ann Clin Transl Neurol 6:2518–2530

30. Irwin DJ, Trojanowski JQ, Grossman M (2013) Cerebrospinal fluid biomarkers for differentiation of frontotemporal lobar degeneration from Alzheimer's disease. Front Aging Neurosci 5:6

31. Bouwman FH, Schoonenboom NS, Verwey NA, van Elk EJ, Kok A, Blankenstein MA et al (2009) CSF biomarker levels in early and late onset Alzheimer's disease. Neurobiol Aging 30:1895–1901

32. Mattsson N, Rosen E, Hansson O, Andreasen N, Parnetti L, Jonsson M et al (2012) Age and diagnostic performance of Alzheimer disease CSF biomarkers. Neurology 78:468–476

33. Pouclet-Courtemanche H, Nguyen TB, Skrobala E, Boutoleau-Bretonniere C, Pasquier F, Bouaziz-Amar E et al (2019) Frontotemporal dementia is the leading cause of "true" a−/T+ profiles defined with Abeta42/40 ratio. Alzheimers Dement (Amst) 11:161–169

34. Verwey NA, van der Flier WM, Blennow K, Clark C, Sokolow S, De Deyn PP et al (2009) A worldwide multicentre comparison of assays for cerebrospinal fluid biomarkers in Alzheimer's disease. Ann Clin Biochem 46:235–240

35. Reijs BL, Teunissen CE, Goncharenko N, Betsou F, Blennow K, Baldeiras I et al (2015) The central biobank and virtual biobank of BIOMARKAPD: a resource for studies on neurodegenerative diseases. Front Neurol 6:216

36. Bian H, Van Swieten JC, Leight S, Massimo L, Wood E, Forman M et al (2008) CSF biomarkers in frontotemporal lobar degeneration with known pathology. Neurology 70:1827–1835

37. Rosso SM, van Herpen E, Pijnenburg YA, Schoonenboom NS, Scheltens P, Heutink P et al (2003) Total tau and phosphorylated tau 181 levels in the cerebrospinal fluid of patients with frontotemporal dementia due to P301L and G272V tau mutations. Arch Neurol 60:1209–1213

38. Irwin DJ, Lleo A, Xie SX, McMillan CT, Wolk DA, Lee EB et al (2017) Ante mortem cerebrospinal fluid tau levels correlate with postmortem tau pathology in frontotemporal lobar degeneration. Ann Neurol 82:247–258

39. Hu WT, Watts K, Grossman M, Glass J, Lah JJ, Hales C et al (2013) Reduced CSF p-Tau181 to tau ratio is a biomarker for FTLD-TDP. Neurology 81:1945–1952

40. Borroni B, Benussi A, Archetti S, Galimberti D, Parnetti L, Nacmias B et al (2015) Csf p-tau181/ tau ratio as biomarker for TDP pathology in frontotemporal dementia. Amyotroph Lateral Scler Frontotemporal Degener. 16:86–91

41. Pijnenburg YA, Verwey NA, van der Flier WM, Scheltens P, Teunissen CE (2015) Discriminative and prognostic potential of cerebrospinal fluid phosphoTau/tau ratio and neurofilaments for frontotemporal dementia subtypes. Alzheimers Dement (Amst). 1:505–512

42. Kuiperij HB, Versleijen AA, Beenes M, Verwey NA, Benussi L, Paterlini A et al (2017) Tau rather than TDP-43 proteins are potential cerebrospinal fluid biomarkers for frontotemporal lobar degeneration subtypes: a pilot study. J Alzheimers Dis 55:585–595

43. Foiani MS, Cicognola C, Ermann N, Woollacott IOC, Heller C, Heslegrave AJ et al (2019) Searching for novel cerebrospinal fluid biomarkers of tau pathology in frontotemporal dementia: an elusive quest. J Neurol Neurosurg Psychiatry 90:740–746

44. Molinuevo JL, Ayton S, Batrla R, Bednar MM, Bittner T, Cummings J et al (2018) Current state of Alzheimer's fluid biomarkers. Acta Neuropathol 136:821–853

45. Zetterberg H, Burnham SC (2019) Blood-based molecular biomarkers for Alzheimer's disease. Mol Brain 12:26

46. Zetterberg H, van Swieten JC, Boxer AL, Rohrer JD (2019) Review: fluid biomarkers for frontotemporal dementias. Neuropathol Appl Neurobiol 45:81–87

47. Gaetani L, Blennow K, Calabresi P, Di Filippo M, Parnetti L, Zetterberg H (2019) Neurofilament light chain as a biomarker in neurological disorders. J Neurol Neurosurg Psychiatry 90:870–881

48. Khalil M, Teunissen CE, Otto M, Piehl F, Sormani MP, Gattringer T et al (2018) Neurofilaments as biomarkers in neurological disorders. Nat Rev Neurol 14:577–589

49. Petzold A (2005) Neurofilament phosphoforms: surrogate markers for axonal injury, degeneration and loss. J Neurol Sci 233:183–198

50. Rosengren LE, Karlsson JE, Sjogren M, Blennow K, Wallin A (1999) Neurofilament protein levels in CSF are increased in dementia. Neurology 52:1090–1093

51. Gisslen M, Price RW, Andreasson U, Norgren N, Nilsson S, Hagberg L et al (2016) Plasma concentration of the Neurofilament light protein (NFL) is a biomarker of CNS injury in HIV infection: a cross-sectional study. EBioMedicine 3:135–140

52. Meeter LH, Dopper EG, Jiskoot LC, Sanchez-Valle R, Graff C, Benussi L et al (2016) Neurofilament light chain: a biomarker for genetic frontotemporal dementia. Ann Clin Transl Neurol 3:623–636

53. Wilke C, Preische O, Deuschle C, Roeben B, Apel A, Barro C et al (2016) Neurofilament light chain in FTD is elevated not only in cerebrospinal fluid, but also in serum. J Neurol Neurosurg Psychiatry 87:1270–1272

54. Kuhle J, Barro C, Andreasson U, Derfuss T, Lindberg R, Sandelius A et al (2016) Comparison of three analytical platforms for quantification of the neurofilament light chain in blood samples: ELISA, electrochemiluminescence immunoassay and Simoa. Clin Chem Lab Med 54:1655–1661

55. Forgrave LM, Ma M, Best JR, DeMarco ML (2019) The diagnostic performance of neurofilament light

chain in CSF and blood for Alzheimer's disease, frontotemporal dementia, and amyotrophic lateral sclerosis: a systematic review and meta-analysis. Alzheimers Dement (Amst) 11:730–743

56. Zhao Y, Xin Y, Meng S, He Z, Hu W (2019) Neurofilament light chain protein in neurodegenerative dementia: a systematic review and network meta-analysis. Neurosci Biobehav Rev 102:123–138

57. Paterson RW, Slattery CF, Poole T, Nicholas JM, Magdalinou NK, Toombs J et al (2018) Cerebrospinal fluid in the differential diagnosis of Alzheimer's disease: clinical utility of an extended panel of biomarkers in a specialist cognitive clinic. Alzheimers Res Ther 10:32

58. Landqvist Waldo M, Frizell Santillo A, Passant U, Zetterberg H, Rosengren L, Nilsson C et al (2013) Cerebrospinal fluid neurofilament light chain protein levels in subtypes of frontotemporal dementia. BMC Neurol 13:54

59. Ljubenkov PA, Staffaroni AM, Rojas JC, Allen IE, Wang P, Heuer H et al (2018) Cerebrospinal fluid biomarkers predict frontotemporal dementia trajectory. Ann Clin Transl Neurol 5:1250–1263

60. Rohrer JD, Woollacott IO, Dick KM, Brotherhood E, Gordon E, Fellows A et al (2016) Serum neurofilament light chain protein is a measure of disease intensity in frontotemporal dementia. Neurology 87:1329–1336

61. Scherling CS, Hall T, Berisha F, Klepac K, Karydas A, Coppola G et al (2014) Cerebrospinal fluid neurofilament concentration reflects disease severity in frontotemporal degeneration. Ann Neurol 75:116–126

62. Steinacker P, Anderl-Straub S, Diehl-Schmid J, Semler E, Uttner I, von Arnim CAF et al (2018) Serum neurofilament light chain in behavioral variant frontotemporal dementia. Neurology 91:e1390–e1401

63. Steinacker P, Semler E, Anderl-Straub S, Diehl-Schmid J, Schroeter ML, Uttner I et al (2017) Neurofilament as a blood marker for diagnosis and monitoring of primary progressive aphasias. Neurology 88:961–969

64. Bridel C, van Wieringen WN, Zetterberg H, Tijms BM, Teunissen CE, and the NFLG, et al (2019) Diagnostic value of cerebrospinal fluid Neurofilament light protein in neurology: a systematic review and meta-analysis. JAMA Neurol 76:1035–1048

65. van der Ende EL, Meeter LH, Poos JM, Panman JL, Jiskoot LC, Dopper EGP et al (2019) Serum neurofilament light chain in genetic frontotemporal dementia: a longitudinal, multicentre cohort study. Lancet Neurol 18:1103–1111

66. Abu-Rumeileh S, Mometto N, Bartoletti-Stella A, Polischi B, Oppi F, Poda R et al (2018) Cerebrospinal fluid biomarkers in patients with frontotemporal dementia Spectrum: a single-center study. J Alzheimers Dis 66:551–563

67. de Jong D, Jansen RW, Pijnenburg YA, van Geel WJ, Borm GF, Kremer HP et al (2007) CSF neurofilament proteins in the differential diagnosis of dementia. J Neurol Neurosurg Psychiatry 78:936–938

68. Skillback T, Farahmand B, Bartlett JW, Rosen C, Mattsson N, Nagga K et al (2014) CSF neurofilament light differs in neurodegenerative diseases and predicts severity and survival. Neurology 83:1945–1953

69. Zerr I, Schmitz M, Karch A, Villar-Pique A, Kanata E, Golanska E et al (2018) Cerebrospinal fluid neurofilament light levels in neurodegenerative dementia: evaluation of diagnostic accuracy in the differential diagnosis of prion diseases. Alzheimers Dement 14:751–763

70. Al Shweiki MR, Steinacker P, Oeckl P, Hengerer B, Danek A, Fassbender K et al (2019) Neurofilament light chain as a blood biomarker to differentiate psychiatric disorders from behavioural variant frontotemporal dementia. J Psychiatr Res 113:137–140

71. Eratne D, Loi SM, Walia N, Farrand S, Li QX, Varghese S et al (2020) A pilot study of the utility of cerebrospinal fluid neurofilament light chain in differentiating neurodegenerative from psychiatric disorders: a 'C-reactive protein' for psychiatrists and neurologists? Aust N Z J Psychiatry 54:57–67

72. Katisko K, Cajanus A, Jaaskelainen O, Kontkanen A, Hartikainen P, Korhonen VE et al (2020) Serum neurofilament light chain is a discriminative biomarker between frontotemporal lobar degeneration and primary psychiatric disorders. J Neurol 267:162–167

73. Vijverberg EG, Dols A, Krudop WA, Del Campo MM, Kerssens CJ, Gossink F et al (2017) Cerebrospinal fluid biomarker examination as a tool to discriminate behavioral variant frontotemporal dementia from primary psychiatric disorders. Alzheimers Dement (Amst) 7:99–106

74. Steinacker P, Huss A, Mayer B, Grehl T, Grosskreutz J, Borck G et al (2017) Diagnostic and prognostic significance of neurofilament light chain NF-L, but not progranulin and S100B, in the course of amyotrophic lateral sclerosis: data from the German MND-net. Amyotroph Lateral Scler Frontotemporal Degener 18:112–119

75. Skillback T, Mattsson N, Blennow K, Zetterberg H (2017) Cerebrospinal fluid neurofilament light concentration in motor neuron disease and frontotemporal dementia predicts survival. Amyotroph Lateral Scler Frontotemporal Degener. 18:397–403

76. Valente ES, Caramelli P, Gambogi LB, Mariano LI, Guimaraes HC, Teixeira AL et al (2019) Phenocopy syndrome of behavioral variant frontotemporal dementia: a systematic review. Alzheimers Res Ther 11:30

77. Varhaug KN, Torkildsen O, Myhr KM, Vedeler CA (2019) Neurofilament light chain as a biomarker in multiple sclerosis. Front Neurol 10:338

78. Bacioglu M, Maia LF, Preische O, Schelle J, Apel A, Kaeser SA et al (2016) Neurofilament light chain in blood and CSF as marker of disease progression

in mouse models and in neurodegenerative diseases. Neuron 91:56–66

79. Arai T, Hasegawa M, Akiyama H, Ikeda K, Nonaka T, Mori H et al (2006) TDP-43 is a component of ubiquitin-positive tau-negative inclusions in frontotemporal lobar degeneration and amyotrophic lateral sclerosis. Biochem Biophys Res Commun 351:602–611

80. Mackenzie IR, Neumann M (2017) Reappraisal of TDP-43 pathology in FTLD-U subtypes. Acta Neuropathol 134:79–96

81. Maekawa S, Leigh PN, King A, Jones E, Steele JC, Bodi I et al (2009) TDP-43 is consistently co-localized with ubiquitinated inclusions in sporadic and Guam amyotrophic lateral sclerosis but not in familial amyotrophic lateral sclerosis with and without SOD1 mutations. Neuropathology 29:672–683

82. Neumann M, Sampathu DM, Kwong LK, Truax AC, Micsenyi MC, Chou TT et al (2006) Ubiquitinated TDP-43 in frontotemporal lobar degeneration and amyotrophic lateral sclerosis. Science 314:130–133

83. Steinacker P, Barschke P, Otto M (2019) Biomarkers for diseases with TDP-43 pathology. Mol Cell Neurosci 97:43–59

84. Ayala YM, Zago P, D'Ambrogio A, Xu YF, Petrucelli L, Buratti E et al (2008) Structural determinants of the cellular localization and shuttling of TDP-43. J Cell Sci 121:3778–3785

85. Buratti E, Baralle FE (2008) Multiple roles of TDP-43 in gene expression, splicing regulation, and human disease. Front Biosci 13:867–878

86. Feneberg E, Gray E, Ansorge O, Talbot K, Turner MR (2018) Towards a TDP-43-based biomarker for ALS and FTLD. Mol Neurobiol 55:7789–7801

87. Junttila A, Kuvaja M, Hartikainen P, Siloaho M, Helisalmi S, Moilanen V et al (2016) Cerebrospinal fluid TDP-43 in frontotemporal lobar degeneration and amyotrophic lateral sclerosis patients with and without the C9ORF72 Hexanucleotide expansion. Dement Geriatr Cogn Dis Extra. 6:142–149

88. Kasai T, Tokuda T, Ishigami N, Sasayama H, Foulds P, Mitchell DJ et al (2009) Increased TDP-43 protein in cerebrospinal fluid of patients with amyotrophic lateral sclerosis. Acta Neuropathol 117:55–62

89. Steinacker P, Hendrich C, Sperfeld AD, Jesse S, von Arnim CA, Lehnert S et al (2008) TDP-43 in cerebrospinal fluid of patients with frontotemporal lobar degeneration and amyotrophic lateral sclerosis. Arch Neurol 65:1481–1487

90. Suarez-Calvet M, Dols-Icardo O, Llado A, Sanchez-Valle R, Hernandez I, Amer G et al (2014) Plasma phosphorylated TDP-43 levels are elevated in patients with frontotemporal dementia carrying a C9orf72 repeat expansion or a GRN mutation. J Neurol Neurosurg Psychiatry 85:684–691

91. Feneberg E, Steinacker P, Lehnert S, Schneider A, Walther P, Thal DR et al (2014) Limited role of free TDP-43 as a diagnostic tool in neurodegenerative diseases. Amyotroph Lateral Scler Frontotemporal Degener 15:351–356

92. Foulds P, McAuley E, Gibbons L, Davidson Y, Pickering-Brown SM, Neary D et al (2008) TDP-43 protein in plasma may index TDP-43 brain pathology in Alzheimer's disease and frontotemporal lobar degeneration. Acta Neuropathol 116:141–146

93. Verstraete E, Kuiperij HB, van Blitterswijk MM, Veldink JH, Schelhaas HJ, van den Berg LH et al (2012) TDP-43 plasma levels are higher in amyotrophic lateral sclerosis. Amyotroph Lateral Scler 13:446–451

94. Broce I, Karch CM, Wen N, Fan CC, Wang Y, Tan CH et al (2018) Immune-related genetic enrichment in frontotemporal dementia: an analysis of genome-wide association studies. PLoS Med 15:e1002487

95. Ferrari R, Hernandez DG, Nalls MA, Rohrer JD, Ramasamy A, Kwok JB et al (2014) Frontotemporal dementia and its subtypes: a genome-wide association study. Lancet Neurol 13:686–699

96. Miller ZA, Rankin KP, Graff-Radford NR, Takada LT, Sturm VE, Cleveland CM et al (2013) TDP-43 frontotemporal lobar degeneration and autoimmune disease. J Neurol Neurosurg Psychiatry 84:956–962

97. Miller ZA, Sturm VE, Camsari GB, Karydas A, Yokoyama JS, Grinberg LT et al (2016) Increased prevalence of autoimmune disease within C9 and FTD/MND cohorts: completing the picture. Neurol Neuroimmunol Neuroinflamm 3:e301

98. Arnold SE, Han LY, Clark CM, Grossman M, Trojanowski JQ (2000) Quantitative neurohistological features of frontotemporal degeneration. Neurobiol Aging 21:913–919

99. Bellucci A, Bugiani O, Ghetti B, Spillantini MG (2011) Presence of reactive microglia and neuroinflammatory mediators in a case of frontotemporal dementia with P301S mutation. Neurodegener Dis 8:221–229

100. Cagnin A, Rossor M, Sampson EL, Mackinnon T, Banati RB (2004) In vivo detection of microglial activation in frontotemporal dementia. Ann Neurol 56:894–897

101. Lant SB, Robinson AC, Thompson JC, Rollinson S, Pickering-Brown S, Snowden JS et al (2014) Patterns of microglial cell activation in frontotemporal lobar degeneration. Neuropathol Appl Neurobiol 40:686–696

102. Bright F, Werry EL, Dobson-Stone C, Piguet O, Ittner LM, Halliday GM et al (2019) Neuroinflammation in frontotemporal dementia. Nat Rev Neurol 15:540–555

103. Bohlen CJ, Bennett FC, Tucker AF, Collins HY, Mulinyawe SB, Barres BA (2017) Diverse requirements for microglial survival, specification, and function revealed by defined-medium cultures. Neuron 94:759–773. e8

104. Grabert K, Michoel T, Karavolos MH, Clohisey S, Baillie JK, Stevens MP et al (2016) Microglial brain region-dependent diversity and selective regional sensitivities to aging. Nat Neurosci 19:504–516

105. Keren-Shaul H, Spinrad A, Weiner A, Matcovitch-Natan O, Dvir-Szternfeld R, Ulland TK et al (2017)

A unique microglia type associated with restricting development of Alzheimer's disease. Cell 169:1276–1290. e17

106. Querol-Vilaseca M, Colom-Cadena M, Pegueroles J, San Martin-Paniello C, Clarimon J, Belbin O et al (2017) YKL-40 (Chitinase 3-like I) is expressed in a subset of astrocytes in Alzheimer's disease and other tauopathies. J Neuroinflammation 14:118

107. Baldacci F, Toschi N, Lista S, Zetterberg H, Blennow K, Kilimann I et al (2017) Two-level diagnostic classification using cerebrospinal fluid YKL-40 in Alzheimer's disease. Alzheimers Dement 13:993–1003

108. Illan-Gala I, Alcolea D, Montal V, Dols-Icardo O, Munoz L, de Luna N et al (2018) CSF sAPPbeta, YKL-40, and NfL along the ALS-FTD spectrum. Neurology 91:e1619–e1628

109. Janelidze S, Mattsson N, Stomrud E, Lindberg O, Palmqvist S, Zetterberg H et al (2018) CSF biomarkers of neuroinflammation and cerebrovascular dysfunction in early Alzheimer disease. Neurology 91:e867–e877

110. Llorens F, Thune K, Tahir W, Kanata E, Diaz-Lucena D, Xanthopoulos K et al (2017) YKL-40 in the brain and cerebrospinal fluid of neurodegenerative dementias. Mol Neurodegener 12:83

111. Thompson AG, Gray E, Thezenas ML, Charles PD, Evetts S, Hu MT et al (2018) Cerebrospinal fluid macrophage biomarkers in amyotrophic lateral sclerosis. Ann Neurol 83:258–268

112. Zhang H, Ng KP, Therriault J, Kang MS, Pascoal TA, Rosa-Neto P et al (2018) Cerebrospinal fluid phosphorylated tau, visinin-like protein-1, and chitinase-3-like protein 1 in mild cognitive impairment and Alzheimer's disease. Transl Neurodegener 7:23

113. Baldacci F, Lista S, Palermo G, Giorgi FS, Vergallo A, Hampel H (2019) The neuroinflammatory biomarker YKL-40 for neurodegenerative diseases: advances in development. Expert Rev Proteomics 16:593–600

114. Abu-Rumeileh S, Steinacker P, Polischi B, Mammana A, Bartoletti-Stella A, Oeckl P et al (2019) CSF biomarkers of neuroinflammation in distinct forms and subtypes of neurodegenerative dementia. Alzheimers Res Ther 12:2

115. Oeckl P, Weydt P, Steinacker P, Anderl-Straub S, Nordin F, Volk AE et al (2019) Different neuroinflammatory profile in amyotrophic lateral sclerosis and frontotemporal dementia is linked to the clinical phase. J Neurol Neurosurg Psychiatry 90:4–10

116. Malaguarnera L, Simpore J, Prodi DA, Angius A, Sassu A, Persico I et al (2003) A 24-bp duplication in exon 10 of human chitotriosidase gene from the sub-Saharan to the Mediterranean area: role of parasitic diseases and environmental conditions. Genes Immun 4:570–574

117. Colangelo AM, Alberghina L, Papa M (2014) Astrogliosis as a therapeutic target for neurodegenerative diseases. Neurosci Lett 565:59–64

118. Heller C, Foiani MS, Moore K, Convery R, Bocchetta M, Neason M et al (2020) Plasma glial fibrillary acidic protein is raised in progranulin-associated frontotemporal dementia. J Neurol Neurosurg Psychiatry 91:263–270

119. Ishiki A, Kamada M, Kawamura Y, Terao C, Shimoda F, Tomita N et al (2016) Glial fibrillar acidic protein in the cerebrospinal fluid of Alzheimer's disease, dementia with Lewy bodies, and frontotemporal lobar degeneration. J Neurochem 136:258–261

120. Colonna M (2003) TREMs in the immune system and beyond. Nat Rev Immunol 3:445–453

121. Cady J, Koval ED, Benitez BA, Zaidman C, Jockel-Balsarotti J, Allred P et al (2014) TREM2 variant p.R47H as a risk factor for sporadic amyotrophic lateral sclerosis. JAMA Neurol 71:449–453

122. Guerreiro R, Wojtas A, Bras J, Carrasquillo M, Rogaeva E, Majounie E et al (2013) TREM2 variants in Alzheimer's disease. N Engl J Med 368:117–127

123. Guerreiro RJ, Lohmann E, Bras JM, Gibbs JR, Rohrer JD, Gurunlian N et al (2013) Using exome sequencing to reveal mutations in TREM2 presenting as a frontotemporal dementia-like syndrome without bone involvement. JAMA Neurol 70:78–84

124. Jonsson T, Stefansson H, Steinberg S, Jonsdottir I, Jonsson PV, Snaedal J et al (2013) Variant of TREM2 associated with the risk of Alzheimer's disease. N Engl J Med 368:107–116

125. Rayaprolu S, Mullen B, Baker M, Lynch T, Finger E, Seeley WW et al (2013) TREM2 in neurodegeneration: evidence for association of the p.R47H variant with frontotemporal dementia and Parkinson's disease. Mol Neurodegener 8:19

126. Su WH, Shi ZH, Liu SL, Wang XD, Liu S, Ji Y (2018) The rs75932628 and rs2234253 polymorphisms of the TREM2 gene were associated with susceptibility to frontotemporal lobar degeneration in Caucasian populations. Ann Hum Genet 82:177–185

127. Wunderlich P, Glebov K, Kemmerling N, Tien NT, Neumann H, Walter J (2013) Sequential proteolytic processing of the triggering receptor expressed on myeloid cells-2 (TREM2) protein by ectodomain shedding and gamma-secretase-dependent intramembranous cleavage. J Biol Chem 288:33027–33036

128. Heslegrave A, Heywood W, Paterson R, Magdalinou N, Svensson J, Johansson P et al (2016) Increased cerebrospinal fluid soluble TREM2 concentration in Alzheimer's disease. Mol Neurodegener 11:3

129. Kleinberger G, Brendel M, Mracsko E, Wefers B, Groeneweg L, Xiang X et al (2017) The FTD-like syndrome causing TREM2 T66M mutation impairs microglia function, brain perfusion, and glucose metabolism. EMBO J 36:1837–1853

130. Piccio L, Deming Y, Del-Aguila JL, Ghezzi L, Holtzman DM, Fagan AM et al (2016) Cerebrospinal fluid soluble TREM2 is higher in Alzheimer disease and associated with mutation status. Acta Neuropathol 131:925–933

131. Woollacott IOC, Nicholas JM, Heslegrave A, Heller C, Foiani MS, Dick KM et al (2018) Cerebrospinal fluid soluble TREM2 levels in frontotemporal dementia differ by genetic and pathological subgroup. Alzheimers Res Ther 10:79

132. van Eijk M, van Roomen CP, Renkema GH, Bussink AP, Andrews L, Blommaart EF et al (2005) Characterization of human phagocyte-derived chitotriosidase, a component of innate immunity. Int Immunol 17:1505–1512

133. Bossu P, Salani F, Alberici A, Archetti S, Bellelli G, Galimberti D et al (2011) Loss of function mutations in the progranulin gene are related to pro-inflammatory cytokine dysregulation in frontotemporal lobar degeneration patients. J Neuroinflammation 8:65

134. Gibbons L, Rollinson S, Thompson JC, Robinson A, Davidson YS, Richardson A et al (1603) Plasma levels of progranulin and interleukin-6 in frontotemporal lobar degeneration. Neurobiol Aging 2015(36):e1–e4

135. Galimberti D, Venturelli E, Fenoglio C, Guidi I, Villa C, Bergamaschini L et al (2008) Intrathecal levels of IL-6, IL-11 and LIF in Alzheimer's disease and frontotemporal lobar degeneration. J Neurol 255:539–544

136. Sjogren M, Folkesson S, Blennow K, Tarkowski E (2004) Increased intrathecal inflammatory activity in frontotemporal dementia: pathophysiological implications. J Neurol Neurosurg Psychiatry 75:1107–1111

137. Galimberti D, Bonsi R, Fenoglio C, Serpente M, Cioffi SM, Fumagalli G et al (2015) Inflammatory molecules in frontotemporal dementia: cerebrospinal fluid signature of progranulin mutation carriers. Brain Behav Immun 49:182–187

138. Galimberti D, Schoonenboom N, Scheltens P, Fenoglio C, Venturelli E, Pijnenburg YA et al (2006) Intrathecal chemokine levels in Alzheimer disease and frontotemporal lobar degeneration. Neurology 66:146–147

139. Galimberti D, Venturelli E, Villa C, Fenoglio C, Clerici F, Marcone A et al (2009) MCP-1 A-2518G polymorphism: effect on susceptibility for frontotemporal lobar degeneration and on cerebrospinal fluid MCP-1 levels. J Alzheimers Dis 17:125–133

140. Cavaillon JM (2001) Pro- versus anti-inflammatory cytokines: myth or reality. Cell Mol Biol (Noisy-le-Grand) 47:695–702

141. Lui H, Zhang J, Makinson SR, Cahill MK, Kelley KW, Huang HY et al (2016) Progranulin deficiency promotes circuit-specific synaptic pruning by microglia via complement activation. Cell 165:921–935

142. Chitramuthu BP, Bennett HPJ, Bateman A (2017) Progranulin: a new avenue towards the understanding and treatment of neurodegenerative disease. Brain 140:3081–3104

143. Smith KR, Damiano J, Franceschetti S, Carpenter S, Canafoglia L, Morbin M et al (2012) Strikingly different clinicopathological phenotypes determined by progranulin-mutation dosage. Am J Hum Genet 90:1102–1107

144. Baker M, Mackenzie IR, Pickering-Brown SM, Gass J, Rademakers R, Lindholm C et al (2006) Mutations in progranulin cause tau-negative frontotemporal dementia linked to chromosome 17. Nature 442:916–919

145. Cruts M, Gijselinck I, van der Zee J, Engelborghs S, Wils H, Pirici D et al (2006) Null mutations in progranulin cause ubiquitin-positive frontotemporal dementia linked to chromosome 17q21. Nature 442:920–924

146. Rademakers R, Neumann M, Mackenzie IR (2012) Advances in understanding the molecular basis of frontotemporal dementia. Nat Rev Neurol 8:423–434

147. Almeida MR, Baldeiras I, Ribeiro MH, Santiago B, Machado C, Massano J et al (2014) Progranulin peripheral levels as a screening tool for the identification of subjects with progranulin mutations in a Portuguese cohort. Neurodegener Dis 13:214–223

148. Carecchio M, Fenoglio C, De Riz M, Guidi I, Comi C, Cortini F et al (2009) Progranulin plasma levels as potential biomarker for the identification of GRN deletion carriers. A case with atypical onset as clinical amnestic mild cognitive impairment converted to Alzheimer's disease. J Neurol Sci 287:291–293

149. Finch N, Baker M, Crook R, Swanson K, Kuntz K, Surtees R et al (2009) Plasma progranulin levels predict progranulin mutation status in frontotemporal dementia patients and asymptomatic family members. Brain 132:583–591

150. Galimberti D, Fumagalli GG, Fenoglio C, Cioffi SMG, Arighi A, Serpente M et al (2018) Progranulin plasma levels predict the presence of GRN mutations in asymptomatic subjects and do not correlate with brain atrophy: results from the GENFI study. Neurobiol Aging 62:245 e9–245e12

151. Ghidoni R, Benussi L, Glionna M, Franzoni M, Binetti G (2008) Low plasma progranulin levels predict progranulin mutations in frontotemporal lobar degeneration. Neurology 71:1235–1239

152. Ghidoni R, Stoppani E, Rossi G, Piccoli E, Albertini V, Paterlini A et al (2012) Optimal plasma progranulin cutoff value for predicting null progranulin mutations in neurodegenerative diseases: a multicenter Italian study. Neurodegener Dis 9:121–127

153. Schofield EC, Halliday GM, Kwok J, Loy C, Double KL, Hodges JR (2010) Low serum progranulin predicts the presence of mutations: a prospective study. J Alzheimers Dis 22:981–984

154. Sleegers K, Brouwers N, Van Damme P, Engelborghs S, Gijselinck I, van der Zee J et al (2009) Serum biomarker for progranulin-associated frontotemporal lobar degeneration. Ann Neurol 65:603–609

155. Meeter LH, Patzke H, Loewen G, Dopper EG, Pijnenburg YA, van Minkelen R et al (2016) Progranulin levels in plasma and cerebrospinal fluid in Granulin mutation carriers. Dement Geriatr Cogn Dis Extra 6:330–340

156. Nicholson AM, Finch NA, Thomas CS, Wojtas A, Rutherford NJ, Mielke MM et al (2014) Progranulin protein levels are differently regulated in plasma and CSF. Neurology 82:1871–1873

157. Wilke C, Gillardon F, Deuschle C, Dubois E, Hobert MA, Muller vom Hagen J et al (2016) Serum levels of Progranulin do not reflect cerebrospinal fluid levels in neurodegenerative disease. Curr Alzheimer Res 13:654–662

158. Han JJ, Yu M, Houston N, Steinberg SM, Kohn EC (2011) Progranulin is a potential prognostic biomarker in advanced epithelial ovarian cancers. Gynecol Oncol 120:5–10

159. Todoric J, Handisurya A, Perkmann T, Knapp B, Wagner O, Tura A et al (2012) Circulating progranulin levels in women with gestational diabetes mellitus and healthy controls during and after pregnancy. Eur J Endocrinol 167:561–567

160. Cruchaga C, Graff C, Chiang HH, Wang J, Hinrichs AL, Spiegel N et al (2011) Association of TMEM106B gene polymorphism with age at onset in granulin mutation carriers and plasma granulin protein levels. Arch Neurol 68:581–586

161. Finch N, Carrasquillo MM, Baker M, Rutherford NJ, Coppola G, Dejesus-Hernandez M et al (2011) TMEM106B regulates progranulin levels and the penetrance of FTLD in GRN mutation carriers. Neurology 76:467–474

162. Hsiung GY, Fok A, Feldman HH, Rademakers R, Mackenzie IR (2011) rs5848 polymorphism and serum progranulin level. J Neurol Sci 300:28–32

163. Byrne S, Elamin M, Bede P, Shatunov A, Walsh C, Corr B et al (2012) Cognitive and clinical characteristics of patients with amyotrophic lateral sclerosis carrying a C9orf72 repeat expansion: a population-based cohort study. Lancet Neurol 11:232–240

164. DeJesus-Hernandez M, Mackenzie IR, Boeve BF, Boxer AL, Baker M, Rutherford NJ et al (2011) Expanded GGGGCC hexanucleotide repeat in non-coding region of C9ORF72 causes chromosome 9p-linked FTD and ALS. Neuron 72:245–256

165. Majounie E, Renton AE, Mok K, Dopper EG, Waite A, Rollinson S et al (2012) Frequency of the C9orf72 hexanucleotide repeat expansion in patients with amyotrophic lateral sclerosis and frontotemporal dementia: a cross-sectional study. Lancet Neurol 11:323–330

166. Pottier C, Ravenscroft TA, Sanchez-Contreras M, Rademakers R (2016) Genetics of FTLD: overview and what else we can expect from genetic studies. J Neurochem 138(Suppl 1):32–53

167. Renton AE, Majounie E, Waite A, Simon-Sanchez J, Rollinson S, Gibbs JR et al (2011) A hexanucleotide repeat expansion in C9ORF72 is the cause of chromosome 9p21-linked ALS-FTD. Neuron 72:257–268

168. van Blitterswijk M, DeJesus-Hernandez M, Rademakers R (2012) How do C9ORF72 repeat expansions cause amyotrophic lateral sclerosis and frontotemporal dementia: can we learn from other noncoding repeat expansion disorders? Curr Opin Neurol 25:689–700

169. Jiang J, Ravits J (2019) Pathogenic mechanisms and therapy development for C9orf72 amyotrophic lateral sclerosis/frontotemporal dementia. Neurotherapeutics 16:1115–1132

170. Mori K, Weng SM, Arzberger T, May S, Rentzsch K, Kremmer E et al (2013) The C9orf72 GGGGCC repeat is translated into aggregating dipeptide-repeat proteins in FTLD/ALS. Science 339:1335–1338

171. Ash PE, Bieniek KF, Gendron TF, Caulfield T, Lin WL, Dejesus-Hernandez M et al (2013) Unconventional translation of C9ORF72 GGGGCC expansion generates insoluble polypeptides specific to c9FTD/ALS. Neuron 77:639–646

172. Gendron TF, Bieniek KF, Zhang YJ, Jansen-West K, Ash PE, Caulfield T et al (2013) Antisense transcripts of the expanded C9ORF72 hexanucleotide repeat form nuclear RNA foci and undergo repeat-associated non-ATG translation in c9FTD/ALS. Acta Neuropathol 126:829–844

173. Zu T, Liu Y, Banez-Coronel M, Reid T, Pletnikova O, Lewis J et al (2013) RAN proteins and RNA foci from antisense transcripts in C9ORF72 ALS and frontotemporal dementia. Proc Natl Acad Sci U S A 110:E4968–E4977

174. Davidson YS, Barker H, Robinson AC, Thompson JC, Harris J, Troakes C et al (2014) Brain distribution of dipeptide repeat proteins in frontotemporal lobar degeneration and motor neurone disease associated with expansions in C9ORF72. Acta Neuropathol Commun 2:70

175. Gomez-Deza J, Lee YB, Troakes C, Nolan M, Al-Sarraj S, Gallo JM et al (2015) Dipeptide repeat protein inclusions are rare in the spinal cord and almost absent from motor neurons in C9ORF72 mutant amyotrophic lateral sclerosis and are unlikely to cause their degeneration. Acta Neuropathol Commun 3:38

176. Mackenzie IR, Arzberger T, Kremmer E, Troost D, Lorenzl S, Mori K et al (2013) Dipeptide repeat protein pathology in C9ORF72 mutation cases: clinico-pathological correlations. Acta Neuropathol 126:859–879

177. Schludi MH, May S, Grasser FA, Rentzsch K, Kremmer E, Kupper C et al (2015) Distribution of dipeptide repeat proteins in cellular models and C9orf72 mutation cases suggests link to transcriptional silencing. Acta Neuropathol 130:537–555

178. Su Z, Zhang Y, Gendron TF, Bauer PO, Chew J, Yang WY et al (2014) Discovery of a biomarker and lead small molecules to target r(GGGGCC)-associated defects in c9FTD/ALS. Neuron 83:1043–1050

179. Gendron TF, Chew J, Stankowski JN, Hayes LR, Zhang YJ, Prudencio M et al (2017) Poly(GP) proteins are a useful pharmacodynamic marker for C9ORF72-associated amyotrophic lateral sclerosis. Sci Transl Med 9 doi:10.1126/scitranslmed. aai7866

180. Lehmer C, Oeckl P, Weishaupt JH, Volk AE, Diehl-Schmid J, Schroeter ML et al (2017) Poly-GP in cerebrospinal fluid links C9orf72-associated dipeptide repeat expression to the asymptomatic phase of ALS/FTD. EMBO Mol Med 9:859–868

181. Gendron TF, Group CONS, Daughrity LM, Heckman MG, Diehl NN, Wuu J et al (2017) Phosphorylated neurofilament heavy chain: a biomarker of survival for C9ORF72-associated amyotrophic lateral sclerosis. Ann Neurol 82:139–146

182. Meeter LHH, Gendron TF, Sias AC, Jiskoot LC, Russo SP, Donker Kaat L et al (2018) Poly(GP), neurofilament and grey matter deficits in C9orf72 expansion carriers. Ann Clin Transl Neurol 5:583–597

183. Baborie A, Griffiths TD, Jaros E, Perry R, McKeith IG, Burn DJ et al (2015) Accumulation of dipeptide repeat proteins predates that of TDP-43 in frontotemporal lobar degeneration associated with hexanucleotide repeat expansions in C9ORF72 gene. Neuropathol Appl Neurobiol 41:601–612

184. Proudfoot M, Gutowski NJ, Edbauer D, Hilton DA, Stephens M, Rankin J et al (2014) Early dipeptide repeat pathology in a frontotemporal dementia kindred with C9ORF72 mutation and intellectual disability. Acta Neuropathol 127:451–458

185. Vatsavayai SC, Yoon SJ, Gardner RC, Gendron TF, Vargas JN, Trujillo A et al (2016) Timing and significance of pathological features in C9orf72 expansion-associated frontotemporal dementia. Brain 139:3202–3216

186. Riemslagh FW. (2019). Molecular mechanisms of C9orf72-linked frontotemporal dementia and amyotrophic lateral sclerosis. Dissertation, Erasmus Universiteit Rotterdam.

187. Oeckl P, Steinacker P, Feneberg E, Otto M (2015) Cerebrospinal fluid proteomics and protein biomarkers in frontotemporal lobar degeneration: current status and future perspectives. Biochim Biophys Acta 1854:757–68

188. van der Ende EL, Meeter HH, Stingl C, van Rooij JGJ, Stoop MP, Nijholt DAT et al (2019) Novel CSF biomarkers in genetic frontotemporal dementia identified by proteomics. Ann Clin Transl Neurol 6:698–707

189. Jahn H, Wittke S, Zurbig P, Raedler TJ, Arlt S, Kellmann M et al (2011) Peptide fingerprinting of Alzheimer's disease in cerebrospinal fluid: identification and prospective evaluation of new synaptic biomarkers. PLoS One 6:e26540

190. Ruetschi U, Zetterberg H, Podust VN, Gottfries J, Li S, Hviid Simonsen A et al (2005) Identification of CSF biomarkers for frontotemporal dementia using SELDI-TOF. Exp Neurol 196:273–281

191. Simonsen AH, McGuire J, Podust VN, Hagnelius NO, Nilsson TK, Kapaki E et al (2007) A novel panel of cerebrospinal fluid biomarkers for the differential diagnosis of Alzheimer's disease versus normal aging and frontotemporal dementia. Dement Geriatr Cogn Disord 24:434–440

192. Teunissen CE, Elias N, Koel-Simmelink MJ, Durieux-Lu S, Malekzadeh A, Pham TV et al (2016) Novel diagnostic cerebrospinal fluid biomarkers for pathologic subtypes of frontotemporal dementia identified by proteomics. Alzheimers Dement (Amst) 2:86–94

193. Barschke P, Oeckl P, Steinacker P, Al Shweiki MR, Weishaupt JH, Landwehrmeyer B, et al. (2020) Different CSF protein profiles in amyotrophic lateral sclerosis and frontotemporal dementia with C9orf72 hexanucleotide repeat expansion. J Neurol Neurosurg Psychiatry 91:503–511

Frontotemporal Dementia: A Cross-Cultural Perspective

Chiadi U. Onyike, Shunichiro Shinagawa, and Ratnavalli Ellajosyula

Introduction

Frontotemporal dementia (FTD) is the term used to indicate clinically and pathologically heterogeneous neurodegenerative syndromes, featuring unrelenting decline in temperament, judgment, conduct, and verbal communication. The onset of FTD occurs most frequently in midlife; most commonly the illness is recognized before age 60. However, when the onset occurs earlier, in youth, the symptoms may mimic a primary psychiatric disorder, for example, schizophrenia or bipolar disorder [1]. Cases arising in the seventh decade of life and later have also been reported [2].

The best known FTD phenotypes are defined, according to the profile of disability and dysfunction, by the behavioral changes, the language deficits, or a combination of cognitive and neurological symptoms. The behavioral phenotype, that is, behavioral FTD, is dominated by dissocial behaviors such as indifference, insensitivity, jocularity, impulsiveness, and compulsive behaviors. Two main language phenotypes are recognized, one, non-fluent FTD, characterized by effortful, dysfluent, non-grammatical speech and difficulty understanding sentences; the other, semantic FTD, by fluent speech, with anomia, agnosia for words and objects, and vacuousness. The features of the behavioral and language phenotypes reflect the degeneration of frontal and temporal cortices. One also encounters FTD syndromes characterized by the association of cognitive, behavioral, or language symptoms with motor dysfunctions that reflect early degeneration of subcortical structures; this combination occurs in diseases such as corticobasal degeneration and progressive supranuclear palsy. It is to be noted that whatever the presenting phenotype may be, FTD progresses to a severe dementia [3].

FTD has been recognized in many countries (see Fig. 1). However, the scope of clinical activity and research varies widely across regions, reflecting, in our view, local expertise, local resources, public health priorities, and sociocultural factors. This chapter attempts to provide a perspective about the international FTD landscape, describing, first, the distribution and demographics; second, the clinical and genetic epidemiology; and, lastly, considering how the diversity of geographic and cultural settings impacts diagnosis, care, and research.

C. U. Onyike (✉)
Division of Geriatric Psychiatry and Neuropsychiatry, The Johns Hopkins University School of Medicine, Baltimore, MD, USA
e-mail: cuo@jhmi.edu

S. Shinagawa
Department of Psychiatry, The Jikei University School of Medicine, Tokyo, Japan

R. Ellajosyula
Department of Neurology, Manipal Clinic, Bangalore, India

B. Ghetti et al. (eds.), *Frontotemporal Dementias*, Advances in Experimental Medicine and Biology 1281, https://doi.org/10.1007/978-3-030-51140-1_10

Fig. 1 Geographical distribution of centers active in research and clinical care focused on frontotemporal dementia. Flags indicates the location of individual centers

Worldwide Distribution of Frontotemporal Dementia

In the past three decades, the frequency of FTD has been described in more than 30 population-based studies from around the world—Australia [4], Brazil [5, 6], Canada [7], Finland [8], Germany [9], India [10], Italy [11–15], Japan [16–20], Netherlands [21], Nigeria [22], Spain [23, 24], South Korea [25], Sweden [26, 27], Turkey [28], the United Kingdom [29–33], and the United States [34].

The most recent systematic review, conducted for the period 2000–2012, summarizes data from studies that were carried out in catchment areas geographically located in 15 countries of the American, European, and Asian continents and showed a point prevalence range of 0.01–4.61/1000 persons [35]. Three-point prevalence rates that fall within that range, from Japan [20], Australia [4], and the United Kingdom [33], were

not included in the review because they became available after 2012. Several studies conducted in India, Korea, Japan, Sweden, and the United Kingdom, during the period 2000–2012, report 1-year cumulative prevalence rates that range 0.16–2.85 per 1000—excluding that from an outlier (31.04/1000), which used an uncommon case definition and a narrow non-representative age range [27]. A study from Nigeria provides a 10-year cumulative prevalence of 0.01/1000 persons, based on archival data, collected from 1998 to 2007, from a large regional neuropsychiatric hospital [22]. However, in the study, the focus on hospital care and the retrospective ascertainment may have resulted in a lower prevalence of FTD than might be found in the reference population.

The worldwide FTD incidence rate of 0.00–0.33/1000 person-years (see Table 1) was estimated from data deriving from a national registry in Denmark [36], as well as catchment area studies conducted in Brazil, Italy, Spain, the United

Table 1 Incidence of frontotemporal dementia (FTD) worldwide

Study/location	Sampling frame and case ascertainment	Age	Years of study	Incidence[a]	95% CI
Knopman et al., 2004 Rochester, USA	City/suburban area; linked and coded records	40–69	4	0.04	0.02–0.11
Nitrini et al., 2004 Cantadeluva, Brazil	Municipality; assessment and diagnostic conference	65+	3	0.28	0.04–1.96
Ravaglia et al., 2005 Conselice, Italy	Provincial town; assessments and diagnostic conference	65+	5	0.33	0.05–2.33
Mercy et al., 2008 Cambridgeshire, UK	Large borough; assessments and diagnostic conference	45–64	6	0.04	0.02–0.06
Garre-Olmo et al., 2010 Girona, Spain	City; assessments and diagnostic conference	30–64	3	0.05	0.04–0.06
Phung et al., 2010 Denmark	Country; national registry, linked and coded records	40+	34	0.00	0.00–0.00
Coyne-Gilchrist et al., 2016 Cambridgeshire and Norfolk, UK	Two large counties; assessments and diagnostic conference	40+	2	0.02	0.01–0.02

[a]Incidence per 1000 person-years

Kingdom, and the United States [6, 11, 24, 32–34, 36].

As noted in the most recent reviews [35, 37], world prevalence and incidence rates for FTD are low and show wide variation. These variations can be explained on methodological grounds. Population-based studies of FTD and other neurodegenerative diseases are technically challenging, due to the difficulty of case definition and ascertainment, the evolving diagnostic rules, as well as the type of expertise and resources that are required [35, 37–39]. Diagnostic criteria have been refined over the past 40 years. Most recently, these refinements were undertaken to address problems related to requirement of the criteria for a multiplicity of symptoms and the rigidity of the algorithms used to define thresholds for diagnosis, as well as a desire to include a ranking for the level of confidence in the diagnosis [40]. The latest criteria [41, 42] addressed these issues, and the next step would be a characterization of interrater reliability, sensitivity, and specificity. The first study to test the interrater reliability of the criteria for behavioral FTD reported an interrater agreement of 82% [43], and another study, using neuropathological data for reference, reported sensitivity and specificity ranging 82–95% and 85–85%, respectively, depending on the assigned level of diagnostic confidence [44]. Positive and negative predictive values were 80–92% and 91–96%, respectively. These estimates are preliminary, as the samples were small or relied on retrospective clinical data. Larger prospective studies are needed to clarify reliability, sensitivity, and specificity of the diagnostic criteria in different clinical and cultural contexts, and to identify areas for refinement.

It is to be noted that geographic regions are diverse with respect to the prevailing cultural and socioeconomic contexts, aspects of which (e.g., poverty, low literacy, lack of infrastructure, and cultural norms) may constitute barriers for research [39]. These challenges explain the variation in the scope, depth, and methodology of FTD surveillance across studies, and, at least partially, the variation in the prevalence and incidence rates. In exceptional circumstances, geographic differences reflect the presence of communities with high rates of mutation carriers [15] or other susceptibility factors. It is also to be noted that current estimates of prevalence and incidence, while of undoubtedly high value for research and policy, are not yet representative of all geographic regions and ethnic groups. The research has been uneven distributed geographically, and there is low ethnic diversity in the studies. For instance, the North American and European cohorts are over 95% Caucasian [37]. In other words, while valuable knowledge has been gained from studies describing FTD distributions in many regions, there is still much to learn.

Genetic Epidemiology

In the past two decades, investigations of familial and hereditary cohorts of FTD have led to the identification of genetic loci for causal dominant mutations, of which the most important are microtubule-associated protein tau (*MAPT*) [45–47], progranulin (*GRN*) [48, 49], and chromosome 9 open reading frame 72 (*C9orf72*) [50, 51]. Mutations of these genes, together, account for a large majority of hereditary FTD in North America and Europe [52], but there are regional variations in the distribution of these mutations. For example, clustering of *GRN* mutations have been observed in northern Italy [53, 54].

The *C9orf72* mutation has a high frequency in North American and European patients who have familial FTD, amyotrophic lateral sclerosis (ALS), or FTD in association with ALS (FTD-ALS)—though it has also been observed in a small proportion of African American, Hispanic American, and Asian patients [55]. *C9orf72* mutations have been also identified in Greek [56] and Turkish [57] cohorts. All *C9orf72* mutation carriers identified in a worldwide epidemiological study (403 and 588 who had FTD or ALS, respectively) were found to have the Finnish founder haplotype by a genome-wide single-nucleotide polymorphism analysis [55]. This finding, which has been replicated many times, points to a founder origin in Northern Europe [58]. The *C9orf72* mutation has been reported as a common cause of hereditary FTD and ALS in Brazil [59], but there were no data on the ethnic background of the mutation carriers. *C9orf72* carriers have also been identified among a small number of Han Chinese patients [60]. These Chinese carriers have been shown to have the same risk haplotype identified in the European cohorts, a finding suggesting the possibility of a shared common founder [61].

Familial and hereditary FTDs appear to be rare in Asian populations. An international Asian collaborative study, that analyzed data from India, Indonesia, Japan, Philippines, and Taiwan, found that few patients had a relative with FTD [62]. Although mutations in the *CHCHD10* gene [63] were found in about 8% of the subjects in a Chinese clinic series [64], a subsequent study found very few carriers of the *MAPT*, *GRN*, and *CHCHD10* mutations [65]. A novel *GRN* mutation was identified recently in one of 116 subjects in a cohort from southern India [66]; no other carriers have yet been identified. Other studies show that mutations in *C9orf72*, *GRN*, and *MAPT* are rare in China [67–69], Japan [70], South Korea [71, 72], and India [73, 74]. There are no data from Africa pertaining to familial or hereditary FTD. A few studies have reported data on the frequency of mutation carriers among African American or Hispanic American subjects, for example [55], but the numbers have been too low for subgroup analyses. However, in an analysis of data collected from ten centers in the United States and Europe and two additional ones in Jamaica and Nigeria, the *C9orf72* mutation was found in three of the 65 FTD subjects of African descent [75]. As the mutation carriers were African American, they may also have had the Finnish founder haplotype.

Clinical Aspects

Cultural Influences

Cultural context may exert strong influences on the experience, expression, or recognition of behavioral dysfunctions, including those of neurodegenerative diseases such as FTD. In western

India, for example, cases tend to present with a severe syndrome [76] in which impulsive and compulsive features are prominent [77]. On the other hand, a report from western Nigeria describes a presentation characterized by abnormal conduct, executive dysfunction, emotional incontinence, and progressive aphasia [78]; however, more data will be needed in order to determine whether other Nigerian cases have similar features. There has been very little description of FTD syndromes in African Americans or Europeans of African descent. Through the analysis of data from the sample of 65 subjects of African descent, it was reported that the behavioral phenotype was most common, and that half the subjects with semantic FTD had behavioral disorder, prosopagnosia, and right-predominant bilateral anteromedial temporal lobe atrophy [75]. Hallucinations were common, even among the cases with language syndromes. Seven subjects had family history of FTD, and three were carriers of the *C9orf72* mutation. None of the four subjects who had motor neuron disease were carriers of the mutation. It must be stated that the degree to which these observations pertain to patients living in Africa is not yet known.

Data from Japan illustrate how cultural factors may influence the *outcomes*, rather than the *features*, of a clinical syndrome. In a study of abnormal eating behaviors in behavioral FTD patients from Japan and the United Kingdom, the symptoms of abnormal eating were similar, whereas weight gain was more common and severe in the United Kingdom patients [79]. This observation was attributed to differences in food culture, including comparatively higher carbohydrate consumption and caloric intake in the United Kingdom. On a historical note, the Japanese construct *Gogi aphasia,* now accepted as corresponding to semantic FTD, was a syndrome defined by word agnosia and preserved phonological and syntactic aspects of language [80].

There have also been interesting observations pertaining to the interaction between the cultural aspects of language and the language phenotypes of FTD. One study from southern India, where multilingualism is ubiquitous, demonstrated strikingly disproportionate loss of the second language in multilingual patients with semantic FTD—suggesting that later-learned languages are more vulnerable to neurodegeneration, and that different languages connect to a common semantic network in the brain [81]. A different line of research, also conducted in India, links multilingualism to concepts of cognitive resilience. Multilingual behavioral FTD patients were found to have older age at illness onset than monolingual subjects, after accounting for literacy and urban exposure, whereas multilingual and monolingual subjects with language phenotypes did not differ in age at illness onset [82]. These data were interpreted as support for the proposal that a robust association between multilingualism and executive functions confers resilience to the cognitive decline associated with neurodegeneration.

Finally, it is not unreasonable to speculate that FTD-ALS is comparatively less common in Asia, on account of the relative infrequency of the *C9orf72* mutation. This may have implications for comparisons of FTD survival across regions, given the rapid progression associated with FTD-ALS. To date, geographic differences have not been shown in the progression of symptoms or in survival with FTD—which has ranged 8–10 years across cohorts [83]. However, most of the data on FTD survival have come from European and North American cohorts.

Clinical Care

Clinical care begins with an accurate and complete diagnosis, wherein dementia status, the specific diagnosis (i.e., recognition of FTD), familial status (if pertinent), severity, and pressing needs are clearly identified. The recognition of FTD syndromes at a late stage appears to be a common problem, with geographic differences that are largely shaped by sociocultural and socioeconomic factors. In developing countries, recognition in the community is low and families tend to report cases late due to low health literacy, not being able to take the time from work, and due to the alternative view (often shared by health providers) that dementia is a stage of life [22].

Recognition is frequently low among healthcare providers, who often lack training and share their community's misconceptions about dementia [84, 85]. In North America, Europe, and Japan, diagnostic delay historically reflected lack of familiarity with FTD and its differential diagnosis but, in these regions, the problem is increasingly mitigated by the utilization of tertiary referrals, professional education, peer-led advocacy, and public education through popular media.

Once a diagnosis has been established, the next step is a plan of care that integrates patient and carer education with pharmacologic prescriptions; behavioral, psychotherapeutic, and rehabilitative interventions; and care management [86]. In the later stages, care requires round-the-clock supervision and hands-on assistance, which usually are delivered by relatives or professional aides (or both), or in custodial care. In many low-income countries, such as in much of Africa, pharmacologic prescriptions are often inaccessible due to cost and supply chain barriers. Residential programs are also uncommon, and there is instead a high reliance on informal care from relatives, which is entrenched in cultural norms. In middle-income countries, such as China and India, behavioral and psychotherapeutic programs are available and often integrated into plans of care [87], whereas residential care remains comparatively uncommon. End-of-life care tends to be less formalized in developing countries, and postmortem diagnosis, whether for clinical or research indications, is uncommon.

Conclusions

FTD has been recognized in many regions of the world. The incidence and prevalence vary widely, but this, at least partly, reflects differences in the methods used to undertake the studies, and disparities in the expertise and resources for research. Some of the epidemiological data point to differences in the distribution of genetic risk factors; mutations in *C9orf72*, *GRN*, and *MAPT* are more frequently reported in patients of European descent, whereas they are less common in those of Asian descent. There appear to be cultural influences on the expression of symptoms, but more research will be needed to clarify the environmental and social mechanisms involved; the interactions of cultural influences with genetic susceptibility factors is not known. The wide variation in care reflects disparities in economic resources and clinical infrastructure, as well as local practices.

There are pressing needs for advancing research on FTD. Population studies are needed in order to fill gaps in our knowledge about FTD frequency and risk factors in developing regions and among minority groups in developed countries, and to facilitate the psychometric characterization of contemporary diagnostic criteria and their translation to different cultural contexts.

The multicentric research collaborations developed in North America (ARTFL-LEFFTDS Longitudinal Frontotemporal Lobar Degeneration, https://www.allftd.org) and Europe (Genetic FTD Initiative, https://www.genfi.org) are yielding important insights from mutation carriers regarding the biological events involved in the development and evolution of symptoms—and the non-Mendelian genetic susceptibility factors that shape the expression and progression of symptoms. Efforts are now underway to translate these insights to sporadic FTD. It is hoped that reflections on FTD from an international perspective will spur an extension of these vibrant multicenter collaborations, to centers in the developing regions of the world. Movement in this direction will depend on advocacy from the International Society for Frontotemporal Dementias, as well as the research community, with an eye to forming strategic partnerships for research and capacity building.

References

1. Velakoulis D, Walterfang M, Mocellin R et al (2009) Frontotemporal dementia presenting as schizophrenia-like psychosis in young people: clinicopathological series and review of cases. Br J Psychiatry 194:298–305

2. Baborie A, Baborie A, Griffiths TD et al (2012) Frontotemporal dementia in elderly individuals. Arch Neurol 69:1052–1060

3. Kertesz A, McMonagle P, Blair M et al (2005) The evolution and pathology of frontotemporal dementia. Brain 128:1996–2005

4. Withall A, Draper B, Seeher K et al (2014) The prevalence and causes of younger onset dementia in eastern Sydney, Australia. Int Psychogeriatr 26:1955–1965

5. Herrera E, Caramelli P, Silveira ASB et al (2002) Epidemiologic survey of dementia in a community-dwelling Brazilian population. Alzheimer Dis Assoc Disord 16:103–108

6. Nitrini R, Caramelli P, Herrera E et al (2004) Incidence of dementia in a community-dwelling Brazilian population. Alzheimer Dis Assoc Disord 18:241–246

7. Feldman HH, Feldman H, Levy AR et al (2003) A Canadian cohort study of cognitive impairment and related dementias (ACCORD): study methods and baseline results. Neuroepidemiology 22:265–274

8. Kivipelto M, Helkala E-L, Laakso MP et al (2002) Apolipoprotein E epsilon4 allele, elevated midlife total cholesterol level, and high midlife systolic blood pressure are independent risk factors for late-life Alzheimer disease. Ann Intern Med 137:149–155

9. Ibach B, Koch H, Koller M et al (2003) Hospital admission circumstances and prevalence of frontotemporal lobar degeneration: a multicenter psychiatric state hospital study in Germany. Dement Geriatr Cogn Disord 16:253–264

10. Banerjee TK, Mukherjee CS, Dutt A et al (2008) Cognitive dysfunction in an urban Indian population – some observations. Neuroepidemiology 31:109–114

11. Ravaglia G, Forti P, Maioli F et al (2005) Incidence and etiology of dementia in a large elderly Italian population. Neurology 64:1525–1530

12. Gilberti N, Turla M, Alberici A et al (2012) Prevalence of frontotemporal lobar degeneration in an isolated population: the Vallecamonica study. Neurol Sci 33:899–904

13. Borroni B, Alberici A, Grassi M et al (2010) Is frontotemporal lobar degeneration a rare disorder? Evidence from a preliminary study in Brescia county, Italy. J Alzheimers Dis 19:111–116

14. Borroni B, Alberici A, Grassi M et al (2011) Prevalence and demographic features of early-onset neurodegenerative dementia in Brescia County, Italy. Alzheimer Dis Assoc Disord 25:341–344

15. Bernardi L, Frangipane F, Smirne N et al (2012) Epidemiology and genetics of frontotemporal dementia: a door-to-door survey in southern Italy. Neurobiol Aging 33:2948.e1–2948.e10

16. Ikeda M, Hokoishi K, Maki N et al (2001) Increased prevalence of vascular dementia in Japan: a community-based epidemiological study. Neurology 57:839–844

17. Yamada T, Hattori H, Miura A et al (2001) Prevalence of Alzheimer's disease, vascular dementia and dementia with Lewy bodies in a Japanese population. Psychiatry Clin Neurosci 55:21–25

18. Wada-Isoe K, Uemura Y, Suto Y et al (2009) Prevalence of dementia in the rural island town of Ama-cho, Japan. Neuroepidemiology 32:101–106
19. Ikejima C, Yasuno F, Mizukami K et al (2009) Prevalence and causes of early-onset dementia in Japan: a population-based study. Stroke 40:2709–2714
20. Wada-Isoe K, Ito S, Adachi T et al (2012) Epidemiological survey of frontotemporal lobar degeneration in Tottori prefecture, Japan. Dement Geriatr Cogn Dis Extra 2:381–386
21. Rosso SM, Donker Kaat L, Baks T et al (2003) Frontotemporal dementia in the Netherlands: patient characteristics and prevalence estimates from a population-based study. Brain 126:2016–2022
22. Amoo G, Akinyemi RO, Onofa LU et al (2011) Profile of clinically-diagnosed dementias in a neuropsychiatric practice in Abeokuta, South-Western Nigeria. Afr J Psychiatry (Johannesbg) 14:377–382
23. Gascón-Bayarri J, Reñé R, Del Barrio JL et al (2007) Prevalence of dementia subtypes in El prat de Llobregat, Catalonia, Spain: the PRATICON study. Neuroepidemiology 28:224–234
24. Garre-Olmo J, Genís Batlle D, del Mar Fernández M et al (2010) Incidence and subtypes of early-onset dementia in a geographically defined general population. Neurology 75:1249–1255
25. Lee DY, Lee JH, Ju Y-S et al (2002) The prevalence of dementia in older people in an urban population of Korea: the Seoul study. J Am Geriatr Soc 50:1233–1239
26. Andreasen N, Blennow K, Sjödin C et al (1999) Prevalence and incidence of clinically diagnosed memory impairments in a geographically defined general population in Sweden. The Piteå Dementia Project Neuroepidemiology 18:144–155
27. Gislason TB, Sjögren M, Larsson L et al (2003) The prevalence of frontal variant frontotemporal dementia and the frontal lobe syndrome in a population-based sample of 85 year olds. J Neurol Neurosurg Psychiatry 74:867–871
28. Gurvit H, Emre M, Tinaz S et al (2008) The prevalence of dementia in an urban Turkish population. Am J Alzheimers Dis Other Dement 23:67–76
29. Ratnavalli E, Ratnavalli E, Brayne C et al (2002) The prevalence of frontotemporal dementia. Neurology 58:1615–1621
30. Stevens T, Livingston G, Kitchen G et al (2002) Islington study of dementia subtypes in the community. Br J Psychiatry 180:270–276
31. Harvey RJ, Skelton-Robinson M, Rossor MN (2003) The prevalence and causes of dementia in people under the age of 65 years. J Neurol Neurosurg Psychiatry 74:1206–1209
32. Mercy L, Hodges JR, Dawson K et al (2008) Incidence of early-onset dementias in Cambridgeshire, United Kingdom. Neurology 71:1496–1499
33. Coyle-Gilchrist ITS, Dick KM, Patterson K et al (2016) Prevalence, characteristics, and survival of frontotemporal lobar degeneration syndromes. Neurology 86:1736–1743
34. Knopman DS, Petersen RC, Edland SD et al (2004) The incidence of frontotemporal lobar degeneration in Rochester, Minnesota, 1990 through 1994. Neurology 62:506–508
35. Hogan DB, Jetté N, Fiest KM et al (2016) The prevalence and incidence of frontotemporal dementia: a systematic review. Can J Neurol Sci 43(Suppl 1):S96–S109
36. Phung TKT, Waltoft BL, Kessing LV et al (2010) Time trend in diagnosing dementia in secondary care. Dement Geriatr Cogn Disord 29:146–153
37. Onyike CU, Diehl-Schmid J (2013) The epidemiology of frontotemporal dementia. Int Rev Psychiatry 25:130–137
38. Knopman DS, Roberts RO (2011) Estimating the number of persons with frontotemporal lobar degeneration in the US population. J Mol Neurosci 45:330–335
39. Lekoubou A, Echouffo-Tcheugui JB, Kengne AP (2014) Epidemiology of neurodegenerative diseases in sub-Saharan Africa: a systematic review. BMC Public Health 14:653
40. Rascovsky K, Hodges JR, Kipps CM et al (2007) Diagnostic criteria for the behavioral variant of frontotemporal dementia (bvFTD): current limitations and future directions. Alzheimer Dis Assoc Disord 21:S14–S18
41. Rascovsky K, Hodges JR, Knopman D et al (2011) Sensitivity of revised diagnostic criteria for the behavioural variant of frontotemporal dementia. Brain 134:2456–2477
42. Gorno-Tempini ML, Hillis AE, Weintraub S et al (2011) Classification of primary progressive aphasia and its variants. Neurology 76:1006–1014
43. Lamarre AK, Rascovsky K, Bostrom A et al (2013) Interrater reliability of the new criteria for behavioral variant frontotemporal dementia. Neurology 80:1973–1977
44. Harris JM, Gall C, Thompson JC et al (2013) Sensitivity and specificity of FTDC criteria for behavioral variant frontotemporal dementia. Neurology 80:1881–1887
45. Poorkaj P, Bird TD, Wijsman E et al (1998) Tau is a candidate gene for chromosome 17 frontotemporal dementia. Ann Neurol 43:815–825
46. Hutton M, Lendon CL, Rizzu P et al (1998) Association of missense and 5′-splice-site mutations in tau with the inherited dementia FTDP-17. Nat Publ Group 393:702–705
47. Spillantini MG, Murrell JR, Goedert M et al (1998) Mutation in the tau gene in familial multiple system tauopathy with presenile dementia. Proc Natl Acad Sci 95:7737–7741
48. Baker M, Mackenzie IR, Pickering-Brown SM et al (2006) Mutations in progranulin cause tau-negative frontotemporal dementia linked to chromosome 17. Nature 442:916–919
49. Cruts M, Gijselinck I, van der Zee J et al (2006) Null mutations in progranulin cause ubiquitin-positive frontotemporal dementia linked to chromosome 17q21. Nature 442:920–924

50. DeJesus-Hernandez M, Mackenzie IR, Boeve BF et al (2011) Expanded GGGGCC hexanucleotide repeat in noncoding region of C9orf72 causes chromosome 9p-linked FTD and ALS. Neuron 72:245–256

51. Renton AE, Majounie E, Waite A et al (2011) A hexanucleotide repeat expansion in C9orf72 is the cause of chromosome 9p21-linked ALS-FTD. Neuron 72:257–268

52. Greaves CV, Rohrer JD (2019) An update on genetic frontotemporal dementia. J Neurol 266:2075–2086

53. Benussi L, Ghidoni R, Pegoiani E et al (2009) Progranulin Leu271LeufsX10 is one of the most common FTLD and CBS associated mutations worldwide. Neurobiol Dis 33:379–385

54. Borroni B, Bonvicini C, Galimberti D et al (2011) Founder effect and estimation of the age of the Progranulin Thr272fs mutation in 14 Italian pedigrees with frontotemporal lobar degeneration. Neurobiol Aging 32:555.e1–555.e8

55. Majounie E, Renton AE, Mok K et al (2012) Frequency of the C9orf72 hexanucleotide repeat expansion in patients with amyotrophic lateral sclerosis and frontotemporal dementia: a cross-sectional study. Lancet Neurol 11:323–330

56. Kartanou C, Karadima G, Koutsis G et al (2018) Screening for the C9orf72 repeat expansion in a greek frontotemporal dementia cohort. Amyotroph Lat Scl Frr 19:152–154

57. Guven G, Lohmann E, Bras J et al (2016) Mutation frequency of the major Frontotemporal dementia genes, MAPT, GRN and C9orf72 in a Turkish cohort of dementia patients. PLoS One 11:e0162592

58. Pliner HA, Mann DM, Traynor BJ (2014) Searching for Grendel: origin and global spread of the C9orf72 repeat expansion. Acta Neuropathol 127:391–396

59. Cintra VP, Bonadia LC, Andrade HMT et al (2018) The frequency of the C9orf72 expansion in a Brazilian population. Neurobiol Aging 66:179.e1–179.e4

60. Jiao B, Tang B, Liu X et al (2014) Identification of C9orf72 repeat expansions in patients with amyotrophic lateral sclerosis and frontotemporal dementia in mainland China. Neurobiol Aging 35:936.e19–936.e22

61. Sirkis DW, Geier EG, Bonham LW et al (2019) Recent advances in the genetics of frontotemporal dementia. Curr Genet Med Rep 7:41–52

62. Fukuhara R, Ghosh A, Fuh J-L et al (2014) Family history of frontotemporal lobar degeneration in Asia – an international multi-center research. Int Psychogeriatr 26:1967–1971

63. Bannwarth S, Ait-El-Mkadem S, Chaussenot A et al (2014) A mitochondrial origin for frontotemporal dementia and amyotrophic lateral sclerosis through CHCHD10 involvement. Brain 137:2329–2345

64. Jiao B, Xiao T, Hou L et al (2016) High prevalence of CHCHD10 mutation in patients with frontotemporal dementia from China. Brain 139:e21–e21

65. Che X-Q, Zhao Q-H, Huang Y et al (2017) Genetic features of MAPT, GRN, C9orf72 and CHCHD10 gene mutations in Chinese patients with Frontotemporal dementia. Curr Alzheimer Res 14:1102–1108

66. Aswathy PM, Jairani PS, Raghavan SK et al (2016) Progranulin mutation analysis: identification of one novel mutation in exon 12 associated with frontotemporal dementia. Neurobiol Aging 39:218.e1–218.e3

67. Jiao B, Guo J-F, Wang Y-Q et al (2013) C9orf72 mutation is rare in Alzheimer's disease, Parkinson's disease, and essential tremor in China. Front Cell Neurosci 7:164

68. Lin C-H, Chen T-F, Chiu M-J et al (2014) Lack of c9orf72 repeat expansion in Taiwanese patients with mixed neurodegenerative disorders. Front Neurol 5:59

69. Tang M, Gu X, Wei J et al (2016) Analyses MAPT, GRN, and C9orf72 mutations in Chinese patients with frontotemporal dementia. Neurobiol Aging 46:235.e11–235.e15

70. Ogaki K, Ogaki K, Li Y et al (2013) Analyses of the MAPT, PGRN, and C9orf72 mutations in Japanese patients with FTLD, PSP, and CBS. Parkinsonism Relat Disord 19:15–20

71. Jang J-H, Kwon M-J, Choi WJ et al (2013) Analysis of the C9orf72 hexanucleotide repeat expansion in Korean patients with familial and sporadic amyotrophic lateral sclerosis. Neurobiol Aging 34:1311.e7–1311.e9

72. Kim E-J, Kwon JC, Park KH et al (2014) Clinical and genetic analysis of MAPT, GRN, and C9orf72 genes in Korean patients with frontotemporal dementia. Neurobiol Aging 35:1213.e13–1213.e17

73. Das G, Sadhukhan T, Sadhukhan D et al (2013) Genetic study on frontotemporal lobar degeneration in India. Parkinsonism Relat Disord 19:487–489

74. Mukherjee O, Das G, Sen S et al (2015) C9orf72 mutations may be rare in frontotemporal lobar degeneration patients in India. Amyotroph Lateral Scler Frontotemporal Degener 17:151–153

75. Josephs K, Hu W, Hillis A et al (2016) Frontotemporal dementia in patients of African descent. J Neurochem 138:229

76. Ghosh A, Dutt A, Ghosh M et al (2013) Using the revised diagnostic criteria for Frontotemporal dementia in India: evidence of an advanced and florid disease. PLoS One 8:e60999

77. Ghosh A, Dutt A (2010) Utilisation behaviour in frontotemporal dementia. J Neurol Neurosurg Psychiatry 81:154–156

78. Akinyemi RO, Akinyemi RO, Owolabi MO et al (2009) Frontotemporal dementia in a Nigerian woman: case report and brief review of the literature. Afr J Med Med Sci 38:71–75

79. Shinagawa S, Ikeda M, Nestor PJ et al (2009) Characteristics of abnormal eating behaviours in frontotemporal lobar degeneration: a cross-cultural survey. J Neurol Neurosurg Psychiatry 80:1413–1414

80. Tanabe H (2007) The uniqueness of Gogi aphasia owing to temporal lobar atrophy. Alzheimer Dis Assoc Disord 21:S12–S13

81. Ellajosyula R, Narayanan J, Patterson K (2020) Striking loss of second language in bilingual patients with semantic dementia. J Neurol 267:551–560

82. Alladi S, Bak TH, Shailaja M et al (2017) Bilingualism delays the onset of behavioral but not aphasic forms of frontotemporal dementia. Neuropsychologia 99:207–212

83. Kansal K, Mareddy M, Sloane KL et al (2016) Survival in Frontotemporal dementia phenotypes: a meta-analysis. Dement Geriatr Cogn Disord 41:109–122

84. Dias A, Patel V (2009) Closing the treatment gap for dementia in India. Indian J Psychiatry 51(Suppl 1):S93–S97

85. Kamoga R, Rukundo GZ, Wakida EK et al (2019) Dementia assessment and diagnostic practices of healthcare workers in rural southwestern Uganda: a cross-sectional qualitative study. BMC Health Serv Res 19:1005

86. Wylie MA, Shnall A, Onyike CU et al (2013) Management of frontotemporal dementia in mental health and multidisciplinary settings. Int Rev Psychiatry 25:230–236

87. Stoner CR, Lakshminarayanan M, Durgante H et al (2019) Psychosocial interventions for dementia in low- and middle-income countries (LMICs): a systematic review of effectiveness and implementation readiness. Aging Ment Health 6:1–12

Progressive Supranuclear Palsy and Corticobasal Degeneration

David G. Coughlin, Dennis W. Dickson, Keith A. Josephs, and Irene Litvan

Introduction

The two most common clinicopathologic subtypes of frontotemporal lobar degeneration (FTLD) are characterized by TDP-43 or tau pathology [1]. Tau is a microtubule-associated protein important for stability and functional properties of microtubules. The gene that encodes tau protein (*MAPT*) is located on chromosome 17, and it undergoes alternative splicing of exons 2, 3, and 10 to generate six isoforms of tau [2]. Alternative splicing of exon 10 generates two major classes of tau protein that contain either three (3R) or four (4R) $\approx$30-amino acid repeats in the microtubule-binding domain of tau. Neurodegenerative tauopathies can be subclassified based upon the predominant type of tau that accumulates in cellular lesions [3]. Pick's dis-

ease, a rare frontotemporal dementia with lobar cortical atrophy and neuronal Pick bodies, is characterized by tau composed predominantly of 3R tau, while neurofibrillary tangles that characterize the pathology in Alzheimer's disease and chronic traumatic encephalopathy are composed of a mixture of 3R and 4R tau with distinct ultrastructural properties [4, 5]. Disorders associated with 4R tau are clinically and pathologically heterogeneous and include aging-related disorders, such as aging-related tau astrogliopathy (ARTAG) [6] and argyrophilic grain disease (AGD) [3, 7]. The most common of the neurodegenerative 4R tauopathies are progressive supranuclear palsy (PSP) and corticobasal degeneration (CBD), which is the focus of this chapter.

Progressive Supranuclear Palsy

PSP was described by Steele, Richardson, and Olszewski in a small autopsy series of patients with postural instability, vertical supranuclear gaze palsy, facial and cervical dystonia, as well as dementia. Despite some clinical variability, they shared distinctive pathologic features, including argyrophilic neurofibrillary tangles in select subcortical and brainstem nuclei. [8]. With the advent of tau biochemistry and molecular biology, the pathologic features of PSP have been expanded to include not only neuronal lesions but also glial lesions [3, 9]. The clinical syndromes

D. G. Coughlin
UC San Diego, Department of Neurosciences,
La Jolla, CA, USA
e-mail: dacoughlin@health.ucsd.edu

D. W. Dickson
Department of Neuroscience, Mayo Clinic,
Jacksonville, FL, USA

K. A. Josephs
Department of Neurology, Mayo Clinic,
Rochester, MN, USA
e-mail: josephs.keith@mayo.edu

I. Litvan (✉)
UC San Diego Department of Neurosciences,
La Jolla, CA, USA
e-mail: ilitvan@health.ucsd.edu

© Springer Nature Switzerland AG 2021
B. Ghetti et al. (eds.), *Frontotemporal Dementias*, Advances in Experimental Medicine and Biology
1281, https://doi.org/10.1007/978-3-030-51140-1_11

associated with the characteristic tau pathology of PSP have also expanded from the original descriptions and is described later in the chapter.

Epidemiology of Progressive Supranuclear Palsy

The prevalence of PSP is thought to be approximately 6/100,000 patients [10–13]; however, there is a growing understanding that PSP pathology is associated with multiple clinical phenotypes, suggesting that the above figure may require revision. Increased awareness of this fact led to increased age-adjusted prevalence estimates in Europe (8.8–10.8/100,000 patients) [11, 14]. Of note, age-adjusted prevalence estimates from the same city in Japan (Yonogo) adjusted to the census of the earlier study increased from 5.8/100,000 patients in 1999 to 17/100,000 patients in 2010 [15, 16]. This is, in part, due to identification of more phenotypes, since the previous studies used the National Institute of Neurologic Disease and Stroke and Society for PSP (NINDS-SPSP) criteria that only identified the classical PSP phenotype (also named PSP-Richardson syndrome [PSP-RS]).

Clinical Features of Progressive Supranuclear Palsy

In addition to the typical presentation described by Richardson and colleagues (PSP-RS), other phenotypes associated with PSP pathology have been described, including an extrapyramidal disorder mimicking Parkinson's disease (PSP-P), corticobasal syndrome (PSP-CBS), dementia with predominantly frontal characteristics (PSP-F), dementia with speech and language disturbances (PSP-SL), and others. Consequently, the newest clinical criteria for PSP, supported by the International Parkinson and Movement Disorder Society (MDS-PSP criteria), include a wider clinical spectrum [17]. Typical age of onset of PSP is in the seventh decade of life [17–19], and average survival is 5–6 years; however, certain phenotypes are associated with much longer disease durations [19, 20].

Several criteria for PSP were proposed based upon clinical case series [21–24], but the first widely used criteria that were based on autopsy-confirmed cases was reported by Litvan et al. [18] and supported by the NINDS-SPSP. The NINDS-SPSP criteria outlined several core features of PSP-RS. Mandatory features included a gradually progressive disorder with age of onset 40 years of age or later, presence of vertical supranuclear gaze palsy, and/or postural instability with falls within the first year of disease. Both features had to be present for a diagnosis of "probable PSP," and only vertical supranuclear gaze palsy or slowing of saccades and postural instability with falls within the first year of disease was consistent with a diagnosis of "possible PSP."

Regarding vertical supranuclear gaze palsy, restricted downward gaze has been considered most specific for PSP because restricted upward gaze can be seen to a lesser degree in aging [25], Parkinson's disease [26], and other conditions [27–31] like severely restricted upward gaze and slowing of vertical saccades. At more advanced stages, horizontal supranuclear gaze palsy may develop, as well [32]. Vertical supranuclear gaze palsy may be preceded by subtle ocular motor abnormalities, including loss of vertical optokinetic nystagmus [33], "stair casing," and the "round the house sign" [34], where horizontal saccadic excursions interrupt vertical eye movements. Other ocular motor movement abnormalities include hypometric saccades, breakdown of smooth pursuit, and square wave jerks [35]. Loss of vergence is observed early and may contribute to frequent complaints of diplopia [36]. Other eye findings include blepharospasm and eyelid-opening apraxia [37], although these are not usually early features.

Early loss of postural reflexes and falls are common and often an early complaint in PSP-RS, usually occurring within the first year of illness. Falls tend to be backwards, but it can occur in any direction and may be compounded by freezing of gait. Falls can result in significant morbidity due to lacerations, fractures, or intracerebral bleeding [32, 38].

While these features define the core clinical features of PSP-RS, a number of other clinical features are often observed. Parkinsonism manifested by symmetric akinesia and rigidity with an axial predominance is common. Neck stiffness with retrocollis has been described in early descriptions of PSP, but it is rare [8]. Facial dystonia produces the so-called PSP stare, with decreased blink rate, furrowed and raised eyebrows, and a look of surprise. Inappropriate laughter and crying episodes are often observed (pseudobulbar affect). Early hypokinetic and spastic dysarthria is a secondary feature, which can progress to anarthria in severe cases [39]. Dysphagia occurs relatively early, and it is frequently implicated as a cause of death due to aspiration pneumonia [40, 41]. Cognitive manifestations associated with PSP overlap with corticobasal syndrome and frontotemporal dementia (FTD). The clinical course of PSP is relentless and nearly always is associated with a frontal-subcortical-type dementia.

PSP-RS phenotype is the clinical syndrome most likely to have PSP pathology at autopsy. Because of this, the NINDS-SPSP criteria proved to be specific for PSP pathology [42, 43], but to have relatively low sensitivity [43–45]. This is because PSP pathology can present with other clinical syndromes, and eye movement abnormalities seen in PSP often occur later in the course of the disease and sometimes not at all [19, 20, 46–57]. In one autopsy series, 76% of pathologically confirmed PSP had a clinical syndrome other than PSP-RS [58].

The most common clinical PSP variant mimics idiopathic Parkinson's disease (PSP-P) and makes up about one-third of pathologically confirmed cases [46, 59–62]. These patients have asymmetric resting tremor and asymmetric appendicular bradykinesia and rigidity, making the distinction between PSP-P and Parkinson's disease challenging [46, 59, 60, 62, 63]. As many as one-third of these patients will respond to levodopa and show greater than 30% reduction in the Unified Parkinson's Disease Rating Scale [46, 64–67]. Some also develop levodopa-induced dyskinesias [46]. Most PSP patients have minimal or no response to levodopa therapy, and if a response occurs, it is typically mild and not sustained [20, 24, 68]. Robust and prolonged response to levodopa therapy is an exclusionary criterion for PSP and makes Parkinson's disease a more likely diagnosis [17]. It can be 3–4 years into the disease course before supranuclear gaze palsy is present to aid in refining the diagnosis in PSP-P [19, 62]. PSP-P patients also have a longer disease duration than PSP-RS, with an average survival of 10–15 years [19, 46, 62].

Other syndromes have been described in autopsy-confirmed PSP. Some present with impulsivity and behavioral changes, including apathy, impulsivity, and social inappropriateness akin to behavioral-variant frontotemporal dementia (PSP-F) [53, 69, 70]. Others present with progressive non-fluent aphasia or apraxia of speech (PSP-SL) [48, 52, 70, 71]. About 10% have a corticobasal syndrome with asymmetrical dystonia, myoclonus, apraxia, and cortical sensory loss (PSP-CBS) [55, 56, 70, 72]. Another rare presentation, but one that is highly predictive of PSP pathology, is pure akinesia with gait freezing (PSP-PAGF) [47, 73, 74]. Early presentations currently considered to be "suggestive" of PSP in MDS-PSP criteria are isolated postural instability (PSP-PI) [19, 75] and isolated oculomotor dysfunction (PSP-OM) [19, 20]. The most uncommon presentations are progressive cerebellar ataxia (PSP-C) [51, 76, 77] and primary lateral sclerosis (PSP-PLS) [50, 57]. It is important to note that while some patients present with discrete syndromes, it is common for considerable overlap, and patients also acquire new signs and symptoms as the disease progresses. Regardless of the initial syndrome, most patients develop vertical supranuclear gaze palsy and postural instability, which are core features of PSP-RS, that make diagnosis obvious, but these may occur only later in the disease course in some of the PSP clinical variants [19].

Recognition of the spectrum of clinical heterogeneity in PSP, led the MDS-PSP criteria to incorporate a broader set of symptoms and signs, as well as levels of certainty that would be associated with PSP pathology [17]. These criteria are more sensitive, but they are less specific than the NINDS-SPSP criteria [78, 79]. The implementation of "multiple allocation extinction" rules (MAX rules) have been necessary to help disentangle patients who may be classified into more than one clinical MDS-PSP category [79]. Even so, these MAX rules may fail to separate up to 40% of patients with PSP-P and PSP-RS overlap syndromes [80]. These issues highlight the ongoing need for specific biomarkers to improve diagnostic accuracy of PSP during life.

Neuropathology of Progressive Supranuclear Palsy

The external appearance of PSP at postmortem evaluation depends upon the clinical syndrome. PSP-RS may have no significant cortical atrophy or mild atrophy affecting the dorsolateral frontal lobe. PSP-F and PSP-CBS usually have more marked frontal atrophy, especially affecting the superior frontal gyrus, while PSP-SL may have more significant frontal atrophy, especially affecting the peri-Sylvian inferior frontal gyrus. Asymmetry, which is not often assessed with research protocols that evaluate only one side of the brain for histology, can be notable in PSP-SL and PSP-CBS. PSP-PLS has focal atrophy affecting the precentral gyrus; it can be asymmetrical as well. The most striking macroscopic finding in PSP-RS (and PSP-P) is midbrain atrophy (Fig. 1a) with loss of neuromelanin pigment on transverse sections of the brainstem (Fig. 1d). The subthalamic nucleus invariably has atrophy (Fig. 1b), and there is also atrophy of the superior cerebellar peduncle (Fig. 1e) and atrophy of the hilus of the cerebellar dentate nucleus (Fig. 1c). Atrophy of subthalamic nucleus and midbrain is usually less severe in PSP-F and PSP-CBS, and often very severe in PSP-PAGF. In the latter, atrophy is frequently accompanied by similar changes in the globus pallidus and with reddish-brown discoloration due to deposition of iron pigment (pallido-nigro-luysial "pigment-spheroid degeneration" [81]).

Histopathologic findings in PSP are similar in the various subtypes. The clinicopathologic sub-types differ in the relative distribution of the neuronal loss and gliosis, and in the density of tau pathology [82]. There are no distinctive cellular pathologies in PSP clinicopathologic variants. The major histopathologic lesions in PSP are neurofibrillary tangles, which often have a globose shape in vulnerable subcortical nuclei, such as the subthalamic nucleus (Fig. 2a) and substantia nigra (Fig. 2b). The morphology and distribution of tangles in PSP is different from the most common disorder with neurofibrillary tangles, Alzheimer's disease (AD), in that subcortical and brainstem nuclei are preferentially affected. The tangles are positive for phospho-tau (Fig. 2d). Using antibodies specific to tau isoforms, the tangles in PSP preferentially accumulate 4R tau (not shown). Tau immunohistochemistry also shows distinctive glial pathology in PSP, including tufted astrocytes (Figs. 2d and 3e) and oligodendroglial coiled bodies (Fig. 2f). Tufted astrocytes are most frequent in neocortex, neostriatum, and midbrain tectum. Coiled bodies are widespread in affected cerebral white matter and vulnerable subcortical fiber tracts in the basal telencephalon, diencephalon, brain stem, and cerebellum. A common neurodegenerative change in the cerebellar dentate nucleus that is not associated with tau pathology is the presence of irregularly swollen cell processes around apical dendrites and cell bodies of cerebellar dentate nucleus neurons (Fig. 2c), a process referred to as grumose degeneration [83]. Glial pathology is increasingly recognized to play a significant role in pathogenesis of neurodegenerative disease, and in PSP microgliosis and astrogliosis parallels the systems affected by neurodegeneration [84], with little evidence to suggest that it precedes tau pathology.

Corticobasal Degeneration

The term corticobasal degeneration was coined by Gibb, Luthert and Marsden [85] to describe the pathology of a rare disorder associated with cognitive and motor features affecting the neocortex and basal ganglia. The clinically defined corticobasal syndrome (CBS) is char-

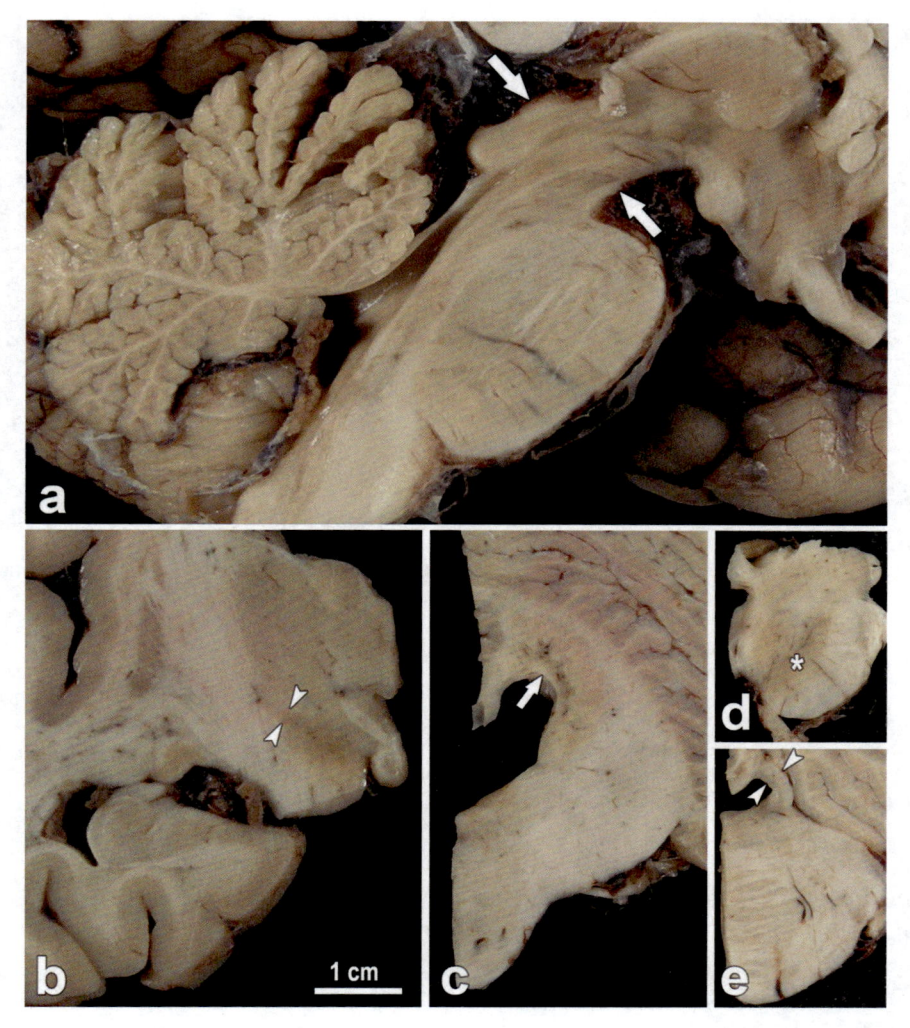

Fig. 1 Macroscopic findings in PSP. (**a**) A sagittal section of the brainstem shows marked atrophy of the midbrain (arrows). (**b**) A coronal section of the diencephalon shows marked atrophy of the subthalamic nucleus (arrowheads). (**c**) A section of the cerebellum at the level of the middle cerebellar peduncle shows marked atrophy and discoloration of the dentate nucleus of the cerebellum (arrow). (**d**) A transverse section of the midbrain shows atrophy and marked neuromelanin pigment loss in the substantia nigra (asterisk). (**e**) A transverse section of the pons shows marked atrophy of the superior cerebellar peduncle (arrowheads)

acterized by progressive cognitive decline associated with asymmetrical rigidity, dystonia, myoclonus, and alien-limb phenomenon. Early autopsy studies reported focal cortical atrophy and swollen achromatic neurons ("ballooned neurons" [86]), as well as neuronal loss in the substantia nigra and cerebellar dentate nucleus—"corticodentatonigral degeneration with neuronal achromasia" [87]. These descriptions did not recognize the tau pathol-ogy in CBD because neuronal lesions in CBD are weakly positive or negative with traditional silver impregnation methods. It was not until the early 1990s that widespread tau pathology in CBD was shown to be distinct from Alzheimer's disease, using immunohistochemistry and ultrastructural methods [88–90]. The pathognomonic astrocytic lesion of CBD ("astrocytic plaques") was described in 1995 [91].

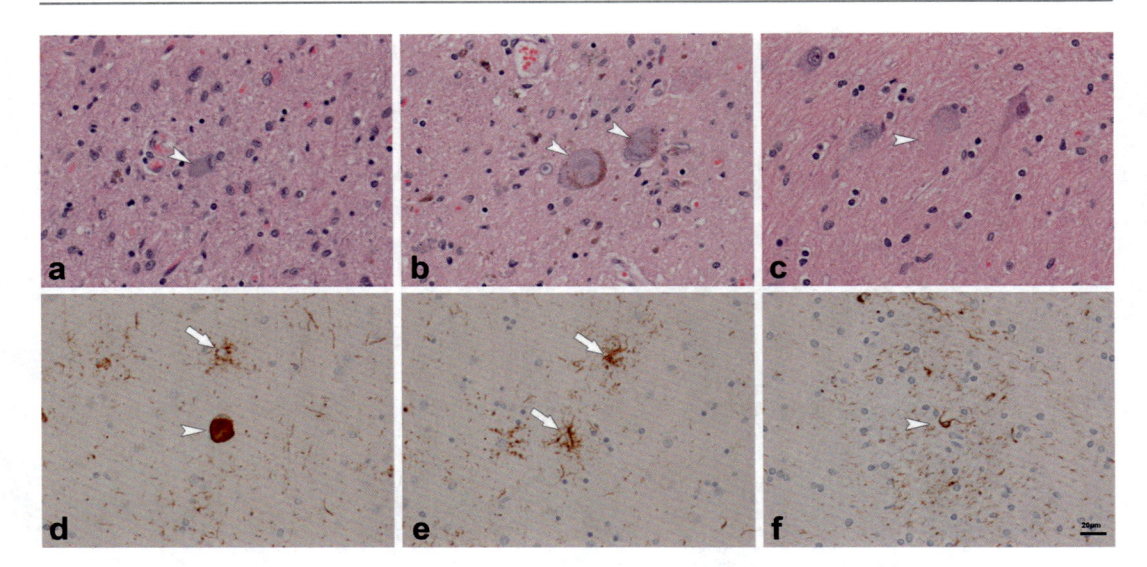

Fig. 2 Microscopic findings in PSP. (**a**) An H&E stained section of the subthalamic nucleus shows severe neuronal loss and astrocytosis, with neurofibrillary tangles (arrow) in residual neurons. (**b**) An H&E stained section of the substantia nigra shows neuronal loss and gliosis with extraneuronal neuromelanin pigment and globose neurofibrillary tangles (arrowheads). (**c**) An H&E stained section of the cerebellar dentate nucleus shows granular eosinophilic swollen cell processes (arrowhead), obscuring the outlines of the neuron, findings characteristic of grumose degeneration (arrow). (**d**) Phospho-tau immunohistochemistry of the caudate nucleus shows a globose neurofibrillary tangle (arrowhead) and a tufted astrocyte (arrow). (**e**) Phospho-tau immunohistochemistry of the caudate nucleus shows several tufted astrocytes (arrows) with morphologic heterogeneity. (**f**) Phospho-tau immunohistochemistry of the internal capsule shows oligodendroglial coiled bodies (arrowhead). All images are of same magnification, bar in (**f**) is 20 μm

Epidemiology of Corticobasal Degeneration

Like PSP, pathologically confirmed CBD has a range of clinical presentations, and CBS may not be the most common. Moreover, the pathologic substrate of CBS is mixed, with PSP being as common as CBD [56, 92], but other disorders, particularly atypical presentations of Alzheimer's disease, can also present with CBS [56, 85, 93–98]. Estimates of prevalence of CBD are inherently flawed. For these reasons, the term corticobasal syndrome (CBS) is now preferred to refer to the clinical presentation described earlier, whereas corticobasal degeneration (CBD) is reserved for the neuropathological diagnosis. The incidence of CBD is estimated to be 0.62–0.92/100,000 [93, 99–101].

Clinical Features of Corticobasal Degeneration Presenting as Corticobasal Syndrome

The onset of CBS is typically in the sixth or seventh decade of life, with a mean survival of about 7 years from diagnosis [93, 99–101]. The motor manifestations of CBS include an asymmetric parkinsonism manifested predominantly by rigidity and bradykinesia [93]. While asymmetry in parkinsonian features is common in Parkinson's disease, the asymmetry in CBS can be striking. There is frequently additional dystonic posturing of the limb. Superimposed may be ideomotor and limb-kinetic apraxia [55, 99, 102]. Alien-limb phenomenon affecting the arm or leg has been described and often results in an unawareness of a levitating hand or leg due to feeling the limb alien, and more rarely, intermanual conflict [103]. Myoclonus is often present, and it may affect limbs or, rarely, the face [99, 104]. Myoclonus is worsened by action, posture, or stimuli [55, 99, 104]. At times, myoclonus can be difficult to differentiate from tremor, although the quality of myoclonic tremor is jerky rather than the smooth oscillatory tremor observed in Parkinson's disease and other parkinsonian disorders [105]. Postural instability and falls are common, but usually later in the disease course than in PSP, unless the symptoms start in lower extremities [93]. Parkinsonism associated with CBS may benefit from levodopa therapy, but improvement in symptoms is rare and levodopa-induced dyskinesias are also rare [55]. Sustained and robust levodopa responsiveness is an exclusionary criterion to the diagnosis of CBS [93, 106].

Several cognitive features and other signs referable to higher-order cortical function are common in CBS. As previously mentioned, apraxia is a core feature. Ideomotor apraxia is usually one of the first disease features. Some patients develop orobuccal apraxia or apraxia of eyelid opening [99, 104, 107]. Cortical sensory loss with astereognosis and agraphesthesia are frequently observed [108, 109]. Visual neglect may be seen, and it is related to parietal lobe dysfunction [95, 107, 110]. A progressive non-fluent aphasia is also described in CBS, with occasional overlay of apraxia of speech from frontal lobe dysfunction [95, 104, 107, 111]. Other features of frontal lobe dysfunction, such as apathy and disinhibition, are common and early [55, 93].

The clinical presentation of autopsy-confirmed CBD is varied, with some presenting with a cognitive syndrome, and some primarily with a motor phenotype. Other neurodegenerative disorders, PSP and Alzheimer's disease in particular, can present with CBS. Unlike PSP, these initial presentations may not necessarily coalesce into a common phenotype over time, making diagnostics even more challenging. Concomitantly, the clinical diagnosis of CBS has relatively poor predictive value for CBD pathology at autopsy compared to other neurodegenerative disorders. The

sensitivity of clinical findings predicting CBD at autopsy is between 26% and 56%. The majority of these studies were performed using older criteria; recently, more specific criteria have not been fully vetted [55, 59, 70, 95]. Current clinical criteria for CBD define a gradual progressive disorder with insidious onset and several possible phenotypes, including CBS, a frontal behavioral-spatial syndrome, a variant of primary non-fluent aphasia, and a PSP syndrome. The clinical syndrome of probable CBS is defined as having two of the following signs: limbs with asymmetric rigidity and akinesia, limb dystonia or limb myoclonus, and two of the following signs and symptoms: orobuccal or limb apraxia, cortical sensory deficits, or alien-limb phenomena. Possible clinical CBS involves having one limb with rigidity or akinesia, limb dystonia, or limb myoclonus with one of the above supportive features. A frontal behavioral spatial syndrome is described with the attendant cognitive features. Non-fluent primary progressive aphasia and a PSP phenotype are recognized but considered as possible CBD. Patients with a PSP clinical syndrome must have at least one additional symptom or sign (limb rigidity/akinesia, limb dystonia or myoclonus, apraxia, and cortical sensory loss) [93].

There are multiple exclusion criteria that, if present, make CBD a less likely cause of the clinical presentation. The most important are the presence of genetic mutations in *GRN*, *FUS*, *TARDBP*, *PSEN1/2*, and *APP* genes. Another exclusionary criterion is a cerebrospinal fluid (CSF) Aβ42/tau ratio consistent with Alzheimer's disease [112]. Classic 4–6 Hz parkinsonian resting tremor, hallucinations, dysautonomia, cerebellar signs, the presence of both upper and motor neuron signs, or the semantic or logopenic variants of primary progressive aphasia are also considered exclusionary; they are more likely to indicate Parkinson's disease, dementia with Lewy bodies, multiple systems atrophy, ALS, or FTLD. Lastly, because there are occasional reports of fulminant presentations of CBD [113, 114], imaging consistent with Creutzfeldt-Jakob disease is also exclusionary.

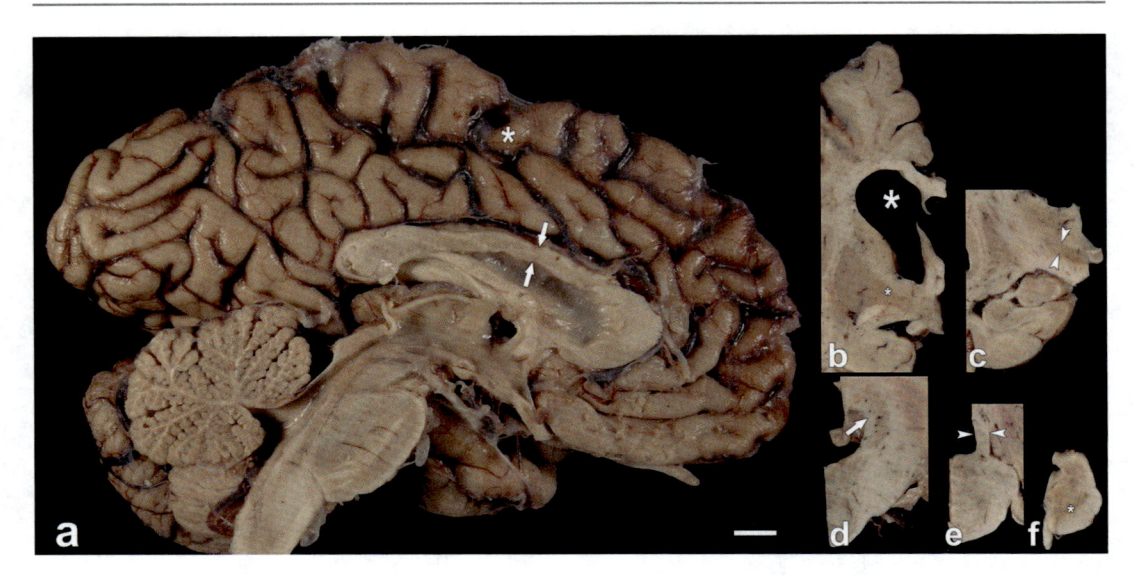

Fig. 3 Macroscopic findings in CBD. (**a**) The medial surface of left hemibrain shows atrophy of the superior frontal gyrus (asterisk indicates area of greatest pathology) and focal atrophy of the corpus callosum (arrows). (**b**) A coronal section of the brain at the level of the fornix shows marked enlargement of the frontal horn of the lateral ventricle (large asterisk). There is also atrophy and discoloration of the globus pallidus (small asterisk). (**c**) A coronal section of the diencephalon and anterior medial temporal lobe shows no hippocampal atrophy and minimal-to-no atrophy of the subthalamic nucleus (arrowheads). (**d**) A section of the cerebellum at the level of the middle cerebellar peduncle shows no atrophy and normal myelin in the hilus of the dentate nucleus (arrow). (**e**) A transverse section of the pons shows no atrophy of the superior cerebellar peduncle (arrowheads). (**f**) A transverse section of the midbrain shows mild atrophy and marked neuromelanin pigment loss in the substantia nigra (asterisk)

Neuropathology of Corticobasal Degeneration

The external appearance of the CBD brain at postmortem evaluation depends upon the clinical syndrome. For patients presenting with CBS or frontotemporal dementia syndromes, there is usually focal atrophy, especially affecting the medial superior frontal gyrus (Fig. 3a). Language-predominant syndromes often have inferior frontal gyrus (peri-Sylvian) atrophy. There is often atrophy of the corpus callosum (Fig. 3a), which tends to parallel the distribution and severity of the focal cortical pathology. Atrophy can be asymmetrical, but this is often difficult to assess at autopsy, given that half the brain is usually frozen for research purposes. Some cases, particularly patients with long tract signs, may have atrophy that extends to the motor cortex. Coronal sections frequently show enlargement of the frontal horn of the lateral ventricle (Fig. 3b). The most common finding in the basal ganglia is atro-

phy and reddish-brown discoloration of the globus pallidus (Fig. 2b). Unlike PSP, there is usually no significant atrophy of the subthalamic nucleus (Fig. 3c). Similarly, the hilus of the cerebellar dentate nucleus (Fig. 3d) and the superior cerebellar peduncle (Fig. 3e) do not have atrophy. Similar to PSP, there is usually loss of neuromelanin pigment in the substantia nigra (Fig. 3f).

Microscopic examination of atrophic cortical sections shows neuronal loss with superficial spongiosis, gliosis, and usually achromatic or ballooned neurons, which are readily detected with routine histology stains, such as hematoxylin-and-eosin (Fig. 4a). Ballooned neurons are found in middle and lower cortical layers of affected neocortices and have diffuse phospho-tau immunoreactivity (Fig. 4d), as well as intense immunoreactivity with antibodies to alpha-B-crystallin, a small heat-shock protein (not shown), and for neurofilament.

In addition to ballooned neurons, the neocortex and neostriatum in CBD have widespread

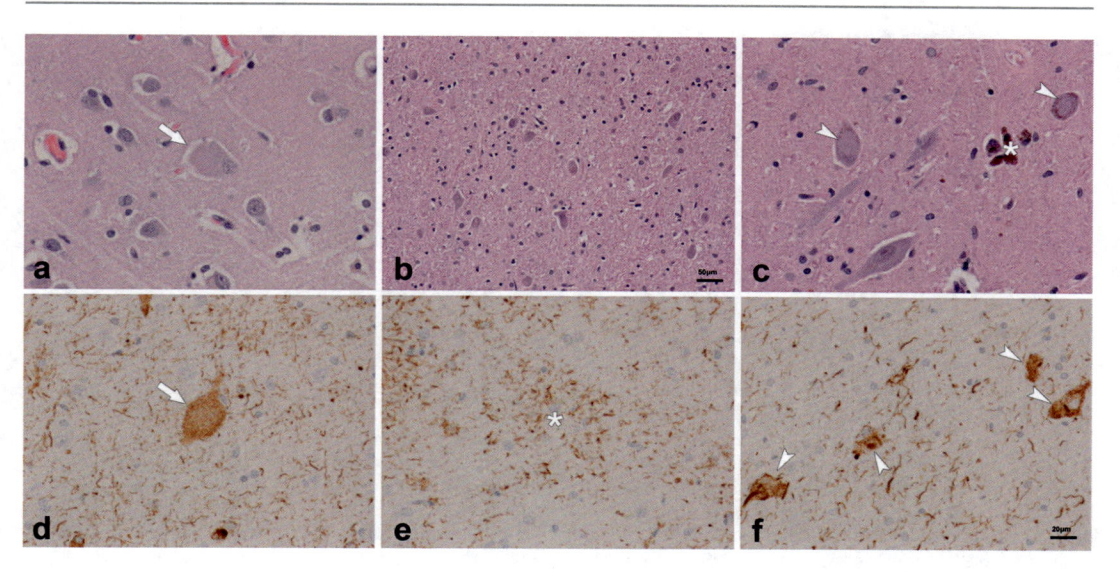

Fig. 4 Microscopic findings in CBD. (**a**) An H&E stained section of superior frontal gyrus shows ballooned neurons (arrow). (**b**) An H&E stained section of the subthalamic nucleus shows mild neuronal loss, but more marked gliosis. (**c**) An H&E stained section of the substantia nigra shows focal neuronal loss (extraneuronal neuromelanin—asterisk) and several neurons with so-called corticobasal bodies (arrowheads). (**d**) Phospho-tau immunohistochemistry of the superior frontal gyrus shows many neuropil threads and a ballooned neuron with diffuse cytoplasmic tau immunoreactivity (arrow). (**e**) Phospho-tau immunohistochemistry of the caudate nucleus shows an astrocytic plaque (asterisk). (**f**) Phospho-tau immunohistochemistry of the subthalamic nucleus shows morphologic heterogeneity of neuronal inclusions (arrowheads). Panels **a** and **c–f** are of same magnification, bar in (**f**) is 20 μm. Panel (**b**) is a lower magnification, bar is 50 μm

deposition of tau in both neurons and glia [3, 9]. Glial inclusions are found in both oligodendroglia and astrocytes. The astrocytic lesions have a characteristic plaque-like morphology ("astrocytic plaques" [91]) (Fig. 4e) that is morphologically distinct from tufted astrocytes of PSP. The pathologic feature that best discriminates PSP from CBD is pervasive thread-like cell processes in affected gray and white matter in CBD, to the extent that the difference can be seen by examining the slide with the naked eye (Fig. 5).

The subthalamic nucleus often has at least mild neuronal loss and gliosis (Fig. 4b), but it is rarely as severe as in PSP. Similarly, the substantia nigra has neuronal loss in CBD, but it can be mild (Fig. 4c). Neurons in the substantia nigra may have so-called corticobasal bodies [85] (Fig. 4c). Cortical neurons in atrophic areas have pleomorphic tau-immunoreactive lesions. In some neurons, tau is densely packed into small irregular inclusion bodies. In other neurons, the inclusions are more diffuse ("pre-tangles"). Neurofibrillary lesions in subcortical nuclei, such as the subthalamic nucleus, also typically have marked morphologic heterogeneity (Fig. 4f), while those in the locus ceruleus and substantia nigra can resemble globose neurofibrillary tangles (Fig. 4c).

Pathogenesis of Progressive Supranuclear Palsy and Corticobasal Degeneration

There is no single cause of PSP or CBD, but several environmental and genetic factors have been investigated. The Environmental Genetic PSP (ENGENE-PSP) study found that lower educational attainment, exposure to well water and industrial wastes, and firearm use were related to higher risk of developing PSP [115, 116]. These findings are also supported by a cluster of PSPs that emerged in northern France in an area of

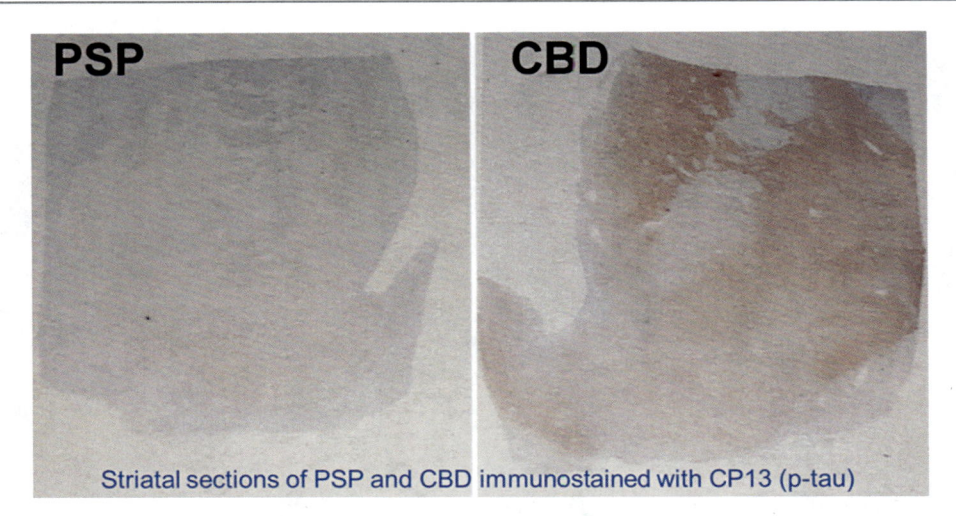

Striatal sections of PSP and CBD immunostained with CP13 (p-tau)

Fig. 5 Comparison of tau burden in PSP and CBD. Sections of the neostriatum in PSP and CBD, immunostained under the same conditions with a sensitive phospho-tau antibody (CP13 from Peter Davies, Feinstein Institute, Long Island, NY), show a clear distinction between PSP and CBD, due to dense tau pathology, mostly thread-like processes (not visible at this magnification), in CBD

high industrial waste contamination [117]. Consumption of high levels of annonacin, a mitochondrial complex I inhibitor, found in the pawpaw fruit was associated with developing PSP or other atypical parkinsonian syndromes in studies in the Caribbean island of Guadeloupe [118, 119]. There may be a slight male predominance within PSP patients [22, 46], and one study documented that increased estrogen exposure in women may be protective against developing PSP [120]. Environmental exposures have not been evaluated in CBD to date.

MAPT mutations may lead to either PSP or CBD [121–124]. Mutations in this gene can also lead to frontotemporal dementia, FTLD with parkinsonism, or primary progressive aphasia [125]. The H1/H1 genotype elevates the risk for developing PSP and CBD [17, 126, 127]. One genome-wide association study in a large cohort of pathologically validated PSP patients additionally identified genetic risk variants at the *MOBP*, *STX6*, and *EIF2AK3* loci [128]. *MOBP*, which encodes for myelin oligodendrocyte-binding protein, is also implicated in CBD and highlights potential importance of white matter [121, 129]. *STX6* encodes for a SNARE protein implicated in fusing vesicles in the Golgi network [130]. *EIF2AK3* encodes for a protein responsible for inhibiting protein synthesis in the face of excess endoplasmic reticulum stress [131, 132]. These genes have been validated in a second genome-wide association study, which additionally identified *SLCO1A2* and *DUSP10* as other genomic loci of interest [133].

Oxidative stress and inflammation can also be demonstrated in PSP and CBD. Mitochondrial enzymatic activity is decreased in both brain tissue and also in skeletal muscle in PSP patients [134–140]. Higher IL-1β and other inflammatory cytokines are found in the brains and CSF of PSP patients and lead to microglial activation [141, 142], which has been implicated in tau deposition [84]. Superoxide dismutase and glutathione, essential antioxidants, are often seen to be elevated in PSP brain tissue, possibly as a defense mechanism [139, 143].

Recent data suggest that misfolded tau oligomers are capable of acting as a template and induce further misfolding of normal monomeric tau leading to larger and larger aggregates, causing cellular damage and ultimately death and likely leading to spreading of disease in a 'prion-like' manner. In vivo animal studies using pre-formed fibrils [144, 145], human diseased brain homogenates [146], and other techniques [147, 148] have shown distal spread of tau pathology via trans-synaptic spread [149, 150]. There may be specific "strains" of tau capable of seeding unique tau pathologies [147, 151, 152].

Biomarkers in Progressive Supranuclear Palsy and Corticobasal Degeneration

The clinicopathologic overlap between PSP and CBD and other neurodegenerative diseases makes the discovery of sensitive and specific biomarkers for these diseases of paramount importance.

Magnetic Resonance Imaging PSP is well described to be associated with several features on structural magnetic resonance imaging (MRI). Most recognized is the presence of midbrain atrophy, resulting in the "hummingbird sign" best seen on the mid-sagittal section (Fig. 6) [153], as well as "morning glory sign [154]", or "Mickey Mouse sign [155]". In one study of an autopsy series of pathologically confirmed cases with PSP, multiple systems atrophy (MSA), or Parkinson's disease (PD), 16/22 (72.7%) of PSP cases were able to be correctly identified by a radiologist reviewing conventional MRI that had been performed during life, and the presence of a hummingbird sign or morning glory sign was

100% specific but was 68.4% sensitive [156]. One study, however, that included different clinical variants of PSP found midbrain atrophy to be a feature of the Richardson syndrome variant, but midbrain atrophy was not found to be a biomarker of PSP pathology [157]. The superior cerebellar peduncle is also frequently atrophied in PSP and, consequently, several different ratios comparing brain stem, pons, superior cerebellar peduncle, and middle cerebellar peduncle measurements have been studied to differentiate PSP from other parkinsonian diseases and from healthy controls. A frequent problem with these measurements is that they are often insensitive, and the radiologic signs will only manifest at later stages of the disease after neurodegeneration has progressed to the point of causing these recognizable patterns [158–163]. A more specific technique to assess the superior cerebellar peduncle is with diffusion tensor imaging (DTI). One DTI study did find the superior cerebellar peduncle to be able to accurately distinguish PSP from normal controls [164]. It is unclear whether atrophy of the superior cerebellar peduncle is a feature of PSP pathology or a feature of Richardson syndrome. Another technique that has also been studied in PSP is resting-state functional magnetic resonance imaging (fMRI). Resting-state fMRI studies have demonstrated disrupted thalamocortical connectivity in PSP [165, 166].

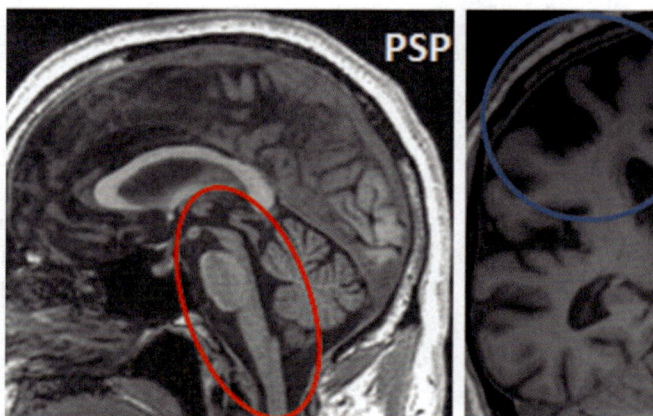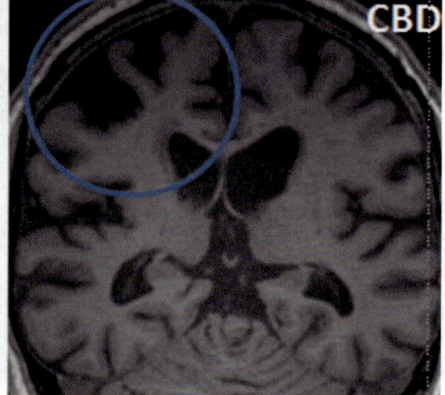

Fig. 6 MRI scan in autopsy-confirmed PSP and CBD. MRI scan in PSP shows the classic hummingbird sign on sagittal MRI, while asymmetric atrophy of the posterior frontal cortex is seen in CBD

Fewer MRI studies have been performed in CBD, but the most frequently cited sign is asymmetric cortical atrophy, affecting the parietal and frontal lobes (Fig. 6) [167–171]. Corpus callosum atrophy is also cited occasionally. Regrettably, neither of these features are specific for CBD to fully differentiate it from other pathologies that cause CBS clinical phenotypes [70, 167, 172]. In addition, symmetric cortical atrophy has been described in autopsy-confirmed cases of CBD [173]. Research studies have utilized voxel-based morphometry to try to distinguish CBD from Alzheimer's disease and other neurodegenerative diseases that present with CBS. These studies have found distinguishing features at the group level [174, 175]. No biomarker exists to distinguish CBD from other neurodegenerative diseases at the single subject level.

Given the prominent white matter degeneration that is common to these conditions, diffusion tensor imaging and white matter volumetric measurements may show more degeneration in PSP and CBD than atypical AD or FTLD TDP-43 that may have overlapping presentations [176–179].

DaTscan A DaTscan is used to detect dopamine transporters on dopamine neurons. DaTscans are typically utilized to differentiate Parkinson's disease from essential tremor. However, DaTscans have been performed in PSP and CBS patients and show a reduction in dopamine transporter receptors. Unfortunately, this finding is nonspecific and can also be seen in other parkinsonian disorders, for example, MSA.

Positron Emission Tomography The most common PET scan is the fluorodeoxyglucose (FDG)-PET scan, which utilizes radioactive glucose to assess for functional integrity of neocortical regions. FDG-PET findings in PSP and CBS tend to mirror findings on MRI. In PSP, hypometabolism is observed in the premotor cortex as well as the midbrain, the latter when present is known as the pimple sign of PSP [180] (Fig. 7). In CBS and CBD, the FDG-PET scan reveals asymmetric frontal and/or parietal hypometabolism (Fig. 7). There are less than a handful of

studies on FDG-PET in autopsy-confirmed PSP, CBD, and other 4R tauopathies. One such study found parietal hypometabolism in CBD and premotor hypometabolism in PSP [181]. Several tracers are currently under investigation that bind to the tau proteins, including ^{18}F-5105, ^{18}F-FDDNP, ^{18}F-THK523, ^{11}C-PBB3, and others [182]. ^{18}F-Flortaucipir (formerly AV-1451 and T807) is the most researched tau tracer to date and appears to bind avidly to paired helical filaments in 3R/4R tauopathies, such as AD [183], and exhibits retention patterns in amnestic AD consistent with Braak tau staging [184, 185] and in posterior cortical regions in posterior cortical atrophy patients [186, 187]. However, ^{18}F-Flortaucipir retention appears to be less robust in 4R tauopathies [183, 188, 189]. Increased retention in the basal ganglia and midbrain can be demonstrated in PSP (Fig. 8), but there is off-site binding, which makes individual patient-level distinctions at early stage difficult [184, 190–193]. Similarly, in CBS, mild increases in retention in cortical regions can be demonstrated (Fig. 8) that correlate with postmortem tau findings [194], although this has been reported to occur predominantly in CBS patients who presented with a motor speech disorder [195]. PET tracers targeting activated microglia (^{11}C-(R) PK11195) may aid in assessing inflammation associated with neurodegeneration in PSP and CBD [196, 197].

Biofluid Biomarkers CSF tau species, including measures of total tau (t-tau) and phosphorylated tau (p-tau) tend not to be elevated in PSP [198–200]. One study reported that a ratio of certain tau fragments may aid in distinguishing PSP from healthy controls and other conditions [201], but the findings could not be replicated [202]. CSF neurofilament light chain (NfL) is an intermediate filament, which can be measured from CSF and is a nonspecific measure of neuronal injury [203], but it shows elevation in PSP, CBD, and other parkinsonian syndromes that can aid in differentiating PSP or CBD from Parkinson's disease [200, 204–207]. The sensitivity of the next-generation single-molecule-array assays has

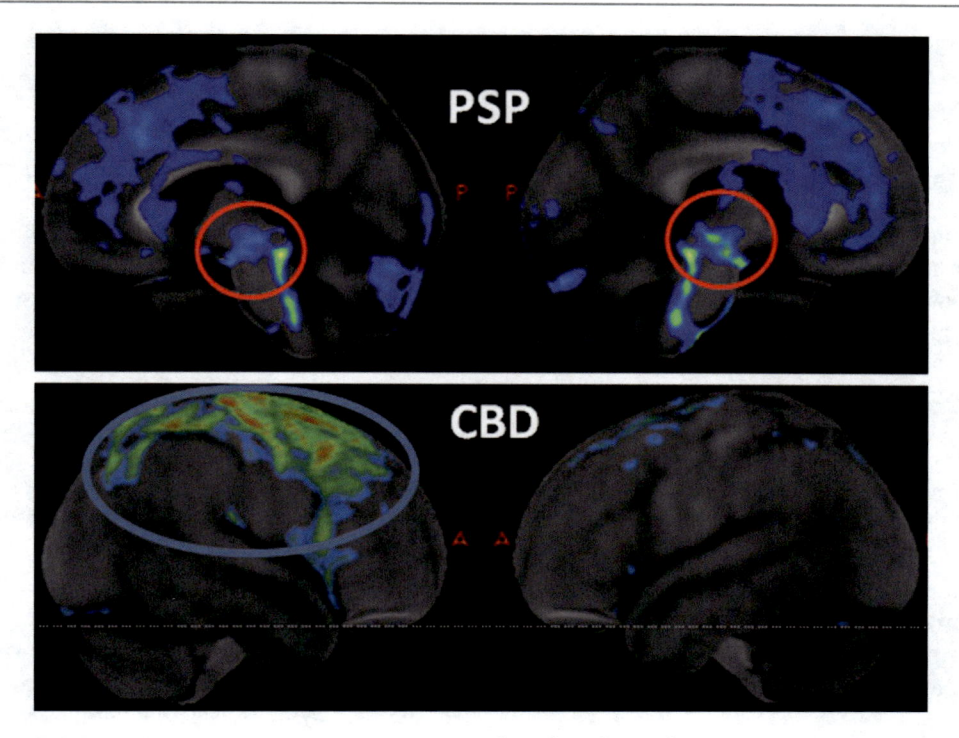

Fig. 7 FDG-PET in autopsy-confirmed PSP and CBD. FDG-PET in PSP shows the classic "pimple sign" (hypometabolism of the midbrain) on mid-sagittal section. Also seen is mild hypometabolism of medial pre- frontal and supplementary motor cortex. In CBD, asymmetric frontoparietal hypometabolism is observed on the lateral view

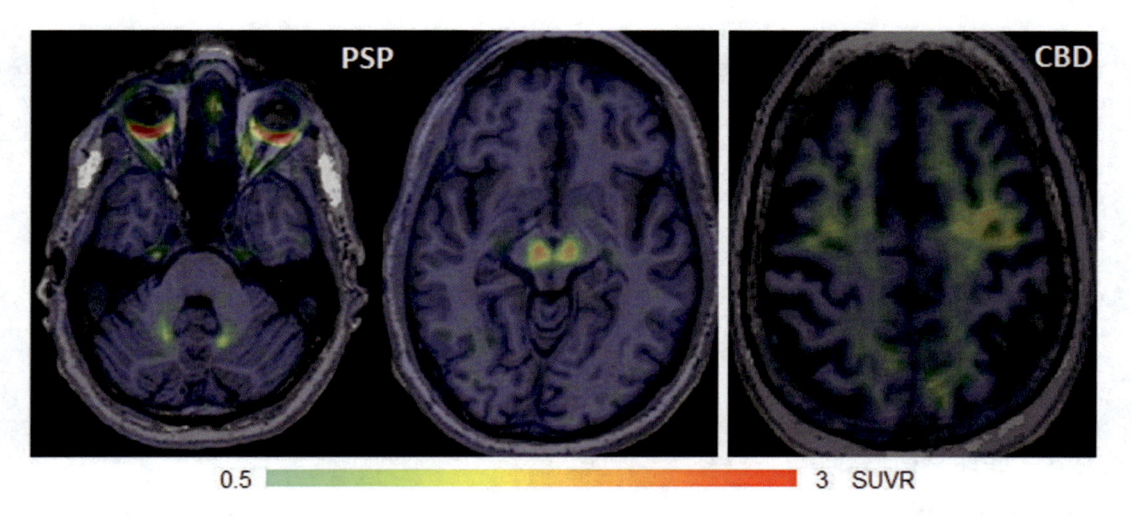

Fig. 8 Flortaucipir PET in autopsy-confirmed PSP and CBD. Flortaucipir PET (AV-1451) in PSP shows increased uptake in the midbrain (substantia nigra) and dentate nucleus of the cerebellum. In a case of CBD that pre- sented with progressive speech apraxia, flortaucipir PET demonstrates asymmetric increased uptake in premotor neocortex

made blood-based NfL measurements possible now as well [208, 209]. Real-time quaking-induced conversion (RT-QuIC) is an emerging assay that was originally developed to aid in diagnosis of Creutzfeldt-Jakob Disease (CJD), where a biologic sample is placed in wells containing monomeric proteins and a fluorescent marker and through polymerization encouraged by sequential shaking steps, can show the presence or absence of a pathologic "seed" from the patient sample. This technique has been adapted to detect alpha-synuclein [210], 3R/4R tau species [211], 3R tau species [212], and a 4R tauopathy assay is under development as well [213], which may offer molecularly specific aid in diagnosis in the near future.

Treatment of Progressive Supranuclear Palsy and Corticobasal Degeneration

Current treatment strategies for both PSP and CBS are supportive and symptomatic as no disease-modulating therapies are currently available for either condition.

Parkinsonism Levodopa preparation may still be trialed to treat the parkinsonism associated with PSP and CBS. In one study of pathologically confirmed PSP patients, approximately one-third of PSP patients showed a significant improvement (> 30% improvement in the Unified Parkinson's Disease Rating Scale) [46], which is a response rate that has been reported in other studies as well [64–67]. Doses of over 1 gm/day of levodopa for 1 month are proposed to elicit responses. Often, however, responses to levodopa are very mild in PSP and CBS, if present at all, and typically wane over time [20, 24, 55, 68, 99, 214]. Dopamine agonists have been trialed in PSP but are generally less effective than levodopa and are more likely to cause side effects [65, 215, 216]. Smaller studies documented improvement in parkinsonism using amantadine or amitriptyline in PSP, but caution is warranted because of possible anticholinergic side effects, including cognitive and psychiatric disturbances, dry

mouth, or difficulty with urination [65, 217–219].

Ocular Symptoms Zolpidem showed mild improvements in saccadic speed in one small study of patients with PSP, but those findings have not been replicated [220–222]. Botulinum toxin may be used to treat blepharospasm and eyelid-opening apraxia, but high doses are often required to achieve benefits [223, 224]. Artificial tears and ophthalmic ointments may be used to treat dry eyes, and sunglasses may be of use to aid in photosensitivity symptoms. Alternating an eye patch is useful for double vision, and, occasionally, prism lenses may be fashioned, if the deficits are fixed.

Spasticity, Dystonia, and Myoclonus Muscle relaxants such as baclofen, tizanadine, and cyclobenzaprine may be considered, but they must be carefully weighed against their possible side effects of somnolence [225]. Botulinum toxin may be used for the disabling focal dystonia of the limbs or neck that occurs in both conditions [223, 225, 226]. Clonazepam or levetiracetam can treat the myoclonus associated with CBS as can valproate [214, 227, 228].

Sialorrhea Again, botulinum toxin may be used to treat sialorrhea [229], as can medications including glycopyrrolate or 1% atropine drops placed sublingually, although the latter, if not carefully applied, can be absorbed systemically and cause anticholinergic side effects [230].

Memory Impairments Acetylcholinesterase inhibitors such as donepezil, rivastigmine, or galantamine may offer some mild improvement in memory function, but studies showed that it may worsen gait and dysphagia in PSP and worsen behavioral symptoms in FTD, so it should be used with caution [227, 231, 232]. No studies of memantine in autopsy-confirmed CBD have been performed, but multiple studies of memantine for memory dysfunction in FTD have failed to show benefits [233, 234].

Mood Changes Selective serotonin reuptake inhibitors or serotonin-norepinephrine reuptake inhibitors may be used to treat depression and anxiety, but they are not helpful for the apathy that can accompany PSP or CBS [227]. Dextromethorphan-quinidine is an effective treatment for pseudobulbar affect as are antidepressants [235].

Nonpharmacological Therapies PSP and CBS patients benefit from multidisciplinary care from providers knowledgeable about these conditions. Physical therapy decreases the likelihood of falls and improves global functioning [227, 236–238]. Weighted walkers are often recommended to aid in safer ambulation. Speech therapy may be employed to strengthen vocal muscles but to also provide strategies for more effective communication [239, 240]. Swallowing evaluations are essential if the patient complains of dysphagia or frequent coughing during meals as food consistency or eating habits may be modified. Safety inspections of the home may be helpful and can often be done my occupational therapists who can suggest changes and modifications to promote safety. Social workers are often needed to aid in utilization of resources that may be available to these patients. Lastly, palliative care consultants can help to manage transitions to less aggressive modalities of care and to promote symptom management and navigate end-of-life decision-making in a way that aids in both the patients and the families' quality of life [241].

Experimental Therapies for Progressive Supranuclear Palsy and Corticobasal Degeneration

Although there are no current disease-modulating treatment for PSP or CBD, several medications are under investigation, many of which target the tau protein by different mechanisms: by decreasing production, stabilizing microtubules, promoting immune system clearance, or modifying post-translational changes.

Tau in PSP and CBD commonly undergoes post-translational phosphorylation and acetylation [242]; unfortunately, trials of the GSK-3β kinase inhibitors lithium, valproate, and Tideglusib failed to show efficacy or were stopped due to poor tolerability [243]. Salsalate inhibits tau acetylation in animal models and is currently under early investigation (NCT02422485) [244]. O-Glc-NAC modification and caspase-mediated cleavage are other potential therapeutic targets [245, 246]

The microtubule-stabilizing agent davunetide failed to show efficacy in a phase IIb/III trial [247], and the taxane derivative TPI-287 inducted anaphylactic reactions, which necessitated trial stoppage [248]. Other compounds still under investigation that are thought to work through this mechanism include epothilone-D and methylene blue [249, 250].

Anti-inflammatory medications have been trialed in PSP, including rasagiline, CoQ10, and riluzole, but studies have failed to show efficacy [251–253], although there was significant benefit in a shorter trial using CoQ10 [254].

Tau immunotherapy is actively under investigation. Specifically, in PSP, the BIIB092 antibody product, directed against the N terminus of extracellular tau [255], showed promise in early trials [256, 257], but a phase II study failed to show efficacy (PASSPORT NCT03068468) [258]. Similarly, ABBV-8E12 had favorable early safety results and good target engagement [259, 260] but failed to show efficacy in larger trials. While these results are discouraging, a number of questions remain regarding this strategy, namely if proper epitopes of tau were selected [261, 262], if oligomeric species or intracellular tau should be prioritized although it is technically more challenging [184, 262–267], or if alternative delivery systems may increase blood–brain barrier penetration of antibody products and improve efficacy [184].

Gene therapy through small interfering RNA (siRNA) or antisense oligonucleotides are currently being investigated in animal models of tauopathies [268–270] and may be of future use in PSP and CBD.

References

1. Josephs KA, Hodges JR, Snowden JS, Mackenzie IR, Neumann M, Mann DM et al (2011) Neuropathological background of phenotypical variability in frontotemporal dementia. Acta Neuropathol 122(2):137–153
2. Spillantini MG, Goedert M (2013) Tau pathology and neurodegeneration. Lancet Neurol 12(6):609–622
3. Dickson DW (2004) Sporadic tauopathies: Pick's disease, corticobasal degeneration, progressive suprnauclear palsy and argyrophilic grain disease. In: Esiri MM, Lee VMY, Trojanowski JQ (eds) The neuropathology of dementia, 2nd edn. Cambridge University Press, Cambridge/New York, pp 227–256
4. Falcon B, Zivanov J, Zhang W, Murzin AG, Garringer HJ, Vidal R et al (2019) Novel tau filament fold in chronic traumatic encephalopathy encloses hydrophobic molecules. Nature 568(7752):420–423
5. Fitzpatrick AWP, Falcon B, He S, Murzin AG, Murshudov G, Garringer HJ et al (2017) Cryo-EM structures of tau filaments from Alzheimer's disease. Nature 547(7662):185–190
6. Kovacs GG, Ferrer I, Grinberg LT, Alafuzoff I, Attems J, Budka H et al (2016) Aging-related tau astrogliopathy (ARTAG): harmonized evaluation strategy. Acta Neuropathol 131(1):87–102
7. Togo T, Sahara N, Yen SH, Cookson N, Ishizawa T, Hutton M et al (2002) Argyrophilic grain disease is a sporadic 4-repeat tauopathy. J Neuropathol Exp Neurol 61(6):547–556
8. Steele JC, Richardson JC, Olszewski J (1964) Progressive supranuclear palsy. A heterogenous degeneration involving the brain stem, basal ganglia and cerebellum with vertical gaze and pseudobulbar palsy, nuchal dystonia and dementia. Arch Neurol 10:333–359
9. Dickson DW, Hauw J-J, Agid Y, Litvan I (2011) Progressive supranuclear palsy and corticobasal degeneration. In: Dickson DW, Weller RO (eds) Neurodegeneration: the molecular pathology of dementia and movement disorders, 2nd edn. Wiley-Blackwell/International Society of Neuropathology, Chichester/West Sussex, p xvii, 477 p.
10. Nath U, Ben-Shlomo Y, Thomson R, Morris HR, Wood N, Lees A et al (2001) The prevalence of progressive supranuclear palsy (Steele–Richardson–Olszewski syndrome) in the UK. Brain 124(7):1438–1449
11. Coyle-Gilchrist IT, Dick KM, Patterson K, Rodríquez PV, Wehmann E, Wilcox A et al (2016) Prevalence, characteristics, and survival of fronto-temporal lobar degeneration syndromes. Neurology 86(18):1736–1743
12. Schrag A, Ben-Shlomo Y, Quinn N (1999) Prevalence of progressive supranuclear palsy and multiple system atrophy: a cross-sectional study. Lancet 354(9192):1771–1775
13. Respondek G, Kurz C, Arzberger T, Compta Y, Englund E, Ferguson LW et al (2017) Which ante mortem clinical features predict progressive supranuclear palsy pathology? Mov Disord 32(7):995–1005
14. Fleury V, Brindel P, Nicastro N, Burkhard PR (2018) Descriptive epidemiology of parkinsonism in the Canton of Geneva, Switzerland. Parkinsonism Relat Disord 54:30–39
15. Kawashima M, Miyake M, Kusumi M, Adachi Y, Nakashima K (2004) Prevalence of progressive supranuclear palsy in Yonago, Japan. Mov Disord: Off J Mov Disord Soc 19(10):1239–1240
16. Takigawa H, Ikeuchi T, Aiba I, Morita M, Onodera O, Shimohata T et al (2016) Japanese longitudinal biomarker study in PSP and CBD (JALPAC): a prospective multicenter PSP/CBD cohort study in Japan. Parkinsonism Relat Disord 22:e120–e1e1
17. Höglinger GU, Respondek G, Stamelou M, Kurz C, Josephs KA, Lang AE et al (2017) Clinical diagnosis of progressive supranuclear palsy: the movement disorder society criteria. Mov Disord:n/a–n/a
18. Litvan I, Agid Y, Calne D, Campbell G, Dubois B, Duvoisin RC et al (1996) Clinical research criteria for the diagnosis of progressive supranuclear palsy (Steele-Richardson-Olszewski syndrome): report of the NINDS-SPSP international workshop. Neurology 47(1):1–9
19. Respondek G, Stamelou M, Kurz C, Ferguson LW, Rajput A, Chiu WZ et al (2014) The phenotypic spectrum of progressive supranuclear palsy: a retrospective multicenter study of 100 definite cases. Mov Disord: Off J Mov Disord Soc 29(14):1758–1766
20. Litvan I, Mangone CA, McKee A, Verny M, Parsa A, Jellinger K et al (1996) Natural history of progressive supranuclear palsy (Steele-Richardson-Olszewski syndrome) and clinical predictors of survival: a clinicopathological study. J Neurol Neurosurg Psychiatry 60(6):615–620
21. Maher E, Lees A (1986) The clinical features and natural history of the Steele-Richardson-Olszewski syndrome (progressive supranuclear palsy). Neurology 36(7):1005
22. Golbe LI, Davis PH, Schoenberg BS, Duvoisin RC (1988) Prevalence and natural history of progressive supranuclear palsy. Neurology 38(7):1031
23. Litvan I, Agid Y (1992) Progressive supranuclear palsy: clinical and research approaches. Oxford University Press, New York
24. Collins S, Ahlskog J, Parisi JE, Maraganore D (1995) Progressive supranuclear palsy: neuropathologically based diagnostic clinical criteria. J Neurol Neurosurg Psychiatry 58(2):167–173

25. Chamberlain W (1971) Restriction in upward gaze with advancing age. Am J Ophthalmol 71(1):341–346

26. Vidailhet M, Rivaud S, Gouider-Khouja N, Pillon B, Bonnet AM, Gaymard B et al (1994) Eye movements in parkinsonian syndromes. Ann Neurol: Off J Am Neurol Assoc Child Neurol Soc 35(4):420–426

27. Gibb W, Esiri M, Lees A (1987) Clinical and pathological features of diffuse cortical Lewy body disease (Lewy body dementia). Brain 110(5): 1131–1153

28. Grant MP, Cohen M, Petersen RB, Halmagyi GM, McDougall A, Tusa RJ et al (1993) Abnormal eye movements in Creutzfeldt–Jakob disease. Ann Neurol: Off J Am Neurol Assoc Child Neurol Soc 34(2):192–197

29. Paulson H, Subramony S (2003) Spinocerebellar Ataxia 3—Machado-Joseph disease (SCA3). In: Genetics of movement disorders. Elsevier, pp 57–69

30. Brüggemann N, Wandinger KP, Gaig C, Sprenger A, Junghanns K, Helmchen C et al (2016) Dystonia, lower limb stiffness, and upward gaze palsy in a patient with IgLON5 antibodies. Mov Disord 31(5):762–764

31. Adams C, McKeon A, Silber MH, Kumar R (2011) Narcolepsy, REM sleep behavior disorder, and supranuclear gaze palsy associated with Ma1 and Ma2 antibodies and tonsillar carcinoma. Arch Neurol 68(4):521–524

32. Boeve BF (2012) Progressive supranuclear palsy. Parkinsonism Relat Disord 18:S192–S1S4

33. Garbutt S, Riley D, Kumar A, Han Y, Harwood M, Leigh R (2004) Abnormalities of optokinetic nystagmus in progressive supranuclear palsy. J Neurol Neurosurg Psychiatry 75(10):1386–1394

34. Quinn N (1996) The "round the houses" sign in progressive supranuclear palsy. Ann Neurol: Off J Am Neurol Assoc Child Neurol Soc 40(6):951

35. Lal V, Truong D (2019) Eye movement in movement disorders. Clin Parkinsonism Relat Disord 1:54–63

36. Kitthaweesin K, Riley DE, Leigh RJ (2002) Vergence disorders in progressive supranuclear palsy. Ann New York Acad Sci 956(1):504–507

37. Yoon WT, Chung EJ, Lee SH, Kim BJ, Lee WY (2005) Clinical analysis of blepharospasm and apraxia of eyelid opening in patients with parkinsonism. J Clin Neurol 1(2):159–165

38. Williams DR, Watt HC, Lees AJ (2006) Predictors of falls and fractures in bradykinetic rigid syndromes: a retrospective study. J Neurol Neurosurg Psychiatry 77(4):468–473

39. Kluin KJ, Foster NL, Berent S, Gilman S (1993) Perceptual analysis of speech disorders in progressive supranuclear palsy. Neurology 43(3 Part 1):563

40. Müller J, Wenning GK, Verny M, McKee A, Chaudhuri KR, Jellinger K et al (2001) Progression of dysarthria and dysphagia in postmortem-confirmed parkinsonian disorders. Arch Neurol 58(2):259–264

41. Papapetropoulos S, Singer C, McCorquodale D, Gonzalez J, Mash DC (2005) Cause, seasonality of death and co-morbidities in progressive supranuclear palsy (PSP). Parkinsonism Relat Disord 11(7):459–463

42. Litvan I, Hauw J, Bartko J, Lantos P, Daniel S, Horoupian D et al (1996) Validity and reliability of the preliminary NINDS neuropathologic criteria for progressive supranuclear palsy and related disorders. J Neuropathol Exp Neurol 55(1):97–105

43. Respondek G, Roeber S, Kretzschmar H, Troakes C, Al-Sarraj S, Gelpi E et al (2013) Accuracy of the National Institute for Neurological Disorders and Stroke/Society for Progressive Supranuclear Palsy and neuroprotection and natural history in Parkinson plus syndromes criteria for the diagnosis of progressive supranuclear palsy. Mov Disord: Off J Mov Disord Soc 28(4):504–509

44. Osaki Y, Ben-Shlomo Y, Lees AJ, Daniel SE, Colosimo C, Wenning G et al (2004) Accuracy of clinical diagnosis of progressive supranuclear palsy. Mov Disord 19(2):181–189

45. Birdi S, Rajput AH, Fenton M, Donat JR, Rozdilsky B, Robinson C et al (2002) Progressive supranuclear palsy diagnosis and confounding features: report on 16 autopsied cases. Mov Disord: Off J Mov Disord Soc 17(6):1255–1264

46. Williams DR, de Silva R, Paviour DC, Pittman A, Watt HC, Kilford L et al (2005) Characteristics of two distinct clinical phenotypes in pathologically proven progressive supranuclear palsy: Richardson's syndrome and PSP-parkinsonism. Brain 128(Pt 6):1247–1258

47. Williams DR, Holton JL, Strand K, Revesz T, Lees AJ (2007) Pure akinesia with gait freezing: a third clinical phenotype of progressive supranuclear palsy. Mov Disord: Off J Mov Disord Soc 22(15):2235–2241

48. Boeve B, Dickson D, Duffy J, Bartleson J, Trenerry M, Petersen R (2003) Progressive nonfluent aphasia and subsequent aphasic dementia associated with atypical progressive supranuclear palsy pathology. Eur Neurol 49(2):72–78

49. Boeve BF, Maraganore D, Parisi JE, Ahlskog J, Graff-Radford N, Caselli RJ et al (1999) Pathologic heterogeneity in clinically diagnosed corticobasal degeneration. Neurology 53(4):795

50. Josephs KA, Katsuse O, Beccano-Kelly DA, Lin W-L, Uitti RJ, Fujino Y et al (2006) Atypical progressive supranuclear palsy with corticospinal tract degeneration. J Neuropathol Exp Neurol 65(4):396–405

51. Koga S, Josephs KA, Ogaki K, Labbé C, Uitti RJ, Graff-Radford N et al (2016) Cerebellar ataxia in progressive supranuclear palsy: an autopsy study of PSP-C. Mov Disord 31(5):653–662

52. Josephs KA, Duffy JR (2008) Apraxia of speech and nonfluent aphasia: a new clinical marker for corticobasal degeneration and progressive supranuclear palsy. Curr Opin Neurol 21(6):688–692

53. Hassan A, Parisi JE, Josephs KA (2012) Autopsy-proven progressive supranuclear palsy presenting as behavioral variant frontotemporal dementia. Neurocase 18(6):478–488

54. Han HJ, Kim H, Park JH, Shin HW, Kim GU, Kim DS et al (2010) Behavioral changes as the earliest clinical manifestation of progressive supranuclear palsy. J Clin Neurol (Seoul, Korea) 6(3):148–151

55. Ling H, O'Sullivan SS, Holton JL, Revesz T, Massey LA, Williams DR et al (2010) Does corticobasal degeneration exist? A clinicopathological re-evaluation. Brain 133(7):2045–2057

56. Ling H, De Silva R, Massey L, Courtney R, Hondhamuni G, Bajaj N et al (2014) Characteristics of progressive supranuclear palsy presenting with corticobasal syndrome: a cortical variant. Neuropathol Appl Neurobiol 40(2):149–163

57. Nagao S, Yokota O, Nanba R, Takata H, Haraguchi T, Ishizu H et al (2012) Progressive supranuclear palsy presenting as primary lateral sclerosis but lacking parkinsonism, gaze palsy, aphasia, or dementia. J Neurol Sci 323(1–2):147–153

58. Respondek G, Höglinger G (2016) The phenotypic spectrum of progressive supranuclear palsy. Parkinsonism Relat Disord 22:S34–SS6

59. Hughes AJ, Daniel SE, Ben-Shlomo Y, Lees AJ (2002) The accuracy of diagnosis of parkinsonian syndromes in a specialist movement disorder service. Brain 125(Pt 4):861–870

60. Hughes AJ, Daniel SE, Kilford L, Lees AJ (1992) Accuracy of clinical diagnosis of idiopathic Parkinson's disease: a clinico-pathological study of 100 cases. J Neurol Neurosurg Psychiatry 55(3):181–184

61. Williams DR, Holton JL, Strand C, Pittman A, de Silva R, Lees AJ et al (2007) Pathological tau burden and distribution distinguishes progressive supranuclear palsy-parkinsonism from Richardson's syndrome. Brain 130(6):1566–1576

62. Williams DR, Lees AJ (2010) What features improve the accuracy of the clinical diagnosis of progressive supranuclear palsy-parkinsonism (PSP-P)? Mov Disord 25(3):357–362

63. Adler CH, Beach TG, Hentz JG, Shill HA, Caviness JN, Driver-Dunckley E et al (2014) Low clinical diagnostic accuracy of early vs advanced Parkinson disease: clinicopathologic study. Neurology 83(5):406–412

64. Richardson J, Steele J, Olszewski J (1963) Supranuclear opthalmoplegia, pseudobulbar palsy, nuchal dystonia and dementia. A clinical report on eight cases of "heterogeneous system degeneration". Trans Am Neurol Assoc 88:25

65. Nieforth KA, Golbe LI (1993) Retrospective study of drug response in 87 patients with progressive supranuclear palsy. Clin Neuropharmacol 16(4):338–346

66. Tan E, Chan L, Wong M (2003) Levodopa-induced oromandibular dystonia in progressive supranuclear palsy. Clin Neurol Neurosurg 105(2):132–134

67. Lang AE (2005) Treatment of progressive supranuclear palsy and corticobasal degeneration. Mov Disord: Off J Mov Disord Soc 20(S12):S83–S91

68. Litvan I, Bhatia KP, Burn DJ, Goetz CG, Lang AE, McKeith I et al (2003) Movement disorders society scientific issues committee report: SIC task force appraisal of clinical diagnostic criteria for parkinsonian disorders. Mov Disord: Off J Mov Disord Soc 18(5):467–486

69. Donker Kaat L, Boon AJ, Kamphorst W, Ravid R, Duivenvoorden HJ, van Swieten JC (2007) Frontal presentation in progressive supranuclear palsy. Neurology 69(8):723–729

70. Josephs KA, Petersen RC, Knopman DS, Boeve BF, Whitwell JL, Duffy JR et al (2006) Clinicopathologic analysis of frontotemporal and corticobasal degenerations and PSP. Neurology 66(1):41–48

71. Mochizuki A, Ueda Y, Komatsuzaki Y, Tsuchiya K, Arai T, Shoji S (2003) Progressive supranuclear palsy presenting with primary progressive aphasia--clinicopathological report of an autopsy case. Acta Neuropathol 105(6):610–614

72. Tsuboi Y, Josephs KA, Boeve BF, Litvan I, Caselli RJ, Caviness JN et al (2005) Increased tau burden in the cortices of progressive supranuclear palsy presenting with corticobasal syndrome. Mov Disord 20(8):982–988

73. Compta Y, Valldeoriola F, Tolosa E, Rey MJ, Martí MJ, Valls-Solé J (2007) Long lasting pure freezing of gait preceding progressive supranuclear palsy: a clinicopathological study. Mov Disord 22(13):1954–1958

74. Facheris MF, Maniak S, Scaravilli F, Schüle B, Klein C, Pramstaller PP (2008) Pure akinesia as initial presentation of PSP: a clinicopathological study. Parkinsonism Relat Disord 14(6):517–519

75. Kurz C, Ebersbach G, Respondek G, Giese A, Arzberger T, Höglinger GU (2016) An autopsy-confirmed case of progressive supranuclear palsy with predominant postural instability. Acta Neuropathol Commun 4(1):120

76. Kanazawa M, Shimohata T, Toyoshima Y, Tada M, Kakita A, Morita T et al (2009) Cerebellar involvement in progressive supranuclear palsy: a clinicopathological study. Mov Disord: Off J Mov Disord Soc 24(9):1312–1318

77. Kanazawa M, Tada M, Onodera O, Takahashi H, Nishizawa M, Shimohata T (2013) Early clinical features of patients with progressive supranuclear palsy with predominant cerebellar ataxia. Parkinsonism Relat Disord 19(12):1149–1151

78. Ali F, Martin PR, Botha H, Ahlskog JE, Bower JH, Masumoto JY et al (2019) Sensitivity and specificity of diagnostic criteria for progressive supranuclear palsy. Mov Disord 34(8):1144–1153

79. Ali F, Botha H, Whitwell JL, Josephs KA (2019) Utility of the movement disorders society criteria for progressive supranuclear palsy in clinical practice. Mov Disord Clin Pract 6(6):436–439

80. Shoeibi A, Litvan I, Juncos JL, Bordelon Y, Riley D, Standaert D et al (2019) Are the International Parkinson disease and Movement Disorder Society progressive supranuclear palsy (IPMDS-PSP) diagnostic criteria accurate enough to differentiate common PSP phenotypes? Parkinsonism Relat Disord. In Press

81. Ahmed Z, Josephs KA, Gonzalez J, DelleDonne A, Dickson DW (2008) Clinical and neuropathologic features of progressive supranuclear palsy with severe pallido-nigro-luysial degeneration and axonal dystrophy. Brain 131(Pt 2):460–472

82. Dickson DW, Ahmed Z, Algom AA, Tsuboi Y, Josephs KA (2010) Neuropathology of variants of progressive supranuclear palsy. Curr Opin Neurol 23(4):394–400

83. Ishizawa K, Lin WL, Tiseo P, Honer WG, Davies P, Dickson DW (2000) A qualitative and quantitative study of grumose degeneration in progressive supranuclear palsy. J Neuropathol Exp Neurol 59(6):513–524

84. Ishizawa K, Dickson DW (2001) Microglial activation parallels system degeneration in progressive supranuclear palsy and corticobasal degeneration. J Neuropathol Exp Neurol 60(6):647–657

85. Gibb WR, Luthert PJ, Marsden CD (1989) Corticobasal degeneration. Brain 112(Pt 5):1171–1192

86. Dickson DW, Yen SH, Suzuki KI, Davies P, Garcia JH, Hirano A (1986) Ballooned neurons in select neurodegenerative diseases contain phosphorylated neurofilament epitopes. Acta Neuropathol 71(3–4):216–223

87. Rebeiz JJ, Kolodny EH, Richardson EP Jr (1967) Corticodentatonigral degeneration with neuronal achromasia: a progressive disorder of late adult life. Trans Am Neurol Assoc 92:23–26

88. Ksiezak-Reding H, Morgan K, Mattiace LA, Davies P, Liu WK, Yen SH et al (1994) Ultrastructure and biochemical composition of paired helical filaments in corticobasal degeneration. Am J Pathol 145(6):1496–1508

89. Mori H, Nishimura M, Namba Y, Oda M (1994) Corticobasal degeneration: a disease with widespread appearance of abnormal tau and neurofibrillary tangles, and its relation to progressive supranuclear palsy. Acta Neuropathol 88(2):113–121

90. Uchihara T, Mitani K, Mori H, Kondo H, Yamada M, Ikeda K (1994) Abnormal cytoskeletal pathology peculiar to corticobasal degeneration is different from that of Alzheimer's disease or progressive supranuclear palsy. Acta Neuropathol 88(4):379–383

91. Feany MB, Dickson DW (1995) Widespread cytoskeletal pathology characterizes corticobasal degeneration. Am J Pathol 146(6):1388–1396

92. Kouri N, Murray ME, Hassan A, Rademakers R, Uitti RJ, Boeve BF et al (2011) Neuropathological features of corticobasal degeneration presenting as corticobasal syndrome or Richardson syndrome. Brain 134(Pt 11):3264–3275

93. Armstrong MJ, Litvan I, Lang AE, Bak TH, Bhatia KP, Borroni B et al (2013) Criteria for the diagnosis of corticobasal degeneration. Neurology 80(5):496–503

94. Boeve BF (2011) The multiple phenotypes of corticobasal syndrome and corticobasal degeneration: implications for further study. J Mol Neuro: MN 45(3):350–353

95. Lee SE, Rabinovici GD, Mayo MC, Wilson SM, Seeley WW, DeArmond SJ et al (2011) Clinicopathological correlations in corticobasal degeneration. Ann Neurol 70(2):327–340

96. Schneider J, Watts R, Gearing M, Brewer R, Mirra S (1997) Corticobasal degeneration neuropathologic and clinical heterogeneity. Neurology 48(4):959–968

97. Hu WT, Rippon GW, Boeve BF, Knopman DS, Petersen RC, Parisi JE et al (2009) Alzheimer's disease and corticobasal degeneration presenting as corticobasal syndrome. Mov Disord: Off J Mov Disord Soc 24(9):1375–1379

98. Chand P, Grafman J, Dickson D, Ishizawa K, Litvan I (2006) Alzheimer's disease presenting as corticobasal syndrome. Mov Disord: Off J Mov Disord Soc 21(11):2018–2022

99. Wenning GK, Litvan I, Jankovic J, Granata R, Mangone CA, McKee A et al (1998) Natural history and survival of 14 patients with corticobasal degeneration confirmed at postmortem examination. J Neurol Neurosurg Psychiatry 64(2):184–189

100. Litvan I, Agid Y, Goetz C, Jankovic J, Wenning GK, Brandel JP et al (1997) Accuracy of the clinical diagnosis of corticobasal degeneration: a clinicopathologic study. Neurology 48(1):119–125

101. Togasaki DM, Tanner CM (2000) Epidemiologic aspects. Adv Neurol 82:53–59

102. Tsuchiya K, Murayama S, Mitani K, Oda T, Arima K, Mimura M et al (2005) Constant and severe involvement of Betz cells in corticobasal degeneration is not consistent with pyramidal signs: a clinicopathological study of ten autopsy cases. Acta Neuropathol 109(4):353–366

103. Josephs KA, Rossor MN (2004) The alien limb. Pract Neurol 4(1):44–45

104. Grimes DA, Lang AE, Bergeron CB (1999) Dementia as the most common presentation of cortical-basal ganglionic degeneration. Neurology 53(9):1969–1974

105. Kompoliti K, Goetz C, Boeve BF, Maraganore D, Ahlskog J, Marsden C et al (1998) Clinical presentation and pharmacological therapy in corticobasal degeneration. Arch Neurol 55(7):957–961

106. Boeve BF, Lang AE, Litvan I (2003) Corticobasal degeneration and its relationship to progressive supranuclear palsy and frontotemporal dementia. Ann Neurol: Off J Am Neurol Assoc Child Neurol Soc 54(S5):S15–S19

107. Murray R, Neumann M, Forman M, Farmer J, Massimo L, Rice A et al (2007) Cognitive and motor assessment in autopsy-proven corticobasal degeneration. Neurology 68(16):1274–1283

108. Jacobs DH, Adair JC, Macauley B, Gold M, Rothi LJG, Heilman KM (1999) Apraxia in corticobasal degeneration. Brain Cogn 40(2):336–354

109. Reich SG, Grill SE (2009) Corticobasal degeneration. Curr Treat Options Neurol 11(3):179

110. Spotorno N, McMillan CT, Powers JP, Clark R, Grossman M (2014) Counting or chunking? Mathematical and heuristic abilities in patients with corticobasal syndrome and posterior cortical atrophy. Neuropsychologia 64:176–183

111. Josephs KA, Duffy JR, Strand EA, Whitwell JL, Layton KF, Parisi JE et al (2006) Clinicopathological and imaging correlates of progressive aphasia and apraxia of speech. Brain 129(Pt 6):1385–1398

112. Shaw LM, Vanderstichele H, Knapik-Czajka M, Clark CM, Aisen PS, Petersen RC et al (2009) Cerebrospinal fluid biomarker signature in Alzheimer's disease neuroimaging initiative subjects. Ann Neurol 65(4):403–413

113. Ling H, Gelpi E, Davey K, Jaunmuktane Z, Mok KY, Jabbari E et al Fulminant corticobasal degeneration: a distinct variant with predominant neuronal tau aggregates. Acta Neuropathol:1–18

114. Rodriguez-Porcel F, Lowder L, Rademakers R, Ravenscroft T, Ghetti B, Hagen MC et al (2016) Fulminant corticobasal degeneration: Agrypnia excitata in corticobasal syndrome. Neurology 86(12):1164–1166

115. Litvan I, Lees PS, Cunningham CR, Rai SN, Cambon AC, Standaert DG et al (2016) Environmental and occupational risk factors for progressive supranuclear palsy: case-control study. Mov Disord 31(5):644–652

116. Kelley KD, Checkoway H, Hall DA, Reich SG, Cunningham C, Litvan I (2018) Traumatic brain injury and firearm use and risk of progressive supranuclear palsy among veterans. Front Neurol 9:474

117. Caparros-Lefebvre D, Golbe LI, Deramecourt V, Maurage CA, Huin V, Buee-Scherrer V et al (2015) A geographical cluster of progressive supranuclear palsy in northern France. Neurology 85(15):1293–1300

118. Caparros-Lefebvre D, Sergeant N, Lees A, Camuzat A, Daniel S, Lannuzel A et al (2002) Guadeloupean parkinsonism: a cluster of progressive supranuclear palsy-like tauopathy. Brain 125(Pt 4):801–811

119. Lannuzel A, Ruberg M, Michel PP (2008) Atypical parkinsonism in the Caribbean island of Guadeloupe: etiological role of the mitochondrial complex I inhibitor annonacin. Mov Disord: Off J Mov Disord Soc 23(15):2122–2128

120. Park HK, Ilango S, Charriez CM, Checkoway H, Riley D, Standaert DG et al (2018) Lifetime exposure to estrogen and progressive supranuclear palsy: environmental and genetic PSP study. Mov Disord 33(3):468–472

121. Kouri N, Ross OA, Dombroski B, Younkin CS, Serie DJ, Soto-Ortolaza A et al (2015) Genome-wide association study of corticobasal degeneration identifies

122. risk variants shared with progressive supranuclear palsy. Nat Commun 6:7247

122. Rohrer JD, Paviour D, Vandrovcova J, Hodges J, De Silva R, Rossor MN (2011) Novel L284R MAPT mutation in a family with an autosomal dominant progressive supranuclear palsy syndrome. Neurodegener Dis 8(3):149–152

123. Ogaki K, Li Y, Takanashi M, Ishikawa K-I, Kobayashi T, Nonaka T et al (2013) Analyses of the MAPT, PGRN, and C9orf72 mutations in Japanese patients with FTLD, PSP, and CBS. Parkinsonism Relat Disord 19(1):15–20

124. Ahmed S, Fairen MD, Sabir MS, Pastor P, Ding J, Ispierto L et al (2019) MAPT p.V363I mutation: a rare cause of corticobasal degeneration. Neurology Genetics 5(4):e347

125. Boeve BF, Hutton M (2008) Refining frontotemporal dementia with parkinsonism linked to chromosome 17: introducing FTDP-17 (MAPT) and FTDP-17 (PGRN). Arch Neurol 65(4):460–464

126. Baker M, Litvan I, Houlden H, Adamson J, Dickson D, Perez-Tur J et al (1999) Association of an extended haplotype in the tau gene with progressive supranuclear palsy. Hum Mol Genet 8(4):711–715

127. Houlden H, Baker M, Morris H, MacDonald N (2001) Pickering–Brown S, Adamson J, et al. Corticobasal degeneration and progressive supranuclear palsy share a common tau haplotype. Neurology 56(12):1702–1706

128. Höglinger GU, Melhem NM, Dickson DW, Sleiman PM, Wang L-S, Klei L et al (2011) Identification of common variants influencing risk of the tauopathy progressive supranuclear palsy. Nat Genet 43(7):699

129. Yokoyama JS, Karch CM, Fan CC, Bonham LW, Kouri N, Ross OA et al (2017) Shared genetic risk between corticobasal degeneration, progressive supranuclear palsy, and frontotemporal dementia. Acta Neuropathol 133(5):825–837

130. Wendler F, Tooze S (2001) Syntaxin 6: the promiscuous behaviour of a SNARE protein. Traffic 2(9):606–611

131. Harding HP, Zhang Y, Bertolotti A, Zeng H, Ron D (2000) Perk is essential for translational regulation and cell survival during the unfolded protein response. Mol Cell 5(5):897–904

132. Yuan SH, Hiramatsu N, Liu Q, Sun XV, Lenh D, Chan P et al (2018) Tauopathy-associated PERK alleles are functional hypomorphs that increase neuronal vulnerability to ER stress. Hum Mol Genet 27(22):3951–3963

133. Sanchez-Contreras MY, Kouri N, Cook CN, Serie DJ, Heckman MG, Finch NA et al (2018) Replication of progressive supranuclear palsy genome-wide association study identifies SLCO1A2 and DUSP10 as new susceptibility loci. Mol Neurodegener 13(1):37

134. Albers DS, Augood SJ, Martin DM, Standaert DG, Vonsattel JPG, Beal MF (1999) Evidence for oxidative stress in the subthalamic nucleus in progressive supranuclear palsy. J Neurochem 73(2):881–884

135. Albers DS, Augood SJ, Park LC, Browne SE, Martin DM, Adamson J et al (2000) Frontal lobe dysfunction in progressive supranuclear palsy: evidence for oxidative stress and mitochondrial impairment. J Neurochem 74(2):878–881

136. Albers DS, Swerdlow RH, Manfredi G, Gajewski C, Yang L, Parker WD Jr et al (2001) Further evidence for mitochondrial dysfunction in progressive supranuclear palsy. Exp Neurol 168(1):196–198

137. Park LC, Albers DS, Xu H, Lindsay JG, Beal MF, Gibson GE (2001) Mitochondrial impairment in the cerebellum of the patients with progressive supranuclear palsy. J Neurosci Res 66(5):1028–1034

138. Albers DS, Beal MF (2002) Mitochondrial dysfunction in progressive supranuclear palsy. Neurochem Int 40(6):559–564

139. Cantuti-Castelvetri I, Keller-McGandy CE, Albers DS, Beal MF, Vonsattel J-P, Standaert DG et al (2002) Expression and activity of antioxidants in the brain in progressive supranuclear palsy. Brain Res 930(1–2):170–181

140. Martinelli P, Scaglione C, Lodi R, Iotti S, Barbiroli B (2000) Deficit of brain and skeletal muscle bioenergetics in progressive supranuclear palsy shown in vivo by phosphorus magnetic resonance spectroscopy. Mov Disord 15(5):889–893

141. Fernandez-Botran R, Ahmed Z, Crespo FA, Gatenbee C, Gonzalez J, Dickson DW et al (2011) Cytokine expression and microglial activation in progressive supranuclear palsy. Parkinsonism Relat Disord 17(9):683–688

142. Starhof C, Winge K, Heegaard NHH, Skogstrand K, Friis S, Hejl A (2018) Cerebrospinal fluid proinflammatory cytokines differentiate parkinsonian syndromes. J Neuroinflammation 15(1):305

143. Sian J, Dexter DT, Lees AJ, Daniel S, Agid Y, Javoy-Agid F et al (1994) Alterations in glutathione levels in Parkinson's disease and other neurodegenerative disorders affecting basal ganglia. Ann Neurol: Off J Am Neurol Assoc Child Neurol Soc 36(3):348–355

144. Iba M, Guo JL, McBride JD, Zhang B, Trojanowski JQ, Lee VM-Y (2013) Synthetic tau fibrils mediate transmission of neurofibrillary tangles in a transgenic mouse model of Alzheimer's-like tauopathy. J Neurosci 33(3):1024–1037

145. Clavaguera F, Lavenir I, Falcon B, Frank S, Goedert M, Tolnay M (2013) "Prion-like" templated misfolding in tauopathies. Brain Pathol 23(3):342–349

146. Clavaguera F, Akatsu H, Fraser G, Crowther RA, Frank S, Hench J et al (2013) Brain homogenates from human tauopathies induce tau inclusions in mouse brain. Proc Natl Acad Sci 110(23):9535–9540

147. Sanders DW, Kaufman SK, DeVos SL, Sharma AM, Mirbaha H, Li A et al (2014) Distinct tau prion strains propagate in cells and mice and define different tauopathies. Neuron 82(6):1271–1288

148. Probst A, Götz J, Wiederhold K, Tolnay M, Mistl C, Jaton A et al (2000) Axonopathy and amyotrophy in mice transgenic for human four-repeat tau protein. Acta Neuropathol 99(5):469–481

149. Dujardin K, Defebvre L, Duhamel A, Lecouffe P, Rogelet P, Steinling M et al (2004) Cognitive and SPECT characteristics predict progression of Parkinson's disease in newly diagnosed patients. J Neurol 251(11):1383–1392

150. Clavaguera F, Hench J, Lavenir I, Schweighauser G, Frank S, Goedert M et al (2014) Peripheral administration of tau aggregates triggers intracerebral tauopathy in transgenic mice. Acta Neuropathol 127(2):299–301

151. Nishimura M, Namba Y, Ikeda K, Oda M (1992) Glial fibrillary tangles with straight tubules in the brains of patients with progressive supranuclear palsy. Neurosci Lett 143(1–2):35–38

152. Yamada T, McGeer P, McGeer E (1992) Appearance of paired nucleated, tau-positive glia in patients with progressive supranuclear palsy brain tissue. Neurosci Lett 135(1):99–102

153. Kato N, Arai K, Hattori T (2003) Study of the rostral midbrain atrophy in progressive supranuclear palsy. J Neurol Sci 210(1–2):57–60

154. Adachi M, KAWANAMI T, OHSHIMA H, Sugai Y, Hosoya T (2004) Morning glory sign: a particular MR finding in progressive supranuclear palsy. Magn Reson Med Sci 3(3):125–132

155. Massey LA, Micallef C, Paviour DC, O'sullivan SS, Ling H, Williams DR et al (2012) Conventional magnetic resonance imaging in confirmed progressive supranuclear palsy and multiple system atrophy. Mov Disord 27(14):1754–1762

156. Massey LA, Micallef C, Paviour DC, O'Sullivan SS, Ling H, Williams DR et al (2012) Conventional magnetic resonance imaging in confirmed progressive supranuclear palsy and multiple system atrophy. Mov Disord: Off J Mov Disord Soc 27(14):1754–1762

157. Whitwell JL, Jack CR Jr, Parisi JE, Gunter JL, Weigand SD, Boeve BF et al (2013) Midbrain atrophy is not a biomarker of progressive supranuclear palsy pathology. Eur J Neurol 20(10):1417–1422

158. Massey LA, Jager HR, Paviour DC, O'Sullivan SS, Ling H, Williams DR et al (2013) The midbrain to pons ratio: a simple and specific MRI sign of progressive supranuclear palsy. Neurology 80(20):1856–1861

159. Quattrone A, Nicoletti G, Messina D, Fera F, Condino F, Pugliese P et al (2008) MR imaging index for differentiation of progressive supranuclear palsy from Parkinson disease and the Parkinson variant of multiple system atrophy. Radiology 246(1):214–221

160. Moller L, Kassubek J, Sudmeyer M, Hilker R, Hattingen E, Egger K et al (2017) Manual MRI morphometry in parkinsonian syndromes. Mov Disord: Off J Mov Disord Soc 32(5):778–782

161. Nigro S, Arabia G, Antonini A, Weis L, Marcante A, Tessitore A et al (2017) Magnetic Resonance Parkinsonism Index: diagnostic accuracy of a fully automated algorithm in comparison with the manual measurement in a large Italian multicentre study in

patients with progressive supranuclear palsy. Eur Radiol 27(6):2665–2675

162. Zanigni S, Calandra-Buonaura G, Manners DN, Testa C, Gibertoni D, Evangelisti S et al (2016) Accuracy of MR markers for differentiating Progressive Supranuclear Palsy from Parkinson's disease. NeuroImage Clinical 11:736–742

163. Hussl A, Mahlknecht P, Scherfler C, Esterhammer R, Schocke M, Poewe W et al (2010) Diagnostic accuracy of the magnetic resonance Parkinsonism index and the midbrain-to-pontine area ratio to differentiate progressive supranuclear palsy from Parkinson's disease and the Parkinson variant of multiple system atrophy. Mov Disord: Off J Mov Disord Soc 25(14):2444–2449

164. Whitwell JL, Master AV, Avula R, Kantarci K, Eggers SD, Edmonson HA et al (2011) Clinical correlates of white matter tract degeneration in progressive supranuclear palsy. Arch Neurol 68(6):753–760

165. Whitwell JL, Avula R, Master A, Vemuri P, Senjem ML, Jones DT et al (2011) Disrupted thalamocortical connectivity in PSP: a resting-state fMRI, DTI, and VBM study. Parkinsonism Relat Disord 17(8):599–605

166. Gardner RC, Boxer AL, Trujillo A, Mirsky JB, Guo CC, Gennatas ED et al (2013) Intrinsic connectivity network disruption in progressive supranuclear palsy. Ann Neurol 73(5):603–616

167. Josephs KA, Whitwell JL, Dickson DW, Boeve BF, Knopman DS, Petersen RC et al (2008) Voxel-based morphometry in autopsy proven PSP and CBD. Neurobiol Aging 29(2):280–289

168. Gröschel K, Hauser T-K, Luft A, Patronas N, Dichgans J, Litvan I et al (2004) Magnetic resonance imaging-based volumetry differentiates progressive supranuclear palsy from corticobasal degeneration. NeuroImage 21(2):714–724

169. Hauser RA, Murtaugh FR, Akhter K, Gold M, Olanow C (1996) Magnetic resonance imaging of corticobasal degeneration. J Neuroimaging 6(4):222–226

170. Koyama M, Yagishita A, Nakata Y, Hayashi M, Bandoh M, Mizutani T (2007) Imaging of corticobasal degeneration syndrome. Neuroradiology 49(11):905–912

171. Boxer AL, Geschwind MD, Belfor N, Gorno-Tempini ML, Schauer GF, Miller BL et al (2006) Patterns of brain atrophy that differentiate corticobasal degeneration syndrome from progressive supranuclear palsy. Arch Neurol 63(1):81–86

172. Josephs KA, Tang-Wai DF, Edland SD, Knopman DS, Dickson DW, Parisi JE et al (2004) Correlation between antemortem magnetic resonance imaging findings and pathologically confirmed corticobasal degeneration. Arch Neurol 61(12):1881–1884

173. Hassan A, Whitwell JL, Boeve BF, Jack CR Jr, Parisi JE, Dickson DW et al (2010) Symmetric corticobasal degeneration (S-CBD). Parkinsonism Relat Disord 16(3):208–214

174. Josephs KA, Whitwell JL, Boeve BF, Knopman DS, Petersen RC, Hu WT et al (2010) Anatomical differences between CBS-corticobasal degeneration and CBS-Alzheimer's disease. Mov Disord: Off J Mov Disord Soc 25(9):1246–1252

175. Whitwell JL, Jack CR Jr, Boeve BF, Parisi JE, Ahlskog JE, Drubach DA et al (2010) Imaging correlates of pathology in corticobasal syndrome. Neurology 75(21):1879–1887

176. McMillan CT, Boyd C, Gross RG, Weinstein J, Firn K, Toledo JB et al (2016) Multimodal imaging evidence of pathology-mediated disease distribution in corticobasal syndrome. Neurology 87(12):1227–1234

177. Whitwell JL, Jack CR, Parisi JE, Knopman DS, Boeve BF, Petersen RC et al (2011) Imaging signatures of molecular pathology in behavioral variant frontotemporal dementia. J Mol Neurosci 45(3):372

178. Whitwell JL, Josephs KA (2011) Neuroimaging in frontotemporal lobar degeneration--predicting molecular pathology. Nat Rev Neurol 8(3):131–142

179. McMillan CT, Irwin DJ, Avants BB, Powers J, Cook PA, Toledo JB et al (2013) White matter imaging helps dissociate tau from TDP-43 in frontotemporal lobar degeneration. J Neurol Neurosurg Psychiatry 84(9):949–955

180. Botha H, Whitwell JL, Madhaven A, Senjem ML, Lowe V, Josephs KA (2014) The pimple sign of progressive supranuclear palsy syndrome. Parkinsonism Relat Disord 20(2):180–185

181. Zalewski N, Botha H, Whitwell JL, Lowe V, Dickson DW, Josephs KA (2014) FDG-PET in pathologically confirmed spontaneous 4R-tauopathy variants. J Neurol 261(4):710–716

182. Villemagne VL, Fodero-Tavoletti MT, Masters CL, Rowe CC (2015) Tau imaging: early progress and future directions. Lancet Neurol 14(1):114–124

183. Marquié M, Normandin MD, Vanderburg CR, Costantino IM, Bien EA, Rycyna LG et al (2015) Validating novel tau positron emission tomography tracer [F-18]-AV-1451 (T807) on postmortem brain tissue. Ann Neurol 78(5):787–800

184. Passamonti L, Vazquez Rodriguez P, Hong YT, Allinson KS, Williamson D, Borchert RJ et al (2017) 18F-AV-1451 positron emission tomography in Alzheimer's disease and progressive supranuclear palsy. Brain 140(3):781–791

185. Pontecorvo MJ, Devous Sr MD, Navitsky M, Lu M, Salloway S, Schaerf FW et al (2017) Relationships between flortaucipir PET tau binding and amyloid burden, clinical diagnosis, age and cognition. Brain 140(3):748–763

186. Nasrallah IM, Chen YJ, Hsieh M-K, Phillips JS, Ternes K, Stockbower GE et al (2018) 18F-Flortaucipir PET/MRI correlations in nonamnestic and amnestic variants of Alzheimer disease. J Nucl Med 59(2):299–306

187. Ossenkoppele R, Schonhaut DR, Schöll M, Lockhart SN, Ayakta N, Baker SL et al (2016) Tau PET pat-

terns mirror clinical and neuroanatomical variability in Alzheimer's disease. Brain 139(5):1551–1567

188. Lowe VJ, Curran G, Fang P, Liesinger AM, Josephs KA, Parisi JE et al (2016) An autoradiographic evaluation of AV-1451 tau PET in dementia. Acta Neuropathol Commun 4(1):58

189. Bevan Jones WR, Cope TE, Passamonti L, Fryer TD, Hong YT, Aigbirhio F et al (2016) [18F]AV-1451 PET in behavioral variant frontotemporal dementia due to MAPT mutation. Ann Clin Trans Neurol 3(12):940–947

190. Cho H, Choi JY, Hwang MS, Lee SH, Ryu YH, Lee MS et al (2017) Subcortical 18 F-AV-1451 binding patterns in progressive supranuclear palsy. Mov Disord: Off J Mov Disord Soc 32(1):134–140

191. Smith R, Schain M, Nilsson C, Strandberg O, Olsson T, Hagerstrom D et al (2017) Increased basal ganglia binding of 18 F-AV-1451 in patients with progressive supranuclear palsy. Mov Disord: Off J Mov Disord Soc 32(1):108–114

192. Whitwell JL, Lowe VJ, Tosakulwong N, Weigand SD, Senjem ML, Schwarz CG et al (2017) [18 F] AV-1451 tau positron emission tomography in progressive supranuclear palsy. Mov Disord: Off J Mov Disord Soc 32(1):124–133

193. Whitwell JL, Tosakulwong N, Botha H, Ali F, Clark HM, Duffy JR et al (2020) Brain volume and flortaucipir analysis of progressive supranuclear palsy clinical variants. NeuroImage: Clinical 25:102152

194. McMillan CT, Irwin DJ, Nasrallah I, Phillips JS, Spindler M, Rascovsky K et al (2016) Multimodal evaluation demonstrates in vivo (18)F-AV-1451 uptake in autopsy-confirmed corticobasal degeneration. Acta Neuropathol 132(6):935–937

195. Ali F, Whitwell JL, Martin PR, Senjem ML, Knopman DS, Jack CR et al (2018) [(18)F] AV-1451 uptake in corticobasal syndrome: the influence of beta-amyloid and clinical presentation. J Neurol 265(5):1079–1088

196. Gerhard A, Trender-Gerhard I, Turkheimer F, Quinn NP, Bhatia KP, Brooks DJ (2006) In vivo imaging of microglial activation with [11C](R)-PK11195 PET in progressive supranuclear palsy. Mov Disord: Off J Mov Disord Soc 21(1):89–93

197. Gerhard A, Watts J, Trender-Gerhard I, Turkheimer F, Banati RB, Bhatia K et al (2004) In vivo imaging of microglial activation with [11C](R)-PK11195 PET in corticobasal degeneration. Mov Disord: Off J Mov Disord Soc 19(10):1221–1226

198. Arai T, Ikeda K, Akiyama H, Shikamoto Y, Tsuchiya K, Yagishita S et al (2001) Distinct isoforms of tau aggregated in neurons and glial cells in brains of patients with Pick's disease, corticobasal degeneration and progressive supranuclear palsy. Acta Neuropathol 101(2):167–173

199. Urakami K, Wada K, Arai H, Sasaki H, Kanai M, Shoji M et al (2001) Diagnostic significance of tau protein in cerebrospinal fluid from patients with corticobasal degeneration or progressive supranuclear palsy. J Neurol Sci 183(1):95–98

200. Hall S, Öhrfelt A, Constantinescu R, Andreasson U, Surova Y, Bostrom F et al (2012) Accuracy of a panel of 5 cerebrospinal fluid biomarkers in the differential diagnosis of patients with dementia and/or parkinsonian disorders. Arch Neurol 69(11):1445–1452

201. Borroni B, Malinverno M, Gardoni F, Alberici A, Parnetti L, Premi E et al (2008) Tau forms in CSF as a reliable biomarker for progressive supranuclear palsy. Neurology 71(22):1796–1803

202. Kuiperij HB, Borroni B, Verbeek MM, Gardoni F, Malinverno M, Padovani A et al (2011) Tau forms in CSF as a reliable biomarker for progressive supranuclear palsy. Neurology 76(16):1443

203. Khalil M, Teunissen CE, Otto M, Piehl F, Sormani MP, Gattringer T et al (2018) Neurofilaments as biomarkers in neurological disorders. Nat Rev Neurol:1

204. Hansson O, Janelidze S, Hall S, Magdalinou N, Lees AJ, Andreasson U et al (2017) Blood-based NfL: a biomarker for differential diagnosis of parkinsonian disorder. Neurology 88(10):930–937

205. Holmberg B, Rosengren L, Karlsson JE, Johnels B (1998) Increased cerebrospinal fluid levels of neurofilament protein in progressive supranuclear palsy and multiple-system atrophy compared with Parkinson's disease. Mov Disord: Off J Mov Disord Soc 13(1):70–77

206. Marques TM, van Rumund A, Oeckl P, Kuiperij HB, Esselink RA, Bloem BR et al (2019) Serum NFL discriminates Parkinson disease from atypical parkinsonisms. Neurology 92(13):e1479–e1486

207. Sako W, Murakami N, Izumi Y, Kaji R (2015) Neurofilament light chain level in cerebrospinal fluid can differentiate Parkinson's disease from atypical parkinsonism: evidence from a meta-analysis. J Neurol Sci 352(1–2):84–87

208. Rojas JC, Karydas A, Bang J, Tsai RM, Blennow K, Liman V et al (2016) Plasma neurofilament light chain predicts progression in progressive supranuclear palsy. Ann Neurol: Off J Am Neurol Assoc Child Neurol Soc 3(3):216–225

209. Kuhle J, Barro C, Andreasson U, Derfuss T, Lindberg R, Sandelius A et al (2016) Comparison of three analytical platforms for quantification of the neurofilament light chain in blood samples: ELISA, electrochemiluminescence immunoassay and Simoa. Clin Chem Lab Med 54(10):1655–1661

210. Groveman BR, Orrù CD, Hughson AG, Raymond LD, Zanusso G, Ghetti B et al (2018) Rapid and ultra-sensitive quantitation of disease-associated α-synuclein seeds in brain and cerebrospinal fluid by αSyn RT-QuIC. Acta Neuropathol Commun 6(1):7

211. Kraus A, Saijo E, Metrick MA 2nd, Newell K, Sigurdson CJ, Zanusso G et al (2019) Seeding selectivity and ultrasensitive detection of tau aggregate conformers of Alzheimer disease. Acta Neuropathol 137(4):585–598

212. Saijo E, Ghetti B, Zanusso G, Oblak A, Furman JL, Diamond MI et al (2017) Ultrasensitive and selective detection of 3-repeat tau seeding activity in Pick dis-

ease brain and cerebrospinal fluid. Acta Neuropathol 133(5):751–765

213. Saijo E, Metrick MA, Koga S, Parchi P, Litvan I, Spina S et al (2020) 4-repeat tau seeds and templating subtypes as brain and CSF biomarkers of frontotemporal lobar degeneration. Acta Neuropathol 139(1):63–77

214. Riley D, Lang A, Ae L, Resch L, Ashby P, Hornykiewicz O et al (1990) Cortical-basal ganglionic degeneration. Neurology 40(8):1203

215. Jackson JA, Jankovic J, Ford J (1983) Progressive supranuclear palsy: clinical features and response to treatment in 16 patients. Ann Neurol: Off J Am Neurol Assoc Child Neurol Soc 13(3):273–278

216. Stowe R, Ives N, Clarke CE, Ferreira J, Hawker RJ, Shah L et al (2008) Dopamine agonist therapy in early Parkinson's disease. Cochrane Database Syst Rev (2):CD006564

217. Kompoliti K, Goetz C, Litvan I, Jellinger K, Verny M (1998) Pharmacological therapy in progressive supranuclear palsy. Arch Neurol 55(8):1099–1102

218. Engel PA (1996) Treatment of progressive supranuclear palsy with amitriptyline: therapeutic and toxic effects. J Am Geriatr Soc 44(9):1072–1074

219. Rajrut A, Uitti R, Fenton M, George D (1997) Amantadine effectiveness in multiple system atrophy and progressive supranuclear palsy. Parkinsonism Relat Disord 3(4):211–214

220. Daniele A, Moro E, Bentivoglio AR (1999) Zolpidem in progressive supranuclear palsy. N Engl J Med 341(7):543–544

221. Cotter C, Armytage T, Crimmins D (2010) The use of zolpidem in the treatment of progressive supranuclear palsy. J Clin Neurosci 17(3):385–386

222. Mayr BJ, Bonelli RM, Niederwieser G, Költringer P, Reisecker F (2002) Zolpidem in progressive supranuclear palsy. Eur J Neurol 9(2):184–185

223. Müller J, Wenning G, Wissel J, Seppi K, Poewe W (2002) Botulinum toxin treatment in atypical parkinsonian disorders associated with disabling focal dystonia. J Neurol 249(3):300–304

224. Piccione F, Mancini E, Tonin P, Bizzarini M (1997) Botulinum toxin treatment of apraxia of eyelid opening in progressive supranuclear palsy: report of two cases. Arch Phys Med Rehabil 78(5):525–529

225. Vanek Z, Jankovic J (2001) Dystonia in corticobasal degeneration. Mov Disord: Off J Mov Disord Soc 16(2):252–257

226. Polo KB, Jabbari B (1994) Botulinum toxin-A improves the rigidity of progressive supranuclear palsy. Ann Neurol: Off J Am Neurol Assoc Child Neurol Soc 35(2):237–239

227. Boeve BF, Josephs KA, Drubach DA (2008) Current and future management of the corticobasal syndrome and corticobasal degeneration. Handb Clin Neurol 89:533–548

228. Kovács T, Farsang M, Vitaszil E, Barsi P, Györke T, Szirmai I et al (2009) Levetiracetam reduces myoclonus in corticobasal degeneration: report of two cases. J Neural Transm 116(12):1631

229. Gómez-Caravaca MT, Cáceres-Redondo MT, Huertas-Fernández I, Vargas-González L, Carrillo F, Carballo M et al (2015) The use of botulinum toxin in the treatment of sialorrhea in parkinsonian disorders. Neurol Sci 36(2):275–279

230. Hyson HC, Johnson AM, Jog MS (2002) Sublingual atropine for sialorrhea secondary to parkinsonism: a pilot study. Mov Disord: Off J Mov Disord Soc 17(6):1318–1320

231. Litvan I, Phipps M, Pharr VL, Hallett M, Grafman J, Salazar A (2001) Randomized placebo-controlled trial of donepezil in patients with progressive supranuclear palsy. Neurology 57(3):467–473

232. Mendez MF, Shapira JS, McMurtray A, Licht E (2007) Preliminary findings: behavioral worsening on donepezil in patients with frontotemporal dementia. Am J Geriatr Psychiatry 15(1):84–87

233. Boxer AL, Knopman DS, Kaufer DI, Grossman M, Onyike C, Graf-Radford N et al (2013) Memantine in patients with frontotemporal lobar degeneration: a multicentre, randomised, double-blind, placebo-controlled trial. Lancet Neurol 12(2):149–156

234. Boxer AL, Lipton AM, Womack K, Merrilees J, Neuhaus J, Pavlic D et al (2009) An open label study of memantine treatment in three subtypes of frontotemporal lobar degeneration. Alzheimer Dis Assoc Disord 23(3):211

235. Pattee GL, Wymer JP, Lomen-Hoerth C, Appel SH, Formella AE, Pope LE (2014) An open-label multicenter study to assess the safety of dextromethorphan/quinidine in patients with pseudobulbar affect associated with a range of underlying neurological conditions. Curr Med Res Opin 30(11):2255–2265

236. Clerici I, Ferrazzoli D, Maestri R, Bossio F, Zivi I, Canesi M et al (2017) Rehabilitation in progressive supranuclear palsy: effectiveness of two multidisciplinary treatments. PLoS One 12(2):e0170927

237. Zampieri C, Di Fabio RP (2008) Balance and eye movement training to improve gait in people with progressive supranuclear palsy: quasi-randomized clinical trial. Phys Ther 88(12):1460–1473

238. Steffen TM, Boeve BF, Mollinger-Riemann LA, Petersen CM (2007) Long-term locomotor training for gait and balance in a patient with mixed progressive supranuclear palsy and corticobasal degeneration. Phys Ther 87(8):1078–1087

239. Henry M, Meese M, Truong S, Babiak M, Miller B, Gorno-Tempini M (2013) Treatment for apraxia of speech in nonfluent variant primary progressive aphasia. Behav Neurol 26(1–2):77–88

240. Farrajota L, Maruta C, Maroco J, Martins IP, Guerreiro M, De Mendonca A (2012) Speech therapy in primary progressive aphasia: a pilot study. Dement Geriatr Cogn Disord Extra 2(1):321–331

241. Wiblin L, Lee M, Burn D (2017) Palliative care and its emerging role in multiple system atrophy and progressive supranuclear palsy. Parkinsonism Relat Disord 34:7–14

242. Buée L, Bussière T, Buée-Scherrer V, Delacourte A, Hof PR (2000) Tau protein isoforms, phosphorylation

and role in neurodegenerative disorders. Brain Res Rev 33(1):95–130

243. Leclair-Visonneau L, Rouaud T, Debilly B, Durif F, Houeto JL, Kreisler A et al (2016) Randomized placebo-controlled trial of sodium valproate in progressive supranuclear palsy. Clin Neurol Neurosurg 146:35–39

244. Min S-W, Chen X, Tracy TE, Li Y, Zhou Y, Wang C et al (2015) Critical role of acetylation in tau-mediated neurodegeneration and cognitive deficits. Nat Med 21(10):1154

245. Yuzwa SA, Shan X, Macauley MS, Clark T, Skorobogatko Y, Vosseller K et al (2012) Increasing O-GlcNAc slows neurodegeneration and stabilizes tau against aggregation. Nat Chem Biol 8(4):393–399

246. Wang AC, Jensen EH, Rexach JE, Vinters HV, Hsieh-Wilson LC (2016) Loss of O-GlcNAc glycosylation in forebrain excitatory neurons induces neurodegeneration. Proc Natl Acad Sci U S A 113(52):15120–15125

247. Boxer AL, Lang AE, Grossman M, Knopman DS, Miller BL, Schneider LS et al (2014) Davunetide in patients with progressive supranuclear palsy: a randomised, double-blind, placebo-controlled phase 2/3 trial. Lancet Neurol 13(7):676–685

248. Boxer AMZ, Tsai R, Koestler M, Rojas J, Ljubenkov P, Rosen H et al (2017) A phase 1B, randomized, double-blind, placebo-controlled, sequential cohort, dose-ranging study of the safety, tolerability, pharmacokinetics, pharmacodynamics, and preliminary efficacy of TPI 287 (abeotaxane) in patients with primary four repeat tauopathies: corticobasal syndrome or progressive supranuclear palsy; or the secondary tauopathy, Alzheimer's disease. J Prev Alz Dis 4(4):282–428

249. Zhang B, Carroll J, Trojanowski JQ, Yao Y, Iba M, Potuzak JS et al (2012) The microtubule-stabilizing agent, epothilone D, reduces axonal dysfunction, neurotoxicity, cognitive deficits, and Alzheimer-like pathology in an interventional study with aged tau transgenic mice. J Neurosci Off J Soc Neurosci 32(11):3601–3611

250. Wischik CM, Staff RT, Wischik DJ, Bentham P, Murray AD, Storey JM et al (2015) Tau aggregation inhibitor therapy: an exploratory phase 2 study in mild or moderate Alzheimer's disease. J Alzheimer's Disease: JAD 44(2):705–720

251. Bensimon G, Ludolph A, Agid Y, Vidailhet M, Payan C, Leigh PN (2009) Riluzole treatment, survival and diagnostic criteria in Parkinson plus disorders: the NNIPPS study. Brain 132(Pt 1):156–171

252. Apetauerova D, Scala SA, Hamill RW, Simon DK, Pathak S, Ruthazer R et al (2016) CoQ10 in progressive supranuclear palsy: a randomized, placebo-controlled, double-blind trial. Neurology(R) Neuroimmunol Neuroinflammation 3(5):e266

253. Nuebling G, Hensler M, Paul S, Zwergal A, Crispin A, Lorenzl S (2016) PROSPERA: a randomized, controlled trial evaluating rasagiline in progressive supranuclear palsy. J Neurol 263(8):1565–1574

254. Stamelou M, Reuss A, Pilatus U, Magerkurth J, Niklowitz P, Eggert KM et al (2008) Short-term effects of coenzyme Q10 in progressive supranuclear palsy: a randomized, placebo-controlled trial. Mov Disord: Off J Mov Disord Soc 23(7):942–949

255. Bright J, Hussain S, Dang V, Wright S, Cooper B, Byun T et al (2015) Human secreted tau increases amyloid-beta production. Neurobiol Aging 36(2):693–709

256. Qureshi IA, Tirucherai G, Ahlijanian MK, Kolaitis G, Bechtold C, Grundman M (2018) A randomized, single ascending dose study of intravenous BIIB092 in healthy participants. Alzheimer's & Dement: Transl Res Clin Interv 4:746–755

257. Boxer AL, Qureshi I, Ahlijanian M, Grundman M, Golbe LI, Litvan I et al (2019) Safety of the tau-directed monoclonal antibody BIIB092 in progressive supranuclear palsy: a randomised, placebo-controlled, multiple ascending dose phase 1b trial. Lancet Neurol 18(6):549–558

258. Dam T, Boxer A, Golbe LI, Höglinger G, Morris HR, Litvan I et al (2018) Efficacy and safety of BIIB092 in patients with progressive supranuclear palsy: passport phase 2 study design (P6. 073). AAN Enterprises

259. Budur K, West T, Braunstein JB, Fogelman I, Bordelon YM, Litvan I et al (2017) Results of a phase 1, single ascending dose, placebo-controlled study of ABBV-8E12 in patients with progressive supranuclear palsy and phase 2 study design in early Alzheimer's disease. Alzheimers Dement 13(7):P599–P600

260. West T, Hu Y, Verghese P, Bateman R, Braunstein J, Fogelman I et al (2017) Preclinical and clinical development of ABBV-8E12, a humanized anti-tau antibody, for treatment of Alzheimer's disease and other tauopathies. J Prev Alzheimers Dis 4(04):236–241

261. Boutajangout A, Ingadottir J, Davies P, Sigurdsson EM (2011) Passive immunization targeting pathological phospho-tau protein in a mouse model reduces functional decline and clears tau aggregates from the brain. J Neurochem 118(4):658–667

262. Chai X, Wu S, Murray TK, Kinley R, Cella CV, Sims H et al (2011) Passive immunization with anti-Tau antibodies in two transgenic models: reduction of Tau pathology and delay of disease progression. J Biol Chem 286(39):34457–34467

263. Sankaranarayanan S, Barten DM, Vana L, Devidze N, Yang L, Cadelina G et al (2015) Passive immunization with phospho-tau antibodies reduces tau pathology and functional deficits in two distinct mouse tauopathy models. PLoS One 10(5):e0125614

264. Ittner A, Bertz J, Suh LS, Stevens CH, Gotz J, Ittner LM (2015) Tau-targeting passive immunization modulates aspects of pathology in tau transgenic mice. J Neurochem 132(1):135–145

265. Castillo-Carranza DL, Gerson JE, Sengupta U, Guerrero-Muñoz MJ, Lasagna-Reeves CA, Kayed R (2014) Specific targeting of tau oligomers in htau mice prevents cognitive impairment and tau toxicity following injection with brain-derived tau oligomeric seeds. J Alzheimers Dis 40(s1):S97–S111

266. Lasagna-Reeves CA, Castillo-Carranza DL, Sengupta U, Guerrero-Munoz MJ, Kiritoshi T, Neugebauer V et al (2012) Alzheimer brain-derived tau oligomers propagate pathology from endogenous tau. Sci Rep 2:700

267. Igawa T, Tsunoda H, Kuramochi T, Sampei Z, Ishii S, Hattori K (2011) Engineering the variable region of therapeutic IgG antibodies. MAbs 3(3):243–252

268. Xu H, Rosler TW, Carlsson T, de Andrade A, Fiala O, Hollerhage M et al (2014) Tau silencing by siRNA in the P301S mouse model of tauopathy. Curr Gene Ther 14(5):343–351

269. Sud R, Geller ET, Schellenberg GD (2014) Antisense-mediated exon skipping decreases tau protein expression: a potential therapy for tauopathies. Mol Ther Nucleic Acids 3:e180

270. Schoch KM, DeVos SL, Miller RL, Chun SJ, Norrbom M, Wozniak DF et al (2016) Increased 4R-tau induces pathological changes in a human-tau mouse model. Neuron 90(5):941–947

Tau Protein and Frontotemporal Dementias

Michel Goedert, Maria Grazia Spillantini,
Benjamin Falcon, Wenjuan Zhang,
Kathy L. Newell, Masato Hasegawa,
Sjors H. W. Scheres, and Bernardino Ghetti

Introduction

Ordered assembly of fewer than ten proteins into filamentous assemblies defines cases of age-related neurodegenerative diseases, including Alzheimer's disease (AD) and Parkinson's disease (PD). Aβ, tau, α-synuclein and TDP-43 are the best known of these proteins. For most diseases, the majority of cases are sporadic, but a small percentage is inherited in a dominant manner. Huntington's disease and other polyglutamine repeat diseases form an exception because all cases are inherited. Chronic traumatic encephalopathy (CTE), by contrast, is probably always environmentally induced. Study of dominantly inherited forms of disease has established a causative role for ordered assembly. By extrapolation, it appears likely that inclusion formation is central to neurodegeneration in all cases of disease. Tau proteinopathies, which are characterised by the assembly of tau protein, are the most common proteinopathies of the human nervous system [1].

Frontotemporal dementias (FTDs), also known as frontotemporal lobar degenerations (FTLDs), are characterised by progressive changes in personality and/or language loss, followed by dementia [2]. Their neuroanatomical substrate is degeneration of frontal and temporal lobes of the cerebral cortex. FTDs have a genetic component that is stronger than for most other neurodegenerative diseases, with mutations in *MAPT*, the tau gene, *GRN*, the progranulin gene and *C9orf72*, the chromosome 9 open reading frame 72 gene, being the most common. Mutations in *MAPT* account for approximately 5% of cases of FTD, with an average age of onset of around 50 years and a duration of disease of approximately 10 years. Some of the clinical and neuropathological features resulting from *MAPT* mutations are reminiscent of sporadic tau proteinopathies, including Pick's disease (PiD), progressive supranuclear palsy (PSP), corticobasal degeneration (CBD), globular glial tauopathy (GGT) and chronic traumatic encephalopathy (CTE). Identification of *MAPT* mutations proved that dysfunction of tau protein is sufficient to cause neurodegeneration and dementia. Here, we first discuss these mutations and their effects, and

M. Goedert (✉) · B. Falcon · W. Zhang · S. H. W. Scheres
MRC Laboratory of Molecular Biology,
Cambridge, UK
e-mail: mg@mrc-lmb.cam.ac.uk

M. G. Spillantini
Department of Clinical Neurosciences, University of Cambridge, Cambridge, UK
e-mail: mgs11@cam.ac.uk

K. L. Newell · B. Ghetti
Department of Pathology and Laboratory Medicine,
Indiana University, Indianapolis, IN, USA

M. Hasegawa
Department of Dementia and Higher Brain Function,
Tokyo Metropolitan Institute of Medical Science,
Tokyo, Japan

B. Ghetti et al. (eds.), *Frontotemporal Dementias*, Advances in Experimental Medicine and Biology
1281, https://doi.org/10.1007/978-3-030-51140-1_12

then focus on sporadic PiD, CBD and CTE and their filament structures.

Tau Protein and Its Isoforms

Tau is an intrinsically disordered protein, which may have many interaction partners. It can be divided into an amino-terminal domain, a proline-rich (PXXP) region, the repeat domain and a carboxy-terminal region. The amino-terminal domain projects away from the microtubule surface and is believed to interact with components of the neuronal plasma membrane. It contains a primate-specific sequence between residues 18 and 28. The PXXP motifs in the proline-rich region are recognised by SH3 domain-containing proteins of the Src family of nonreceptor tyrosine kinases, such as Fyn [3].

The repeat region and some adjacent sequences mediate interactions between tau and microtubules. Electron cryo-microscopy (cryo-EM) has shown that each tau repeat binds to the outer microtubule surface and adopts an extended structure along protofilaments, interacting with α- and β-tubulins [4, 5]. Single-molecule tracking revealed a kiss-and-hop mechanism, with a dwell time of tau on individual microtubules of approximately 40 ms [6, 7]. Isoform differences do not influence this interaction. Despite these rapid dynamics, tau promotes microtubule assembly. It remains to be seen if microtubules are also stabilised. Tau is most abundant in the labile domain of microtubules, which has led to the suggestion that it may not stabilise microtubules, but it may enable them to have long labile domains [8, 9]. Less is known about the function of the carboxy-terminal region, which may inhibit assembly into filaments.

Despite lacking a typical low-complexity domain, full-length tau can undergo liquid-liquid phase separation through electrostatic and hydrophobic interactions [10, 11], which has been found in conjunction with amyloid aggregation, at least in vitro. Although liquid-liquid phase separation and amyloid aggregation of tau are independent processes, they may be able to influence each other.

Six tau isoforms ranging from 352 to 441 amino acids are expressed in adult human brain from a single *MAPT* gene [12] (Fig. 1). They differ by the presence or absence of inserts of 29 and 58 amino acids (encoded by exons 2 and 3, with exon 3 being only transcribed with exon 2) in the amino-terminal half, and the inclusion, or not, of the 31 amino acid microtubule-binding repeat, encoded by exon 10, in the carboxy-terminal half. Inclusion of exon 10 results in the production of three isoforms with four repeats (4R) and its exclusion in a further three isoforms with three repeats (3R). The repeats comprise residues 244–368, in the numbering of the 441 amino acid isoform. In adult human brain, similar levels of 3R and 4R tau are expressed [13]; the finding that a correct isoform ratio is essential for preventing neurodegeneration and dementia came as a surprise. The 2 N isoforms are underrepresented in comparison with isoforms that include exon 2 or exclude both exons 2 and 3; 2 N, 1 N and 0 N tau isoforms make up 9%, 54% and 37%, respectively. Big tau, which carries an additional large exon in the amino-terminal half, is only expressed in the peripheral nervous system.

Isoform expression is not conserved between species. Thus, in adult mouse brain, 4R tau isoforms are almost exclusively present, whereas adult chicken brain expresses 3R, 4R and 5R tau isoforms [14]. However, the presence of one hyperphosphorylated 3R tau isoform lacking amino-terminal inserts is characteristic of developing vertebrates. In mice, the switch from 3R to 4R tau occurs between postnatal days 9 and 18, with tau phosphorylation decreasing over time. However, isoform switching and phosphorylation are regulated differently [15]. Adult 4R tau isoforms are better at promoting microtubule assembly and at binding to microtubules than the 3R tau isoform expressed during development.

Tau Assemblies

Full-length tau assembles into filaments [1, 16]. Negative-stain immuno-electron microscopy showed that antibodies specific for the N- and C-termini of tau decorate filaments. This was not

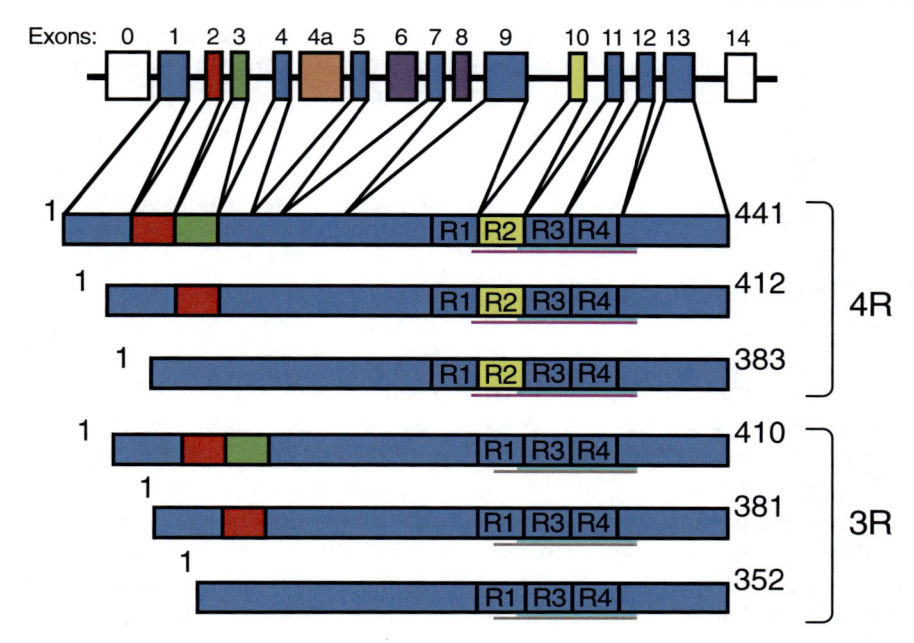

Fig. 1 Human brain tau isoforms. *MAPT* and the six tau isoforms expressed in adult human brain. *MAPT* consists of 14 exons (E). Alternative mRNA splicing of E2 (red), E3 (green) and E10 (yellow) gives rise to six tau isoforms (352–441 amino acids). The constitutively spliced exons (E1, E4, E5, E7, E9, E11, E12 and E13) are shown in blue. E6 and E8 (violet) are not transcribed in human brain. E4a (orange) is only expressed in the peripheral nervous system. The repeats (R1–R4) are shown, with three isoforms having four repeats (4R) and three isoforms with three repeats (3R). The core sequences of tau filaments from chronic traumatic encephalopathy (K274/S305-R379) determined by cryo-EM are underlined (in blue); the core sequences of tau filaments from Pick's disease (K254-F378 of 3R tau) are underlined (in grey); and the core sequences of tau filaments from corticobasal degeneration (K274-E380 of 4R tau) are underlined (in cyan)

the case of antibodies directed against R3 and R4 of tau because their epitopes are occluded in the filaments [17–19]. Together with biochemical studies, this work established that tau filaments consist of a core region and a fuzzy coat. Tau filaments have the biophysical characteristics of amyloid [20]. Because the region in tau that binds to microtubules also forms the filament cores, physiological function and pathological assembly may be mutually exclusive.

Phosphorylation negatively regulates the ability of tau to interact with microtubules, and filamentous tau is abnormally hyperphosphorylated [21]. It remains to be seen if phosphorylation is necessary and/or sufficient for the assembly of tau into filaments. Alternatively, a change in conformation as part of the assembly process may lead to tau hyperphosphorylation. Because tau is hydrophilic, it is not surprising that unmodified full-length protein requires cofactors, such as heparin, to assemble into filaments [22–25]. Cofactors other than heparin and/or post-translational modifications may cause the assembly of tau in human brain [26, 27].

Besides phosphorylation, other modifications may also be involved. Thus, acetylation, methylation, glycation, isomerisation, O-GlcNAcylation, nitration, sumoylation, ubiquitination and truncation of assembled tau have been described. In particular, acetylation of lysine residues has come to the fore in recent years. It reduces charge, which may play a role in filament assembly of tau. Site-specific acetylation of K280 has been shown to enhance tau aggregation, while reducing microtubule assembly [28]. Twenty-one lysine residues are present between residues 244 and 380 of tau.

In AD, CTE, tangle-only dementia and many other tauopathies, all six tau isoforms are present in disease filaments. Pick bodies of PiD are made of only 3R tau. In CBD, PSP, argyrophilic grain disease (AGD), GGT and several other diseases, 4R tau isoforms make up the filaments. The morphologies of tau filaments vary in the different diseases, even when they are made of the same isoforms.

Genetics of Microtubule-Associated Protein Tau

The relevance of tau dysfunction for neurodegeneration became clear in June 1998, when dominantly inherited mutations in *MAPT* were shown to cause a form of frontotemporal dementia that can be associated with parkinsonism, frontotemporal dementia and parkinsonism linked to chromosome 17 and caused by mutations in the tau gene (FTDP-17 T, also known as familial FTLD-tau) [29–31]. In FTDP-17 T, abundant filamentous tau inclusions are present either in nerve cells or in both nerve cells and glial cells. Aβ deposits, a defining feature of AD, are not present. This work established that a pathological pathway, leading from monomeric to assembled tau, is sufficient to cause neurodegeneration and dementia.

Sixty-five mutations in *MAPT* have been identified in FTDP-17 T (Fig. 2). Filamentous inclusions are composed of either 3R, 4R or 3R + 4R tau [2]. *MAPT* mutations are concentrated in exons 9–12 (encoding R1–R4) and the introns flanking exon 10, with a smaller number of disease-causing mutations in exon 13. Two mutations (R5H and R5L) are present in exon 1 of *MAPT*. Mutations can be divided into those with a primary effect at the protein level and those affecting the alternative messenger ribonucleic acid (mRNA) splicing of tau pre-mRNA.

The architecture of *MAPT* on chromosome 17q21.31 is characterised by two haplotypes as the result of a 900 kb inversion (H1) or noninversion (H2) polymorphism [32]. Inheritance of the H1 haplotype of *MAPT* is a risk factor for PSP, CBD, PD and amyotrophic lateral sclerosis (ALS), but not for PiD [33–38]. The H2 haplotype is associated with increased expression of exon 3 of *MAPT* in grey matter, suggesting that

inclusion of exon 3 may protect against PSP, CBD, PD and ALS [39]. In experimental studies, exon 3-containing tau isoforms have been found to aggregate less than those lacking exon 3 [40].

Disease-causing mutations in *MAPT* have made it possible to produce transgenic rodent lines that form tau filaments and show neurodegeneration [41–43]. Aggregation of tau correlates with neurodegeneration [44]. Reducing aggregation and increasing degradation of aggregates are therefore therapeutic objectives. It has been reported that the removal of senescent brain cells leads to a reduction in both tau aggregates and neurodegeneration in transgenic mice [45].

Transgenic mouse lines were also essential for identification of the prion-like properties of assembled tau. Aggregation of hyperphosphorylated tau was induced following intracerebral injection of tau seeds from mice transgenic for human mutant 0N4R P301S tau into transgenic mice expressing wild-type non-aggregated 2N4R tau and, to a lesser extent, following intracerebral injection into wild-type mice [46]. Tauopathy then spread to connected brain regions, indicative of seed endocytosis, seeded aggregation, intracellular transport, and release of tau seeds. This work was complemented by studies in cells [47]. It was subsequently shown that in brain extracts from mice transgenic for human P301S tau, short filaments had the greatest seeding activity [48]. These findings may be mechanistically related to the observation that in the process leading to AD, seed-competent tau inclusions first appear in transentorhinal cortex, followed by the hippocampal formation and large parts of the neocortex [49, 50].

Conformers of assembled tau seem to exist that influence the pattern of spread in brain, reminiscent of prion strains [51–53]. They may explain the variety of human tauopathies. Inclusions formed and spread of pathology occurred after intracerebral injection of brain homogenates from cases of AD, tangle-only dementia, PSP, CBD and AGD into a mouse line transgenic for wild-type human 4R tau and, to a lesser extent, following intracerebral injection into non-transgenic mice [51]. PiD, the filamentous inclusions of which are made of 3R tau only,

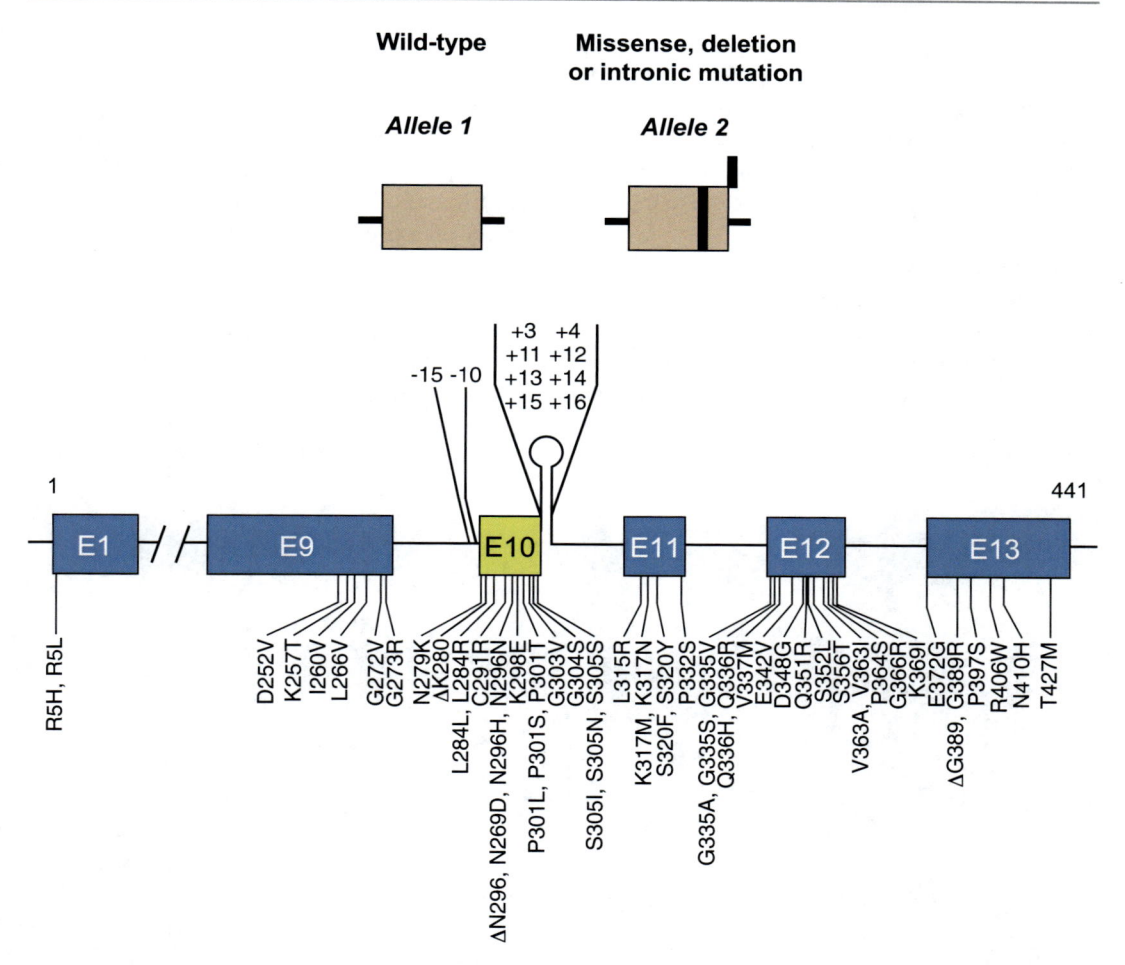

Fig. 2 Mutations in *MAPT* in FTDP-17 T. Missense, deletion and intronic mutations are dominantly inherited. Fifty-five coding region and ten intronic mutations are shown

was an exception. However, seeds from PiD brain induced inclusion formation and spreading in a mouse line, expressing equal amounts of human 3R and 4R tau, in the absence of mouse tau [53].

The tau sequence and, possibly, non-tau molecular requirements for seeded aggregation in vivo remain to be defined. Tau assemblies reminiscent of those in the corresponding human diseases were observed, following the injection of brain homogenates from patients with PSP, CBD and AGD, which are 4R tau proteinopathies [51] and PiD, a 3R tau proteinopathy [53]. Although these findings are consistent with the existence of distinct tau aggregate conformers, structural information is required to prove their existence.

Neuropathological Phenotypes of FTDP-17T

Cases of FTDP-17 T are characterised by the presence of filamentous tau inclusions in nerve cells or in both nerve cells and glial cells [1, 2]. Cases with glial inclusions only have not been described. Tau inclusions are most abundant in hippocampal formation and cerebral cortex.

Inclusions similar to Pick bodies are often observed in the brains of individuals with mutations in exons 9, 11, 12 and 13 of *MAPT*. Similar to sporadic PiD, inclusions associated with mutations G272V in exon 9 and ΔK280 in exon 10 are made of 3R tau and are not phosphorylated at S262 [54–56]. For other mutations, such as G389R in exon 13, variable amounts of 4R tau and some phosphorylation of S262 are seen in Pick-like bodies [57] (Figs. 3 and 8). Mutation N410H in exon 13 phenocopies the tau pathology of CBD [58].

In the study mentioned earlier, tau deposits are found predominantly in neurons, whereas mutations in exon 1 and exon 10, as well as in the introns following exon 9 and exon 10, are associated with abundant neuronal and glia tau inclusions [2]. Glial pathology is in the form of coiled bodies in oligodendroglia, as well as

tufted astrocytes and astrocytic plaques reminiscent of PSP and CBD. Mutations in exon 10 cause the formation of inclusions made of 4R tau; most of these mutations affect exon 10 pre-mRNA splicing, altering the ratio of 3R/4R tau. *MAPT* mutations P301L, P301S and P301T, the primary effects of which are at the protein level, are exceptions (Figs. 4 and 8). They continue to be important for the generation of experimental models of tauopathy and illustrate the clinical and pathological heterogeneity associated with *MAPT* mutations. Although most individuals with mutations P301L and P301S develop behavioural-variant FTD, cases of primary pro-

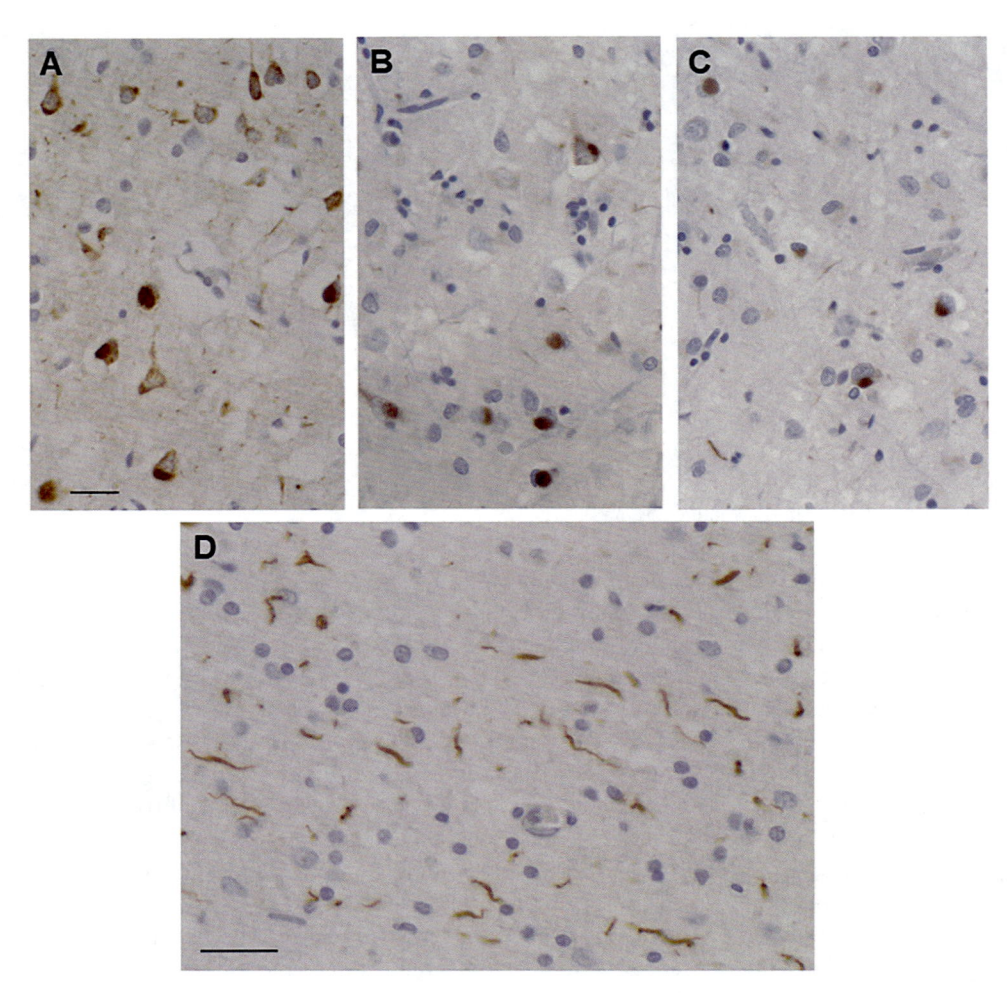

Fig. 3 Tau pathology in the frontal cortex of a patient with the G389R mutation in *MAPT*. Pick-like bodies in grey matter and neuropil threads in white matter are labelled by anti-tau antibodies AT8 (**a**, **d**), RD3 (**b**) and RD4 (**c**). More Pick-like bodies were labelled with RD3 than RD4. Scale bar, 25 μm

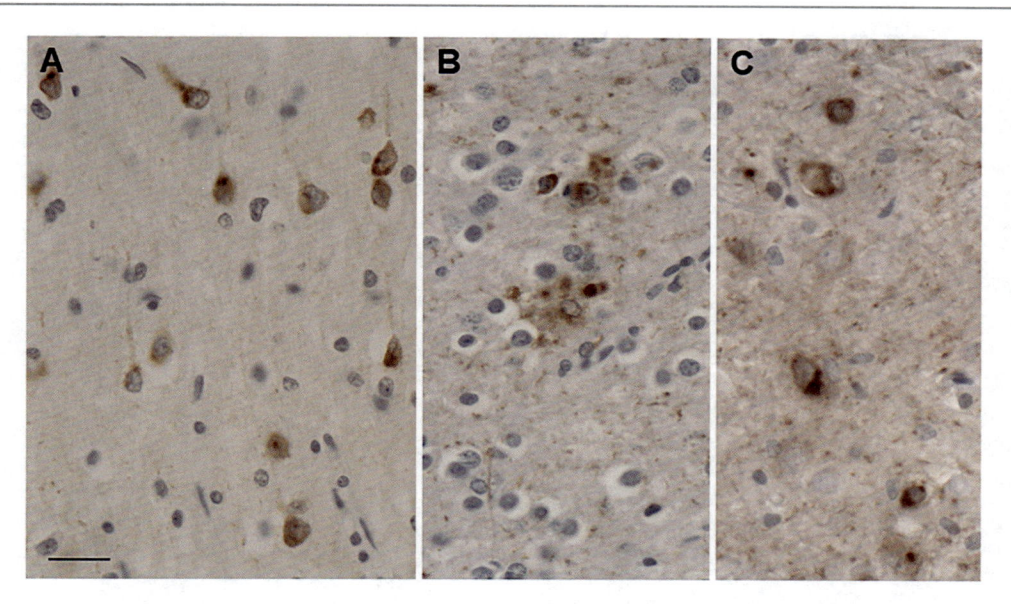

Fig. 4 Tau pathology in the frontal cortex of a patient with the P301L mutation in *MAPT*. Tau inclusions in nerve cells and astrocytes are labelled by anti-tau antibod- ies AT8 (**a**, **b**) and RD4 (**c**). These inclusions were not labelled by RD3. Scale bar, 25 μm

gressive aphasia have been described [59]. A P301S carrier presented with corticobasal syndrome [60]. A P301L patient had GGT, as had individuals with mutation P301T [59, 61]. GGT has emerged as a common disease associated with mutations in *MAPT*. Mutations in codon 301 affect only 20–25% of tau molecules, with 75–80% being wild type, arguing against a simple loss-of-function mechanism as an important disease determinant [62].

Intronic mutations in *MAPT* and most mutations in exon 10 affect the ratio of 3R/4R tau, which is normally 1:1, without changing the amount of total tau (Figs. 5, 6 and 8). For most mutations, this results in the relative overproduction of wild-type 4R tau and its assembly into filamentous inclusions. Tau filaments appear as twisted ribbons or half ribbons. Although these mutations often give rise to behavioural-variant FTD, cases of atypical PSP have also been described [63]. For other mutations, such as V337M (exon 12) [64] and R406W (exon 13) (Figs. 7 and 8) [65], tau inclusions resemble those of AD, and filaments are made of all six brain tau isoforms.

Structures of Tau Filaments from Pick's Disease

PiD accounts for approximately 20% cases of FTLD-tau. Behavioural-variant frontotemporal dementia and progressive non-fluent aphasia are its most common clinical manifestations. Arnold Pick described the clinical picture and macroscopic findings in 1892 [66], and Alois Alzheimer reported the microscopic features in 1911 [67]. The presence of tau protein in Pick bodies was shown in 1985 [68, 69].

Nerve cell loss predominates in cerebral cortex (frontal > temporal > parietal), followed by hippocampal formation and amygdala, with subcortical structures being affected to variable extents [70]. The substantia nigra may be affected in some cases, while the nucleus basalis of Meynert is mostly unaffected. In frontal and anterior temporal lobes, severe circumscribed (knifeedge) atrophy is commonly seen. Microscopically, the Pick body, which consists of assembled, hyperphosphorylated 3R tau, is the pathognomonic inclusion of PiD (Fig. 9a, b) [71]. Biochemical studies have also suggested the

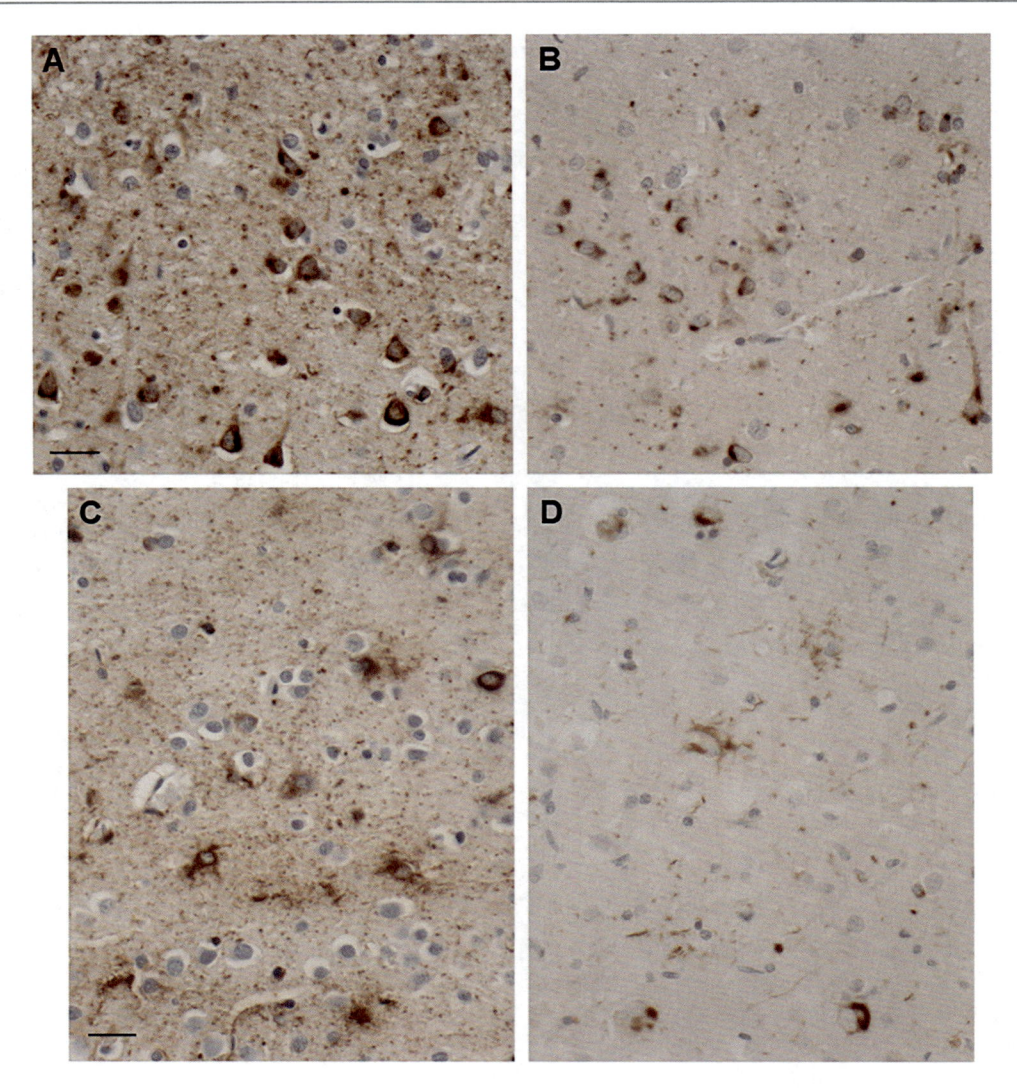

Fig. 5 Tau pathology in the frontal cortex of a patient with the IVS10 + 16 mutation in *MAPT*. Tau inclusions in nerve cells and astrocytes are labelled by anti-tau antibodies AT8 (**a, c**) and RD4 (**b, d**). These inclusions were not labelled by RD3. Scale bar, 25 μm

presence of 4R tau pathology. However, this probably reflects coexisting pathologies [72] or the presence of a *MAPT* mutation. Pick bodies predominate in hippocampus and cerebral cortex. Fewer assemblies are present in glial cells (Fig. 9c). The glial tau pathology of PiD consists of ramified astrocytes and globular glial inclusions in oligodendrocytes. By Western blotting, assembled tau from PiD brain runs as a doublet of 60 and 64 kDa, which reveals the presence of 3R tau upon dephosphorylation [73].

By negative stain electron microscopy of sarkosyl-insoluble filaments from PiD brain, we observed narrow (Type I) and wide (Type II) tau filaments [74]. Narrow filaments had previously been described as straight, but they have a helical twist with a crossover distance of approximately 1000 Å and widths of 50–150 Å. Wide filaments have a similar crossover distance, but their widths vary from 150 to 300 Å. Immunogold negative-stain electron microscopy showed that most filaments are Type I, with a minority of Type II

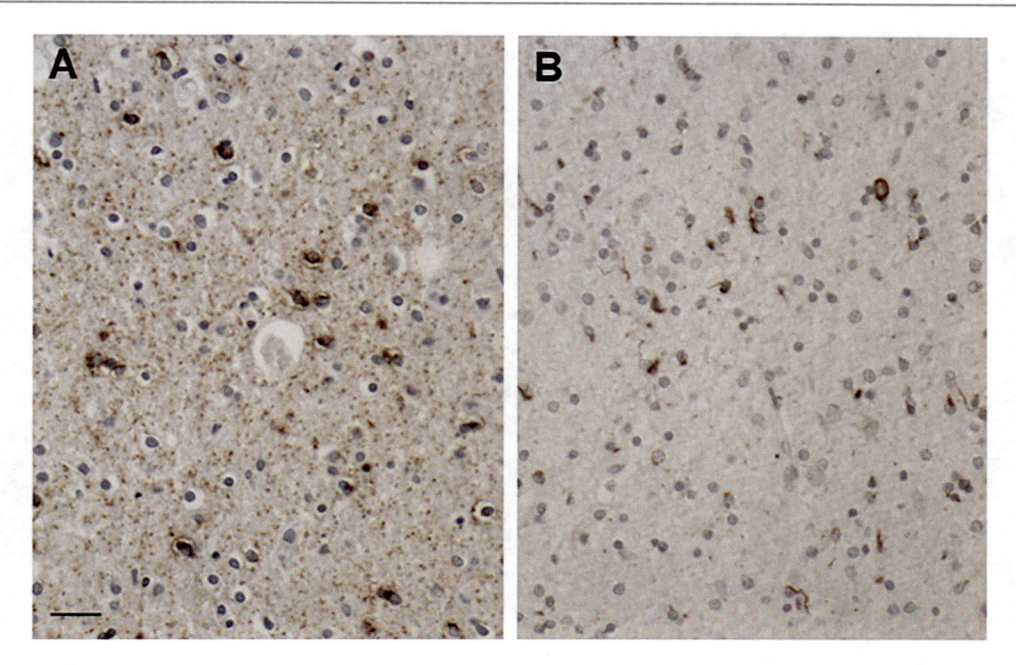

Fig. 6 Tau pathology in the subcortical white matter of the frontal lobe in a patient with the IVS10 + 16 mutation in *MAPT*. Tau inclusions in oligodendrocytes in white matter are labelled by anti-tau antibodies AT8 (**a**) and RD4 (**b**). These inclusions were not labelled by RD3. Scale bar, 25 μm

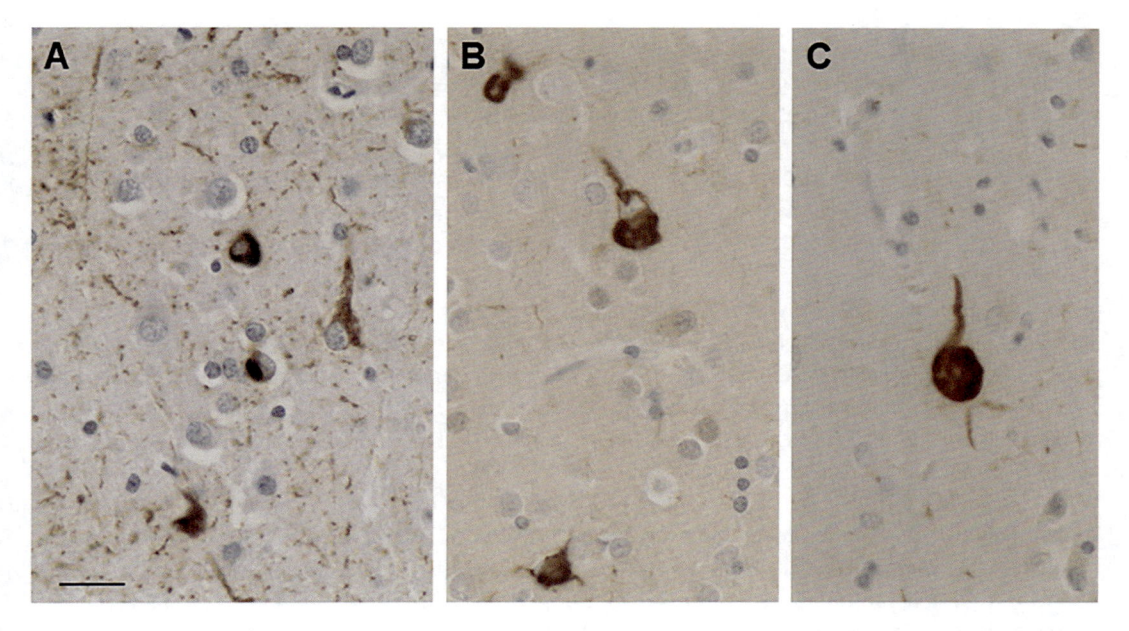

Fig. 7 Tau pathology in the frontal cortex of a patient with the R406W mutation in *MAPT*. Neurofibrillary tangles and neuropil threads are labelled by anti-tau antibodies AT8 (**a**), RD3 (**b**) and RD4 (**c**). Scale bar, 25 μm

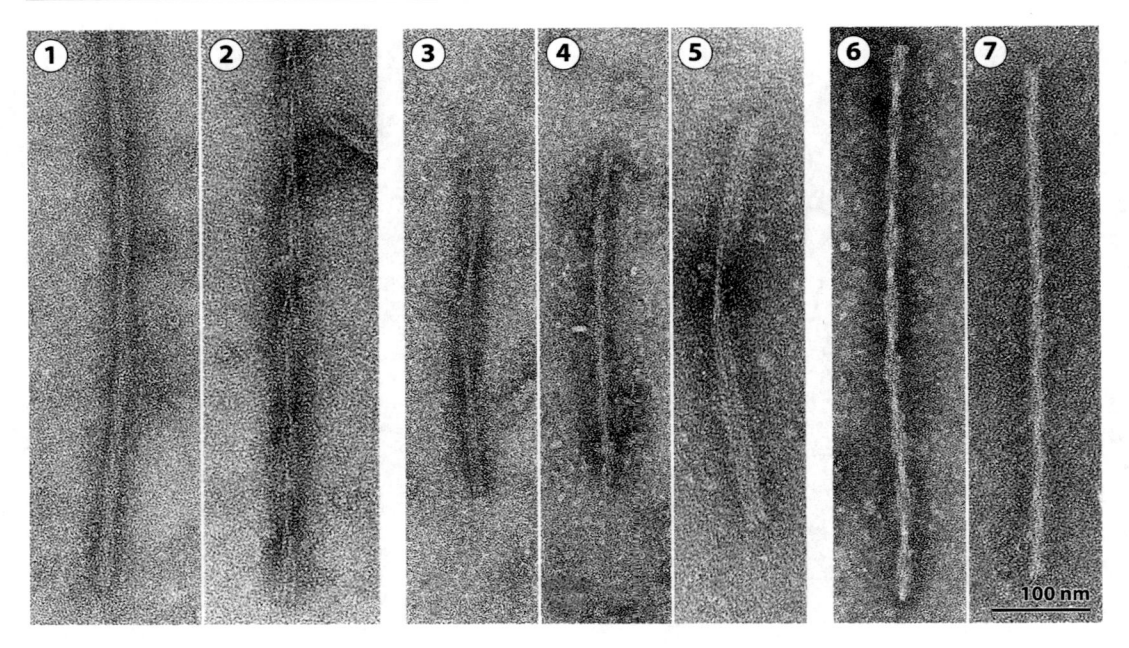

Fig. 8 Negative-stain electron microscopy of tau filaments from cases of frontotemporal dementia and parkinsonism linked to chromosome 17 caused by *MAPT* mutations (FTDP-17 T). (1, 2), Tau filaments from a case with abundant Pick body–like inclusions and a G389R mutation. (1) Straight filaments form the majority species and (2) strongly stranded filaments are in the minority. (3–5). Tau filaments from cases with neuronal and glial inclusions and a P301L mutation or an IVS10 mutation. (3) Narrow twisted ribbons and (4) occasional rope-like filaments. (5) Wide twisted ribbons. (6, 7) Paired helical and straight tau filaments as in AD are present in cases with mutations V337M and R406W in *MAPT*

filaments. Filaments were not decorated by antibodies specific for R1, R3 or R4 of tau, indicating that these repeats form part of the ordered filament core.

By cryo-EM, structures of tau filaments were determined from combined frontal and temporal cortices of an individual with PiD (Fig. 10) [74]. The core of Type I filaments is made of a single protofilament that consists of residues K254-F378 of 3R tau (93 amino acids), which adopt an elongated, J-shaped, cross-β structure (Fig. 10a, c). Type II filaments are formed by the association of two Type I filaments at the distal tips of the J, where they form tight contacts through van der Waals interactions (Fig. 10b). We determined a 3.2 Å resolution map of the ordered cores of Type I filaments; the map of Type II filaments was limited to 8 Å. Each protofilament comprises nine β-strands, which are arranged into four cross-β packing stacks and are connected by turns and arcs. R1 provides two β-strands, and R3 and R4 three β-strands each. The stacks pack together in a hairpin-like fashion: β1 against one side of β8, β2 against β7, β3 against β6 and β4 against β5. The final strand, β9, is formed from the ten amino acids after R4 and packs against the other side of β8.

Three regions of less well-resolved density bordering the solvent-exposed faces of β4, β5 and β9 are apparent in Type I and Type II filaments. They may represent less ordered, heterogeneous and/or transiently occupied structures. The density bordering β4 is similarly located, but more extended, than that found to interact with the side chains of K317, T319 and K321 in tau filaments from AD.

Unlike tau filaments of CBD, CTE and AD, Pick body filaments are not phosphorylated at S262 [75, 76]. The reasons for this differential phosphorylation are unknown. The cryo-EM structure shows that the tight turn at G261 prevents phosphorylation of S262 in the ordered core of PiD filaments, whereas phosphorylated S262 is outside the ordered cores of tau filaments

Fig. 9 Tau pathology in the frontal cortex of a patient with Pick's disease. Tau inclusions in nerve cells and glia in grey matter (**a**, **b**), as well as oligodendrocytes in white matter (**c**) labelled by anti-tau antibodies AT8 (**a**, **c**) and RD3 (**b**). These inclusions were not labelled by RD4. Scale bar, 25 μm

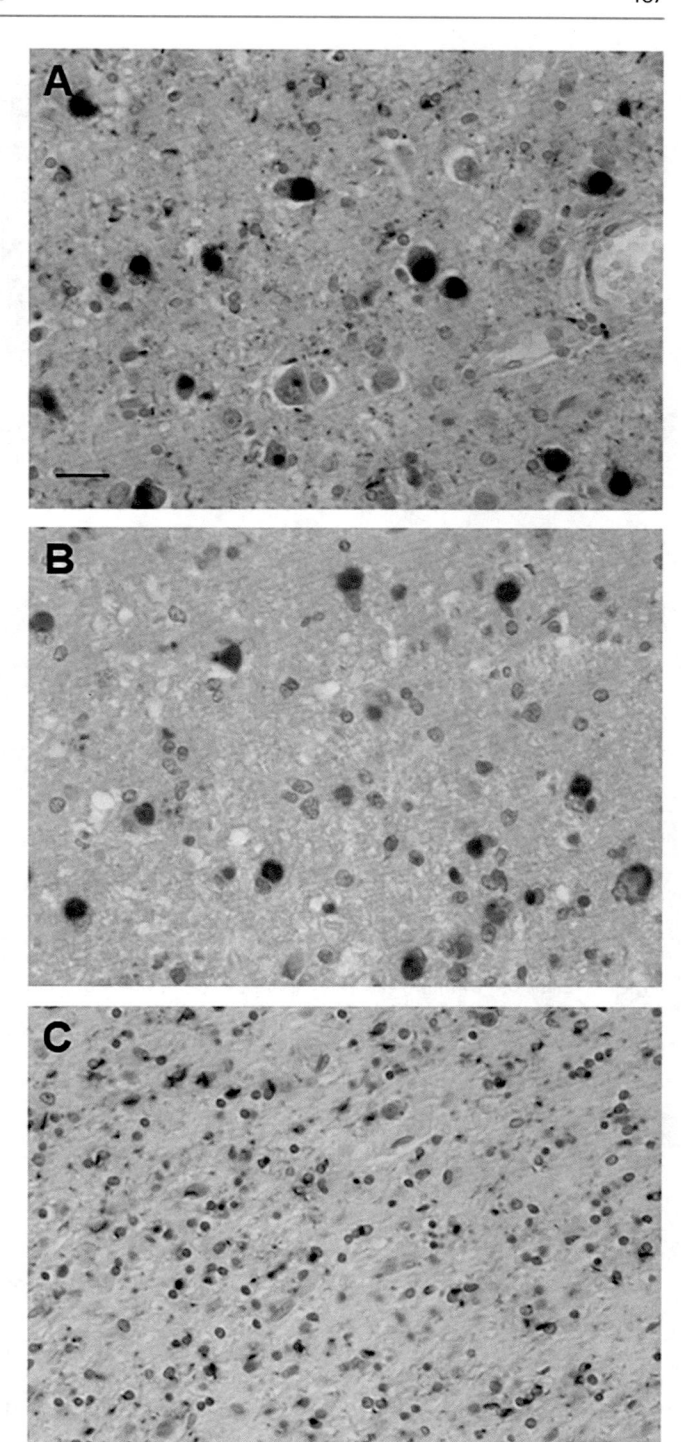

A Type I Pick filament   **B** Type II Pick filament

C Pick fold (3R Tau)

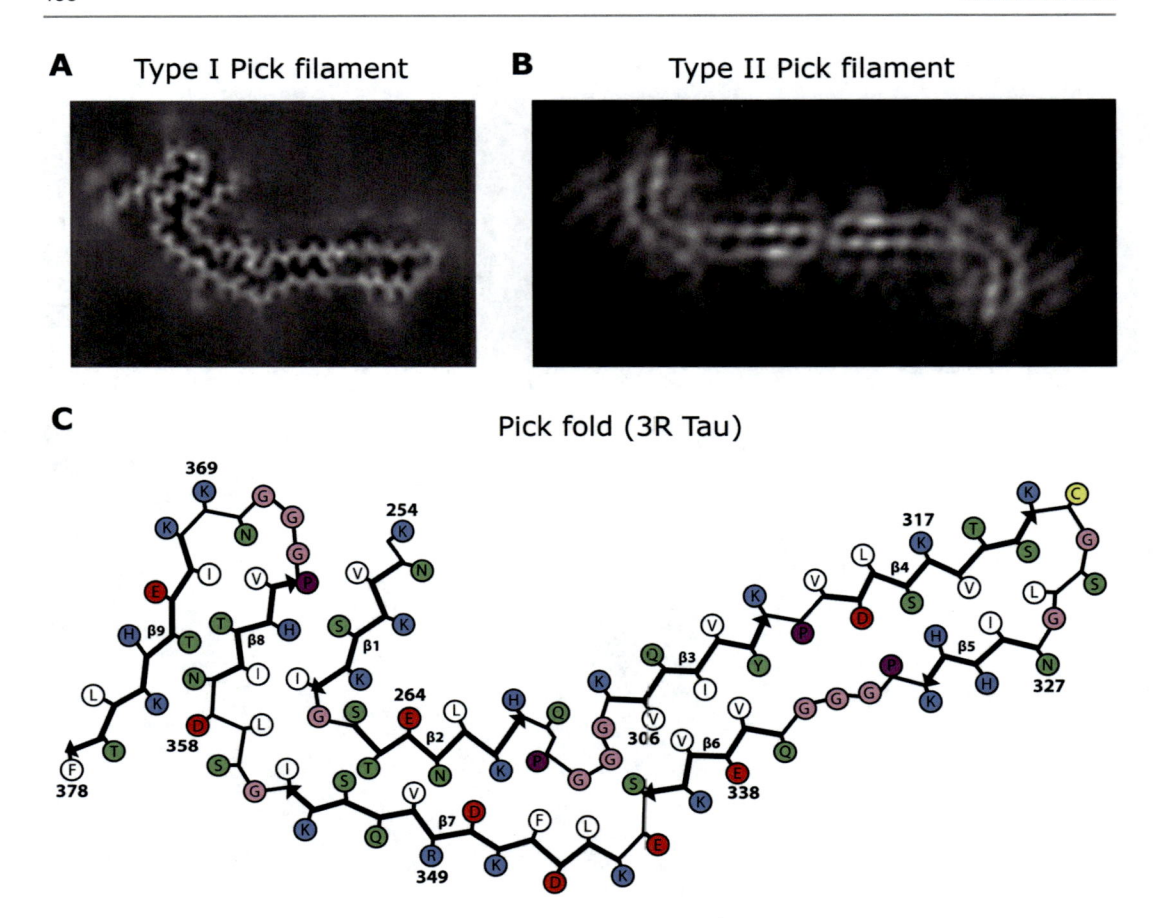

Fig. 10 Structures of tau filaments from Pick's disease. Type I and Type II tau filaments are characteristic, with Type I filaments forming the vast majority (**a, b**), Unsharpened cryo-EM densities of Type I (**a**) and Type II (**b**) filaments. Type I Pick filaments contain a single proto-filament, whereas in Type II filaments, two identical protofilaments pack against each other symmetrically through Van der Waals interactions at the tip of the J. (**c**), Schematic view of the tau protofilament core of PiD. The observed nine β-strands (β1–β9) are shown as arrows

from CBD, CTE and AD. This may explain the differential phosphorylation and raises the question of whether phosphorylation at S262 may protect against PiD.

It was not previously known why only 3R tau, which lacks R2, is present in Pick body filaments. The above shows that despite sequence homology, the structure formed by K254-K274 of R1 is inaccessible to the corresponding residues from R2 (S285-S305). In support, tau filaments extracted from the brain of the patient with PiD used for cryo-EM seeded the aggregation of recombinant 3R, but not 4R, tau. Such templated misfolding may explain the selective incorporation of 3R tau in Pick body filaments.

Structures of Tau Filaments from Corticobasal Degeneration

CBD typically presents as corticobasal syndrome, which includes cortical signs, asymmetric apraxia, rigidity, myoclonus and alien limb phenomenon. It can also present as behavioural-variant FTD, Richardson's syndrome and posterior cortical atrophy [77]. In 1925, Lhermitte et al. probably described cases of what is now known as CBD [78]. In 1968, Rebeiz et al. reported the disease as 'corticonigral degeneration with neuronal achromasia' [79]. The term CBD was introduced by Gibb et al. in 1989 [80]. The presence of tau protein in the inclusions of CBD was shown in 1990 [81].

Neuropathologically, CBD is characterised by asymmetric focal cortical atrophy and depigmentation of the substantia nigra. Nerve cells show diffuse cytoplasmic tau immunoreactivity, abundant neuropil threads in grey and white matter, as well as pathognomonic astrocytic plaques, mainly in affected cortical areas and in striatum (Fig. 11) [82, 83]. By Western blotting, assembled tau from CBD brains runs as a doublet of 64 kDa and 68 kDa, which consists of 4R tau upon dephosphorylation [84]. In addition, two closely related tau bands of approximately 37 kDa are typical of CBD [85].

By negative stain electron microscopy of sarkosyl-insoluble filaments from CBD brains, we observed narrow (Type I) and wide (Type II) tau filaments [86], in agreement with previous findings [87]. Narrow filaments have a helical twist with a crossover distance of approximately 1000 Å and widths of 80–130 Å. Wide filaments have a crossover distance of approximately 1400 Å and widths of 130–260 Å. Immunogold negative-stain electron microscopy showed that Type I and Type II filaments are present in similar amounts in some cases of CBD, with Type II filaments being more abundant in others. Filaments were not decorated by antibodies specific for R2, R3 or R4 of tau, indicating that these repeats form part of the ordered filament cores.

Structures of tau filaments were determined by cryo-EM from the frontal cortex of three individuals with CBD (Fig. 12) [86]. The core of Type I

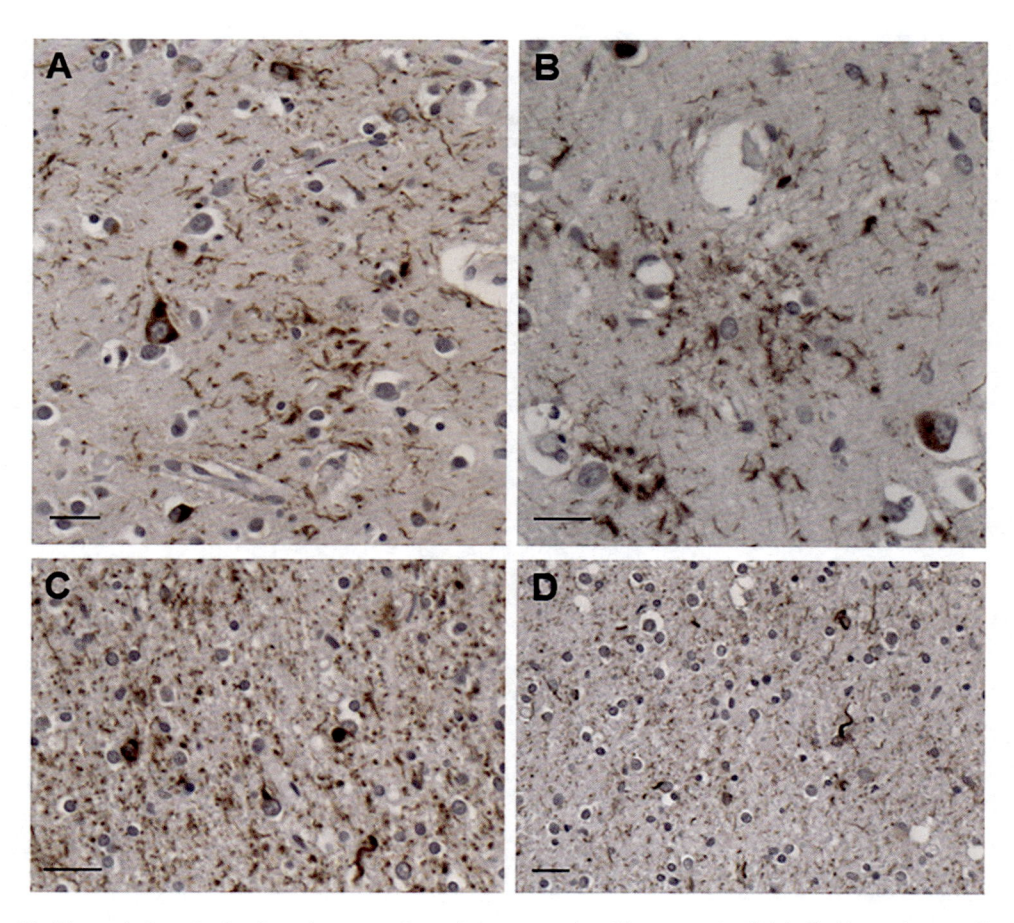

Fig. 11 Tau pathology in the frontal cortex of a patient with corticobasal degeneration. Tau inclusions in nerve cells and glia in grey matter (**a, b**), as well as oligodendrocytes in white matter (**c, d**) labelled by anti-tau antibodies AT8 (**a, c**) and RD4 (**b, d**). These inclusions were not labelled by RD3. Scale bars, 25 μm

A   Type I CBD filament

B   Type II CBD filament

C   CBD fold (4R Tau)

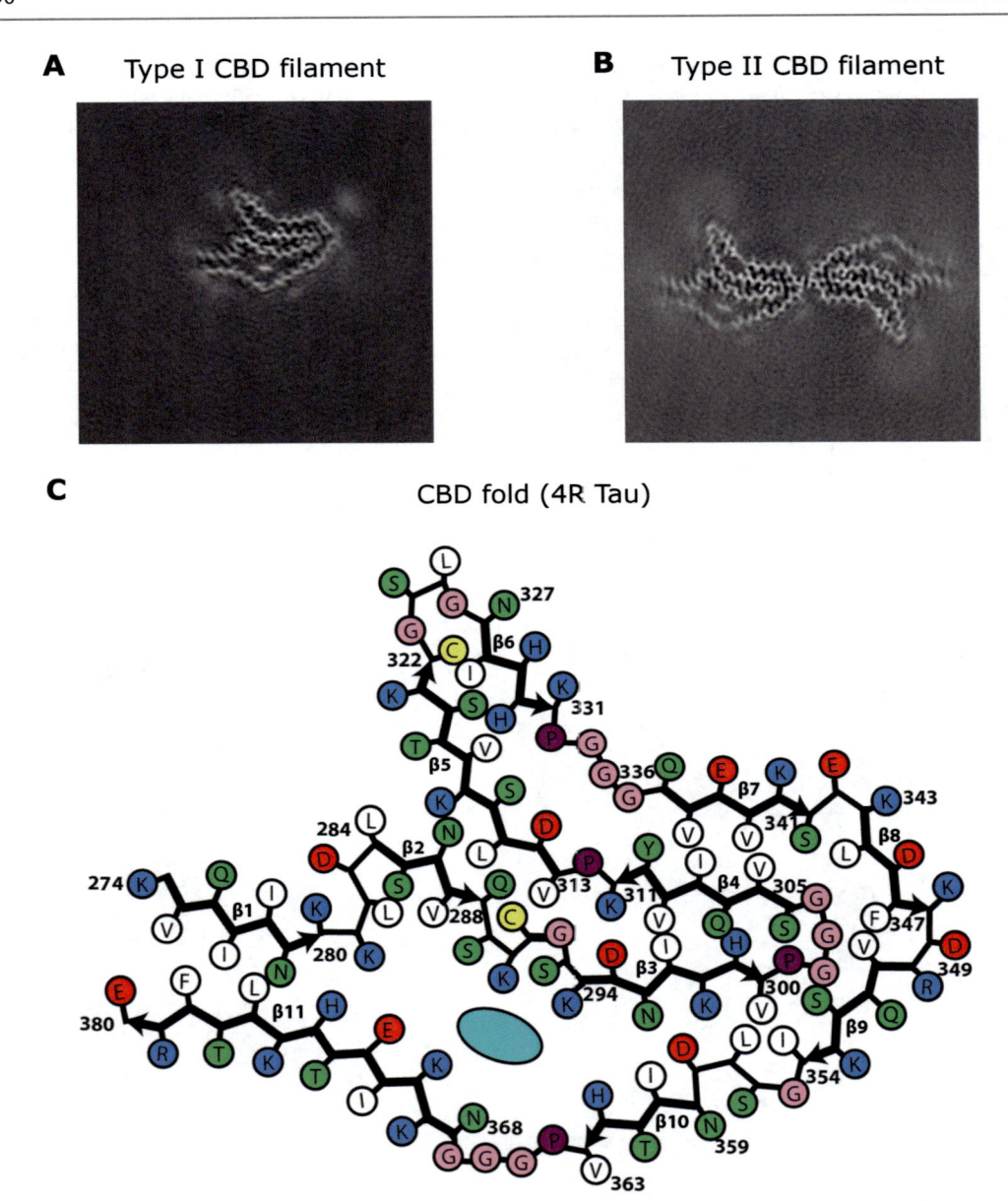

Fig. 12 Structures of tau filaments from corticobasal degeneration. Type I and Type II tau filaments are characteristic, with Type II filaments being more numerous in some cases. (**a**, **b**), Unsharpened cryo-EM densities of Type I (**a**) and Type II (**b**) filaments. Type I filaments contain a single protofilament, whereas two symmetrically packed protofilaments are present in Type II filaments. The protofilament interface is formed by anti-parallel stacking of [343]KLDFKDR[349]. (**c**), Schematic view of the tau protofilament core of CBD. The observed 11 β-strands (β1–β11) are shown as arrows. The central non-proteinaceous density is shown in blue

filaments is made of a single protofilament that consists of residues K274-E380 of 4R tau (107 amino acids; Fig. 12a, c). It encompasses the last residue of R1; all of R2, R3 and R4; as well as 12 amino acids after R4. In the core, there are 11 β-strands (β1–β11): three from R2 (β1–β3), three from R3 (β4–β6), four from R4 (β7–β10) and one from the sequence after R4 (β11). Each protofilament of CBD contains an additional density that is surrounded by the density of tau protein within a positively charged environment. The molecular identities of this density, as well as of those present on the outside of filament structures, remain to be identified. It has been suggested that they may correspond to post-translational modifications of tau [88]. Type II filaments consist of pairs of identical protofilaments of Type I (Fig. 12b). We obtained maps of Type I and Type II filaments at overall resolutions of 3.2 Å and 3.0 Å.

The 11 β-strands of each protofilament are connected by arcs and turns and form a four-layered structure. The central four layers are formed by β7, β4, β3 and β10. Strands β3 and β4 are connected by a sharp turn, whereas β7 and β10 are connected through β8 and β9, which wrap around the turn. On the other side, β2, β5 and β6 form a three-layered structure. β2 packs against one end of β5, and β6 packs against the other end. The first and the last strands, β1 and β11, pack against each other and close a hydrophilic cavity formed by residues from β2, β3, β10, β11 and the connections between β1 and β2, as well as between β2 and β3.

Each tau repeat contains a PGGG (proline-glycine-glycine-glycine) motif. In the CBD fold, that of R1 (residues 270–273) is located just outside the structured core. The PGGG motif of R2 (residues 301–304) forms a tight turn between β3 and β4, which is essential for the formation of the four-layered cross-β packing. The PGGG motif of R3 (residues 332–335) adopts an extended conformation between β6 and β7, compensating for the shorter lengths of these strands compared to the opposing β4 and β5 connected by P312. The PGGG motif of R4 (residues 364–367) adopts a similar extended conformation, forming part of the hydrophilic cavity.

In CBD Type II filaments, protofilaments are related by C2 symmetry. Their interface is formed by anti-parallel stacking of [343]KLDFKDR[349]. Besides van der Waals interactions between the anti-parallel side chains of K347 from each protofilament, the side chain of K347 is positioned to form hydrogen bonds with the carboxyl group of D348 and the backbone carbonyl of K347 on the opposite protofilament.

CBD is characterised by abundant neuronal and glial inclusions of 4R tau. It remains to be determined if Type I and Type II filaments are differentially distributed between neuronal and glial inclusions. This notwithstanding, a single tau protofilament is characteristic of these inclusions.

Structures of Tau Filaments from Chronic Traumatic Encephalopathy

CTE is associated with repetitive head impacts or exposure to blast waves. Described as punch-drunk syndrome by Martland in 1928 [89] and dementia pugilistica by Millspaugh in 1937 [90], CTE has since been identified in former participants of other contact sports, ex-military personnel and after physical abuse. Critchley used the term in a book chapter in 1949 [91]. CTE is the best-known example of an environmentally induced neurodegenerative disease.

Clinically, CTE is characterised by behavioural, mood, cognitive and motor impairments [92]. Initial mood and behavioural changes that progress to marked cognitive impairment are often seen. For this and other reasons, we decided to include CTE in the present discussion, even though it is not generally classified under the umbrella of FTD. Motor impairments, including parkinsonism and cerebellar ataxia, have been described mostly in retired boxers.

The neuropathological concept of CTE was emphasised by Corsellis et al. in 1973, who identified generalised cerebral atrophy and widespread cortical neurofibrillary lesions in some retired boxers [93]. Antigenic similarities between the neurofibrillary lesions of CTE and Alzheimer's disease were noted in 1988 [94]; this was followed by the description of tau inclusions using immunohistochemistry [95]. CTE is defined by an abundance of hyperphosphorylated tau in neurons, astrocytes and cell processes around small blood vessels (Figs. 13 and 14). Together with the accumulation of tau inclusions in cortical layers II and III [96], this distinguishes CTE from Alzheimer's disease and other tauopa-

Fig. 13 Tau pathology in the temporal cortex of a patient (former American football player) with chronic traumatic encephalopathy. Tau inclusions in nerve cells and neuropil threads labelled by anti-tau antibody AT8. Scale bar, 25 μm

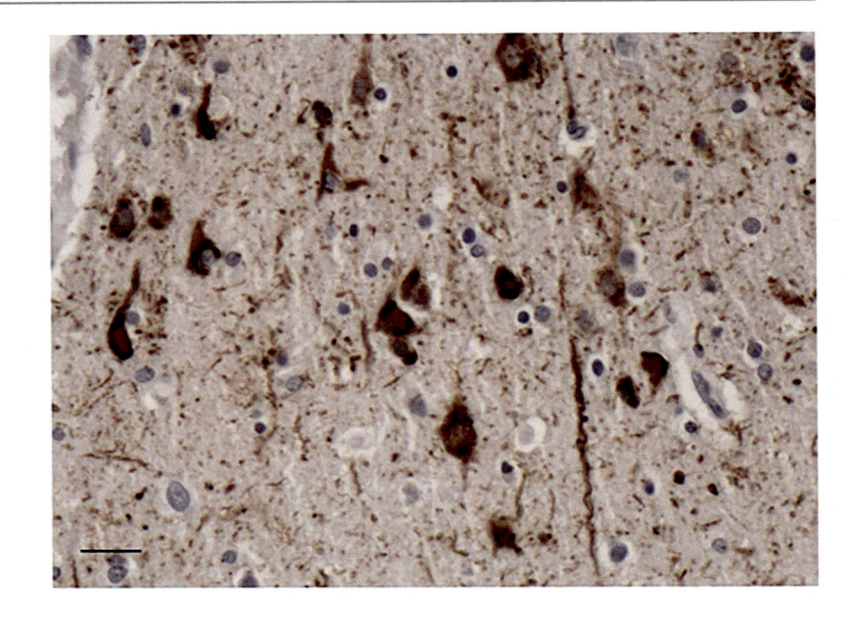

thies. By Western blotting, assembled tau from CTE runs as major bands of 60 kDa, 64 kDa and 68 kDa, which consist of all six brain tau isoforms upon dephosphorylation [97].

By negative-stain electron microscopy of sarkosyl-insoluble material from CTE brains, it was shown that Type I tau filaments make up about 90% of filaments [98]. They differ from tau filaments of PiD, CBD and AD [74, 86, 99]. Widths are 20–25 nm and crossover spacings 65–80 nm. The remaining filaments (Type II) resemble paired helical filaments of Alzheimer's disease; they have pronounced helical twists that result in projected widths of 15–30 nm.

Structures of tau filaments were determined by cryo-EM from the frontal cortex of three individuals with CTE (one former American football player and two ex-boxers) (Fig. 15). The core of Type I filaments is made of pairs of identical protofilaments that consist of residues K274/S305-R379 of tau (74 amino acids) (Fig. 15a). The protofilament structure (CTE fold) is similar to the C-shaped Alzheimer fold [99], but it adopts a more open conformation (Fig. 15c). Most notably, additional density—which is not present in the Alzheimer fold—is surrounded by the density of tau protein within the ordered core. Analysis of the minority Type II filaments revealed the presence of two kinds of filament, something that

was not apparent by negative staining. Approximately 75% of these filaments (Type II) were composed of pairs of the same protofilament as in Type I tau filaments (including the extra density) but with a different protofilament interface. CTE Type I and Type II filaments are thus ultrastructural polymorphs that have different protofilament interfaces, but a common protofilament structure. The remaining filaments were identical to paired helical filaments of Alzheimer's disease. This shows that cryo-EM was able to resolve what looked like paired helical filaments by negative staining into CTE Type II filaments and paired helical filaments. By cryo-EM, paired helical filaments made up 1–2% of filaments.

Each CTE protofilament is C-shaped and contains eight β-strands, five of which give rise to two regions of anti-parallel β-sheets, with the other three forming a β-helix. The carboxy-terminal residues of R1 and R2 form part of the first β-strand. R3 contributes three and R4 four β-strands, with the final β-strand being formed by the 11 amino acids after the end of R4; β1 and β2 pack against β8, β3 packs against β7, with β4, β5 and β6 giving rise to the C-shaped β-helix. The CTE fold is similar to the Alzheimer fold [98, 99], with the main differences being present at the tip of the C, where the packing of β4–β6 coin-

Fig. 14 Tau pathology in the temporal cortex of a patient (ex-boxer) with chronic traumatic encephalopathy. Tau inclusions in nerve cells and glia adjacent to small blood vessels labelled with anti-tau antibodies AT8 (**a**), RD3 (**b**) and RD4 (**c**). Scale bar, 25 µm

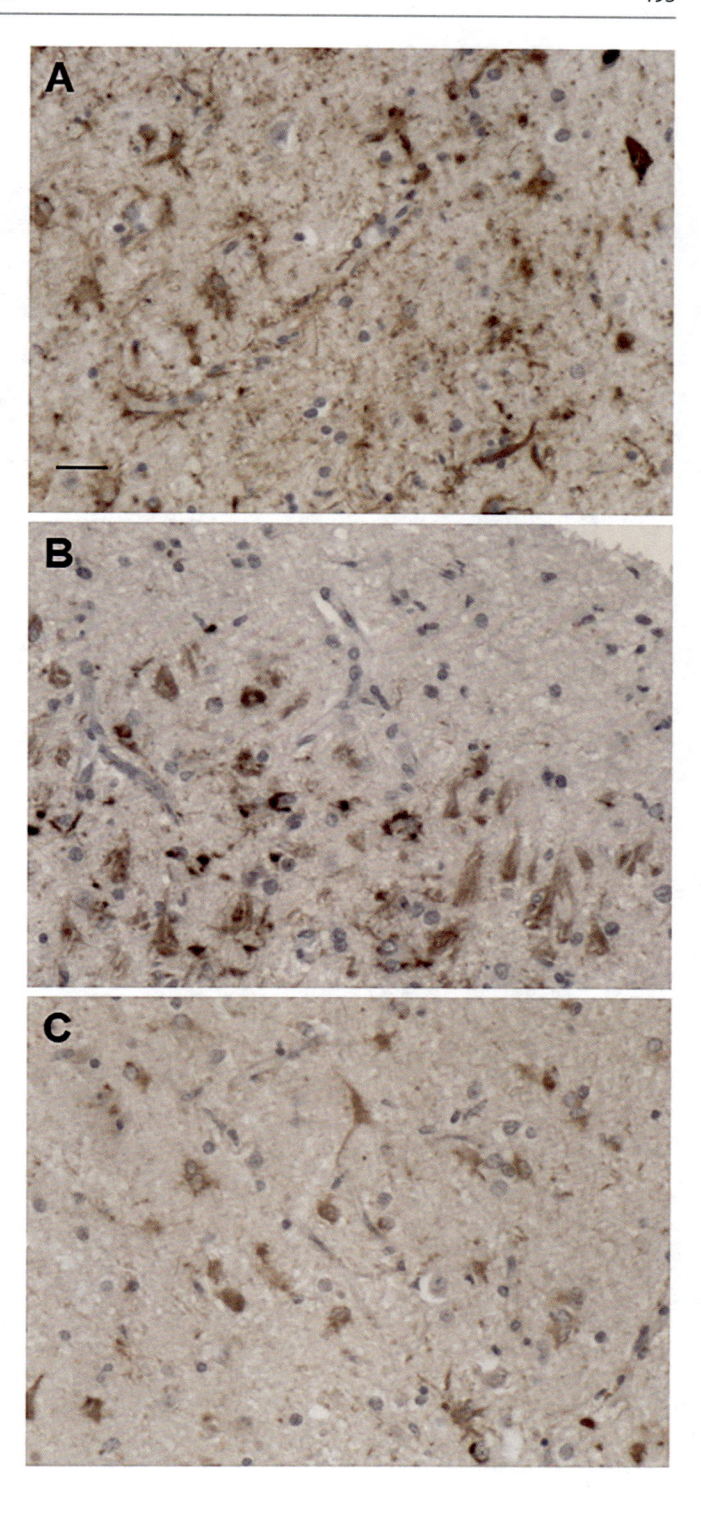

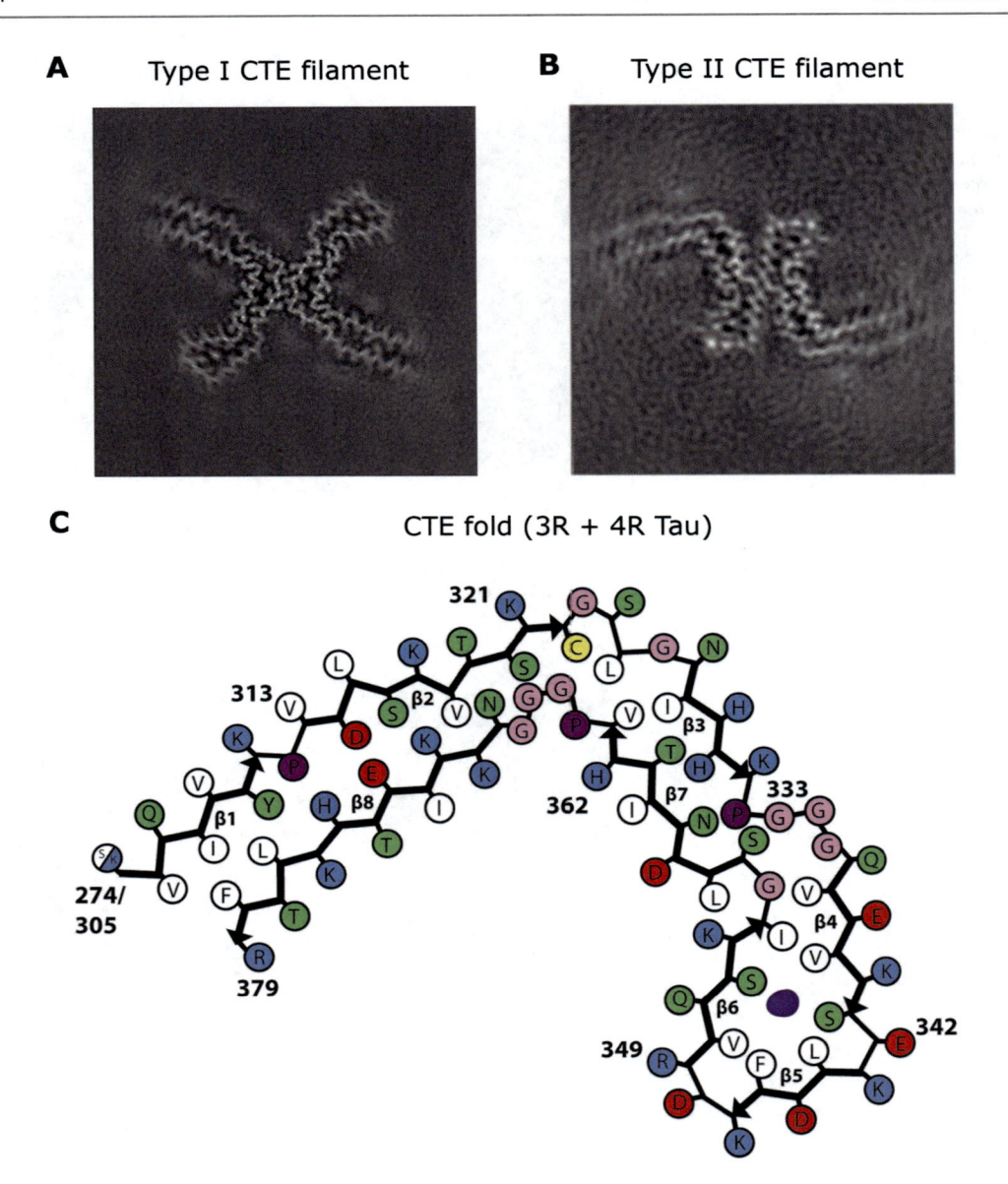

A Type I CTE filament **B** Type II CTE filament

C CTE fold (3R + 4R Tau)

Fig. 15 Structures of tau filaments from chronic traumatic encephalopathy. Type I and Type II tau filaments are characteristic, with Type I filaments forming the vast majority. (**a**, **b**), Unsharpened cryo-EM densities of Type I (**a**) and Type II (**b**) filaments. The Type I filament was resolved to 2.3 Å and the Type II filament to 3.4 Å. Both filament types show identical pairs of protofilaments. They differ in their inter-protofilament packing (ultrastructural polymorphs). In CTE Type I filaments, protofilaments pack through an anti-parallel steric zipper formed by residues [324]SLGNIH[329]. The interface in CTE Type II filaments comprises residues [332]PGGGQ[336]. (**c**), Schematic view of the tau protofilament core of CTE. The observed eight β-strands (β1–β8) are shown as arrows. The central non-proteinaceous density is shown in violet

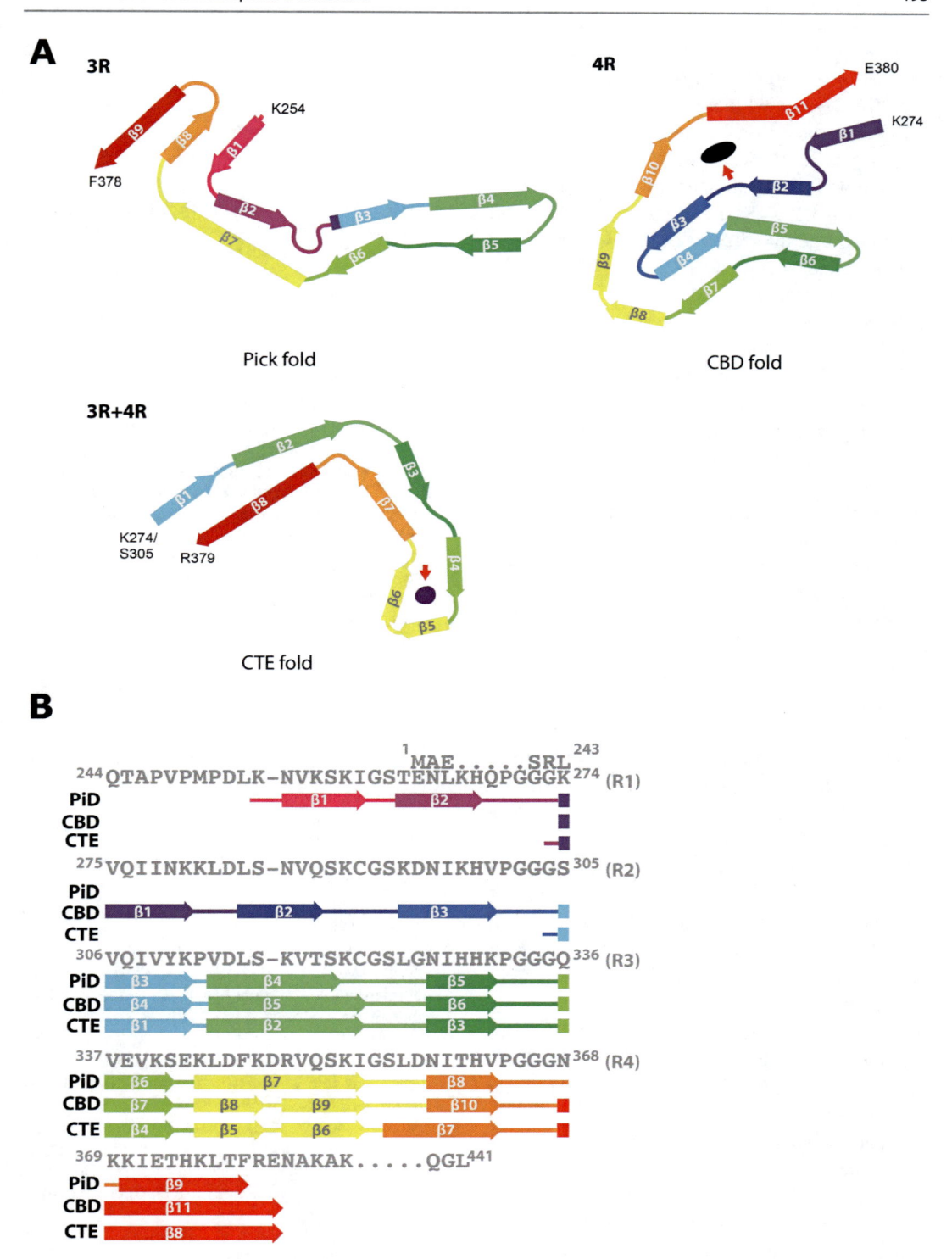

Fig. 16 Structures of tau filament cores from human brain. (**a**) Protofilament from Pick's disease (Pick fold), a 3R tauopathy; protofilament from corticobasal degeneration (CBD fold), a 4R tauopathy; protofilament from chronic traumatic encephalopathy (CTE fold), a 3R + 4R tauopathy. Red arrows point to the internal densities in CBD and CTE folds. β-Strands are marked by thick arrows (11 in the CBD fold, 9 in the Pick fold and 8 in the CTE fold). (**b**), Schematic depicting the microtubule-binding repeats (R1-R4) of tau and the sequence after R4, with β-strands found in the cores of tau filaments marked by thick arrows. Colours of individual β-strands are the same in (**a**) and (**b**)

cides with an opening up of the C-shape, and a reversal in the orientation of residues S356 and L357. In CTE Type I filaments, two identical protofilaments pack in a staggered manner through an anti-parallel steric zipper formed by residues [324]SLGNIH[329]. The interface in CTE Type II filaments is also staggered and comprises the same residues as the interface in Alzheimer's disease–paired helical filaments ([332]PGGGQ[336]), but a kinked conformation reduces the number of hydrogen bonds across the interface.

The above-mentioned findings establish CTE as different from Alzheimer's disease, even though tau inclusions of both diseases are made of all six brain isoforms. In contrast to Alzheimer's disease, CTE is also characterised by an abundant glial tau pathology. The presence of a single CTE tau fold implies that the glial and neuronal tau inclusions are made of the same protofilament. The presence of identical CTE tau folds in the brains of a former American footballer and two ex-boxers establishes the presence of the same disease.

Conclusion

Assembled tau protein has been known to form the filamentous inclusions of a number of frontotemporal dementias since the 1980s. The finding that the same protein can be found in the inclusions of multiple diseases led some to conclude that the formation of tau inclusions is an epiphenomenon of little significance. The identification of mutations in *MAPT* in FTDP-17 T changed all that. To date, 65 disease-causing mutations have been identified. Most are missense mutations, but some change the ratio of 3R/4R tau. Clinicopathological studies have shown links between some mutations in *MAPT* and sporadic tauopathies.

Ongoing work has shown that the structures of tau filaments from sporadic PiD, CBD and CTE are different. Thus, the same protein takes on distinct structures in different diseases (Fig. 16). So far, in individuals with the same disease, be it PiD, CBD or CTE, filament structures were identical. It remains to be seen how the structures of tau filaments from the brains of individuals with *MAPT* mutations compare to each other and to those from sporadic diseases.

References

1. Goedert M, Eisenberg DS, Crowther RA (2017) Propagation of tau aggregates and neurodegeneration. Annu Rev Neurosci 40:189–210
2. Ghetti B, Oblak AL, Boeve BF, Johnson KA, Dickerson DC, Goedert M (2015) Frontotemporal dementia caused by microtubule-associated protein tau gene (*MAPT*) mutations: a chameleon for neuropathology and neuroimaging. Neuropathol Appl Neurobiol 41:24–46
3. Lee G, Newman ST, Gard DL, Band H, Panchamoorthy G (1998) Tau interacts with src-family non-receptor tyrosine kinases. J Cell Sci 111:3167–3177
4. Al-Bassam J, Ozer RS, Safer D, Halpain DS, Milligan RA (2002) MAP 2 and tau bind longitudinally along the outer ridges of microtubule protofilaments. J Cell Biol 157:1187–1196
5. Kellogg EH, Hejab NMA, Poepsel S, Downing KH, DiMaio F, Nogales E (2018) Near-atomic model of microtubule-tau interactions. Science 360:1242–1246
6. Janning D, Igaev M, Sündermann F, Brühmann J, Beutel O, Heinisch JJ et al (2014) Single-molecule tracking of tau reveals fast kiss-and-hop interaction with microtubules in living neurons. Mol Biol Cell 25:3541–3551
7. Niewidok B, Igaev M, Sündermann F, Janning DF, Bakota L, Brandt R (2016) Presence of a carboxy-terminal pseudorepeat and disease-like pseudohyperphosphorylation critically influence tau's interaction with microtubules in axon-like processes. Mol Biol Cell 27:3537–3549
8. Black MM, Slaughter T, Moshiach S, Obrocka M, Fischer I (1996) Tau is enriched on dynamic microtubules in the distal region of growing axons. J Neurosci 16:3601–3619
9. Qiang L, Sun X, Austin TO, Muralidharan H, Jean DC, Liu M et al (2018) Tau does not stabilize axonal microtubules but rather enables them to have long labile domains. Curr Biol 28:2181–2189
10. Boyka S, Qi X, Chen TH, Surewicz K, Surewicz WT (2019) Liquid-liquid phase separation of tau protein: the crucial role of electrostatic interactions. J Biol Chem 294:11054–11059
11. Lin Y, Fichou Y, Zeng Z, Hu NY, Han S (2020) Electrostatically driven complex coacervation and amyloid aggregation of tau are independent processes with overlapping conditions. ACS Chem Neurosci 11:615–627
12. Goedert M, Spillantini MG, Jakes R, Rutherford D, Crowther RA (1989) Multiple isoforms of human microtubule-associated protein tau: sequences and localization in neurofibrillary tangles of Alzheimer's disease. Neuron 3:519–526

13. Goedert M, Jakes R (1990) Expression of separate isoforms of human tau protein: correlation with the tau pattern in brain and effects on tubulin polymerization. EMBO J 9:4225–4230

14. Yoshida H, Goedert M (2002) Molecular cloning and functional characterization of chicken brain tau: isoforms with up to five tandem repeats. Biochemistry 41:15203–15211

15. Tuerde D, Kimura T, Miyasaka T, Furusawa K, Shimozawa A, Hasegawa M et al (2018) Isoform-independent and –dependent phosphorylation of microtubule-associated protein tau in mouse brain during postnatal development. J Biol Chem 293:1781–1793

16. Brion JP, Passareiro H, Nunez J, Flament-Durand J (1985) Mise en évidence immunologique de la protéine tau au niveau des lésions de dégénérescence neurofibrillaire de la maladie d'Alzheimer. Arch Biol 95:229–235

17. Goedert M, Wischik CM, Crowther RA, Walker JE, Klug A (1988) Cloning and sequencing of the cDNA encoding a core protein of the paired helical filament of Alzheimer disease: identification as the microtubule-associated protein tau. Proc Natl Acad Sci U S A 85:4051–4055

18. Wischik CM, Novak M, Thøgersen HC, Edwards PC, Runswick MJ, Jakes R et al (1988) Isolation of a fragment of tau derived from the core of the paired helical filament of Alzheimer disease. Proc Natl Acad Sci U S A 85:4506–4510

19. Wischik CM, Novak M, Edwards PC, Klug A, Tichelaar W, Crowther RA (1988) Structural characterization of the core of the paired helical filament of Alzheimer disease. Proc Natl Acad Sci U S A 85:4884–4888

20. Berriman J, Serpell LC, Oberg KA, Fink AL, Goedert M, Crowther RA (2003) Tau filaments from human brain and from in vitro assembly of recombinant protein show cross-beta structure. Proc Natl Acad Sci U S A 100:9034–9038

21. Iqbal K, Liu F, Gong CX (2016) Tau and neurodegenerative disease: the story so far. Nat Rev Neurol 12:15–27

22. Goedert M, Jakes R, Spillantini MG, Hasegawa M, Smith MJ, Crowther RA (1996) Assembly of microtubule-associated protein tau into Alzheimer-like filaments induced by sulphated glycosaminoglycans. Nature 383:550–553

23. Kampers T, Friedhoff P, Biernat J, Mandelkow EM, Mandelkow E (1996) RNA stimulates aggregation of microtubule-associated protein tau into Alzheimer-like paired helical filaments. FEBS Lett 399:344–349

24. Pérez M, Valpuesta JM, Medina M, Montejo de Garcini E, Avila J (1996) Polymerization of tau into filaments in the presence of heparin: the minimal sequence required for tau-tau interaction. J Neurochem 67:1183–1190

25. Wilson DM, Binder LI (1997) Free fatty acids stimulate the polymerization of tau and amyloid beta peptides. Am J Pathol 150:2181–2195

26. Falcon B, Zhang W, Schweighauser M, Murzin AG, Vidal R, Garringer HJ et al (2018) Tau filaments from multiple cases of sporadic and inherited Alzheimer's disease adopt a common fold. Acta Neuropathol 136:699–708

27. Fichou Y, Lin Y, Rauch JN, Vigers M, Zeng Z, Srivasta M et al (2018) Cofactors are essential constituents of stable and seeding-active tau fibrils. Proc Natl Acad Sci U S A 115:13234–13239

28. Haj-Yahya M, Lashuel HA (2018) Protein semisynthesis provides access to tau disease-associated post-translational modifications (PTMs) and paves the way to deciphering the tau PTM code in health and diseased states. J Am Chem Soc 140:6611–6621

29. Poorkaj P, Bird TD, Wijsman E, Nemens E, Garruto RM, Anderson L et al (1998) Tau is a candidate gene for chromosome 17 frontotemporal dementia. Ann Neurol 43:815–825

30. Hutton M, Lendon CL, Rizzu P, Baker M, Froelich S, Houlden H et al (1998) Association of missense and 5'-splice site mutations in tau with the inherited dementia FTDP-17. Nature 393:702–705

31. Spillantini MG, Murrell JR, Goedert M, Farlow MR, Klug A, Ghetti B (1998) Mutation in the tau gene in familial multiple system tauopathy with familial presenile dementia. Proc Natl Acad Sci U S A 95:7737–7741

32. Stefansson H, Helgason A, Thorleifsson G, Steinthorsdottir V, Masson G, Barnard J et al (2005) A common inversion under selection in Europeans. Nat Genet 37:129–137

33. Conrad C, Andreadis A, Trojanowski JQ, Dickson DW, Kang D, Chen X et al (1997) Genetic evidence for the involvement of tau in progressive supranuclear palsy. Ann Neurol 41:277–281

34. Baker M, Litvan I, Houlden H, Adamson J, Dickson D, Perez-Tur J et al (1999) Association of an extended haplotype in the tau gene with progressive supranuclear palsy. Hum Mol Genet 8:711–715

35. Di Maria E, Tabaton M, Vigo T, Abbruzzese G, Bellone E, Donati C et al (2000) Corticobasal degeneration shares a common genetic background with progressive supranuclear palsy. Ann Neurol 47:374–377

36. Pastor P, Ezquerra M, Munoz E, Marti MJ, Blersa R, Tolosa E et al (2000) Significant association between the tau gene A0/A0 genotype and Parkinson's disease. Ann Neurol 47:242–245

37. Morris HR, Baker M, Yasojima K, Houlden H, Khan MN, Wood NW et al (2002) Analysis of tau haplotypes in Pick's disease. Neurology 59:443–445

38. Zhang CC, Zhu JX, Wan Y, Tan L, Wang HF, Yu JT et al (2017) Meta-analysis of the association between variants in MAPT and neurodegenerative diseases. Oncotarget 8:4494–4507

39. Caffrey TM, Joachim C, Wade-Martins R (2008) Haplotype-specific expression of the N-terminal exons 2 and 3 at the human MAPT locus. Neurobiol Aging 29:1923–1929

40. Zhong Q, Congdon EE, Nagaraja HN, Kuret J (2012) Tau isoform composition influences rate and extent of filament formation. J Biol Chem 287:20711–20719

41. Lewis J, McGowan E, Rockwood J, Melrose H, Nacharaju P, van Slegtenhorst M et al (2000) Neurofibrillary tangles, amyotrophy and progressive motor disturbance in mice expressing mutant (P301L) tau protein. Nature Genet 25:402–405

42. Allen B, Ingram E, Takao M, Smith MJ, Jakes R, Virdee K et al (2002) Abundant tau filaments an nonapoptotic neurodegeneration in transgenic mice expressing human P301S tau protein. J Neurosci 22:9340–9351

43. Götz J, Bodea LG, Goedert M (2018) Rodent models for Alzheimer disease. Nat Rev Neurosci 19:583–598

44. Macdonald JA, Bronner IF, Drynan L, Fan J, Curry A, Fraser G et al (2019) Assembly of transgenic human P301S tau is necessary for neurodegeneration in murine spinal cord. Acta Neuropathol Commun 7:44

45. Bussian TJ, Aziz A, Meyer CF, Swenson BL, Van Deursen JM, Baker DJ (2018) Clearance of senescent glial cells prevents tau-dependent pathology and cognitive decline. Nature 562:578–582

46. Clavaguera F, Bolmont T, Crowther RA, Abramowski D, Frank S, Probst A et al (2009) Transmission and spreading of tauopathy in transgenic mouse brain. Nat Cell Biol 11:909–913

47. Frost B, Jacks RL, Diamond MI (2009) Propagation of tau misfolding from the outside to the inside of a cell. J Biol Chem 284:12845–12852

48. Jackson SJ, Kerridge C, Cooper J, Cavallini A, Falcon B, Cella CV et al (2016) Short fibrils constitute the major species of seed-competent tau in the brains of mice transgenic for human P301S tau. J Neurosci 36:762–772

49. Braak H, Braak E (1991) Neuropathological staging of Alzheimer-related changes. Acta Neuropathol 82:239–259

50. Kaufman SK, Del Tredici K, Thomas TL, Braak H, Diamond MI (2018) Tau seeding activity begins in transentorhinal/entorhinal regions and anticipates phospho-tau pathology in Alzheimer's disease and PART. Acta Neuropathol 136:57–67

51. Clavaguera F, Akatsu H, Fraser G, Crowther RA, Frank S, Hench J et al (2013) Brain homogenates from human tauopathies induce tau inclusions in mouse brain. Proc Natl Acad Sci U S A 110:9535–9540

52. Sanders DW, Kaufman SK, DeVos SL, Sharma AM, Mirhaba H, Li A et al (2014) Distinct tau prion strains propagate in cells and mice and define different tauopathies. Neuron 82:1271–1288

53. He Z, McBride JD, Xu H, Changolkar L, Kim SJ, Zhang B et al (2020) Transmission of tauopathy strains is independent of their isoform composition. Nat Commun 11:7

54. Spillantini MG, Crowther RA, Kamphorst W, Heutink P, Van Swieten JC (1998) Tau pathology in two Dutch families with mutations in the microtubule-binding region of tau. Am J Pathol 153:1359–1363

55. Bronner IF, Ter Meulen BC, Azmani A, Severijnen LA, Willemsen R, Kamphorst W et al (2005) Hereditary Pick's disease with the G272V mutation shows predominant three-repeat tau pathology. Brain 128:2645–2653

56. Van Swieten JC, Bronner IF, Azmani A, Severijnen LA, Kamphorst W, Ravid R et al (2007) The deltaK280 mutation in *MAPT* favors exon 10 skipping *in vivo*. J Neuropathol Exp Neurol 66:17–25

57. Murrell JR, Spillantini MG, Zolo P, Guazzelli M, Smith MJ, Hasegawa M et al (1999) Tau gene mutation G389R causes a tauopathy with abundant Pick body-like inclusions and axonal deposits. J Neuropathol Exp Neurol 58:1207–1226

58. Kouri N, Carlomagno Y, Baker M, Liesinger AM, Caselli RJ, Wszolek ZK et al (2014) Novel mutation in *MAPT* exon 13 (p.N410H) causes corticobasal degeneration. Acta Neuropathol 127:271–282

59. Tacik P, Sanchez-Contreras M, DeTure M, Murray ME, Rademakers R, Ross OA et al (2017) Clinicopathologic heterogeneity in frontotemporal dementia and parkinsonism linked to chromosome 17 (FTDP-17) due to microtubule-associated protein tau (MAPT) p.P301L mutation, including a patient with globular glial tauopathy. Neuropathol Appl Neurobiol 43:200–214

60. Bugiani O, Murrell JR, Giaccone G, Hasegawa M, Ghigo G, Tabaton M et al (1999) Frontotemporal dementia and corticobasal degeneration in a family with a P301S mutation in *Tau*. J Neuropathol Exp Neurol 58:667–677

61. Erro ME, Zelaya MV, Mendioroz M, Larumbe R, Ortega-Cubero S, Lanciego JL et al (2019) Globular glial tauopathy caused by *MAPT* P301T mutation: clinical and neuropathological findings. J Neurol 266:2396–2405

62. Goedert M (2016) The ordered assembly of tau is the gain of toxic function that causes human tauopathies. Alzheimers Dement 12:1040–1050

63. Spina S, Farlow MR, Unverzagt FW, Kareken DA, Murrell JR, Fraser G et al (2008) The tauopathy associated with mutation +3 in intron 10 of *Tau*: characterization of the MSTD family. Brain 131:72–89

64. Spillantini MG, Crowther RA, Goedert M (1996) Comparison of the neurofibrillary pathology in Alzheimer's disease and familial presenile dementia with tangles. Acta Neuropathol 92:42–48

65. Reed LA, Grabowski TJ, Schmidt ML, Morris JC, Goate A, Solodkin A et al (1997) Autosomal dominant dementia with widespread neurofibrillary tangles. Ann Neurol 42:564–572

66. Pick A (1892) Über die Beziehungen der senilen Hirnatrophie zur Aphasie. Prager Med Wochenschr 17:165–167

67. Alzheimer A (1911) Über eigenartige Krankheitsfälle des späteren Alters. Z ges Neurol Psychiat 22:146–148

68. Rasool CG, Selkoe DJ (1985) Sharing of specific antigens by degenerating neurons in Pick's disease and Alzheimer's disease. N Engl J Med 312:700–705

69. Pollock NJ, Mirra SS, Binder LI, Hansen LA, Wood JG (1986) Filamentous aggregates in Pick's disease, progressive supranuclear palsy, and Alzheimer's dis-

ease share antigenic determinants with microtubule-associated protein tau. Lancet 328:1211

70. Kertesz A, Munoz DG (eds) (1998) Pick's disease and Pick complex. Wiley-Liss, Weinheim

71. Kovacs GG, Rozemuller AJM, Van Swieten JC, Gelpi E, Majtenyi K, Al-Sarraj S et al (2013) Neuropathology of the hippocampus in FTLD-tau with Pick bodies: a study of the brain net Europe consortium. Neuropathol Appl Neurobiol 39:166–178

72. Motoi Y, Iwamoto H, Itaya M, Kobayashi T, Hasegawa M, Yasuda M et al (2005) Four-repeat tau-positive Pick body-like inclusions are distinct from classic Pick bodies. Acta Neuropathol 110:431–433

73. Delacourte A, Robitaille Y, Sergeant N, Buée L, Hof PR, Wattez A et al (1996) Specific pathological tau protein variants characterize Pick's disease. J Neuropathol Exp Neurol 55:159–168

74. Falcon B, Zhang W, Murzin AG, Murshudov G, Garringer HJ, Vidal R et al (2018) Structures of filaments from Pick's disease reveal a novel tau protein fold. Nature 561:137–140

75. Probst A, Tolnay M, Langui D, Goedert M, Spillantini MG (1996) Pick's disease: Hyperphosphorylated tau protein segregates to the somatoaxonal compartment. Acta Neuropathol 92:588–596

76. Delacourte A, Sergeant N, Wattez A, Gauvreau D, Robitaille Y (1998) Vulnerable neuronal subsets in Alzheimer's and Pick's disease are distinguished by their tau isoform distribution and phosphorylation. Ann Neurol 43:193–204

77. Lee SE, Rabinovici GD, Mayo MC, Wilson SM, Seeley WW, DeArmond SJ et al (2001) Clinicopathological correlations in corticobasal degeneration. Ann Neurol 70:327–340

78. Lhermitte J, Lévy G, Kyriaco N (1925) Les perturbations de la représentation spatiale chez les apraxiques. Rev Neurol (Paris) 2:586–600

79. Rebeiz JJ, Kolodny EH, Richardson EP (1968) Corticodentatonigral degeneration with neuronal achromasia. Arch Neurol 18:20–33

80. Gibb WRG, Luthert PJ, Marsden CD (1989) Corticobasal degeneration. Brain 112:1171–1192

81. Paulus W, Selim M (1990) Corticonigral degeneration with neuronal achromasia and basal neurofibrillary tangles. Acta Neuropathol 81:89–94

82. Feany MB, Dickson DW (1995) Widespread cytoskeletal pathology characterizes corticobasal degeneration. Am J Pathol 146:1388–1396

83. Kouri N, Whitwell JL, Josephs KA, Radekmakers R, Dickson DW (2011) Corticobasal degeneration: a pathologically distinct 4R tauopathy. Nat Rev Neurol 7:263–272

84. Sergeant N, Wattez A, Delacourte A (1999) Neurofibrillary degeneration in progressive supranuclear palsy and corticobasal degeneration: tau pathologies with exclusively "exon 10" isoforms. J Neurochem 72:1243–1249

85. Arai T, Ikeda K, Akiyama H, Nonaka T, Hasegawa M, Ishiguro K et al (2004) Identification of amino-terminally cleaved tau fragments that distinguish progressive supranuclear palsy from corticobasal degeneration. Ann Neurol 55:72–79

86. Zhang W, Tarutani A, Newell KL, Murzin AG, Matsubara T, Falcon B et al (2020) Novel tau filament fold in corticobasal degeneration. Nature 580:283–287

87. Ksiezak-Reding H, Tracz E, Yang LS, Dickson DW, Simon M, Wall JS (1996) Ultrastructural instability of paired helical filaments from corticobasal degeneration as examined by scanning transmission electron microscopy. Am J Pathol 149:639–651

88. Arakhamia T, Lee CE, Carlomagno Y, Duong DD, Kundinger SR, Wang K et al (2020) Posttranslational modifications mediate the structural diversity of tauopathy strains. Cell 180:633–644

89. Martland HS (1928) Punch drunk. J Am Med Assoc 91:1103–1107

90. Millspaugh JA (1937) Dementia pugilistica. US Nav Med Bull 35:297–303

91. Critchley M (1949) Punch drunk syndromes: the chronic traumatic encephalopathy of boxers. In: Hommage à Clovis Vincent. Maloine, Paris, pp 131–145

92. McKee AC, Cantu RC, Nowinski CJ, Stern RA, Daneshvar DH, Alvarez VE et al (2013) The spectrum of disease in chronic traumatic encephalopathy. Brain 136:43–64

93. Corsellis JAN, Bruton CJ, Freeman-Browne D (1973) The aftermath of boxing. Psychol Med 3:270–303

94. Roberts GW (1988) Immunocytochemistry of neurofibrillary tangles in dementia pugilistica and Alzheimer's disease: evidence for common genesis. Lancet 232:1456–1458

95. Tokuda T, Ikeda S, Yanagisawa N, Ihara Y, Glenner GG (1991) Re-examination of ex-boxers' brains using immunohistochemistry with antibodies to amyloid β-protein and tau protein. Acta Neuropathol 82:280–285

96. Hof PR, Bouras C, Buée L, Delacourte A, Perl DP, Morrison JH (1995) Differential distribution of neurofibrillary tangles in the cerebral cortex of dementia pugilistica and Alzheimer's disease cases. Acta Neuropathol 85:23–30

97. Schmidt ML, Zhukareva V, Newell KL, Lee VMY, Trojanowski JQ (2002) Tau isoform profile and phosphorylation state in dementia pugilistica recapitulate Alzheimer's disease. Acta Neuropathol 101:518–524

98. Falcon B, Zivanov J, Zhang W, Murzin AG, Garringer HJ, Vidal R et al (2019) Novel tau filament fold in chronic traumatic encephalopathy encloses hydrophobic molecules. Nature 568:420–423

99. Fitzpatrick AWP, Falcon B, He S, Murzin AG, Murshudov G, Garringer HJ et al (2017) Cryo-EM structures of tau filaments from Alzheimer's disease. Nature 54:185–190

Frontotemporal Lobar Degeneration TDP-43-Immunoreactive Pathological Subtypes: Clinical and Mechanistic Significance

Manuela Neumann, Edward B. Lee, and Ian R. Mackenzie

Introduction

Frontotemporal dementia (FTD) is a heterogeneous clinical syndrome, characterized by progressive changes in behavior, personality, and/or language, with relative preservation of memory [1]. Major clinical subtypes include the behavioral-variant FTD (bvFTD) and two forms of primary progressive aphasia (PPA); the nonfluent/agrammatic and semantic variants (nfvPPA and svPPA, respectively). In addition, FTD is often associated with motor features, either an extrapyramidal movement disorder (atypical par-

M. Neumann
Department of Neuropathology, University of Tübingen, Tübingen, Germany

DZNE, German Center for Neurodegenerative Diseases, Tübingen, Germany
e-mail: Manuela.Neumann@dzne.de

E. B. Lee
Department of Pathology and Laboratory Medicine, and Center for Neurodegenerative Disease Research, University of Pennsylvania, Philadelphia, PA, USA
e-mail: edward.lee@pennmedicine.upenn.edu

I. R. Mackenzie (✉)
Department of Pathology, University of British Columbia, Vancouver, BC, Canada

Department of Pathology, Vancouver General Hospital, Vancouver, BC, Canada
e-mail: ian.mackenzie@vch.ca

kinsonism or corticobasal syndrome—CBS) or motor neuron disease (MND; usually classical amyotrophic lateral sclerosis—ALS). A family history is present in 25–50% of cases, with autosomal dominant FTD caused by mutations in several different genes [2].

The neuropathology underlying clinical FTD is also heterogeneous. Relatively selective degeneration of the frontal and temporal lobes is a consistent feature and "frontotemporal lobar degeneration" (FTLD) is used as the generic term for those pathologies that commonly present as clinical FTD [3, 4]. As with many other neurodegenerative conditions, the pathology of most cases of FTD includes the abnormal intracellular aggregation and accumulation of some pathological protein(s). Until quite recently, the vast majority of FTLD cases fell into two broad categories—those characterized by cellular inclusions composed of the microtubule-associated protein tau (FTLD-tau) and those with tau-negative inclusions that could only be detected with immunohistochemistry (IHC) against the nonspecific marker of pathological protein accumulation, ubiquitin (FTLD-U) [5]. In 2006, two publications each described three distinct patterns of FTLD-U pathology, based on the anatomical distribution and morphology of ubiquitin immunoreactive (-ir) neuronal inclusions in the cerebral cortex

© Springer Nature Switzerland AG 2021
B. Ghetti et al. (eds.), *Frontotemporal Dementias*, Advances in Experimental Medicine and Biology
1281, https://doi.org/10.1007/978-3-030-51140-1_13

[6, 7]. Importantly, the pathological features that defined each of the subtypes in these two independent studies were almost identical, providing powerful validation of the results. The significance and legitimacy of the pathological subtypes were further supported by the finding of relatively specific correlations with different clinical phenotypes [6] and with the subsequent recognition that most of the newly identified genetic causes of FTD were each consistently associated with a specific type of FTLD-U pathology, including a novel (fourth) pattern that is only found in cases caused by mutations in the valosin-containing protein gene (*VCP*) [8–10]. A major breakthrough occurred, later in 2006, when the transactive response DNA-binding protein with M_r 43 kD (TDP-43) was identified as the ubiquitinated pathological protein in most cases of FTLD-U (which now became FTLD-TDP) and in sporadic ALS, strengthening the concept that FTD and ALS are closely related conditions with overlapping pathogenesis [11, 12]. Subsequent studies confirmed TDP-43 as the pathological protein in most clinical and genetic subtypes of FTLD-U, and the same criteria were adopted for the pathological subclassification of FTLD-TDP, with only minor modifications [13–15].

Over the past decade, the concept and utility of the current FTLD-TDP subtyping system has gained wide acceptance and has been repeatedly validated through its application in new case series and by the discovery of additional clinical, genetic, and pathological correlations. Moreover, recent studies have demonstrated that cases with each of the different pathological subtypes are associated with different genetic risk factors, and that the insoluble protein extracted from postmortem brain tissue has differing physical and biochemical properties [16–18]. These findings suggest that accurate pathological subtyping of cases and a better understanding of their biochemical basis will likely be important to advance the development of biomarkers and targeted therapies for FTD.

Major Frontotemporal Lobar Degeneration TDP-43-Immunoreactive Pathological Subtypes

Although the studies that originally described the FTLD-U subtypes-evaluated ubiquitin-ir pathology in neocortex, hippocampus, and (in one study) striatum [6, 7], the diagnostic criteria that are now commonly used to subclassify FTLD-TDP cases are based exclusively on neocortical features (Table 1). Several studies have shown that these criteria are equally applicable and give comparable results regardless of whether the antibody used recognizes phosphorylated or phosphorylation-independent TDP-43 [15, 19]. The two discordant numbering systems introduced in the original papers have since been replaced with the harmonized alphabetic classification that is used later in the chapter [20].

Neocortical Features

Frontotemporal Lobar Degeneration TDP-43-Immunoreactive Pathology Type A

Type A cases are characterized by abundant TDP-43-ir neuronal cytoplasmic inclusions (NCIs) and short thick dystrophic neurites (DN), which are concentrated in the superficial cortical layers (Fig. 1). The NCIs are mostly compact (cNCIs) and have an oval or crescentic shape. Lentiform neuronal intranuclear inclusions (NIIs) are also usually present, but they are much less abundant.

Frontotemporal Lobar Degeneration TDP-43-Immunoreactive Pathology Type B

Type B cases have at least moderate numbers of NCI in both superficial and deep cortical layers, with relatively few DN and no NII. Most of the NCIs have a diffuse granular morphology (dNCI), sometimes referred to as "pre-inclusions." Importantly, some cases also have a background of delicate and small, TDP-43-ir threads and dots (ThD), which, when concentrated in layer II, may

Table 1 FTLD-TDP subtypes: distinguishing pathological features*, associated phenotypes, and causal mutations

	Type A	Type B	Type C	Type D
TDP-ir pathology				
Neocortex	II: cNCI, DN, NII	II-VI: dNCI	II-VI: long DN	II-VI:DN, NII
Hippocampus	den: NII CA1: threads	den: dNCI	den: cNCI	
Subcortical	WM: threads BG: DN, NII SN: DN	WM: GCI BG: dNCI, GCI SN: dNCI, GCI LMN: NCI	BG: cNCI	BG: DN, NII SN: DN, NII
Phenotypes	bvFTD, nfvPPA	bvFTD, nfvPPA, ALS	svPPA	IBMPFD, ALS
Mutations	*GRN* *C9orf72, TBK1*	*C9orf72, TBK1*		*VCP*

*See main text for full description of regional pathology. *II* cortical lamina II; *II–VI* cortical laminae II to VI; *ALS* amyotrophic lateral sclerosis; *BG* basal ganglia; *bvFTD* behavioral variant frontotemporal dementia; *C9orf72* chromosome 9 open reading frame 72 gene; *CA1* cornu ammonis region 1; *cNCI* compact neuronal cytoplasmic inclusions; *den* dentate lamina of hippocampus; *DN* dystrophic neurites (short unless otherwise specified); *dNCI* diffuse NCI; *GCI* glial cytoplasmic inclusions; *GRN* granulin gene; *IBMPFD* inclusion body myopathy with Paget disease of bone and frontotemporal dementia; *LMN* lower motor neurons; *NII* neuronal intranuclear inclusions; *nfvPPA* non-fluent-variant primary progressive aphasia; *SN* substantia nigra; *svPPA* semantic variant PPA; *TBK1* TANK binding kinase 1 gene; *TDP-ir* TDP-43 immunoreactive; *VCP* valosin containing protein gene; *WM* white matter

resemble the superficial laminar distribution that is typical of type A cases; however, this ThD pathology is neither consistent nor specific for type B cases.

Frontotemporal Lobar Degeneration TDP-43-Immunoreactive Pathology Type C

Type C cases have a predominance of DN with few, if any, NCI and no NII. DNs are somewhat more abundant in superficial cortical layers, and many have a unique long, tortuous morphology.

Frontotemporal Lobar Degeneration TDP-43-Immunoreactive Pathology Type D

The characteristic feature of FTLD-TDP type D pathology is an abundance of lentiform NII and delicate short DN, which are somewhat concentrated in superficial laminae. cNCI are rare in this subtype.

Hippocampal and Subcortical Pathology

In addition to the characteristic neocortical features, most cases of FTLD-TDP are also found to have significant TDP-43-ir pathology in limbic and subcortical anatomical regions (Table 1) [8, 21, 22]. Although not included in the diagnostic criteria, each of the neocortical subtypes shows a highly consistent pattern of subcortical involvement, which may be helpful when classifying difficult cases, and which may help to explain the range of associated clinical features [22].

Frontotemporal Lobar Degeneration TDP-43-Immunoreactive Pathology Type A

A highly characteristic feature of type A cases is the presence of delicate TDP-43-ir threads in hippocampal CA1 region, which is often associated with significant pyramidal cell loss (hippocampal sclerosis) (Fig. 1). Type A cases also tend to have abundant white matter threads, a predominance of DN in subcortical grey matter regions, and small numbers of NIIs in the hippocampus and striatum. Diffuse and compact NCIs are also present in the hippocampal dentate and striatum, but they tend to be less abundant than in type B or C cases.

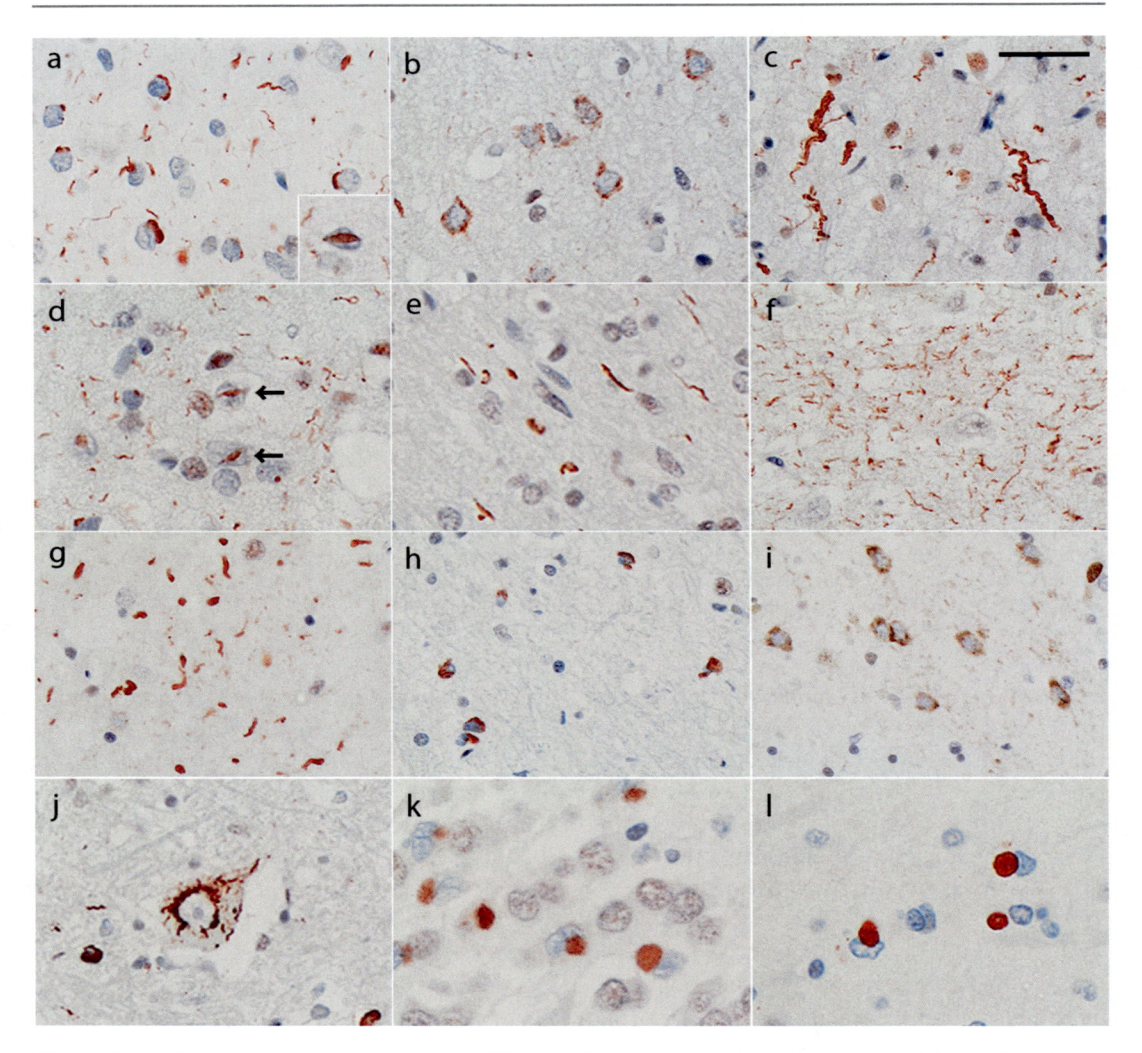

Fig. 1 TDP-43 immunoreactive pathology in different FTLD-TDP subtypes. Subtypes are defined by the pattern in the neocortex: type A has compact neuronal cytoplasmic inclusions (cNCIs), short dystrophic neurites (DNs), and some lentiform neuronal intranuclear inclusions (NIIs, insert) concentrated in layer II (**a**); type B has diffuse granular NCIs (dNCIs) throughout the neocortex (**b**); type C has DNs, many of which are long and tortuous DNs (**c**), and type D has numerous NIIs (arrows) and delicate short DNs (**d**). Each subtype also shows a characteristic pattern of pathology in the hippocampus and subcortical regions. Type A cases have thread pathology in the subcortical white matter (**e**), delicate wispy threads in hippocampal CA1 (**f**), and a predominance of DN and occasional NII in striatum and other subcortical grey matter regions (**g**). Type B cases have glial cytoplasmic inclusions in the subcortical white matter (**h**), a predominance of dNCI in subcortical grey matter (**i**), and NCI in lower motor neurons of the medulla and spinal cord (**j**). Type C cases have compact "Pick body-like" NCI in dentate granule cells of the hippocampus (**k**) and striatum (**l**). Bar: 40 μm (**a–c**, **f–j**), 10 μm (**a**, insert), 30 μm (**d**, **e**, **l**), 25 μm (**k**). TDP-43 immunohistochemistry

Frontotemporal Lobar Degeneration TDP-43-Immunoreactive Pathology Type B

The most defining subcortical feature of type B cases is frequent NCI in lower motor neurons

(LMN) of the hypoglossal nucleus and spinal cord, which may have diffuse, compact, or filamentous morphology. Moderate numbers of TDP-43-ir glial cytoplasmic inclusions (GCIs) are present in the cerebral white matter. Many subcortical gray matter regions have abundant NCIs, which are predominantly diffuse, with more modest numbers of GCI.

Frontotemporal Lobar Degeneration TDP-43-Immunoreactive Pathology Type C

The hippocampal dentate gyrus and striatum consistently show numerous cNCIs that have a unique "Pick body-like" morphology with uniform solid consistency and smooth round contour (in contrast to the cNCI found in some type A and type B cases, which usually appeared as a compact aggregate of coarse granules). The cerebral white matter is not involved, and most other subcortical structures show only occasional DN.

Frontotemporal Lobar Degeneration TDP-43-Immunoreactive Pathology Type D

Modest numbers of DN and NII are present in the amygdala, basal ganglia, nucleus basalis, thalamus, and midbrain. The pons, medulla, and cerebellum are consistently spared. Notably, the dentate granule cells of the hippocampus are free of NCI.

Clinical Correlations

There is significant overlap in the clinical features associated with each of the different major protein classes of FTLD (FTLD-tau, FTLD-TDP, and FTLD-FET) and among the subtypes within each class [1, 23–25]. Moreover, in cases within all the pathological groups, a patient's phenotype often evolves as their disease progresses to include additional clinical features. In general, cases of svPPA and FTD combined with ALS are usually found to have underlying FTLD-TDP pathology; those with nfvPPA or prominent extrapyramidal features (particularly sporadic cases) more often have FTLD-tau, whereas bvFTD can be associated with any of the FTLD pathologies. Within the FTLD-TDP group, each of the subtypes shows a number of important clinical correlations.

Frontotemporal Lobar Degeneration TDP-43-Immunoreactive Pathology Type A

Most cases present with features of bvFTD, often with prominent apathy and social withdrawal. An aphasic presentation is less common and may be nfvPPA or more difficult to classify. Executive dysfunction and some degree of memory impairment are not uncommon, particularly with older age at presentation. Neuropsychiatric manifestations (delusions, hallucinations, or obsessive behaviors) are particularly common in those with an underlying *GRN* or *C9orf72* mutation. Extrapyramidal features are reported in up to half of the cases but are rarely the presenting or predominant feature, whereas ALS is highly unusual.

Frontotemporal Lobar Degeneration TDP-43-Immunoreactive Pathology Type B

This pathology underlies the vast majority of cases in which FTD occurs in combination with clinical features of ALS. The presenting dementia syndrome is most often bvFTD, while language problems usually develop later. Psychosis is particularly common in those caused by the *C9orf72* repeat expansion, where they may be the presenting feature in one-third [26]. Extrapyramidal features develop in at least half.

Frontotemporal Lobar Degeneration TDP-43-Immunoreactive Pathology Type C

There is a particularly strong correlation between this pathology and svPPA, with most cases of clinical svPPA having FTLD-TDP type C pathology. There are often some associated behavioral changes, and cases with predominant *right* temporal involvement may present with loss of sympathy/empathy, hyposexuality, prosopagnosia, and obsessive/compulsive behavior. Psychiatric features and extrapyramidal movement disorders are much less common than with the other subtypes. Although these cases do not develop ALS, they may have some upper motor neuron features. Patients with this pathology also tend to have a

M. Neumann et al.

slower disease progression and older age at death compared to those with the other FTLD-TDP subtypes.

Frontotemporal Lobar Degeneration TDP-43-Immunoreactive Pathology Type D

This pathology is exclusively found in familial cases with *VCP* mutations in which there is variable penetrance of inclusion body myopathy (90%), Paget disease of bone (45%), FTD (30%), and ALS (10%) [27]. The FTD syndrome is usually bvFTD, with language dysfunction and extrapyramidal motor features being relatively uncommon.

Genetic Correlations

Patients with FTD due to mutations in *GRN* are consistently found to have FTLD-TDP type A pathology at autopsy [10, 13, 15], while those with *VCP* mutations always have type D (Table 1) [8, 28]. In contrast, the *C9orf72* repeat expansion has more variable TDP-43-ir pathology, with most studies reporting some cases with type A and others with type B FTLD-TDP [26, 29–32]. Moreover, two recent studies found that only half of *C9orf72* mutation cases had either typical type A or type B pathology, while the largest group had the combined pathological features of both type A and type B (type A + B, see later) [15, 22]. Although there are currently few reports describing the pathology in cases of FTD caused by mutations in the TANK-binding kinase 1 gene (*TBK1*), these also seem to include both type A and type B cases [33–36]. There are a number of other rare genetic causes of FTD that have been reported to have TDP-43 pathology but for which there is currently insufficient information to define the specific pattern (e.g., *TARDBP*, *CHCHD10*, *OPTN*, *SQSTM1*) [2]. Finally, in addition to causal mutations, genetic risk factors have been identified for FTLD-TDP, some of which are associated with a specific pathological subtype (e.g., a variant in *UNC13A* was found to be associated with FTLD-TDP type B cases but not A or C) [17].

Other Frontotemporal Lobar Degeneration TDP-43-Immunoreactive Pathology Subtypes and Patterns of Pathology

Although the subtyping of FTLD-TDP cases has proven to be useful and the current criteria generally accepted, several reports have identified cases that are difficult to classify, either because the pattern of pathology does not to fit with any of the existing subtypes or because it shows overlapping features of more than one subtype [15, 22, 37–42]. Although these cases represent a small minority in most series, they highlight some of the technical and interpretive differences that exist among neuropathologists in applying the current FTLD-TDP classification criteria.

Frontotemporal Lobar Degeneration TDP-43-Immunoreactive Pathology Cases with Overlapping Pathological Features

In a series of 30 FTLD-TDP cases selected for a BrainNet Europe study, an initial panel of five neuropathologists designated three cases as "atypical" type B, four cases as having features of both type A and type B (A + B), and two cases that had insufficient TDP-43 pathology for typing [37]. A follow-up analysis of this case series, involving a much larger group of investigators, found relatively poor agreement among the reviewers in assigning FTLD-TDP subtypes (~62%), with the worst agreement observed for FTLD-TDP type B cases. However, agreement was better (up to 85%) when raters were asked to simply dichotomize between types A or B and type C, suggesting that the major difficulty was in differentiating between type A and type B. An earlier study by Armstrong et al., that used principal component analysis, and that included a combination of TDP-43-ir pathology and additional changes that are not part of the standard subtyping criteria (neuronal loss, neuronal enlargement, neuropil vacuolation, oligodendroglial inclusions) also found significant overlap among FTLD-TDP subtypes, particularly between type A and type B [43]. Finally, a small series of four cases of FTD with delusions also reported two as having mixed type A + B pathology and two which were unclassifiable [39]. Importantly, all four of these cases harbored the *C9orf72* repeat expansion.

The issue of combined subtypes was addressed more specifically in a study designed to compare the pathological features that define the subtypes, based on the original ubiquitin-based criteria versus TDP-43 IHC [15]. In this series of 78 FTLD-TDP cases, the majority (81%) were easily classified as types A, B, or C; however, 15 cases demonstrated mixed features of both FTLD-TDP type A and type B. These mixed cases were characterized by NII, NCI, and short DN in layer II (type A features), as well as granular NCI in deeper neocortical layers that were at least as numerous as in layer II (type B features) (Fig. 2). Importantly, 12 of the 15 type A + B cases carried the *C9orf72* repeat expansion, while the remaining three cases had clinical or pathologic evidence of MND. In fact, half of the *C9orf72* mutation cases in this study had FTLD-TDP type A + B pathology, while the other half were classified as pure type B.

A similar analysis of 89 cases by another group found that a higher proportion of cases (96%) could be readily subtyped as A, B, or C, whereas five cases were judged to have features that crossed FTLD-TDP subtypes, all of which also had concomitant MND pathology [44]. One case with the *C9orf72* mutation exhibited type B features with NII (type A + B), while another *C9orf72* case exhibited a mixed type B + C pattern. The other three were non-*C9orf72* cases and included one type C with NII (type A + C), and two type B with long DN (type B + C).

Although the current subtyping criteria are based solely on pathological findings in neocortical sections, each of the different FTLD-TDP subtypes has also been reported to be associated with distinctive patterns of TDP-43 pathology in limbic and subcortical regions (see above) [21, 22]. In a recent study, Mackenzie and Neumann investigated whether including pathological data from subcortical anatomical regions would allow for better classification of cases with a mixed pattern of neocortical TDP-43-ir pathology [22]. Using standard observational assessment of neocortical sections, all of the non-*C9orf72* mutation cases could be readily classified as type A, B or C, and these results were validated using non-biased hierarchical clustering analysis (HCA). Furthermore, HCA of the pathological data from subcortical regions found that these cases again formed three distinct clusters, which perfectly matched the neocortical type A, B, and C groups. In contrast, using the neocortical data, only half of the *C9orf72* mutation cases clustered with either the type A or type B cases, and the remaining 14 formed a distinct cluster exhibiting mixed features of type A and type B. When the same group of *C9orf72* mutation cases was analyzed using the limbic and subcortical TDP-43 pathology data, more of the cases segregated as type A or type B; however, five cases remained as a separate mixed A + B cluster.

The results of these studies indicate that, although the vast majority of FTLD-TDP cases can be readily subclassified, based on the current criteria, there exists a minority that are difficult to assign because they have a combination of pathological features that characterize more than one subtype. Interestingly, these mixed patterns of pathology seem to be particularly common in cases with the *C9orf72* repeat expansion and sporadic cases that have features of both FTD and ALS [15, 22, 39, 41, 44], suggesting that there may be something unique about the mechanism of TDP-43 mis-metabolism in these clinical and genetic groups that result in greater pathological heterogeneity.

Novel Frontotemporal Lobar Degeneration TDP-43-Immunoreactive Pathology Subtypes

In 2017, Lee et al. described a series of seven cases that were difficult to categorize, based on the 2011 harmonized FTLD-TDP classification,

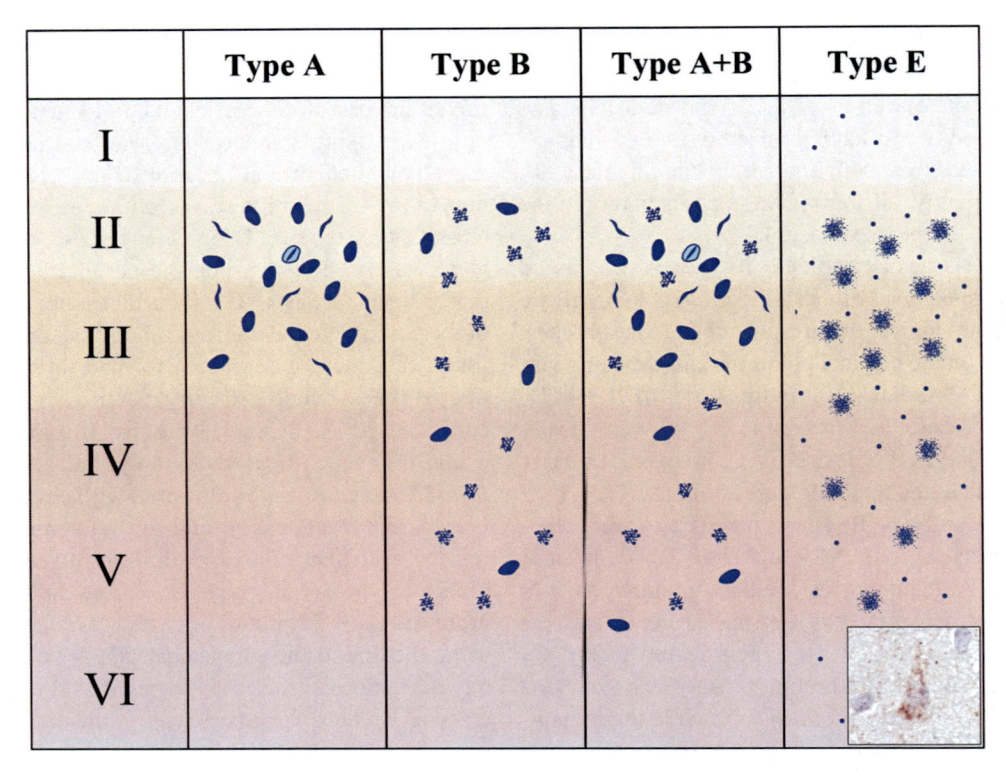

Fig. 2 Schematic representation of TDP-43 inclusion morphologies and distribution in cases with mixed (A + B) and novel (E) subtypes. Type A + B cases show the characteristic features of type A (compact neuronal cytoplasmic inclusions (NCI), short dystrophic neurites, and neuronal intranuclear inclusions, concentrated in layer II), as well as the characteristic features of type B (compact and diffuse granular NCI in deep and superficial layers). Type E cases exhibit granulofilamentous neuronal cytoplasmic inclusions and a background of and fine grains (inset photo) throughout the neocortex. (Modified from Lee et al. 2017 [40])

that they felt represented a unique subtype, which they designated as type E [40]. The neocortical TDP-43 pathology involved all cortical layers and consisted of weakly staining granulofilamentous neuronal cytoplasmic inclusions (GFNIs) set in a background of very fine grain-like deposits (Fig. 2). In contrast to the NCI found in other FTLD-TDP subtypes, these GFNIs were negative for ubiquitin and mostly negative for p62. GFNI and grain pathology, as well as TDP-43-ir oligodendroglial inclusions, were also present in a wide range of neocortical and subcortical regions, sparing only of the occipital neocortex and cerebellum. Motor neuron involvement was a consistent feature, although only one case was associated with clinical features of ALS. Interestingly, these FTLD-TDP type E cases were consistently asso-

ciated with a rapid clinical course of 1–3 years' duration.

Some additional reports have described cases with pathology similarity to the type E of Lee et al. Takeuchi et al. reported a subset of sporadic ALS cases with NCI, granular or dot like DN, and a high density of GCI, involving motor cortex, other neocortical regions, basal ganglia, and spinal cord [41]. Ubiquitin and p62 IHC were not performed. The authors interpreted these findings as distinct from FTLD-TDP types A–D. More recently, two cases with 1-year duration of PPA and ALS were reported to exhibit FTLD-TDP type E pathology, consisting of TDP-43-ir, p62-negative GFNI, and grains [42]. Finally, a case of rapidly progressive Foix-Chavany-Marie syndrome (FCMS) has been reported to exhibit FTLD-TDP type E [45]. FCMS, also known as bilateral opercular syndrome, is characterized by prominent motor dysfunction, involving muscles of the face, tongue, and pharynx. While the etiology of FCMS is diverse, sometimes being associated with bilateral opercular infarcts, progressive forms of FCMS share clinical similarities to FTD [46].

While FTLD-TDP type E may represent a distinct subtype, the association with ALS, in some cases, and the similarities with the pathological features of type B cases, raise the possibility that types E and B represent a continuum. Indeed, FTLD-TDP type B has been described as often having a predominance of granular rather than compact NCIs, a synaptic pattern of neuropil inclusions, and abundant threads and dots [15, 21, 32]. Given the relatively short disease duration of most cases with FTLD-TDP type E, one possibility is that these represent "early-stage" disease when the TDP-43 inclusions are still immature and have not yet coalesced into a more typical FTLD-TDP type B morphology and become ubiquitinated. Alternatively, FTLD-TDP type E could represent a more virulent pathology, which spreads through the brain and spinal cord quickly, resulting in rapid clinical disease progression and relatively immature inclusions.

Unique Patterns of TDP-43 Pathology in Rare Disorders

Unique patterns of TDP-43 proteinopathy have been described in a few rare neurodegenerative diseases, not typically classified as FTLD or ALS. A screen of non-neurodegenerative disease neuropathology specimens revealed that Rosenthal fibers and eosinophilic granular bodies, which may be present in reactive gliosis and in some low-grade astrocytic brain tumors, label with TDP-43 IHC [47]. Rosenthal fibers are protein aggregates within astrocytes that are composed primarily of glial fibrillary acidic protein (GFAP) and are also the defining pathological feature of Alexander disease, a leukodystrophy associated with *GFAP* mutations

[48]. A subsequent study demonstrated that Rosenthal fibers in Alexander disease are also TDP-43-ir [49]. Thus, Alexander disease represents a unique TDP-43 proteinopathy, in which neurodegeneration is associated exclusively with astrocytic inclusions.

Another unique pattern of TDP-43 proteinopathy is found in Perry syndrome, a progressive neurodegenerative disease characterized by parkinsonism, psychiatric symptoms, and hypoventilation, caused by mutations in the gene encoding dynactin-1 (*DCTN1*) [50]. In addition to modest numbers of NCI composed of the dynactin subunit p50, cases of Perry syndrome exhibit TDP-43-ir NCI, DN, oligodendroglial GCI, axonal spheroids, and perivascular astrocytic inclusions [51]. Based on the very limited number of cases reported ($n = 3$), the pattern of TDP-43 pathology in Perry syndrome seems to be distinct from FTLD-TDP, with a predisposition for the substantia nigra and other subcortical regions with only mild and inconsistent involvement of the cerebral cortex.

In both Alexander disease and Perry syndrome, TDP-43 protein aggregation is likely secondary to the accumulation and dysfunction of other proteins (GFAP and dynactin, respectively). Nonetheless, these conditions are informative by demonstrating that TDP-43 proteinopathy may result from diverse mechanisms.

TDP-43 Pathology in Aging and Common Neurodegenerative Disorders

Finally, it is important to recognize that some degree of TDP-43-ir pathology is a common finding in the limbic structures of the mesial temporal lobe in aging and in association with many common neurodegenerative disorders, including Alzheimer's disease and Lewy body disease [52]. The clinical relevance of this pathology and its relationship to FTLD-TDP is currently the topic of tremendous interest and controversy [53, 54], but it is beyond the scope of this chapter.

Biochemical Basis of Frontotemporal Lobar Degeneration TDP-43-Immunoreactive Pathology Subtypes

Biochemical Properties of TDP-43 Aggregates and Disease-Associated Modifications

Aggregated TDP-43 isolated from human postmortem FTLD-TDP brain tissue is poorly detergent soluble and subject to a variety of disease-associated posttranslational modifications (PTMs). These result in a highly characteristic biochemical banding pattern by immunoblot analysis, with the presence of disease-specific bands of ~25 kDa, ~45 kDa, and a high molecular smear, in addition to the ~43 kDa band corresponding to normal TDP-43 (Fig. 3a) [11, 12]. PTM of TDP-43 include N-terminal truncation, phosphorylation, ubiquitination, acetylation, cysteine oxidation, and sumoylation [55]. The characterization of the various PTMs and their functional consequences are still poorly understood and not fully validated in human postmortem tissue; however, there is increasing evidence

that TDP-43 PTM may play a crucial role in disease pathogenesis, and that modulation of disease-relevant PTM might be a promising avenue for future therapeutic approaches.

N-Terminal Truncation and C-Terminal Fragments

The presence of short TDP-43 fragments of ~25 kDa is a hallmark feature of FTLD-TDP [11, 12]. They are composed of N-terminally truncated TDP-43 species as demonstrated by absent labeling with antibodies raised against the N-terminus (amino acids 6–24) but detection with antibodies against the extreme C-terminus of TDP-43 [56]. N-terminal sequencing of fragments isolated from human postmortem tissue has revealed arginine at position 208 [57], and mass spectrometry analysis of tryptic digests of isolated fragments has demonstrated aspartic acid residues at positions 219 and 247 [58] as potential cleavage sites. Although these experiments clearly demonstrate that these short species contain N-terminally truncated fragments that extend to the extreme C-terminus, it is still unclear whether they all include the entire C-terminal region. Moreover, the origins and pathomechanistic relevance of the C-terminal

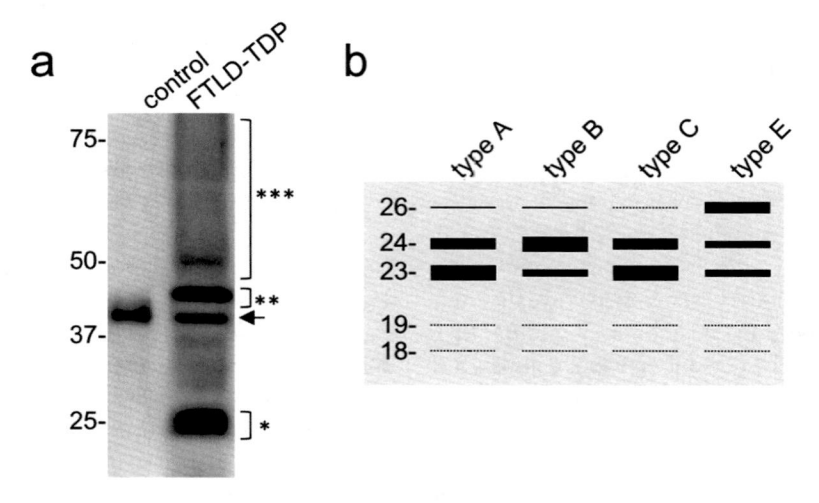

Fig. 3 Immunoblot analysis of sarcosyl-insoluble protein fractions from FTLD-TDP shows the disease-specific biochemical signature of TDP-43 with pathological bands −25 kDa (*), −45 kDa (**), and a high molecular smear (***), in addition to the physiological TDP-43 band (arrow) also present in control brains (**a**). Schematic representation of distinct banding patterns of C-terminal fragments among FTLD-TDP subtypes (types A, B, and C based on Kawakami et al. 2018; type E based on Lee et al. 2017) [40, 84] (**b**)

fragments (CTFs) remain to be fully established. Most studies propose proteolytic cleavage/degradation by caspases [59, 60], asparaginyl endopeptidase [61], or calpains [62]; although other explanations include alternative splicing events or usage of alternate translational start sites [63, 64]. However, several proposed cleavage sites and generated fragments/isoforms in these studies do not match well with the fragments observed in human postmortem tissue, suggesting that additional enzymes and/or mechanisms might exist.

The potential role of TDP-43 CTF in disease pathogenesis is supported by findings of cellular toxicity upon overexpression of CTF in some cellular and animal models [57, 65]; however, in several other model systems, the correlation is less clear [66]. Moreover, while enrichment for CTF over full-length TDP-43 is a characteristic feature of most types of cellular inclusions in the cerebral cortex, CTFs are less abundant or absent in inclusions in spinal cord LMN in FTLD-TDP/ALS [56] and in cortical pre-inclusions [67], thereby suggesting that the formation of CTF might not be mandatory for aggregation and toxicity.

Phosphorylation

Aberrant phosphorylation of TDP-43 has been recognized as one of the major PTMs of pathological TDP-43 since its initial discovery as the disease protein in FTLD-TDP and ALS [11, 12]. The fact that the majority of pathogenic *TARDBP* mutations either introduce or disrupt potential serine/threonine phosphorylation sites or introduce phosphomimic residues (glutamate/aspartate) suggest that alterations in the phosphorylation status of TDP-43 play a crucial role in the pathogenesis of TDP-43 proteinopathies [55]. TDP-43 has 41 serine, 15 threonine, and 8 tyrosine residues acting as potential phosphorylation sites. Mass spectrometry analysis of recombinant TDP-43 treated with casein kinase 1 and of aggregated TDP-43 isolated from human postmortem tissue has revealed several phosphorylated residues [68–70]; however, so far, only five sites at the C-terminus of TDP-43 (pS379, pS403, pS404, pS409, pS410) have been validated in pathological TDP-43 inclusions in human postmortem tissue with phosphorylation-site-specific antibodies [19, 68]. Phosphorylation at these C-terminal serine residues (with pS409/410 as most studied sites) is a highly consistent and specific feature of aggregated TDP-43 in all types of pathological TDP-43 inclusions, in all sporadic and familial FTLD-TDP subtypes, and is considered an abnormal event due to the lack of phosphorylation of these sites under physiological conditions [15, 19, 68, 70, 71]. The functional consequences of TDP-43 C-terminal phosphorylation are not fully resolved. While some experimental studies have described an association with decreased solubility of TDP-43 and greater toxicity [68, 72], others have reported the opposite effects with phosphomimicking mutants showing increased solubility and reduced toxicity [73, 74]. Further insights into the role of TDP-43 phosphorylation, in regulating its physiological functions (e.g., RNA binding, dimerization) and the impact of abnormal phosphorylation events through the identification of the involved kinases and phosphatases, will be crucial steps to elucidate the pathological processes in TDP-proteinopathies.

Ubiquitination

Ubiquitination of TDP-43 aggregates is a key feature in FTLD-TDP; however, insights into the specific lysin residues that are ubiquitinated in human FTLD-TDP tissues and the functional consequences are still limited. The detection of TDP-43 Lys-48- and Lys-63-linked polyubiquitin chains in cellular models is suggestive of proteasomal and autophagosomal degradation of TDP-43 [75]. Lysine residues 84, 95, 102, 114, 121, 140, 145, 160, 176, 181, and 263 have been identified as ubiquitinated TDP-43 residues in cellular models, however, with some variability among studies, most likely reflecting the complexity of ubiquitin-proteasome regulation of TDP-43 in a highly context-dependent manner [76–79]. Notably, ubiquitination of lysin 84 has been postulated as an important modifier of nuclear import of TDP-43 in mutagenesis experiments, and a complex interplay between TDP-43 ubiquitination at distinct sites and phosphorylation at pS409/410 has been observed [77]. However, validation of any ubiquitination site in human postmortem FTLD-TDP tissue is lacking, and, to date, the only ubiquitinated residue identified by mass spectrometry of insoluble protein extracts from postmortem tissue (of an ALS patient) is lysine 79 [80].

Acetylation

Another modification of lysine residue is acetylation. So far, two acetylation sites have been identified in cellular models at lysine 145 (located in RRM1) and lysine 192 (located in RRM2) [81]. However, since mutation of TDP-43 at these two sites did not completely abrogate acetylation, additional acetylated lysine residues may be present. Acetylation at lysine 145 and lysine 192 has been shown to impair the binding of TDP-43 to RNA and to promote TDP-43 phosphorylation at pS409/410 [81]. The potential role of this modification in disease was demonstrated using an antibody specific for TDP-43 acetylated at lysine 145, which revealed acetylated TDP-43 as a biochemical component of the TDP-43 inclusions in ALS/FTD spinal cord, which are known to be composed of the full-length protein, but not the inclusions in cerebral cortex, which are composed primarily of CTFs that lack the epitope recognized by the antibody [81].

Sumoylation

Evidence for sumoylation of TDP-43 comes mainly from a proteomics approach that revealed SUMO-2/3 in complex with insoluble TDP-43 in a cellular model system overexpressing a CTF [75]; however, sumoylation of TDP-43 has not yet been directly demonstrated in human disease tissue.

Cysteine Oxidation

Upon exposure to oxidative stressors, TDP-43 has been reported to undergo cysteine oxidation and disulfide cross-linking in vitro and in cellular models, resulting in enhanced TDP-43 aggregation and alterations in subcellular distribution [82]. TDP-43 has six cysteine residues, and there is experimental evidence that all sites contribute to proper folding, self-assembly, and oligomer-

ization of TDP-43 [55]. While increased levels of cross-linked TDP-43 species are present in FTLD-TDP brains [82], the pathomechanistic role of cysteine oxidation and cross-linking remains to be fully determined.

Biochemical Diversity of TDP-43 Aggregates in Frontotemporal Lobar Degeneration TDP-43-Immunoreactive Pathology Subtypes and Evidence for TDP-43 Strains

A crucial open question in the FTLD-TDP research field is the molecular basis behind the huge clinical and neuropathological phenotypic variability, as well as the selective vulnerability in FTLD-TDP subtypes and ALS. The concept that distinct self-propagating conformers of an aggregated protein ("strains") represent the basis for phenotypic diversity in a neurodegenerative disease was first established in prion diseases [83]. By analogy, a popular hypothesis to explain the heterogeneity in FTLD-TDP is the presence of different conformational types of misfolded TDP-43 ("TDP-43 strains") that can propagate in a prion-like manner [84]. In fact, there is a growing body of evidence supporting this idea.

Biochemical heterogeneity of aggregated TDP-43 has already been recognized in the initial report on the discovery of TDP-43 as the disease protein [11]. Briefly, monoclonal antibodies (clones 182 and 406) generated against insoluble protein fractions from FTLD-TDP brains, each labeled distinct bands of the N-terminally truncated TDP-43 species by immunoblot, specific for either type A or type B FTLD-TDP cases (then referred to as FTLD-U type 3 or type 1, respectively). This suggested that each antibody was recognizing either a specific conformation or a specific pattern of PTM of aggregated TDP-43 species, each being specific for a different FTLD-TDP subtype. Several studies have been performed since then to further characterize and correlate an immunoblot banding pattern of TDP-43 CTF with distinct FTLD-TDP subtypes, with most employing antibodies against pS409/410 [19, 40, 68, 85]. Using high-percentage polyacrylamide gel electrophoresis, distinct CTF with up to three major bands (23 kDa, 24 kDa, and 26 kDa) and two minor bands (18 kDa and 19 kDa) can be present in sarkosyl-insoluble lysates of FTLD-TDP brains, with some studies demonstrating subtle differences in the banding pattern among FTLD-TDP subtypes (Fig. 3b) [40, 68, 85]. Briefly, in type A, the most intense major band is at 23 kDa; in type B, it is at 24 kDa; type C lacks the 26 kDa band and has a more prominent 23 kDa band; and type E shows three major bands with the most intense at 26 kDa. However, significant variability within and overlap between subtypes exists [19]; so, the biochemical classification of subtypes remains challenging, and more sensitive methods of detection, quantification, and analysis of various CTFs and their PTMs are required.

Nevertheless, it is tempting to speculate that the different banding patterns in FTLD-TDP may correspond to different conformational species of abnormal TDP-43. In strong support of this idea, protease treatment of insoluble TDP-43 aggregates has revealed different patterns of protease-resistant cores among FTLD-TDP subtypes, highly suggestive of different conformers [18]. More recently, a new extraction method, termed "SarkoSpin," has been developed that allows extraction of pathological TDP-43 species from postmortem tissue with improved separation from physiological TDP-43, compared to the previous sequential extraction protocols [16]. This approach has revealed additional insights into distinct biophysical properties of aggregated TDP-43 among the TDP-43 proteinopathies, with TDP-43 from FTLD-TDP type C found to exhibit a higher intrinsic density and protease-resistant CTF core compared to that from cases of type A or ALS (type B not examined).

In addition to the observed biochemical/structural differences, crucial support for the idea that distinct pathological TDP-43 species may (at least partially) explain the clinical and pathological variability in FTLD-TDP comes from the observations that TDP-43 extracted from different FTLD-TDP subtypes exhibits different levels of seeding activity and toxicity in vitro and in vivo. The first such evidence was provided by Nokanko et al. who reported that seeding activity of TDP-43 extracted from human postmortem tissue in a cell culture model was more efficient when using extracts from type A and type B cases compared to type C [86]. Interestingly, the banding pattern of insoluble CTF extracted from the seeded cell lysates resembled that from the corresponding FTLD-TDP subject used as the seed, suggestive of a prion-like self-templating process of TDP-43 aggregation. These results were validated and expanded in a report where TDP-43 aggregates extracted using the SarkoSpin protocol from FTLD-TDP type A cases demonstrated templated seeding and toxicity in cultured primary neurons, while those from subtype C seemed inert [16]. While in these studies no differences between sporadic and genetic cases were mentioned, Porta et al. reported that lysates from *GRN* mutation carriers had the highest seeding activity in their cellular screening assay, followed by *C9orf72* mutation carriers and sporadic FTLD-TDP type A and type B cases [87]. Biochemical analyses of the lysates revealed a correlation between the presence of two minor CTF bands of 18 kDa and 19 kDa and seeding activity, suggesting that distinct fragments and/or conformational TDP-43 species seem to be more potent [87]. Most importantly, this study provided the first in vivo evidence for propagation of TDP-43 pathology in a prion-like manner by demonstrating the induction and spreading of de novo TDP-43 pathology, following the intracerebral injection of FTLD-TDP aggregates isolated from human FTLD-TDP type A tissue into transgenic mice expressing cytoplasmic human TDP-43 and non-transgenic mice [87].

Therefore, current insights are consistent with the idea that the progression of FTLD-TDP pathology involves self-templating seeded aggregation and cell-to-cell spreading of pathological TDP-43 that exists in different conformations. However, more extensive biochemical, biophysical, and seeding studies are needed to strengthen the hypothesis that different TDP-43 conformers/species, indeed, contribute to the phenotypic heterogeneity in FTLD-TDP patients (e.g., by demonstrating whether distinct FTLD-TDP subtype-derived TDP-43 aggregates can reproduce their distinct clinical and neuropathological characteristics in animal models).

Finally, in addition to biochemical differences of TDP-43 itself, co-aggregation of other proteins into TDP-43 inclusions might contribute to the diversity among FTLD-TDP subtypes. This hypothesis is supported by double-label immuno-histochemical findings with co-localization of hnRNP E2 and TDP-43 in FTLD-TDP subtype C and subsets of FTLD-TDP type A inclusions, but not in type B cases [88, 89]. However, in-depth biochemical characterization of the protein composition of TDP-43 inclusions is required to further address this.

Summary

The current criteria for the pathological subclassification of FTLD-TDP are widely accepted and show a number of highly relevant clinical and genetic associations. However, the presence of a small proportion of cases with novel patterns of TDP-43-ir pathology indicates the need for additional correlative studies. Investigations, to date, suggest that the basis for the different subtypes is, at least partially, biochemical and/or conformational variation in the aggregating protein. Further studies to more fully elucidate the nature of the subtype-specific pathological species of TDP-43 will be crucial to the development of useful biomarkers and targeted therapies.

Acknowledgments This work was supported by the NOMIS foundation (MN) and the Canadian Institutes of Health Research (IRM).

References

1. Woollacott IO, Rohrer JD (2016) The clinical spectrum of sporadic and familial forms of frontotemporal dementia. J Neurochem 138(Suppl 1):6–31
2. Pottier C, Ravenscroft TA, Sanchez-Contreras M, Rademakers R (2016) Genetics of FTLD: overview and what else we can expect from genetic studies. J Neurochem 138(Suppl 1):32–53
3. Mackenzie IR, Neumann M, Bigio EH, Cairns NJ, Alafuzoff I, Kril J et al (2009) Nomenclature for neuropathologic subtypes of frontotemporal lobar degeneration: consensus recommendations. Acta Neuropathol 117(1):15–18
4. Mackenzie IR, Neumann M, Bigio EH, Cairns NJ, Alafuzoff I, Kril J et al (2010) Nomenclature and nosology for neuropathologic subtypes of frontotemporal lobar degeneration: an update. Acta Neuropathol 119(1):1–4
5. Mackenzie IR, Shi J, Shaw CL, Duplessis D, Neary D, Snowden JS et al (2006) Dementia lacking distinctive histology (DLDH) revisited. Acta Neuropathol 112(5):551–559
6. Mackenzie IR, Baborie A, Pickering-Brown S, Du Plessis D, Jaros E, Perry RH et al (2006) Heterogeneity of ubiquitin pathology in frontotemporal lobar degeneration: classification and relation to clinical phenotype. Acta Neuropathol 112(5):539–549
7. Sampathu DM, Neumann M, Kwong LK, Chou TT, Micsenyi M, Truax A et al (2006) Pathological heterogeneity of frontotemporal lobar degeneration with ubiquitin-positive inclusions delineated by ubiquitin immunohistochemistry and novel monoclonal antibodies. Am J Pathol 169(4):1343–1352
8. Forman MS, Mackenzie IR, Cairns NJ, Swanson E, Boyer PJ, Drachman DA et al (2006) Novel ubiquitin neuropathology in frontotemporal dementia with valosin-containing protein gene mutations. J Neuropathol Exp Neurol 65(6):571–581
9. Holm IE, Englund E, Mackenzie IR, Johannsen P, Isaacs AM (2007) A reassessment of the neuropathology of frontotemporal dementia linked to chromosome 3. J Neuropathol Exp Neurol 66(10):884–891
10. Mackenzie IR, Baker M, Pickering-Brown S, Hsiung GY, Lindholm C, Dwosh E et al (2006) The neuropathology of frontotemporal lobar degeneration caused by mutations in the progranulin gene. Brain 129(Pt 11):3081–3090
11. Neumann M, Sampathu DM, Kwong LK, Truax AC, Micsenyi MC, Chou TT et al (2006) Ubiquitinated TDP-43 in frontotemporal lobar degeneration and amyotrophic lateral sclerosis. Science 314(5796):130–133
12. Arai T, Hasegawa M, Akiyama H, Ikeda K, Nonaka T, Mori H et al (2006) TDP-43 is a component of ubiquitin-positive tau-negative inclusions in frontotemporal lobar degeneration and amyotrophic lateral sclerosis. Biochem Biophys Res Commun 351:602–611
13. Cairns NJ, Neumann M, Bigio EH, Holm IE, Troost D, Hatanpaa KJ et al (2007) TDP-43 in familial and sporadic frontotemporal lobar degeneration with ubiquitin inclusions. Am J Pathol 171(1):227–240
14. Davidson Y, Kelley T, Mackenzie IR, Pickering-Brown S, Du Plessis D, Neary D et al (2007) Ubiquitinated pathological lesions in frontotemporal lobar degeneration contain the TAR DNA-binding protein, TDP-43. Acta Neuropathol 113(5):521–533
15. Mackenzie IR, Neumann M (2017) Reappraisal of TDP-43 pathology in FTLD-U subtypes. Acta Neuropathol 134(1):79–96

16. Laferriere F, Maniecka Z, Perez-Berlanga M, Hruska-Plochan M, Gilhespy L, Hock EM et al (2019) TDP-43 extracted from frontotemporal lobar degeneration subject brains displays distinct aggregate assemblies and neurotoxic effects reflecting disease progression rates. Nat Neurosci 22(1):65–77

17. Pottier C, Ren Y, Perkerson RB 3rd, Baker M, Jenkins GD, van Blitterswijk M et al (2019) Genome-wide analyses as part of the international FTLD-TDP whole-genome sequencing consortium reveals novel disease risk factors and increases support for immune dysfunction in FTLD. Acta Neuropathol 137(6):879–899

18. Tsuji H, Arai T, Kametani F, Nonaka T, Yamashita M, Suzukake M et al (2012) Molecular analysis and biochemical classification of TDP-43 proteinopathy. Brain 135(Pt 11):3380–3391

19. Neumann M, Kwong LK, Lee EB, Kremmer E, Flatley A, Xu Y et al (2009) Phosphorylation of S409/410 of TDP-43 is a consistent feature in all sporadic and familial forms of TDP-43 proteinopathies. Acta Neuropathol 117(2):137–149

20. Mackenzie IR, Neumann M, Baborie A, Sampathu DM, Du Plessis D, Jaros E et al (2011) A harmonized classification system for FTLD-TDP pathology. Acta Neuropathol 122(1):111–113

21. Josephs KA, Stroh A, Dugger B, Dickson DW (2009) Evaluation of subcortical pathology and clinical correlations in FTLD-U subtypes. Acta Neuropathol 118(3):349–358

22. Mackenzie IR, Neumann M (2020) Subcortical TDP-43 pathology patterns validate cortical FTLD-TDP subtypes and demonstrate unique aspects of C9orf72 mutation cases. Acta Neuropathol 139(1):83–98

23. Josephs KA, Hodges JR, Snowden JS, Mackenzie IR, Neumann M, Mann DM et al (2011) Neuropathological background of phenotypical variability in frontotemporal dementia. Acta Neuropathol 122(2):137–153

24. Perry DC, Brown JA, Possin KL, Datta S, Trujillo A, Radke A et al (2017) Clinicopathological correlations in behavioural variant frontotemporal dementia. Brain 140(12):3329–3345

25. Rohrer JD, Lashley T, Holton J, Revesz T, Urwin H, Isaacs AM et al (2011) The clinical and neuroanatomical phenotype of FUS associated frontotemporal lobar degeneration. J Neurol Neurosurg Psychiatry 82(12):1405–1407

26. Snowden JS, Rollinson S, Thompson JC, Harris JM, Stopford CL, Richardson AM et al (2012) Distinct clinical and pathological characteristics of frontotemporal dementia associated with C9ORF72 mutations. Brain 135(Pt 3):693–708

27. Al-Obeidi E, Al-Tahan S, Surampalli A, Goyal N, Wang AK, Hermann A et al (2018) Genotype-phenotype study in patients with valosin-containing protein mutations associated with multisystem proteinopathy. Clin Genet 93(1):119–125

28. Neumann M, Mackenzie IR, Cairns NJ, Boyer PJ, Markesbery WR, Smith CD et al (2007) TDP-43 in the ubiquitin pathology of frontotemporal dementia with VCP gene mutations. J Neuropathol Exp Neurol 66(2):152–157

29. Bigio EH, Weintraub S, Rademakers R, Baker M, Ahmadian SS, Rademaker A et al (2013) Frontotemporal lobar degeneration with TDP-43 proteinopathy and chromosome 9p repeat expansion in C9ORF72: clinicopathologic correlation. Neuropathology 33(2):122–133

30. Boeve BF, Boylan KB, Graff-Radford NR, DeJesus-Hernandez M, Knopman DS, Pedraza O et al (2012) Characterization of frontotemporal dementia and/or amyotrophic lateral sclerosis associated with the GGGGCC repeat expansion in C9ORF72. Brain 135(Pt 3):765–783

31. Mahoney CJ, Beck J, Rohrer JD, Lashley T, Mok K, Shakespeare T et al (2012) Frontotemporal dementia with the C9ORF72 hexanucleotide repeat expansion: clinical, neuroanatomical and neuropathological features. Brain 135(Pt 3):736–750

32. Murray ME, Dejesus-Hernandez M, Rutherford NJ, Baker M, Duara R, Graff-Radford NR et al (2011) Clinical and neuropathologic heterogeneity of c9FTD/ALS associated with hexanucleotide repeat expansion in C9ORF72. Acta Neuropathol 122(6):673–690

33. Gijselinck I, Van Mossevelde S, van der Zee J, Sieben A, Philtjens S, Heeman B et al (2015) Loss of TBK1 is a frequent cause of frontotemporal dementia in a Belgian cohort. Neurology 85(24):2116–2125

34. Hirsch-Reinshagen V, Alfaify OA, Hsiung GR, Pottier C, Baker M, Perkerson RB 3rd et al (2019) Clinicopathologic correlations in a family with a TBK1 mutation presenting as primary progressive aphasia and primary lateral sclerosis. Amyotroph Lateral Scler Frontotemporal Degener 20:568–575

35. Koriath CA, Bocchetta M, Brotherhood E, Woollacott IO, Norsworthy P, Simon-Sanchez J et al (2017) The clinical, neuroanatomical, and neuropathologic phenotype of TBK1-associated frontotemporal dementia: a longitudinal case report. Alzheimers Dement (Amst) 6:75–81

36. Lamb R, Rohrer JD, Real R, Lubbe SJ, Waite AJ, Blake DJ et al (2019) A novel TBK1 mutation in a family with diverse frontotemporal dementia spectrum disorders. Cold Spring Harb Mol Case Stud 5(3):a003913

37. Alafuzoff I, Pikkarainen M, Neumann M, Arzberger T, Al-Sarraj S, Bodi I et al (2015) Neuropathological assessments of the pathology in frontotemporal lobar degeneration with TDP43-positive inclusions: an inter-laboratory study by the BrainNet Europe consortium. J Neural Transm 122(7):957–972

38. Pikkarainen M, Hartikainen P, Alafuzoff I (2008) Neuropathologic features of frontotemporal lobar degeneration with ubiquitin-positive inclusions visualized with ubiquitin-binding protein p62 immunohistochemistry. J Neuropathol Exp Neurol 67(4):280–298

39. Shinagawa S, Naasan G, Karydas AM, Coppola G, Pribadi M, Seeley WW et al (2015) Clinicopathological study of patients with C9ORF72-associated fron-

totemporal dementia presenting with delusions. J Geriatr Psychiatry Neurol 28(2):99–107

40. Lee EB, Porta S, Michael Baer G, Xu Y, Suh E, Kwong LK et al (2017) Expansion of the classification of FTLD-TDP: distinct pathology associated with rapidly progressive frontotemporal degeneration. Acta Neuropathol 134(1):65–78

41. Takeuchi R, Tada M, Shiga A, Toyoshima Y, Konno T, Sato T et al (2016) Heterogeneity of cerebral TDP-43 pathology in sporadic amyotrophic lateral sclerosis: evidence for clinico-pathologic subtypes. Acta Neuropathol Commun 4(1):61

42. Tan RH, Guennewig B, Dobson-Stone C, Kwok JBJ, Kril JJ, Kiernan MC et al (2019) The underacknowledged PPA-ALS: a unique clinicopathologic subtype with strong heritability. Neurology 92(12):e1354–e1e66

43. Armstrong RA, Ellis W, Hamilton RL, Mackenzie IR, Hedreen J, Gearing M et al (2010) Neuropathological heterogeneity in frontotemporal lobar degeneration with TDP-43 proteinopathy: a quantitative study of 94 cases using principal components analysis. J Neural Transm (Vienna) 117(2):227–239

44. Nishihira Y, Gefen T, Mao Q, Appin C, Kohler M, Walker J et al (2019) Revisiting the utility of TDP-43 immunoreactive (TDP-43-ir) pathology to classify FTLD-TDP subtypes. Acta Neuropathol 138(1):167–169

45. Clark CN, Quaegebeur A, Nirmalananthan N, MacKinnon AD, Revesz T, Holton JL et al (2019) Foix-Chavany-Marie syndrome due to type E TDP43 pathology. Neuropathol Appl Neurobio l46(3):292–295.

46. Ihori N, Araki S, Ishihara K, Kawamura M (2006) A case of frontotemporal lobar degeneration with progressive dysarthria. Behav Neurol 17(2):97–104

47. Lee EB, Lee VM, Trojanowski JQ, Neumann M (2008) TDP-43 immunoreactivity in anoxic, ischemic and neoplastic lesions of the central nervous system. Acta Neuropathol 115(3):305–311

48. Brenner M, Johnson AB, Boespflug-Tanguy O, Rodriguez D, Goldman JE, Messing A (2001) Mutations in GFAP, encoding glial fibrillary acidic protein, are associated with Alexander disease. Nat Genet 27(1):117–120

49. Walker AK, Daniels CM, Goldman JE, Trojanowski JQ, Lee VM, Messing A (2014) Astrocytic TDP-43 pathology in Alexander disease. J Neurosci 34(19):6448–6458

50. Farrer MJ, Hulihan MM, Kachergus JM, Dachsel JC, Stoessl AJ, Grantier LL et al (2009) DCTN1 mutations in Perry syndrome. Nat Genet 41(2):163–165

51. Mishima T, Koga S, Lin WL, Kasanuki K, Castanedes-Casey M, Wszolek ZK et al (2017) Perry syndrome: a distinctive type of TDP-43 Proteinopathy. J Neuropathol Exp Neurol 76(8):676–682

52. Josephs KA, Murray ME, Whitwell JL, Parisi JE, Petrucelli L, Jack CR et al (2014) Staging TDP-43 pathology in Alzheimer's disease. Acta Neuropathol 127(3):441–450

53. Josephs KA, Mackenzie I, Frosch MP, Bigio EH, Neumann M, Arai T et al (2019) LATE to the PART-y. Brain 142(9):e47

54. Nelson PT, Dickson DW, Trojanowski JQ, Jack CR, Boyle PA, Arfanakis K et al (2019) Limbic-predominant age-related TDP-43 encephalopathy (LATE): consensus working group report. Brain 142(6):1503–1527

55. Buratti E (2018) TDP-43 post-translational modifications in health and disease. Expert Opin Ther Targets 22(3):279–293

56. Igaz LM, Kwong LK, Xu Y, Truax AC, Uryu K, Neumann M et al (2008) Enrichment of C-terminal fragments in TAR DNA-binding protein-43 cytoplasmic inclusions in brain but not in spinal cord of frontotemporal lobar degeneration and amyotrophic lateral sclerosis. Am J Pathol 173(1):182–194

57. Igaz LM, Kwong LK, Chen-Plotkin A, Winton MJ, Unger TL, Xu Y et al (2009) Expression of TDP-43 C-terminal fragments in vitro recapitulates pathological features of TDP-43 Proteinopathies. J Biol Chem 284(13):8516–8524

58. Nonaka T, Kametani F, Arai T, Akiyama H, Hasegawa M (2009) Truncation and pathogenic mutations facilitate the formation of intracellular aggregates of TDP-43. Hum Mol Genet 18(18):3353–3364

59. Zhang YJ, Xu YF, Dickey CA, Buratti E, Baralle F, Bailey R et al (2007) Progranulin mediates caspase-dependent cleavage of TAR DNA binding protein-43. J Neurosci 27(39):10530–10534

60. Dormann D, Capell A, Carlson AM, Shankaran SS, Rodde R, Neumann M et al (2009) Proteolytic processing of TAR DNA binding protein-43 by caspases produces C-terminal fragments with disease defining properties independent of progranulin. J Neurochem 110(3):1082–1094

61. Herskowitz JH, Gozal YM, Duong DM, Dammer EB, Gearing M, Ye K et al (2012) Asparaginyl endopeptidase cleaves TDP-43 in brain. Proteomics 12(15–16):2455–2463

62. Yamashita T, Hideyama T, Hachiga K, Teramoto S, Takano J, Iwata N et al (2012) A role for calpain-dependent cleavage of TDP-43 in amyotrophic lateral sclerosis pathology. Nat Commun 3:1307

63. Xiao S, Sanelli T, Chiang H, Sun Y, Chakrabartty A, Keith J et al (2015) Low molecular weight species of TDP-43 generated by abnormal splicing form inclusions in amyotrophic lateral sclerosis and result in motor neuron death. Acta Neuropathol 130(1):49–61

64. Nishimoto Y, Ito D, Yagi T, Nihei Y, Tsunoda Y, Suzuki N (2010) Characterization of alternative isoforms and inclusion body of the TAR DNA-binding protein-43. J Biol Chem 285(1):608–619

65. Zhang YJ, Xu YF, Cook C, Gendron TF, Roettges P, Link CD et al (2009) Aberrant cleavage of TDP-43 enhances aggregation and cellular toxicity. Proc Natl Acad Sci U S A 106(18):7607–7612

66. Berning BA, Walker AK (2019) The pathobiology of TDP-43 C-terminal fragments in ALS and FTLD. Front Neurosci 13:335

67. Josephs KA, Zhang YJ, Baker M, Rademakers R, Petrucelli L, Dickson DW (2019) C-terminal and full length TDP-43 specie differ according to FTLD-TDP lesion type but not genetic mutation. Acta Neuropathol Commun 7(1):100

68. Hasegawa M, Arai T, Nonaka T, Kametani F, Yoshida M, Hashizume Y et al (2008) Phosphorylated TDP-43 in frontotemporal lobar degeneration and amyotrophic lateral sclerosis. Ann Neurol 64(1):60–70

69. Kametani F, Nonaka T, Suzuki T, Arai T, Dohmae N, Akiyama H et al (2009) Identification of casein kinase-1 phosphorylation sites on TDP-43. Biochem Biophys Res Commun 382(2):405–409

70. Inukai Y, Nonaka T, Arai T, Yoshida M, Hashizume Y, Beach TG et al (2008) Abnormal phosphorylation of Ser409/410 of TDP-43 in FTLD-U and ALS. FEBS Lett 582(19):2899–2904

71. Tan RH, Shepherd CE, Kril JJ, McCann H, McGeachie A, McGinley C et al (2013) Classification of FTLD-TDP cases into pathological subtypes using antibodies against phosphorylated and non-phosphorylated TDP43. Acta Neuropathol Commun 1:33

72. Kim KY, Lee HW, Shim YM, Mook-Jung I, Jeon GS, Sung JJ (2015) A phosphomimetic mutant TDP-43 (S409/410E) induces Drosha instability and cytotoxicity in neuro 2A cells. Biochem Biophys Res Commun 464(1):236–243

73. Brady OA, Meng P, Zheng Y, Mao Y, Hu F (2011) Regulation of TDP-43 aggregation by phosphorylation and p62/SQSTM1. J Neurochem 116(2):248–259

74. Li HY, Yeh PA, Chiu HC, Tang CY, Tu BP (2011) Hyperphosphorylation as a defense mechanism to reduce TDP-43 aggregation. PLoS One 6(8):e23075

75. Seyfried NT, Gozal YM, Dammer EB, Xia Q, Duong DM, Cheng D et al (2010) Multiplex SILAC analysis of a cellular TDP-43 proteinopathy model reveals protein inclusions associated with SUMOylation and diverse polyubiquitin chains. Mol Cell Proteomics 9(4):705–718

76. Dammer EB, Fallini C, Gozal YM, Duong DM, Rossoll W, Xu P et al (2012) Coaggregation of RNA-binding proteins in a model of TDP-43 proteinopathy with selective RGG motif methylation and a role for RRM1 ubiquitination. PLoS One 7(6):e38658

77. Hans F, Eckert M, von Zweydorf F, Gloeckner CJ, Kahle PJ (2018) Identification and characterization of ubiquitinylation sites in TAR DNA-binding protein of 43 kDa (TDP-43). J Biol Chem 293(41):16083–16099

78. Kim W, Bennett EJ, Huttlin EL, Guo A, Li J, Possemato A et al (2011) Systematic and quantitative assessment of the ubiquitin-modified proteome. Mol Cell 44(2):325–340

79. Wagner SA, Beli P, Weinert BT, Nielsen ML, Cox J, Mann M et al (2011) A proteome-wide, quantitative survey of in vivo ubiquitylation sites reveals widespread regulatory roles. Mol Cell Proteomics 10(10):M111 013284

80. Kametani F, Obi T, Shishido T, Akatsu H, Murayama S, Saito Y et al (2016) Mass spectrometric analysis of accumulated TDP-43 in amyotrophic lateral sclerosis brains. Sci Rep 6:23281

81. Cohen TJ, Hwang AW, Restrepo CR, Yuan CX, Trojanowski JQ, Lee VM (2015) An acetylation switch controls TDP-43 function and aggregation propensity. Nat Commun 6:5845

82. Cohen TJ, Hwang AW, Unger T, Trojanowski JQ, Lee VM (2012) Redox signalling directly regulates TDP-43 via cysteine oxidation and disulphide cross-linking. EMBO J 31(5):1241–1252

83. Aguzzi A, Heikenwalder M, Polymenidou M (2007) Insights into prion strains and neurotoxicity. Nat Rev Mol Cell Biol 8(7):552–561

84. Kawakami I, Arai T, Hasegawa M (2019) The basis of clinicopathological heterogeneity in TDP-43 proteinopathy. Acta Neuropathol 138(5):751–770

85. Hasegawa M, Nonaka T, Tsuji H, Tamaoka A, Yamashita M, Kametani F et al (2011) Molecular dissection of TDP-43 proteinopathies. J Mol Neurosci 45(3):480–485

86. Nonaka T, Masuda-Suzukake M, Arai T, Hasegawa Y, Akatsu H, Obi T et al (2013) Prion-like properties of pathological TDP-43 aggregates from diseased brains. Cell Rep 4(1):124–134

87. Porta S, Xu Y, Restrepo CR, Kwong LK, Zhang B, Brown HJ et al (2018) Patient-derived frontotemporal lobar degeneration brain extracts induce formation and spreading of TDP-43 pathology in vivo. Nat Commun 9(1):4220

88. Davidson YS, Robinson AC, Flood L, Rollinson S, Benson BC, Asi YT et al (2017) Heterogeneous ribonuclear protein E2 (hnRNP E2) is associated with TDP-43-immunoreactive neurites in semantic dementia but not with other TDP-43 pathological subtypes of frontotemporal lobar degeneration. Acta Neuropathol Commun 5(1):54

89. Kattuah W, Rogelj B, King A, Shaw CE, Hortobagyi T, Troakes C (2019) Heterogeneous nuclear ribonucleoprotein E2 (hnRNP E2) is a component of TDP-43 aggregates specifically in the a and C pathological subtypes of frontotemporal lobar degeneration. Front Neurosci 13:551

Lysosomal Dysfunction and Other Pathomechanisms in FTLD: Evidence from Progranulin Genetics and Biology

Xiaolai Zhou, Thomas Kukar,
and Rosa Rademakers

Introduction

Frontotemporal lobar degeneration (FTLD) is a complex disease, characterized by progressive degeneration of frontal and temporal lobes and extensive neuroinflammation, which manifests with a range of clinical disorders and inevitably leads to death [1]. The most common clinical presentation is behavioral variant frontotemporal dementia (bvFTD) characterized by progressive deterioration of personality, social behavior with disinhibition, and cognition [2]. However, in other FTLD patients, language dysfunction in the form of primary progressive aphasia is the predominant feature [3]. FTLD spectrum disorders are a leading cause of early-onset dementia with most patients presenting first symptoms around 60 years of age; however, a range from 25 to

X. Zhou
Department of Neuroscience, Mayo Clinic,
Jacksonville, FL, USA
e-mail: Zhou.xiaolai@mayo.edu

T. Kukar
Department of Pharmacology and Chemical Biology,
Center for Neurodegenerative Disease, Emory
University School of Medicine, Atlanta, GA, USA
e-mail: tkukar@emory.edu

R. Rademakers (✉)
Department of Neuroscience, Mayo Clinic,
Jacksonville, FL, USA

VIB Center for Molecular Neurology, University of
Antwerp-CDE, Antwerp, Belgium
e-mail: rosa.rademakers@uantwerpen.vib.be

90 years has been reported [4]. Importantly, more than 40% of FTLD patients have a positive family history of FTLD or related neurodegenerative disorders, sometimes with an autosomal dominant pattern of inheritance, which speaks to the strong genetic component of the disease [5–7].

In 1998, mutations in the microtubule-associated protein tau gene (*MAPT*) were identified as the first genetic cause of FTLD in a set of families with bvFTD and parkinsonism [8–10]. The subsequent identification of several FTLD families that lacked mutations or rearrangements in *MAPT*, despite genetic linkage to the same chromosomal region, suggested the presence of another genetic cause for FTLD close to the *MAPT* locus on chromosome 17q21 [11]. Intriguingly, these families also had pathology distinct from the *MAPT* carriers: they showed pathological inclusions positive for ubiquitin but negative for the tau protein. This remained a conundrum in the field until 2006 when systematic sequencing of candidate genes in a 6 Mb critical region, defined by the linked families, led to the identification of heterozygous progranulin gene (*GRN*) mutations as the second cause of autosomal dominant FTLD [12, 13]. In the same year, the TAR DNA-binding protein 43 (TDP-43) was found to be the main component of the ubiquitin inclusions in the *GRN* families, and FTLD with TDP-43 pathology (FTLD-TDP) was discovered to be the most common type of FTLD pathology [14, 15]. We now know that *GRN*

© Springer Nature Switzerland AG 2021
B. Ghetti et al. (eds.), *Frontotemporal Dementias*, Advances in Experimental Medicine and Biology
1281, https://doi.org/10.1007/978-3-030-51140-1_14

mutation carriers always present with FTLD-TDP type A, a specific FTLD-TDP subtype defined based on the distribution, cellular localization, and shape of the TDP-43 inclusions [16].

Progranulin (PGRN), encoded by *GRN*, is a conserved 593-amino-acid secreted glycoprotein. It has an unusual structure with seven full-length and one half-length granulin domains connected by linker regions and can be proteolytically cleaved to release individual 6 kDa granulin peptides [17] (Fig. 1). Multiple proteases are able to generate granulins from PGRN including neutrophil elastase [18, 19], proteinase 3 (a neutrophil protease) [19], matrix metalloproteinase 12

(MMP-12) [20], MMP-14 [21], and a disintegrin and metalloproteinase with thrombospondin motifs 7 (ADAMTS-7) [22]. On the other hand, PGRN can be stabilized from proteolysis by secretory leukocyte protease inhibitor (SLPI). Notably, in vitro assays showed that the cleavage of PGRN by proteases does not always result in the release of solely 6–12 kDa granulin fragments; instead, multiple intermediate-sized granulin products are also produced [18–20].

PGRN is highly expressed in epithelial cells such as those in the intestinal crypt, skin, kidney, and reproductive tracts, as well as immune cells within the lymphoid tissue of the lung, gut, and

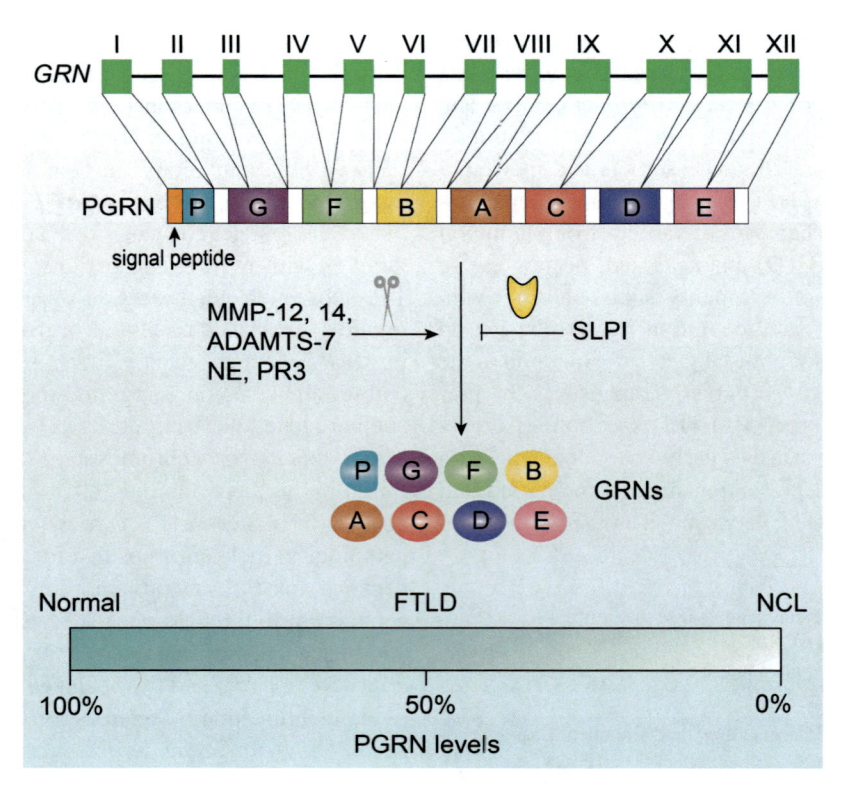

Fig. 1 PGRN, granulins, and associated disease phenotypes. Schematic of a part of the genomic structure of the progranulin gene (*GRN*) with 12 coding exons represented by green boxes. Following mRNA transcription and translation, the precursor protein progranulin (PGRN) is generated consisting of a signal-peptide, seven full-length granulin domains (granulins G, F, B, A, C, D, E) and one half-length granulin domain (granulin P). PGRN can be further cleaved by multiple enzymes to generate individual granulins. The cleavage of PGRN can be inhib-

ited through its binding to secretory leukocyte protease inhibitor (SLPI). Heterozygous loss-of-function *GRN* mutations which reduce PGRN levels to 50% of normal levels cause frontotemporal lobar degeneration (FTLD), whereas homozygous loss-of-function *GRN* mutations with no residual PGRN expression cause neuronal ceroid lipofuscinosis (NCL). *NE* neutrophil elastase; *PR3* proteinase 3; *MMP* matrix metalloproteinase; *ADAMTS-7* a disintegrin and metalloproteinase with thrombospondin motifs 7

spleen [23, 24]. In the brain, PGRN is mainly expressed in microglia and in different types of neurons including Purkinje cells, hippocampus pyramidal cells, and cerebral cortical neurons [23, 25, 26]. Both PGRN and granulins have been implicated in diverse functional processes. Specifically, early work focused on the role of PGRN in cell cycle progression and cell migration in a range of tissue remodeling processes including development, wound repair/inflammation, and tumorigenesis [27]. It was later determined that PGRN and one of the granulins, granulin E, exhibited neurotrophic properties [28] and most recently PGRN and granulins were shown to act as a key regulator of lysosomal health [29]. How *GRN* mutations affect these various biological processes and which mechanism is most important for the development and progression of FTLD remains an area of active investigation.

In this chapter, we briefly summarize the *GRN* mutational spectrum and its associated phenotypes, followed by an in-depth discussion on possible *GRN*-related disease mechanisms with emphasis on the recent evidence implicating PGRN and granulins in lysosomal function and dysfunction.

PGRN Mutational Spectrum and Associated Phenotypes

Heterozygous Loss-of-Function Mutations in *GRN* Cause FTLD

Through sequencing studies in FTLD and early-onset dementia populations, *GRN* mutations are now estimated to account for 5–20% of patients with a positive family history and 1–5% of apparently sporadic FTLD patients [30]. Mutations are mostly small insertions, deletions, or duplications affecting the *GRN* reading frame, splice-site mutations, or nonsense mutations, all leading to a premature termination codon and degradation of the mutant *GRN* mRNA transcript through nonsense-mediated decay. Larger partial or complete gene deletions have also been reported [31–33]. Mutations affecting the signal-peptide

sequence of PGRN, such as p.W7R and p.A9D, are also considered pathogenic because these mutants are unable to recruit the signal recognition particle, preventing secretion, leading to degradation of mutant *GRN* mRNA [34–36]. Two recent international studies summarized the different *GRN* mutations and number of families reported, showing at least 140 different loss-of-function mutations in more than 400 unrelated families (more families were reported by Moore et al., but genomic information was not used to determine cryptic relationships) [4, 37]. The most common mutation is c.813_816del (p.T272Sfs*10), with hundreds of affected patients from a founder population in Italy [38]. Other common mutations include c.1477C > T (p.R493*), c.709-1G > A (p.?), and c.26C > (p.A9D), all of which are more geographically distributed [4].

FTLD patients are heterozygous carriers; thus, a loss of 50% PGRN is the uniform consequence of all known pathogenic mutations resulting in PGRN haploinsufficiency. Because PGRN is a secreted protein, the reduction in PGRN can be detected in plasma or cerebrospinal fluid (CSF) samples from *GRN* mutation carriers and used as a diagnostic biomarker [39–42]. Together with in vitro functional assays, these PGRN measurements in human biofluids have also proven useful in the study of *GRN* missense variants which were identified through routine screening of FTLD patients but for which the pathogenicity is less obvious [35, 43–45]. For a select few missense mutations (including p.C105R, p.C139R, and p.C521Y affecting critical conserved cysteine residues), compelling evidence has now been gathered to support an effect on PGRN; however, most of these mutations do not completely eliminate PGRN expression and/or function and thus may represent FTLD risk factors rather than clear pathogenic mutations. The notion that a partial loss of PGRN (resulting in less than 100% but more than 50% remaining expression) could function as an FTLD risk factor is already demonstrated by rs5848, a common variant in the 3' untranslated region of *GRN* which was first described as a risk factor for FTLD-TDP in 2008 and was shown to partially reduce PGRN expression [46]. A highly significant association of this variant with risk to develop FTLD-TDP type A (indistinguishable from the pathology seen in *GRN* mutation carriers) was recently confirmed in a large international study [47]. Interestingly, this same variant has been implicated in other neurodegenerative disorders, including Alzheimer's disease (AD) and hippocampal sclerosis of aging, which may point to the fact that a partial loss of PGRN leads to a general increase in neurodegenerative disease risk [48–51].

Homozygous Loss-of-Function Mutations in *GRN* Cause Neuronal Ceroid Lipofuscinosis

Unexpectedly, homozygous loss-of-function mutations in *GRN* were reported in 2012 as the cause of neuronal ceroid lipofuscinosis (NCL) type 11 [52]. NCLs are neurodegenerative disorders characterized by the accumulation of abnormal lipopigment in lysosomes and clinical features of (usually) childhood-onset visual failure, cerebellar ataxia, seizures, and progressive decline in cognitive and motor functions [53]. The discovery of homozygous *GRN* mutations in patients with a lysosomal storage disorder marked a landmark finding providing novel and strong evidence for a functional role of PGRN within lysosomes. To date, eight different families with a total of 11 homozygous *GRN* mutation carriers have been reported (summarized in [54]). Strikingly, while most patients presented with classical NCL symptoms with a juvenile onset, three patients developed behavioral and cognitive symptoms that would allow the diagnosis of probable bvFTD [2], with one patient only developing symptoms at 56 years of age. This suggests that FTLD and NCL are extreme phenotypes on a spectrum with as yet unknown factors contributing to the phenotypic presentation. Residual expression of PGRN in homozygous *GRN* mutation carriers, as a result of hypomorphic variants that still synthesize some PGRN, may explain the bvFTD phenotype in some patients, but other factors likely play a role. Importantly, neuropathological examination in one patient homozygous for *GRN* mutations showed typical hallmarks of neuronal ceroid lipofuscinosis but no TDP-43 inclusions similar to those observed in FTLD [54].

Genetic Modifiers of FTLD-*GRN*

The large variability in age at disease onset among pathogenic *GRN* mutation carriers, even within single families [55], recently prompted an unbiased two-stage genome-wide association study using more than 400 patients from unrelated *GRN* families [37]. No genome-wide significant association with age at onset was identified. However, when symptomatic *GRN* carriers were compared to healthy individuals (in an attempt to identify possible protective factors), a genome-wide significant association was reported for genetic variants at the *TMEM106B* locus (rs1990622) and the *GFRA2* locus (rs36196656). These findings imply that even pathogenic *GRN* mutations are not fully penetrant and provide hope that TMEM106B-related and/or GFRA2-related pathways might be future targets for treatments for FTLD. The current biological knowledge on these candidate proteins in relation to PGRN is discussed in sections "PGRN Neurotrophic Receptors and Signaling Pathways" (GFRA2) and "Lessons from TMEM106B". (TMEM106B).

PGRN Deficiency Leads to a Loss of Neurotrophic Support

Neurotrophic Effect of PGRN and Granulins

Before the link of PGRN with FTLD, its function in cell growth had been extensively studied in the cancer biology field. Increased expression of PGRN was reported in several types of cancer including liver, breast, kidney, prostate, and ovarian cancer and was found to be associated with poor prognosis (for review, see [56, 57]). In vitro studies found PGRN functions as a growth factor. Treatment with PGRN induced cell proliferation

[58, 59] and prevented the apoptosis of tumor cells [18, 60–63]. In vivo, a reduction in PGRN expression greatly reduced tumor formation [64–67].

Prompted by the discovery of *GRN* mutations in FTLD patients, it was subsequently shown that PGRN was able to regulate survival and neurite outgrowth of different types of neurons. Primary cultured cortical and hippocampal neurons derived from *Grn*$^{-/-}$ mice showed deficits in neurite outgrowth and branching, significantly reduced neuronal survival, and increased caspase-mediated apoptosis [68, 69]. PGRN knockdown in NSC-34 motor neurons and human neural cells, differentiated from NHNP cells (a human neural progenitor cell line), also significantly reduced survival [70, 71], whereas PGRN-deficient hippocampal slices were susceptible to glucose deprivation [72]. On the other hand, either overexpression of PGRN or treatment with recombinant PGRN protein increased neurite outgrowth and the survival of primary cortical, hippocampal, and motor neurons [28, 70, 73]. Moreover, in vivo studies using zebrafish showed that PGRN knockdown decreased axonal outgrowth inducing motor neuron deficits, which could be rescued by overexpression of PGRN [74, 75]. Interestingly, overexpression of human PGRN mRNA also rescued human TDP-43-induced axon growth deficits in zebrafish [76].

It is known that PGRN can be cleaved into mature ~6 kDa granulin peptides as well as intermediate-length cleavage products (as mentioned in the introduction). Whereas the function of intermediate progranulin products in this context remains to be determined, diverse effects of granulins have been reported. Granulin A was shown to either induce cell growth or inhibit cell proliferation in different cell lines, while granulin B presented with inhibitory or antagonistic effects to granulin A [77–79]. Granulin D has been shown to regulate DNA synthesis in cultured astrocytes and glioblastoma cells [80]. Granulins C and E have also been shown to have neurotrophic properties. In hippocampal neurons, granulin C was shown to have comparable neurotrophic effects to granulin E [69], whereas in another study in primary motor neurons and

cortical neurons, granulin E but not granulin C had an effect [81]. Moreover, granulin AaE (equivalent to human granulin E in zebrafish) was shown to promote the survival of neuronal cells in zebrafish [82]. Interestingly, deletion of granulin E from PGRN completely abolished the neurotrophic effect of PGRN suggesting that granulin E may be the key domain or region involved in the neurotrophic effect of PGRN [82]. In line with these findings, inhibition of PGRN processing (by SLPI) abolishes PGRN-enhanced survival and neurite outgrowth in cortical neurons [28].

PGRN Neurotrophic Receptors and Signaling Pathways

In both cancer cells and primary neurons, PGRN has been shown to stimulate cell proliferation and promote cell survival through the activation of typical growth factor signal transduction pathways such as extracellular regulated kinase (ERK1/2) and the phosphatidylinositol-3 kinase (PI3K)/protein kinase B (Akt) cell survival pathways [59, 60, 68, 83–87]. One study revealed that PGRN treatment stimulated the phosphorylation of glycogen synthase kinase-3 beta (GSK-3β) in cultured neurons and knockdown of PGRN in SH-SY5Y cells impaired retinoic acid-induced differentiation and reduced the level of phosphorylated GSK-3β [73]. In addition, loss of PGRN in a human neural progenitor cell line led to an increase in Wnt/β-catenin signaling [71]. The involvement of a wide range of signaling cascades suggests PGRN might function through different neurotrophic receptors. However, thus far, the nature of the neurotrophic receptor(s) in the CNS remains unclear.

Sortilin (SORT1), a member of the vacuolar protein sorting 10 protein (VPS10P) domain receptor family [88], is one of the best-studied cell receptors for PGRN. Like PGRN, SORT1 is highly expressed in neurons in the frontal cortex, one of the most vulnerable brain regions in FTLD-*GRN*, and SORT1 also has a high binding affinity to PGRN [89]. SORT1 is known to be involved in the trafficking and signaling of several neurotrophins [90]. For instance, SORT1, forming a receptor complex with the common neurotrophin receptor (p75NTR), binds to the pro-form of nerve growth factor-β (proNGF) and triggers cell death signaling [91]. However, SORT1 solely functions as a sorting receptor for PGRN [89]. Indeed, multiple studies have shown the neurotrophic effect of PGRN and granulins is independent of SORT1. Either pharmacologic inhibition of the granulin E-SORT1 interaction or deletion of the SORT1 binding site of granulin E failed to abolish the neurotrophic function of granulin E [81]. In support of this notion, knockout or knockdown of SORT1 in mouse and zebrafish does not cause axonal outgrowth defects [81], and loss of SORT1 fails to abrogate the neurotrophic effect of PGRN in cultured neurons [69].

What about other candidate neurotrophic receptors? By using an unbiased antibody-based screen for differential tyrosine phosphorylation levels of 49 different human receptor tyrosine kinases, Neill et al. recently found that PGRN rapidly increased tyrosine phosphorylation of ephrin type A receptor 2 (EphA2) in a human urinary bladder carcinoma cell line [92]. PGRN binds to EphA2 with an affinity comparable to SORT1 (both of them are around the nanomolar range) [89]. PGRN binds to EphA2 on the cell surface and activates both mitogen-activated protein kinase and Akt and promotes capillary morphogenesis (Fig. 2). Separately, proteomic analysis of transgenic mice with inducible neuronal PGRN overexpression predicted activation of Notch signaling pathways in this model [93], and additional experiments confirmed that PGRN can bind to all four Notch receptors through the extracellular domain. PGRN also co-localized with Notch1 in primary dorsal root ganglia (DRG) neurons. Interestingly, upon nerve injury, the expression of *Hey1* and *Hes* (two Notch target genes) increased in *Grn* overexpression and decreased in *Grn*−/− mouse DRG neurons compared to wild-type mice. These findings indicate that PGRN can activate Notch and EphA2 in the peripheral nervous system. Given the established survival support roles of EphA2 and Notch, further studies are warranted to determine if basal levels of PGRN activate EphA2 or Notch in the brain.

Finally, we recently identified genetic variants at the *GFRA2* locus as novel modifiers of the disease risk in FTLD patients carrying a *GRN* mutation [37]. GDNF family receptor alpha 2 (GFRA2) is a member of the glial cell line-derived neurotrophic factor (GDNF) receptor family and is known to function as the preferred receptor for neurturin (NRTN) [94]. GFRA2 binds with NRTN and further recruits and activates a transmembrane tyrosine kinase receptor known as RET, which can activate the mitogen-activated protein kinase (MAPK) and Akt signaling pathways (Fig. 2). The risk haplotype at the *GFRA2* locus is associated with lower mRNA levels of *GFRA2* in brain tissue as compared to the protective haplotype. Moreover, we determined that PGRN can directly interact with GFRA2 [37]. Notably, GFRA2 is also abundantly expressed in different brain regions, especially in the frontal cortex [37], a vulnerable brain region in FTLD-*GRN*. While more studies are needed, this work suggests that GFRA2 could potentially function as a signaling receptor for PGRN in the CNS and upregulation of GFRA2 could be considered as a therapeutic strategy.

Role of Inflammation in FTLD-*GRN*

Overview of PGRN and Inflammation

Although PGRN is widely expressed throughout the body, the expression of PGRN is enriched in the spleen and cells of the hematopoietic lineage, supporting the idea that PGRN is involved in the function and maintenance of the immune system. In particular, *GRN* expression is enriched in monocytes, dendritic cells, and granulocytes within the blood and microglia in the brain [95]. Further, early work isolated and identified peptide fragments of PGRN from inflammatory cells leading to speculation that PGRN may be involved in inflammation and wound healing [96].

Subsequent studies have found increased levels of PGRN in tissue and biofluids from many types of inflammatory states and conditions, ranging from bacterial and viral infections [97–101], insulin resistance and type 2 diabetes [102–

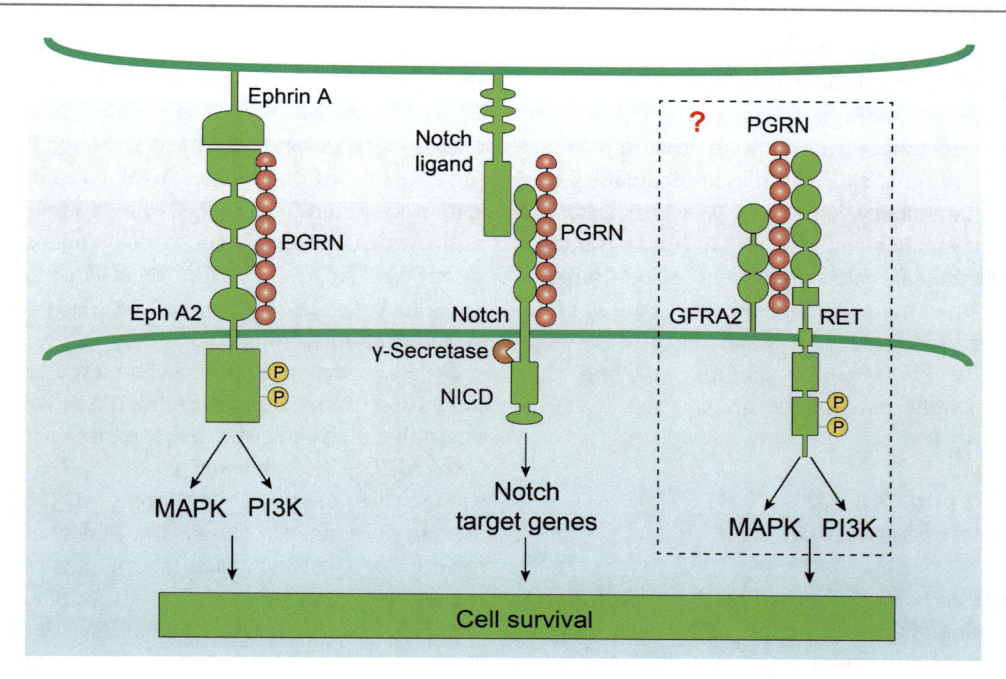

Fig. 2 Possible PGRN neurotrophic receptors and signaling pathways. Progranulin (PGRN) may bind to ephrin type A receptor 2 (EphA2) on the neuronal cell surface in the presence of ephrinA (from an adjacent cell) and activate its receptor tyrosine kinase, initiating neurotrophic signaling through MAPK/ERK and PI3K-Akt signaling pathways. PGRN might bind to Notch on the neuronal cell surface in the presence of Notch ligand (from adjacent cell) and trigger its cleavage to release the intracellular domain of the Notch protein (NICD), which then moves to the nucleus and increases the expression of genes involved in cell survival. PGRN might bind to GDNF family receptor alpha 2 (GFRA2) and further recruit and activate a transmembrane tyrosine kinase receptor RET, which can activate MAPK/ERK and PI3K-Akt signaling pathways. *MAPK* mitogen-activated protein kinase; *ERK* extracellular regulated kinase; *PI3K* phosphatidylinositol-3 kinase; *Akt* protein kinase B

104], liver dysfunction [105], and arthritis [106, 107]. Further, the *GRN* promoter has multiple binding sites for transcription factors related to inflammation, such as phorbol esters and multiple cytokines involved in the inflammatory response [108]. Indeed, treatment of murine embryo fibroblasts with interleukin-1 (IL-1) or tumor necrosis factor (TNF), two pro-inflammatory cytokines, leads to robust upregulation of *GRN* expression [109].

These observations led investigators to study whether the upregulation of PGRN was a consequence of inflammation or if PGRN can directly modulate inflammation. Work from the Bateman lab demonstrated that PGRN expression was upregulated during the wound response and PGRN likely functioned as a growth factor to facilitate wound healing [110]. In particular, they found that delivery of extracellular PGRN increased neutrophils, macrophages, fibroblasts, and blood vessel formation within the wound, leading them to conclude PGRN helped stimulate inflammation that is necessary for wound repair. Other labs, however, have found PGRN has the opposite effect in different models of inflammation. In 2002, PGRN (also called proepithelin or PEPI) was reported to have anti-inflammatory activity through blocking TNF activation of neutrophils [18]. In contrast, one of the granulins, granulin B (also called epithelin B), was pro-inflammatory in multiple assays. For example, granulin B inhibited proliferation of epithelial cells and induced the release of IL-8, a chemokine that attracts neutrophils. Work from Kessenbrock et al. found that application of recombinant PGRN can also reduced the influx of neutrophils following immune complex (IC)-stimulated inflammation in vivo [19].

Although some of these observations are conflicting, when considered together, these studies provide compelling evidence that PGRN expression is correlated with inflammation in different model systems and may play a modulatory role in inflammatory pathways. Nevertheless, the precise mechanism(s) how PGRN and granulins might mediate such pleiotropic effects on specific inflammatory cascades is much less clear. Future studies to understand the precise functions of PGRN and granulins will hopefully shine a light on these questions.

PGRN and Central Neuroinflammation

While early work focused on the role of PGRN on inflammation in peripheral tissues, the discovery of *GRN* mutations in FTLD prompted researchers to investigate whether PGRN was involved in inflammation in the central nervous system (CNS). PGRN is expressed in multiple neuronal populations as well as microglia throughout the brain [23, 111, 112], whereas astrocytes and oligodendrocytes do not appear to express PGRN at significant levels in vivo [113]. Multiple studies have discovered that the expression of PGRN by microglia is dramatically upregulated following an injury or other insults in animal models. For example, spinal cord injury [114], traumatic brain injury [115], injection of the neurotoxin quinolinic acid [116], or the endotoxin lipopolysaccharide [117] all activate microglia and lead to robust upregulation of *GRN* mRNA and PGRN protein. Most relevant to FTLD is the fact that PGRN levels are elevated across a wide variety of neurodegenerative diseases, which are often associated with neuroinflammation. Indeed, microglial PGRN itself is elevated in FTLD cases not caused by *GRN* mutations [118]. In contrast, FTLD cases with *GRN* mutations have reduced immunoreactivity for neuronal and microglia PGRN, supporting the idea that haploinsufficiency of PGRN and granulins extends to multiple cell types in the brain [119, 120]. In amyotrophic lateral sclerosis, PGRN expression is upregulated in the spinal cord, most likely due to activation of microglia [121]. Expression profiling of microglia from a mouse model of Creutzfeldt-Jakob disease (CJD) found increased levels of PGRN among a host of other genes involved in inflammation, interferon response, and complement pathways. Unbiased expression studies also identified changes in PGRN levels in multiple lysosomal storage disease models, likely driven by activated microglia [122].

Studies have also found increased PGRN expression in Alzheimer's disease (AD), which is especially relevant given the association of the *GRN* SNP rs5848 that decreases PGRN levels and increases the risk of developing AD [50, 51]. Increased immunoreactivity for PGRN in AD brain tissue labels activated microglia surrounding amyloid plaques as well as dystrophic neurites [13, 112, 113]. This observation extends to mouse models of AD where multiple groups have found increased levels of PGRN associated with amyloid plaques [123–125]. In clinical late-onset AD, higher CSF PGRN levels were associated with more advanced disease stages and cognitive impairment which was thought to reflect microglial activation during disease [126]. Importantly, a recent detailed immunohistochemical analysis of PGRN in AD brains replicated earlier findings that PGRN levels are increased with AD disease status but concluded that the increased PGRN signal is derived primarily from extracellular PGRN associated directly with amyloid beta (Aβ). Thus, further work is needed to understand the specific role of PGRN in AD pathology and pathogenesis [127].

Potential Mechanisms of Neuroinflammation in FILD-*GRN*

The clearest evidence that PGRN has an important role in central and peripheral inflammation comes from experiments examining the phenotype of *Grn*-deficient mice (reviewed in depth elsewhere) [57, 128]. Five unique *Grn*$^{-/-}$ and a novel *Grn*$^{R493X/R493X}$ knock-in mouse models have been developed thus far [72, 129–133]. All six models share a consistent age-dependent microgliosis and astrogliosis throughout the brain including the cortex, hippocampus, and thalamus [133–136]. Moreover, macrophages and microglia isolated from $Grn^{-/-}$ or $Grn^{R493X/R493X}$ mice have an exacerbated inflammatory response when challenged with pro-inflammatory molecules [137–140]. Further, loss of PGRN leads to microglial upregulation of multiple lysosomal genes, increased production of cytokines and complement, and enhanced synaptic pruning by microglia in $Grn^{-/-}$ mice [138]. The authors suggest that PGRN may normally function as a "brake" to suppress aberrant activation of microglia during aging by facilitating proper phagocytosis and lysosome function. Taken together, it is clear that PGRN plays an important role in decreasing, or modulating, neuroinflammation; however, the precise mechanism(s) by which this is accomplished still needs further investigation. Next, we will examine a few possible mechanisms that could help explain how PGRN may modulate neuroinflammation and how loss of PGRN function contributes to FTLD pathogenesis.

PGRN Anti-inflammatory Activity Through Signaling

A key unresolved question is whether PGRN has inherent anti-inflammatory activity. It is well established that administration of exogenous PGRN has pleiotropic effects in cells, some of which can be considered anti-inflammatory. These observations led many investigators to search for PGRN receptors that may mediate these effects. In 2002, PGRN was reported to decrease inflammation through inhibition of TNF signaling in neutrophils, likely downstream after TNF binding to its receptors [18]. Subsequent work supported an anti-inflammatory effect of PGRN, potentially mediated through the TNF pathway. Extracellular PGRN was found to moderately reduce secretion of IL-8 from human aortic smooth muscle cells induced by TNF treatment [141]. In 2011, Tang et al. reported that PGRN bound directly to the TNF receptors (TNFRSF1A and TNFRSF1A) and functioned as a TNF antagonist [142]. Intriguingly, PGRN bound to TNFRSF1A and TNFRSF1A with a higher affinity compared to TNF, the known ligand. Further,

recombinant PGRN was found to block TNF-induced inflammation in multiple cell culture assays and mouse models of TNF activity [142]. These results were initially greeted with excitement by the FTLD community because it might explain how decreased levels of PGRN lead to a pro-inflammatory state and ultimately set the stage for neurodegeneration. Unfortunately, following the original report, multiple pharmaceutical companies (*personal communication*) and academic labs have been unable to replicate the ability of PGRN to antagonize TNF binding or function [143–148]. The reasons for these discrepancies are unknown. Furthermore, from a broader perspective, it was never clear how an excess of TNF activity, theoretically caused by decreased levels of PGRN, could be the fundamental driver of either FTLD or NCL.

Although the potential anti-inflammatory activity of PGRN is fascinating, the preponderance of evidence suggests that it is unlikely to be mediated through antagonism of TNF activity. Alternatively, the binding of PGRN to other cell surface receptors may modulate inflammation. PGRN has been reported to bind directly, or indirectly, to a number of transmembrane receptors including delta homolog 1 (DLK1) [149], SORT1 [89], Toll-like receptor 9 (TLR9) [150], Notch1 [93], EphA2 [92], prosaposin (PSAP) mediated binding to the cation-independent mannose-6-phosphate receptor (M6PR), low-density lipoprotein receptor-related protein 1 (LRP1) [151], tyrosine-protein kinase receptor (Tyro3) [147], and GFRA2 [37]. Besides SORT1, most of these interactions have only been reported once and have not been extensively investigated, especially in the CNS. Moreover, the role of PGRN binding to any of these receptors and downstream effects on inflammation is speculative and needs to be validated and more thoroughly investigated.

The PGRN:Granulin Balance

Granulins are thought to be pro-inflammatory and could be another potential player in inflammation related to decreases in PGRN. The name "granulin" was derived from the fact that they were enriched in granules isolated from granulocytes and speculated to be cytokines [96].

Seminal work by Zhu et al. in 2002 found that PGRN could be cleaved in the extracellular space by elastase released from white blood cells, blocking PGRN's anti-inflammatory activity [18]. In contrast, the cleaved and released granulins reduced cell growth and produced a pro-inflammatory response. How might this observation be involved in FTLD? Some have speculated that the ratio of circulating PGRN to granulins is altered in FTD-*GRN* carriers leading to an imbalance, an increase in granulins, and subsequent inflammation. This idea has not been formally tested and awaits the development of antibodies specific to granulins, which is ongoing [120]. Although compelling, this hypothesis does not explain the even greater neurodegeneration and neuroinflammation that occurs in mice and humans completely deficient in PGRN and granulins [52, 54, 134, 152, 153]. Further, the precise pro-inflammatory mechanism for granulins, such as a signaling receptor, is unknown. Additionally, granulins have also been reported to have many beneficial effects, such as enhancing survival of motor neurons in culture [28], inducing neuronal outgrowth and branching [69], enhancing neuron survival and axon growth [154], and protecting retinal photoreceptor cell degeneration [155]. Finally, we, and other labs, have found that granulins are a common, endogenous protein produced in the lysosome of many cells [120, 156, 157]. This would suggest that granulins have a normal, homeostatic function inside the cell and aren't necessarily pro-inflammatory. In summary, granulins may play divergent roles depending on their location, and unraveling their function both inside of lysosomes and outside of the cell is an important focus of future research.

Contribution of Lysosomal Dysfunction in FTLD-*GRN*

Involvement of Lysosomal Dysfunction in FTLD-*GRN*

While the neurotrophic and anti-neuroinflammatory effects of PGRN have been well documented and studied for some time, recent evidence

suggested a previously unrecognized but important function of PGRN within lysosomes. Firstly, although PGRN is a secreted protein, its main localization within the cells is in lysosomes [89, 114]. Moreover, at the transcriptional level, *GRN* is co-regulated with other lysosomal genes such as cathepsin D (*CTSD*) by transcription factor EB (TFEB), a master regulator of lysosomal biogenesis [158]. Finally, as discussed in the section "Homozygous Loss-of-Function Mutations in *GRN* Cause Neuronal Ceroid Lipofuscinosis", a complete loss of PGRN due to homozygous loss-of-function *GRN* mutations leads to NCL, a lysosomal storage disease characterized by the accumulation of autofluorescent storage material (lipofuscin) [52]. Interestingly, lipofuscin accumulation is also consistently seen in the brains of different $Grn^{-/-}$ mouse lines [130, 132, 134, 159, 160]. Accumulation of ubiquitin [72, 130, 132, 134–136] and p62-positive [132, 161, 162] protein aggregates and increased levels of lysosomal proteins such as CTSD and LAMP1/2 [162–164] have also been detected in $Grn^{-/-}$ mice.

But what evidence suggests that even a partial loss of PGRN, such as is the case in FTLD-*GRN* patients, is sufficient to develop lysosomal dysfunction or NCL-like pathology? First, sphingolipid activator protein (saposin) D and subunit c of mitochondrial ATP synthase (SCMAS), two major protein components of lipofuscin [165, 166], are elevated in patients with FTLD-*GRN* [163]. In addition, preclinical retinal lipofuscinosis was detected in retinas of heterozygous loss-of-function *GRN* mutation carriers, and increased lipofuscinosis and intracellular NCL-like storage material also occurred in circulating lymphoblasts as well as postmortem cortex of these patients. Interestingly, the NCL-like pathological changes found in lymphoblasts from heterozygous *GRN* mutation carriers could be fully rescued by normalizing PGRN expression [167]. Similarly, FTLD-*GRN* patient induced pluripotent stem cell (iPSC)-derived cortical neurons have been shown to develop NCL-like pathologies including enlarged vesicles and lipofuscin accumulation [168]. Together, these findings strongly suggest that PGRN plays vital roles in lysosomes and dysfunction of the lysosomes due to (even a partial) loss of PGRN may be an important disease mechanism in FTLD-*GRN*. To provide further context to these recent developments, we will next summarize current insights into lysosomal trafficking of PGRN, its processing into granulins within lysosomes, and the functional evidence of the involvement of PGRN and granulins in the regulation of a growing list of lysosomal enzymes.

Lysosomal Trafficking of PGRN

Using an alkaline phosphatase-mediated cell surface binding assay, Hu et al. identified SORT1 as a high-affinity binding cell surface receptor for PGRN [89]. PGRN binds to the beta-propeller region of SORT1 in the extracellular domain through its last three amino acids (QLL) [89, 169]. SORT1 is a well-known sorting receptor [170]. The cytoplasmic tail of SORT1 encodes two sorting motifs, a tyrosine-based YSVL motif and an acidic dileucine cluster motif, which facilitate both endocytosis and intracellular trafficking of SORT1 [171]. Further studies found SORT1 functions as a sorting receptor for PGRN. It binds to PGRN both intracellularly and extracellularly and facilitates its lysosomal trafficking from the biosynthetic pathway and from the extracellular space [89] (Fig. 3). Overexpression of SORT1 reduces extracellular PGRN levels, whereas downregulation of SORT1 or abolishing the binding between PGRN and SORT1 increases extracellular PGRN levels [89, 169, 172, 173]. Interestingly, genetic knockout of *Sort1* in *Grn*[+/−] mice completely corrects PGRN serum levels from the haploinsufficiency state back to normal [89]. Notably, Carrasquillo et al. took a human genetic-based approach and performed a genome-wide association analysis of common genetic variants with human plasma PGRN levels. Genetic variants at the *SORT1* locus were found to be the most significantly associated with plasma PGRN levels, suggesting that differences in *SORT1* expression (predicted to result from these variants) also regulate extracellular levels of PGRN in vivo [173].

Importantly, while a complete loss of *Sort1* in mice leads to a robust accumulation of Pgrn in serum, a substantial amount of Pgrn (~50%) can still be detected in lysosomes in cortical neurons derived from these mice suggesting the existing of alternative lysosomal pathway(s) for PGRN, in addition to SORT1 [89]. Using an unbiased proteomic approach, Zhou et al. identified PSAP as a strong binding partner of PGRN [151]. Like PGRN, PSAP is also a secreted glycoprotein that is predominantly localized to lysosomes [174]. Similar to the binding to SORT1, PSAP binds to PGRN within the cell as well as in the extracellular space. Through further binding to two trafficking receptors (M6PR and LRP1), PSAP allows PGRN a piggyback ride and delivers it into lysosomes [151] (Fig. 3). Disruption of the binding of PGRN and PSAP completely abolishes the PSAP-mediated lysosomal trafficking of PGRN from both the biosynthetic pathway and extracellular space [175]. Loss of *Psap* in mice increases serum Pgrn level to a similar extent as loss of *Sort1* (~five- to sixfold) [89, 151]. Interestingly, Nicholson et al. independently demonstrated a physical interaction between PGRN and PSAP in functional follow-up experiments after genetic variants at the *PSAP* locus were found to be associated with human plasma PGRN levels [176].

Notably, PSAP-mediated lysosomal trafficking of PGRN is independent from the SORT1 pathway since loss of SORT1 failed to abolish PSAP-mediated lysosomal trafficking of PGRN. Moreover, the deficits of lysosomal trafficking of PGRN that resulted from the loss of PSAP can be fully rescued by overexpression of SORT1 [151]. Thus, the SORT1 and PSAP pathways are two complementary pathways that regulate the lysosomal trafficking of PGRN. The contribution that each of these two pathways plays in the lysosomal trafficking of PGRN might be determined by the expression levels or abundance of the pathway components in different types of tissues and different developmental stages. Indeed, SORT1 is almost undetectable in mouse fibroblasts, and as a consequence, the lysosomal trafficking is completely dependent

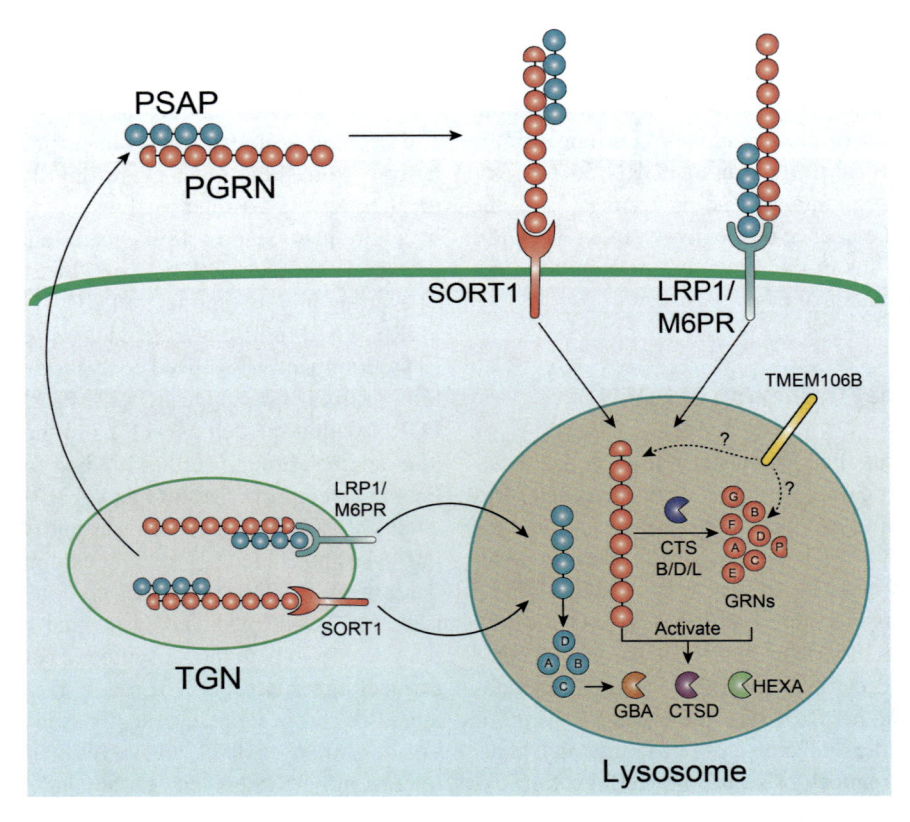

Fig. 3 Lysosomal trafficking and function of PGRN. Progranulin (PGRN) is targeted into lysosomes through either sortilin 1 (SORT1) or PSAP-LRP1/M6PR pathways from the extracellular space and trans-Golgi network (TGN). Lysosomal targeted PGRN is further processed into stable granulin peptides (GRNs) by lysosomal enzymes including cathepsin L (CTSL), cathepsin B (CTSB), and cathepsin D (CTSD). Recent work provided some first insights into the role of PGRN in lysosomes. It was suggested that PGRN may indirectly regulate lysosomal function by controlling the lysosomal trafficking and processing of prosaposin (PSAP) or may directly regulate lysosomal enzymes such as CTSD, glucocerebrosidase (GBA), and β-hexosaminidase A (HEXA) in concert with GRNs. TMEM106B might either directly or indirectly interact with PGRN and GRNs to co-regulate lysosomal function. *M6PR* mannose-6-phosphate receptor; *LRP1* low-density lipoprotein receptor-related protein 1

on PSAP, whereas in neurons where both SORT1 and PSAP/LRP1/M6PR are highly expressed, both pathways contribute to the lysosomal trafficking of PGRN. Taken together, these findings suggest SORT1 and PSAP pathways coordinate with each other to regulate the lysosomal trafficking of PGRN in a spatiotemporal-dependent manner. Further studies on examining lysosomal localization of PGRN in *Psap* and *Sort1* double-knockout mice will be important to reveal whether other pathway(s) might still exist.

Lysosomal Processing of PGRN

PGRN and PSAP share multiple biological features including lysosomal localization and trafficking mechanisms. In addition, PGRN and PSAP are both precursor proteins which can be further processed into a group of smaller mature functional proteins. Whereas PGRN had been shown to be proteolytically processed into seven and a half granulin peptides in the extracellular space [18], PSAP is proteolytically processed into four saposin peptides within lysosomes

[177]. Given these similarities, it was hypothesized that PGRN, like PSAP, could be processed into granulin peptides within lysosomes [178], which was eventually experimentally demonstrated by three independent groups in 2017 [120, 156, 157] (Fig. 3). Multiple lines of evidence were provided to support this important discovery: first, mature granulins were detected in multiple different cell lines including Hela, HEK293T, H4, SH-SY5Y, SW13, and primary fibroblasts as well as multiple different tissues such as brain, liver, spleen, kidney, and heart [120, 157]. Second, either disruption of the lysosomal trafficking of PGRN or inhibition of lysosomal activities abolished the generation of granulins suggesting that the intracellular processing of PGRN indeed occurs within lysosomes [120, 157]. Finally, in vitro cleavage assays directly showed multiple lysosomal proteases such as cathepsin L, B, and D are able to cleave PGRN into granulins [120, 156, 157]. The cleavage effects of these lysosomal proteases were further verified within cells by using different cathepsin knockout mouse fibroblasts [157]. Of note, in addition to stable granulins, multiple intermediate products, like di- or multi-granulin peptides, were also produced [120, 156, 157]. The observation of different cleavage patterns of PGRN upon incubation with different lysosomal proteases [157] suggested that the lysosomal processing of PGRN might require the coordination of different lysosomal proteases, further highlighting the complexity of PGRN processing and its regulation. Most important in the context of FTLD is the fact that haploinsufficiency of PGRN in $Grn^{+/-}$ mice and FTLD-GRN patients was shown to lead to a comparable reduction of the granulin peptides [120, 157]. Taken together, these findings clearly demonstrated that PGRN is converted into granulins within lysosomes. Further study toward the understanding of the function of individual granulins as well as its processing intermediates in lysosomes might ultimately reveal the disease mechanism of both NCL and FTLD caused by GRN mutation.

Lysosomal Function of PGRN

Effect on PSAP

In the section "Lysosomal Trafficking of PGRN", we described how PSAP binds to PGRN, thereby offering PGRN a way into lysosomes through PSAP receptors: LRP1 and M6PR [151]. Interestingly, the reverse also appears to take place. Specifically, PGRN can facilitate the lysosomal trafficking of PSAP through its receptor SORT1 [26]. Loss of Grn in mice compromised the neuronal lysosomal targeting of Psap leading to a reduction of neuronal lysosomal Psap and saposins and an increase of Psap in serum. Loss of $Sort1$ also leads to a comparable increase of Psap in serum as compared to what is seen upon Grn loss. Similarly, neuronal PSAP and saposins were found to be decreased in FTLD-GRN but not control individuals or FTLD patients with tau pathology. Furthermore, it is known that loss of PSAP or saposins can also cause lysosomal storage disease [177]. Moreover, loss of $Psap$ in mice results in FTLD-like behavioral phenotypes as well as FTLD-like pathologies including accumulation of phospho-TDP-43 (pTDP-43) and massive gliosis [26]. Together, these findings suggest that PGRN indirectly influences lysosomal function by controlling the lysosomal level of PSAP and saposins and that reduced levels of neuronal lysosomal PSAP and saposins due to the haploinsufficiency of PGRN may be a contributing factor in the development or progression of FTLD-GRN.

Effect on CTSD

Recently, multiple groups independently demonstrated a surprising role of PGRN and granulins in the regulation of cathepsin D (CTSD) enzymatic activity [167, 168, 180, 181]. CTSD is an important aspartyl protease responsible for the degradation of proteins in lysosomes. PGRN directly interacts with CTSD [168, 180, 181] and increases its enzymatic activity [168, 180]. Furthermore, granulin E is also able to bind CTSD [181], and co-incubation of granulin E with CTSD is sufficient to stimulate the proteolytic activity of CTSD in vitro [168, 180]. In support of the function of PGRN in CTSD activity regulation, loss of *Grn* in different mouse tissues such as brain, liver, and spleen [180, 181] as well as the partial loss of PGRN in human fibroblasts [167] and iPSC-derived human cortical neurons from FTLD-*GRN* patients [168] resulted in a reduction of CTSD activity. In fact, elevated CTSD protein level was detected in both postmortem brains of FTLD-*GRN* patients [163] and *Grn*-deficient mouse tissues [163, 180], likely due to a feedback loop resulting from the reduced CTSD activity. Most recently, in vitro experiments found PGRN may enhance the conversion of the CTSD precursor to mature CTSD in a concentration-dependent manner [182]. Combined with the fact that loss of *Ctsd* in mice was shown to induce pTDP-43 aggregates [163], these findings suggest that reduced CTSD activity due to PGRN haploinsufficiency might play a role in the development of FTLD-*GRN*.

Effect on GBA

An important association of PGRN with glucocerebrosidase (GBA), a lysosomal enzyme involved in the glucocerebroside degradation, has also been revealed [183–187]. Homozygous *GBA* mutations cause Gaucher disease (GD), a common lysosomal storage disease [188], whereas heterozygous *GBA* mutations are associated with Parkinson's disease and Lewy-body dementia [189]. Jian et al. recently reported an association of decreased serum PGRN levels with GD [183]. In the animal study, they showed that under challenging conditions such as ovalbumin-induced chronic inflammation or during aging, *Grn*$^{-/-}$ mice develop GD-like phenotypes, including typical Gaucher-like cells in lung, spleen, and bone marrow as well as GD-like lysosomal morphological changes [183]. Mechanistically, they speculated that loss of PGRN leads to disruption of lysosomal trafficking of GBA, but the enzymatic activity of GBA was not affected [183, 184]. Inconsistently, loss of *Grn* in mice has been shown to result in a significant reduction of GBA activity in multiple different tissues including liver, spleen, and brain [185]. The reduced GBA enzymatic activity has been further confirmed in

postmortem brains from FTLD-*GRN* patients [187]. Importantly, comparable amounts of GBA were detected in the lysosomal fractions of wild-type and *Grn*$^{-/-}$ mouse tissues strongly arguing that this *Grn* deficiency-mediated reduction in GBA activity is unlikely due to its lysosomal trafficking deficits [185]. Notably, although PGRN and granulins bind to GBA, addition of recombinant PGRN or granulins fails to increase GBA activity in vitro suggesting the contribution of indirect mechanisms. In this regard, it is known that saposins, the processing products from PSAP, positively regulate GBA activity [190]. CTSD is the major protease for PSAP processing [191], and its activity is further regulated by PGRN as described above [167, 168, 180, 181]. Thus, it is possible that PGRN regulates GBA activity through its control on the CTSD-PSAP-saposin axis. Indeed, a recent study showed loss of PGRN impairs the processing of PSAP to saposin C and the treatment of saposin C rescued the reduction of GBA activity in PGRN-deficient cells [186].

Effect on HexA

Most recently, PGRN has been associated with β-hexosaminidase A (HexA) [192], a lysosomal enzyme that is involved in GM2 ganglioside degradation. Loss of Hex A results in GM2 ganglioside accumulation, leading to Tay-Sachs disease (TSD), a typical lysosomal storage disease [193]. PGRN binds to HexA and increases the enzymatic activity and lysosomal delivery of HexA. Both aged and ovalbumin-challenged adult *Grn*-deficient mice were shown to have significant GM2 ganglioside accumulation and the appearance of typical TSD cells containing zebra bodies [192]. Treatment of either recombinant PGRN or Pcgin, an engineered PGRN derivative, reversed PGRN deficiency-induced lysosomal accumulation of GM2 ganglioside.

Lessons from TMEM106B

Genetic studies have clearly established *TMEM106B* variants as genetic modifiers of disease risk in *GRN* mutation carriers [37, 194, 195]. While the functional variant(s) responsible for the risk-modifying effect remain largely unknown, multiple studies have suggested that the "risk" haplotype is associated with higher levels of transmembrane protein 106B (TMEM106B) as compared to the "protective" haplotype [194, 196]. Mechanistically, it has been suggested that a noncoding variant could change the transcription of *TMEM106B* by altering chromatin architecture [196]; however, a role of the coding p.T185S variant cannot be excluded [197–199].

Given that TMEM106B is a type II transmembrane protein with its main intracellular localization at lysosomes [198, 200, 201], this genetic finding provides independent support for the important role of PGRN in lysosomes and its possible dysfunction in FTLD-*GRN*. Overexpression of TMEM106B in vitro results in multiple lysosomal dysfunctions, including enlarged lysosomal size, reduced lysosomal pH, decreased degradation capacity of endocytic cargo, and deficits of endolysosomal trafficking [198–202], eventually leading to cell death [203]. Elevated levels of TMEM106B have been found in postmortem brains of FTLD-TDP patients [200, 204] as well as brains of $Grn^{-/-}$ mice [160]. Furthermore, increased TMEM106B exacerbated

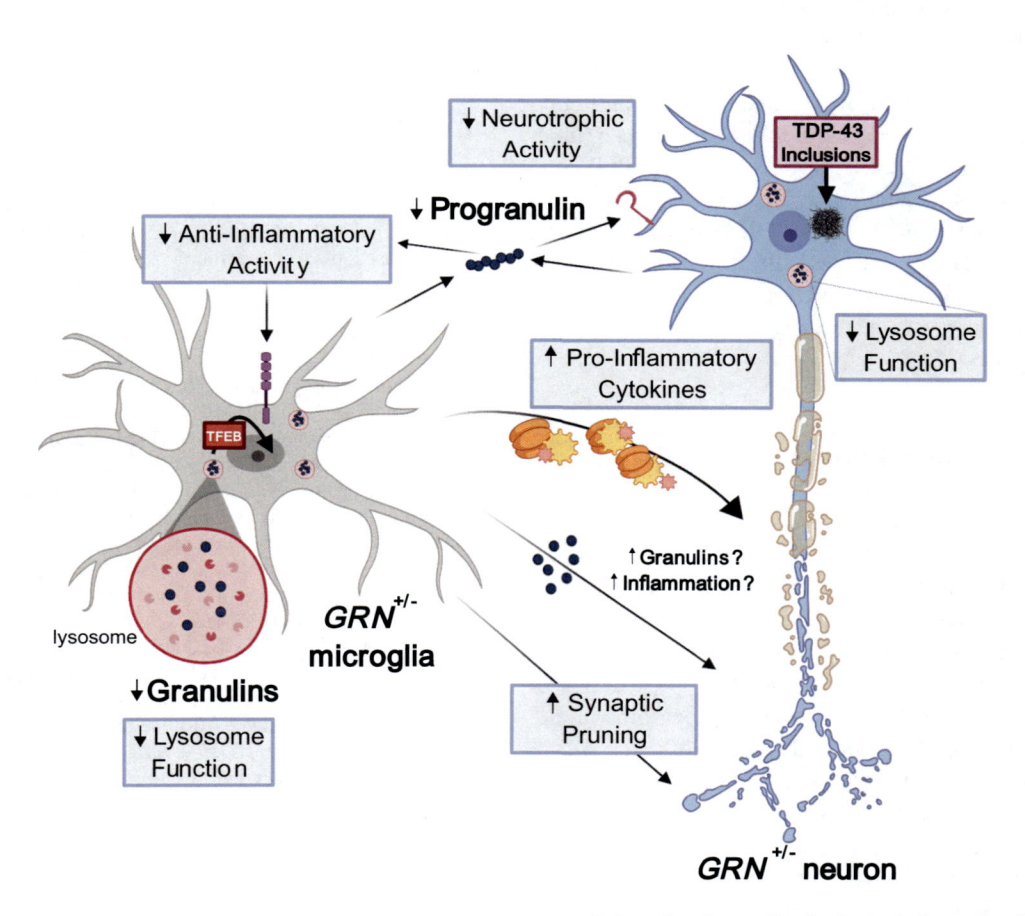

Fig. 4 Summary of potential disease mechanisms in FTLD-*GRN*. Progranulin (PGRN) is secreted by neurons and microglia in the brain, which can have beneficial activity through neurotrophic support and/or suppressing inflammation, both of which are decreased by PGRN haploinsufficiency. PGRN is normally trafficked to the lysosome in neurons and microglia, processed into granulins, which are thought to have a homeostatic function. The extracellular role of granulins in the brain is unclear but may increase inflammation. In neurons, decreased lysosomal function leads to defects in protein homeostasis and accumulation of ubiquitinated TDP-43 inclusions. In microglia, lysosome dysfunction can activate TFEB, which may exacerbate the release of pro-inflammatory cytokines and increase synaptic pruning, contributing to neuronal toxicity and degeneration

the FTLD-related pathologies such as lipofuscin and lysosome dysfunction in $Grn-/-$ brains at old age [160]. Intriguingly, loss of Tmem106b in $Grn^{-/-}$ mice was shown to ameliorate both the lysosomal and FTLD-related phenotypes in young $Grn^{-/-}$ mice [205]. Other studies, however, failed to observe noticeable benefits from the loss of Tmem106b in heterozygous $Grn^{+/-}$ mice and in C9orf72-repeat overexpressing mice, a mouse model for another type of FTD-TDP, where genetic studies had also identified a disease-modifying effect for *TMEM106B* haplotypes [206, 207]. Moreover, our unpublished work shows loss of *Tmem106b* results in myelination deficits and further loss of *Tmem106b* in $Grn^{-/-}$ mice exacerbates its FTLD-related pathologies leading to severe motor deficits (*personal communication*). These studies underscore a functional interaction between TMEM106B and PGRN, but additional mechanistic insight into the biology of either one of these proteins remains to be learned. They also illustrate that more work is needed before lowering TMEM106B can be considered as a therapeutic strategy in *GRN* carriers.

Concluding Remarks

Almost 14 years after the initial discovery of *GRN* mutations in FTLD patients, important new insights into its function and dysfunction have emerged. At least three independent disease mechanisms have been proposed to contribute to the development of FTLD-*GRN*: a loss of neurotrophic support, an increase in neuroinflammation, and lysosomal dysfunction (Fig. 4). In individual patients, a combination of these pathways may well be involved, potentially modified by additional genetic and/or environmental factors. In parallel to this increase in knowledge, global efforts have emerged to prepare the field for PGRN-related clinical trials by focusing on the identification of cohorts of mutation carriers and the development of robust biomarkers of disease onset and progression [208, 209]. It is the hope that the significant progress in this field will lay the foundation for the future development of

successful therapies for FTLD-*GRN*. However, until then, key outstanding questions remain to be answered in relation to the normal function of PGRN and its role in disease, including but not limited to: (1) What is the lysosomal function of PGRN and/or granulins? (2) Which receptors are most critical for the neurotrophic and inflammatory activities of PGRN and granulins? (3) How does the PGRN-granulin balance affect disease development or progression? (4) What is the functional interaction between PGRN and granulins and TMEM106B within lysosomes? Future studies should focus on these important topics.

References

1. Graff-Radford NR, Woodruff BK (2007) Frontotemporal dementia. Semin Neurol 27(1):48–57
2. Rascovsky K, Hodges JR, Knopman D, Mendez MF, Kramer JH, Neuhaus J et al (2011) Sensitivity of revised diagnostic criteria for the behavioural variant of frontotemporal dementia. Brain 134(Pt 9):2456–2477
3. Gorno-Tempini ML, Hillis AE, Weintraub S, Kertesz A, Mendez M, Cappa SF et al (2011) Classification of primary progressive aphasia and its variants. Neurology 76(11):1006–1014
4. Moore KM, Nicholas J, Grossman M, McMillan CT, Irwin DJ, Massimo L et al (2020) Age at symptom onset and death and disease duration in genetic frontotemporal dementia: an international retrospective cohort study. Lancet Neurol 19(2):145–156
5. Ratnavalli E, Brayne C, Dawson K, Hodges JR (2002) The prevalence of frontotemporal dementia. Neurology 58(11):1615–1621
6. Rosso SM, Donker Kaat L, Baks T, Joosse M, de Koning I, Pijnenburg Y et al (2003) Frontotemporal dementia in The Netherlands: patient characteristics and prevalence estimates from a population-based study. Brain 126(Pt 9):2016–2022
7. Goldman JS, Farmer JM, Wood EM, Johnson JK, Boxer A, Neuhaus J et al (2005) Comparison of family histories in FTLD subtypes and related tauopathies. Neurology 65(11):1817–1819
8. Hutton M, Lendon CL, Rizzu P, Baker M, Froelich S, Houlden H et al (1998) Association of missense and 5'-splice-site mutations in tau with the inherited dementia FTDP-17. Nature 393(6686):702–705
9. Poorkaj P, Bird TD, Wijsman E, Nemens E, Garruto RM, Anderson L et al (1998) Tau is a candidate gene for chromosome 17 frontotemporal dementia. Ann Neurol 43(6):815–825

10. Spillantini MG, Murrell JR, Goedert M, Farlow MR, Klug A, Ghetti B (1998) Mutation in the tau gene in familial multiple system tauopathy with presenile dementia. Proc Natl Acad Sci U S A 95(13):7737–7741

11. Rademakers R, Cruts M, Dermaut B, Sleegers K, Rosso SM, Van den Broeck M et al (2002) Tau negative frontal lobe dementia at 17q21: significant fine-mapping of the candidate region to a 4.8 cM interval. Mol Psychiatry 7(10):1064–1074

12. Cruts M, Gijselinck I, van der Zee J, Engelborghs S, Wils H, Pirici D et al (2006) Null mutations in progranulin cause ubiquitin-positive frontotemporal dementia linked to chromosome 17q21. Nature 442(7105):920–924

13. Baker M, Mackenzie IR, Pickering-Brown SM, Gass J, Rademakers R, Lindholm C et al (2006) Mutations in progranulin cause tau-negative frontotemporal dementia linked to chromosome 17. Nature 442(7105):916–919

14. Neumann M, Sampathu DM, Kwong LK, Truax AC, Micsenyi MC, Chou TT et al (2006) Ubiquitinated TDP-43 in frontotemporal lobar degeneration and amyotrophic lateral sclerosis. Science 314(5796):130–133

15. Arai T, Hasegawa M, Akiyama H, Ikeda K, Nonaka T, Mori H et al (2006) TDP-43 is a component of ubiquitin-positive tau-negative inclusions in frontotemporal lobar degeneration and amyotrophic lateral sclerosis. Biochem Biophys Res Commun 351(3):602–611

16. Mackenzie IR, Neumann M, Baborie A, Sampathu DM, Du Plessis D, Jaros E et al (2011) A harmonized classification system for FTLD-TDP pathology. Acta Neuropathol 122(1):111–113

17. Bateman A, Bennett HP (1998) Granulins: the structure and function of an emerging family of growth factors. J Endocrinol 158(2):145–151

18. Zhu J, Nathan C, Jin W, Sim D, Ashcroft GS, Wahl SM et al (2002) Conversion of proepithelin to epithelins: roles of SLPI and elastase in host defense and wound repair. Cell 111(6):867–878

19. Kessenbrock K, Frohlich L, Sixt M, Lammermann T, Pfister H, Bateman A et al (2008) Proteinase 3 and neutrophil elastase enhance inflammation in mice by inactivating antiinflammatory progranulin. J Clin Invest 118(7):2438–2447

20. Suh HS, Choi N, Tarassishin L, Lee SC (2012) Regulation of progranulin expression in human microglia and proteolysis of progranulin by matrix metalloproteinase-12 (MMP-12). PLoS One 7(4):e35115

21. Butler GS, Dean RA, Tam EM, Overall CM (2008) Pharmacoproteomics of a metalloproteinase hydroxamate inhibitor in breast cancer cells: dynamics of membrane type 1 matrix metalloproteinase-mediated membrane protein shedding. Mol Cell Biol 28(15):4896–4914

22. Bai XH, Wang DW, Kong L, Zhang Y, Luan Y, Kobayashi T et al (2009) ADAMTS-7, a direct target of PTHrP, adversely regulates endochondral bone growth by associating with and inactivating GEP growth factor. Mol Cell Biol 29(15):4201–4219

23. Daniel R, He Z, Carmichael KP, Halper J, Bateman A (2000) Cellular localization of gene expression for progranulin. J Histochem Cytochem 48(7):999–1009

24. Daniel R, Daniels E, He Z, Bateman A (2003) Progranulin (acrogranin/PC cell-derived growth factor/granulin-epithelin precursor) is expressed in the placenta, epidermis, microvasculature, and brain during murine development. Dev Dyn 227(4):593–599

25. Zhang Y, Chen K, Sloan SA, Bennett ML, Scholze AR, O'Keeffe S et al (2014) An RNA-sequencing transcriptome and splicing database of glia, neurons, and vascular cells of the cerebral cortex. J Neurosci 34(36):11929–11947

26. Zhou X, Sun L, Bracko O, Choi JW, Jia Y, Nana AL et al (2017) Impaired prosaposin lysosomal trafficking in frontotemporal lobar degeneration due to progranulin mutations. Nat Commun 8:15277

27. Bateman A, Bennett HP (2009) The granulin gene family: from cancer to dementia. BioEssays 31(11):1245–1254

28. Van Damme P, Van Hoecke A, Lambrechts D, Vanacker P, Bogaert E, van Swieten J et al (2008) Progranulin functions as a neurotrophic factor to regulate neurite outgrowth and enhance neuronal survival. J Cell Biol 181(1):37–41

29. Paushter DH, Du H, Feng T, Hu F (2018) The lysosomal function of progranulin, a guardian against neurodegeneration. Acta Neuropathol 136(1):1–17

30. Rademakers R, Neumann M, Mackenzie IR (2012) Advances in understanding the molecular basis of frontotemporal dementia. Nat Rev Neurol 8(8):423–434

31. Clot F, Rovelet-Lecrux A, Lamari F, Noel S, Keren B, Camuzat A et al (2014) Partial deletions of the GRN gene are a cause of frontotemporal lobar degeneration. Neurogenetics 15(2):95–100

32. Rovelet-Lecrux A, Deramecourt V, Legallic S, Maurage CA, Le Ber I, Brice A et al (2008) Deletion of the progranulin gene in patients with frontotemporal lobar degeneration or Parkinson disease. Neurobiol Dis 31(1):41–45

33. Gijselinck I, van der Zee J, Engelborghs S, Goossens D, Peeters K, Mattheijssens M et al (2008) Progranulin locus deletion in frontotemporal dementia. Hum Mutat 29(1):53–58

34. Pinarbasi ES, Karamyshev AL, Tikhonova EB, Wu IH, Hudson H, Thomas PJ (2018) Pathogenic signal sequence mutations in progranulin disrupt SRP interactions required for mRNA stability. Cell Rep 23(10):2844–2851

35. Shankaran SS, Capell A, Hruscha AT, Fellerer K, Neumann M, Schmid B et al (2008) Missense mutations in the progranulin gene linked to frontotemporal lobar degeneration with ubiquitin-immunoreactive inclusions reduce progranulin production and secretion. J Biol Chem 283(3):1744–1753

36. Saracino D, Sellami L, Clot F, Camuzat A, Lamari F, Rucheton B et al (2020) The missense p.Trp7Arg mutation in GRN gene leads to progranulin haploinsufficiency. Neurobiol Aging 85:154 e9–154e11

37. Pottier C, Zhou X, Perkerson RB 3rd, Baker M, Jenkins GD, Serie DJ et al (2018) Potential genetic modifiers of disease risk and age at onset in patients with frontotemporal lobar degeneration and GRN mutations: a genome-wide association study. Lancet Neurol 17(6):548–558

38. Benussi L, Rademakers R, Rutherford NJ, Wojtas A, Glionna M, Paterlini A et al (2013) Estimating the age of the most common Italian GRN mutation: walking back to Canossa times. J Alzheimers Dis 33(1):69–76

39. Ghidoni R, Benussi L, Glionna M, Franzoni M, Binetti G (2008) Low plasma progranulin levels predict progranulin mutations in frontotemporal lobar degeneration. Neurology 71(16):1235–1239

40. Finch N, Baker M, Crook R, Swanson K, Kuntz K, Surtees R et al (2009) Plasma progranulin levels predict progranulin mutation status in frontotemporal dementia patients and asymptomatic family members. Brain 132(Pt 3):583–591

41. Sleegers K, Brouwers N, Van Damme P, Engelborghs S, Gijselinck I, van der Zee J et al (2009) Serum biomarker for progranulin-associated frontotemporal lobar degeneration. Ann Neurol 65(5):603–609

42. Galimberti D, Fumagalli GG, Fenoglio C, Cioffi SMG, Arighi A, Serpente M et al (2018) Progranulin plasma levels predict the presence of GRN mutations in asymptomatic subjects and do not correlate with brain atrophy: results from the GENFI study. Neurobiol Aging 62:245 e9–245e12

43. Karch CM, Ezerskiy L, Redaelli V, Giovagnoli AR, Tiraboschi P, Pelliccioni G et al (2016) Missense mutations in progranulin gene associated with frontotemporal lobar degeneration: study of pathogenetic features. Neurobiol Aging 38:215 e1–215e12

44. Wang J, Van Damme P, Cruchaga C, Gitcho MA, Vidal JM, Seijo-Martinez M et al (2010) Pathogenic cysteine mutations affect progranulin function and production of mature granulins. J Neurochem 112(5):1305–1315

45. Kleinberger G, Capell A, Brouwers N, Fellerer K, Sleegers K, Cruts M et al (2016) Reduced secretion and altered proteolytic processing caused by missense mutations in progranulin. Neurobiol Aging 39:220 e17–220 e26

46. Rademakers R, Eriksen JL, Baker M, Robinson T, Ahmed Z, Lincoln SJ et al (2008) Common variation in the miR-659 binding-site of GRN is a major risk factor for TDP43-positive frontotemporal dementia. Hum Mol Genet 17(23):3631–3642

47. Pottier C, Ren Y, Perkerson RB 3rd, Baker M, Jenkins GD, van Blitterswijk M et al (2019) Genome-wide analyses as part of the international FTLD-TDP whole-genome sequencing consortium reveals novel disease risk factors and increases support for immune dysfunction in FTLD. Acta Neuropathol 137(6):879–899

48. Hokkanen SRK, Kero M, Kaivola K, Hunter S, Keage HAD, Kiviharju A et al (2019) Putative risk alleles for LATE-NC with hippocampal sclerosis in population-representative autopsy cohorts. Brain Pathol 30(2):364–372

49. Nho K, Saykin AJ, Alzheimer's Disease Neuroimaging I, Nelson PT (2016) Hippocampal sclerosis of aging, a common Alzheimer's disease 'mimic': risk genotypes are associated with brain atrophy outside the temporal lobe. J Alzheimers Dis 52(1):373–383

50. Sheng J, Su L, Xu Z, Chen G (2014) Progranulin polymorphism rs5848 is associated with increased risk of Alzheimer's disease. Gene 542(2):141–145

51. Xu HM, Tan L, Wan Y, Tan MS, Zhang W, Zheng ZJ et al (2017) PGRN is associated with late-onset Alzheimer's disease: a case-control replication study and meta-analysis. Mol Neurobiol 54(2):1187–1195

52. Smith KR, Damiano J, Franceschetti S, Carpenter S, Canafoglia L, Morbin M et al (2012) Strikingly different clinicopathological phenotypes determined by progranulin-mutation dosage. Am J Hum Genet 90(6):1102–1107

53. Mole SE, Anderson G, Band HA, Berkovic SF, Cooper JD, Kleine Holthaus SM et al (2019) Clinical challenges and future therapeutic approaches for neuronal ceroid lipofuscinosis. Lancet Neurol 18(1):107–116

54. Huin V, Barbier M, Bottani A, Lobrinus JA, Clot F, Lamari F et al (2020) Homozygous GRN mutations: new phenotypes and new insights into pathological and molecular mechanisms. Brain 143(1):303–319

55. Rademakers R, Baker M, Gass J, Adamson J, Huey ED, Momeni P et al (2007) Phenotypic variability associated with progranulin haploinsufficiency in patients with the common 1477C-->T (Arg493X) mutation: an international initiative. Lancet Neurol 6(10):857–868

56. Toh H, Chitramuthu BP, Bennett HP, Bateman A (2011) Structure, function, and mechanism of progranulin; the brain and beyond. J Mol Neurosci 45(3):538–548

57. Chitramuthu BP, Bennett HPJ, Bateman A (2017) Progranulin: a new avenue towards the understanding and treatment of neurodegenerative disease. Brain 140(12):3081–3104

58. He Z, Bateman A (1999) Progranulin gene expression regulates epithelial cell growth and promotes tumor growth in vivo. Cancer Res 59(13):3222–3229

59. Lu R, Serrero G (2001) Mediation of estrogen mitogenic effect in human breast cancer MCF-7 cells by PC-cell-derived growth factor (PCDGF/granulin precursor). Proc Natl Acad Sci U S A 98(1):142–147

60. He Z, Ismail A, Kriazhev L, Sadvakassova G, Bateman A (2002) Progranulin (PC-cell-derived growth factor/acrogranin) regulates invasion and cell survival. Cancer Res 62(19):5590–5596

61. Tangkeangsirisin W, Hayashi J, Serrero G (2004) PC cell-derived growth factor mediates tamoxifen resistance and promotes tumor growth of human breast cancer cells. Cancer Res 64(5):1737–1743

62. Kim WE, Serrero G (2006) PC cell-derived growth factor stimulates proliferation and confers Trastuzumab resistance to Her-2-overexpressing breast cancer cells. Clin Cancer Res 12(14 Pt 1):4192–4199

63. Pizarro GO, Zhou XC, Koch A, Gharib M, Raval S, Bible K et al (2007) Prosurvival function of the granulin-epithelin precursor is important in tumor progression and chemoresponse. Int J Cancer 120(11):2339–2343

64. Zhang H, Serrero G (1998) Inhibition of tumorigenicity of the teratoma PC cell line by transfection with antisense cDNA for PC cell-derived growth factor (PCDGF, epithelin/granulin precursor). Proc Natl Acad Sci U S A 95(24):14202–14207

65. Lu R, Serrero G (2000) Inhibition of PC cell-derived growth factor (PCDGF, epithelin/granulin precursor) expression by antisense PCDGF cDNA transfection inhibits tumorigenicity of the human breast carcinoma cell line MDA-MB-468. Proc Natl Acad Sci U S A 97(8):3993–3998

66. Chen XY, Li JS, Liang QP, He DZ, Zhao J (2008) Expression of PC cell-derived growth factor and vascular endothelial growth factor in esophageal squamous cell carcinoma and their clinicopathologic significance. Chin Med J 121(10):881–886

67. Cheung ST, Wong SY, Leung KL, Chen X, So S, Ng IO et al (2004) Granulin-epithelin precursor overexpression promotes growth and invasion of hepatocellular carcinoma. Clin Cancer Res 10(22):7629–7636

68. Kleinberger G, Wils H, Ponsaerts P, Joris G, Timmermans JP, Van Broeckhoven C et al (2010) Increased caspase activation and decreased TDP-43 solubility in progranulin knockout cortical cultures. J Neurochem 115(3):735–747

69. Gass J, Lee WC, Cook C, Finch N, Stetler C, Jansen-West K et al (2012) Progranulin regulates neuronal outgrowth independent of sortilin. Mol Neurodegener 7:33

70. Ryan CL, Baranowski DC, Chitramuthu BP, Malik S, Li Z, Cao M et al (2009) Progranulin is expressed within motor neurons and promotes neuronal cell survival. BMC Neurosci 10:130

71. Rosen EY, Wexler EM, Versano R, Coppola G, Gao F, Winden KD et al (2011) Functional genomic analyses identify pathways dysregulated by progranulin deficiency, implicating Wnt signaling. Neuron 71(6):1030–1042

72. Yin F, Banerjee R, Thomas B, Zhou P, Qian L, Jia T et al (2010) Exaggerated inflammation, impaired host defense, and neuropathology in progranulin-deficient mice. J Exp Med 207(1):117–128

73. Gao X, Joselin AP, Wang L, Kar A, Ray P, Bateman A et al (2010) Progranulin promotes neurite outgrowth and neuronal differentiation by regulating GSK-3beta. Protein Cell 1(6):552–562

74. Chitramuthu BP, Baranowski DC, Kay DG, Bateman A, Bennett HP (2010) Progranulin modulates zebrafish motoneuron development in vivo and rescues truncation defects associated with knockdown of survival motor neuron 1. Mol Neurodegener 5:41

75. Laird AS, Van Hoecke A, De Muynck L, Timmers M, Van den Bosch L, Van Damme P et al (2010) Progranulin is neurotrophic in vivo and protects against a mutant TDP-43 induced axonopathy. PLoS One 5(10):e13368

76. Chitramuthu BP, Kay DG, Bateman A, Bennett HP (2017) Neurotrophic effects of progranulin in vivo in reversing motor neuron defects caused by over or under expression of TDP-43 or FUS. PLoS One 12(3):e0174784

77. Shoyab M, McDonald VL, Byles C, Todaro GJ, Plowman GD (1990) Epithelins 1 and 2: isolation and characterization of two cysteine-rich growth-modulating proteins. Proc Natl Acad Sci U S A 87(20):7912–7916

78. Plowman GD, Green JM, Neubauer MG, Buckley SD, McDonald VL, Todaro GJ et al (1992) The epithelin precursor encodes two proteins with opposing activities on epithelial cell growth. J Biol Chem 267(18):13073–13078

79. Culouscou JM, Carlton GW, Shoyab M (1993) Biochemical analysis of the epithelin receptor. J Biol Chem 268(14):10458–10462

80. Liau LM, Lallone RL, Seitz RS, Buznikov A, Gregg JP, Kornblum HI et al (2000) Identification of a human glioma-associated growth factor gene, granulin, using differential immuno-absorption. Cancer Res 60(5):1353–1360

81. De Muynck L, Herdewyn S, Beel S, Scheveneels W, Van Den Bosch L, Robberecht W et al (2013) The neurotrophic properties of progranulin depend on the granulin E domain but do not require sortilin binding. Neurobiol Aging 34(11):2541–2547

82. Wang P, Chitramuthu B, Bateman A, Bennett HPJ, Xu P, Ni F (2018) Structure dissection of zebrafish progranulins identifies a well-folded granulin/epithelin module protein with pro-cell survival activities. Protein Sci 27(8):1476–1490

83. Feng JQ, Guo FJ, Jiang BC, Zhang Y, Frenkel S, Wang DW et al (2010) Granulin epithelin precursor: a bone morphogenic protein 2-inducible growth factor that activates Erk1/2 signaling and JunB transcription factor in chondrogenesis. FASEB J 24(6):1879–1892

84. Zanocco-Marani T, Bateman A, Romano G, Valentinis B, He ZH, Baserga R (1999) Biological activities and signaling pathways of the granulin/epithelin precursor. Cancer Res 59(20):5331–5340

85. Monami G, Gonzalez EM, Hellman M, Gomella LG, Baffa R, Iozzo RV et al (2006) Proepithelin promotes migration and invasion of 5637 bladder cancer cells through the activation of ERK1/2 and the formation of a paxillin/FAK/ERK complex. Cancer Res 66(14):7103–7110

86. Ong CH, Bateman A (2003) Progranulin (granulin-epithelin precursor, PC-cell derived growth factor, acrogranin) in proliferation and tumorigenesis. Histol Histopathol 18(4):1275–1288

87. Xu J, Xilouri M, Bruban J, Shioi J, Shao Z, Papazoglou I et al (2011) Extracellular progranulin protects cortical neurons from toxic insults by activating survival signaling. Neurobiol Aging 32(12):2326 e5–2326 16

88. Jansen P, Giehl K, Nyengaard JR, Teng K, Lioubinski O, Sjoegaard SS et al (2007) Roles for the pro-neurotrophin receptor sortilin in neuronal development, aging and brain injury. Nat Neurosci 10(11):1449–1457

89. Hu F, Padukkavidana T, Vaegter CB, Brady OA, Zheng Y, Mackenzie IR et al (2010) Sortilin-mediated endocytosis determines levels of the frontotemporal dementia protein, progranulin. Neuron 68(4):654–667

90. Nykjaer A, Willnow TE (2012) Sortilin: a receptor to regulate neuronal viability and function. Trends Neurosci 35(4):261–270

91. Nykjaer A, Lee R, Teng KK, Jansen P, Madsen P, Nielsen MS et al (2004) Sortilin is essential for proNGF-induced neuronal cell death. Nature 427(6977):843–848

92. Neill T, Buraschi S, Goyal A, Sharpe C, Natkanski E, Schaefer L et al (2016) EphA2 is a functional receptor for the growth factor progranulin. J Cell Biol 215(5):687–703

93. Altmann C, Vasic V, Hardt S, Heidler J, Haussler A, Wittig I et al (2016) Progranulin promotes peripheral nerve regeneration and reinnervation: role of notch signaling. Mol Neurodegener 11(1):69

94. Airaksinen MS, Saarma M (2002) The GDNF family: signalling, biological functions and therapeutic value. Nat Rev Neurosci 3(5):383–394

95. Uhlen M, Karlsson MJ, Zhong W, Tebani A, Pou C, Mikes J et al (2019) A genome-wide transcriptomic analysis of protein-coding genes in human blood cells. Science 366(6472):eaax9198

96. Bateman A, Belcourt D, Bennett H, Lazure C, Solomon S (1990) Granulins, a novel class of peptide from leukocytes. Biochem Biophys Res Commun 173(3):1161–1168

97. Gong Y, Zhan T, Li Q, Zhang G, Tan B, Yang X et al (2016) Serum progranulin levels are elevated in patients with chronic hepatitis B virus infection, reflecting viral load. Cytokine 85:26–29

98. Wei F, Jiang Z, Sun H, Pu J, Sun Y, Wang M et al (2019) Induction of PGRN by influenza virus inhibits the antiviral immune responses through downregulation of type I interferons signaling. PLoS Pathog 15(10):e1008062

99. Zou S, Luo Q, Song Z, Zhang L, Xia Y, Xu H et al (2017) Contribution of progranulin to protective lung immunity during bacterial pneumonia. J Infect Dis 215(11):1764–1773

100. Suh HS, Gelman BB, Lee SC (2014) Potential roles of microglial cell progranulin in HIV-associated CNS pathologies and neurocognitive impairment. J Neuroimmune Pharmacol 9(2):117–132

101. Suh HS, Lo Y, Choi N, Letendre S, Lee SC (2014) Evidence of the innate antiviral and neuroprotective properties of progranulin. PLoS One 9(5):e98184

102. Alissa EM, Sutaih RH, Kamfar HZ, Alagha AE, Marzouki ZM (2017) Serum progranulin levels in relation to insulin resistance in childhood obesity. J Pediatr Endocrinol Metab 30(12):1251–1256

103. Abella V, Pino J, Scotece M, Conde J, Lago F, Gonzalez-Gay MA et al (2017) Progranulin as a biomarker and potential therapeutic agent. Drug Discov Today 22(10):1557–1564

104. Korolczuk A, Beltowski J (2017) Progranulin, a new adipokine at the crossroads of metabolic syndrome, diabetes, dyslipidemia and hypertension. Curr Pharm Des 23(10):1533–1539

105. Tanaka Y, Takahashi T, Tamori Y (2014) Circulating progranulin level is associated with visceral fat and elevated liver enzymes: significance of serum progranulin as a useful marker for liver dysfunction. Endocr J 61(12):1191–1196

106. Yamamoto Y, Takemura M, Serrero G, Hayashi J, Yue B, Tsuboi A et al (2014) Increased serum GP88 (Progranulin) concentrations in rheumatoid arthritis. Inflammation 37(5):1806–1813

107. Cerezo LA, Kuklova M, Hulejova H, Vernerova Z, Kasprikova N, Veigl D et al (2015) Progranulin is associated with disease activity in patients with rheumatoid arthritis. Mediat Inflamm 2015:740357

108. Bhandari V, Daniel R, Lim PS, Bateman A (1996) Structural and functional analysis of a promoter of the human granulin/epithelin gene. Biochem J 319(Pt 2):441–447

109. Li X, Massa PE, Hanidu A, Peet GW, Aro P, Savitt A et al (2002) IKKalpha, IKKbeta, and NEMO/IKKgamma are each required for the NF-kappa B-mediated inflammatory response program. J Biol Chem 277(47):45129–45140

110. He Z, Ong CH, Halper J, Bateman A (2003) Progranulin is a mediator of the wound response. Nat Med 9(2):225–229

111. Mackenzie IR, Baker M, Pickering-Brown S, Hsiung GY, Lindholm C, Dwosh E et al (2006) The neuropathology of frontotemporal lobar degeneration caused by mutations in the progranulin gene. Brain 129(Pt 11):3081–3090

112. Ahmed Z, Mackenzie IR, Hutton ML, Dickson DW (2007) Progranulin in frontotemporal lobar degeneration and neuroinflammation. J Neuroinflammation 4:7

113. Mukherjee O, Pastor P, Cairns NJ, Chakraverty S, Kauwe JS, Shears S et al (2006) HDDD2 is a familial frontotemporal lobar degeneration with ubiquitin-positive, tau-negative inclusions caused by a missense mutation in the signal peptide of progranulin. Ann Neurol 60(3):314–322

114. Naphade SB, Kigerl KA, Jakeman LB, Kostyk SK, Popovich PG, Kuret J (2010) Progranulin expression

is upregulated after spinal contusion in mice. Acta Neuropathol 119(1):123–133

115. Tanaka Y, Matsuwaki T, Yamanouchi K, Nishihara M (2013) Exacerbated inflammatory responses related to activated microglia after traumatic brain injury in progranulin-deficient mice. Neuroscience 231:49–60

116. Petkau TL, Neal SJ, Orban PC, MacDonald JL, Hill AM, Lu G et al (2010) Progranulin expression in the developing and adult murine brain. J Comp Neurol 518(19):3931–3947

117. Ma Y, Matsuwaki T, Yamanouchi K, Nishihara M (2017) Progranulin protects hippocampal neurogenesis via suppression of neuroinflammatory responses under acute immune stress. Mol Neurobiol 54(5):3717–3728

118. Mao Q, Wang D, Li Y, Kohler M, Wilson J, Parton Z et al (2017) Disease and region specificity of granulin immunopositivities in Alzheimer disease and frontotemporal lobar degeneration. J Neuropathol Exp Neurol 76(11):957–968

119. Mao Q, Zheng X, Gefen T, Rogalski E, Spencer CL, Rademakers R et al (2019) FTLD-TDP with and without GRN mutations cause different patterns of CA1 pathology. J Neuropathol Exp Neurol 78(9):844–853

120. Holler CJ, Taylor G, Deng Q, Kukar T (2017) Intracellular proteolysis of progranulin generates stable, lysosomal granulins that are haploinsufficient in patients with frontotemporal dementia caused by GRN mutations. eNeuro 4(4):ENEURO.0100-17

121. Malaspina A, Kaushik N, de Belleroche J (2001) Differential expression of 14 genes in amyotrophic lateral sclerosis spinal cord detected using gridded cDNA arrays. J Neurochem 77(1):132–145

122. Ohmi K, Greenberg DS, Rajavel KS, Ryazantsev S, Li HH, Neufeld EF (2003) Activated microglia in cortex of mouse models of mucopolysaccharidoses I and IIIB. Proc Natl Acad Sci U S A 100(4):1902–1907

123. Minami SS, Min SW, Krabbe G, Wang C, Zhou Y, Asgarov R et al (2014) Progranulin protects against amyloid beta deposition and toxicity in Alzheimer's disease mouse models. Nat Med 20(10):1157–1164

124. Gowrishankar S, Yuan P, Wu Y, Schrag M, Paradise S, Grutzendler J et al (2015) Massive accumulation of luminal protease-deficient axonal lysosomes at Alzheimer's disease amyloid plaques. Proc Natl Acad Sci U S A 112(28):E3699–E3708

125. Pereson S, Wils H, Kleinberger G, McGowan E, Vandewoestyne M, Van Broeck B et al (2009) Progranulin expression correlates with dense-core amyloid plaque burden in Alzheimer disease mouse models. J Pathol 219(2):173–181

126. Suarez-Calvet M, Capell A, Araque Caballero MA, Morenas-Rodriguez E, Fellerer K, Franzmeier N et al (2018) CSF progranulin increases in the course of Alzheimer's disease and is associated with sTREM2, neurodegeneration and cognitive decline. EMBO Mol Med 10(12):e9712

127. Mendsaikhan A, Tooyama I, Bellier JP, Serrano GE, Sue LI, Lue LF et al (2019) Characterization of lysosomal proteins progranulin and prosaposin and their interactions in Alzheimer's disease and aged brains: increased levels correlate with neuropathology. Acta Neuropathol Commun 7(1):215

128. Kleinberger G, Capell A, Haass C, Van Broeckhoven C (2013) Mechanisms of granulin deficiency: lessons from cellular and animal models. Mol Neurobiol 47(1):337–360

129. Kayasuga Y, Chiba S, Suzuki M, Kikusui T, Matsuwaki T, Yamanouchi K et al (2007) Alteration of behavioural phenotype in mice by targeted disruption of the progranulin gene. Behav Brain Res 185(2):110–118

130. Petkau TL, Neal SJ, Milnerwood A, Mew A, Hill AM, Orban P et al (2012) Synaptic dysfunction in progranulin-deficient mice. Neurobiol Dis 45(2):711–722

131. Kao AW, Eisenhut RJ, Martens LH, Nakamura A, Huang A, Bagley JA et al (2011) A neurodegenerative disease mutation that accelerates the clearance of apoptotic cells. Proc Natl Acad Sci U S A 108(11):4441–4446

132. Wils H, Kleinberger G, Pereson S, Janssens J, Capell A, Van Dam D et al (2012) Cellular ageing, increased mortality and FTLD-TDP-associated neuropathology in progranulin knockout mice. J Pathol 228(1):67–76

133. Nguyen AD, Nguyen TA, Zhang J, Devireddy S, Zhou P, Karydas AM et al (2018) Murine knockin model for progranulin-deficient frontotemporal dementia with nonsense-mediated mRNA decay. Proc Natl Acad Sci U S A 115(12):E2849–E2E58

134. Ahmed Z, Sheng H, Xu YF, Lin WL, Innes AE, Gass J et al (2010) Accelerated lipofuscinosis and ubiquitination in granulin knockout mice suggest a role for progranulin in successful aging. Am J Pathol 177(1):311–324

135. Ghoshal N, Dearborn JT, Wozniak DF, Cairns NJ (2012) Core features of frontotemporal dementia recapitulated in progranulin knockout mice. Neurobiol Dis 45(1):395–408

136. Yin F, Dumont M, Banerjee R, Ma Y, Li H, Lin MT et al (2010) Behavioral deficits and progressive neuropathology in progranulin-deficient mice: a mouse model of frontotemporal dementia. FASEB J 24(12):4639–4647

137. Martens LH, Zhang J, Barmada SJ, Zhou P, Kamiya S, Sun B et al (2012) Progranulin deficiency promotes neuroinflammation and neuron loss following toxin-induced injury. J Clin Invest 122(11):3955–3959

138. Lui H, Zhang J, Makinson SR, Cahill MK, Kelley KW, Huang HY et al (2016) Progranulin deficiency promotes circuit-specific synaptic pruning by microglia via complement activation. Cell 165(4):921–935

139. Krabbe G, Minami SS, Etchegaray JI, Taneja P, Djukic B, Davalos D et al (2017) Microglial NFkappaB-TNFalpha hyperactivation induces

obsessive-compulsive behavior in mouse models of progranulin-deficient frontotemporal dementia. Proc Natl Acad Sci U S A 114(19):5029–5034

140. Gotzl JK, Brendel M, Werner G, Parhizkar S, Sebastian Monasor L, Kleinberger G et al (2019) Opposite microglial activation stages upon loss of PGRN or TREM2 result in reduced cerebral glucose metabolism. EMBO Mol Med 11(6):e9711

141. Kojima Y, Ono K, Inoue K, Takagi Y, Kikuta K, Nishimura M et al (2009) Progranulin expression in advanced human atherosclerotic plaque. Atherosclerosis 206(1):102–108

142. Tang W, Lu Y, Tian QY, Zhang Y, Guo FJ, Liu GY et al (2011) The growth factor progranulin binds to TNF receptors and is therapeutic against inflammatory arthritis in mice. Science 332(6028):478–484

143. Chen X, Chang J, Deng Q, Xu J, Nguyen TA, Martens LH et al (2013) Progranulin does not bind tumor necrosis factor (TNF) receptors and is not a direct regulator of TNF-dependent signaling or bioactivity in immune or neuronal cells. J Neurosci 33(21):9202–9213

144. Etemadi N, Webb A, Bankovacki A, Silke J, Nachbur U (2013) Progranulin does not inhibit TNF and lymphotoxin-alpha signalling through TNF receptor 1. Immunol Cell Biol 91(10):661–664

145. Hu Y, Xiao H, Shi T, Oppenheim JJ, Chen X (2014) Progranulin promotes tumour necrosis factor-induced proliferation of suppressive mouse CD4(+) Foxp3(+) regulatory T cells. Immunology 142(2):193–201

146. Stubert J, Waldmann K, Dieterich M, Richter DU, Briese V (2014) Progranulin shows cytoprotective effects on trophoblast cells in vitro but does not antagonize TNF-alpha-induced apoptosis. Arch Gynecol Obstet 290(5):867–873

147. Fujita K, Chen X, Homma H, Tagawa K, Amano M, Saito A et al (2018) Targeting Tyro3 ameliorates a model of PGRN-mutant FTLD-TDP via tau-mediated synaptic pathology. Nat Commun 9(1):433

148. Lang I, Fullsack S, Wajant H (2018) Lack of evidence for a direct interaction of progranulin and tumor necrosis factor receptor-1 and tumor necrosis factor receptor-2 from cellular binding studies. Front Immunol 9:793

149. Baladron V, Ruiz-Hidalgo MJ, Bonvini E, Gubina E, Notario V, Laborda J (2002) The EGF-like homeotic protein dlk affects cell growth and interacts with growth-modulating molecules in the yeast two-hybrid system. Biochem Biophys Res Commun 291(2):193–204

150. Park B, Buti L, Lee S, Matsuwaki T, Spooner E, Brinkmann MM et al (2011) Granulin is a soluble cofactor for toll-like receptor 9 signaling. Immunity 34(4):505–513

151. Zhou X, Sun L, Bastos de Oliveira F, Qi X, Brown WJ, Smolka MB et al (2015) Prosaposin facilitates sortilin-independent lysosomal trafficking of progranulin. J Cell Biol 210(6):991–1002

152. Almeida MR, Macario MC, Ramos L, Baldeiras I, Ribeiro MH, Santana I (2016) Portuguese family with the co-occurrence of frontotemporal lobar degeneration and neuronal ceroid lipofuscinosis phenotypes due to progranulin gene mutation. Neurobiol Aging 41:200 e1–200 e5

153. Kamate M, Detroja M, Hattiholi V (2019) Neuronal ceroid lipofuscinosis type-11 in an adolescent. Brain and Development 41(6):542–545

154. Hyung S, Im SK, Lee BY, Shin J, Park JC, Lee C et al (2019) Dedifferentiated Schwann cells secrete progranulin that enhances the survival and axon growth of motor neurons. Glia 67(2):360–375

155. Tanaka M, Kuse Y, Nakamura S, Hara H, Shimazawa M (2019) Potential effects of progranulin and granulins against retinal photoreceptor cell degeneration. Mol Vis 25:902–911

156. Lee CW, Stankowski JN, Chew J, Cook CN, Lam YW, Almeida S et al (2017) The lysosomal protein cathepsin L is a progranulin protease. Mol Neurodegener 12(1):55

157. Zhou X, Paushter DH, Feng T, Sun L, Reinheckel T, Hu F (2017) Lysosomal processing of progranulin. Mol Neurodegener 12(1):62

158. Belcastro V, Siciliano V, Gregoretti F, Mithbaokar P, Dharmalingam G, Berlingieri S et al (2011) Transcriptional gene network inference from a massive dataset elucidates transcriptome organization and gene function. Nucleic Acids Res 39(20):8677–8688

159. Filiano AJ, Martens LH, Young AH, Warmus BA, Zhou P, Diaz-Ramirez G et al (2013) Dissociation of frontotemporal dementia-related deficits and neuroinflammation in progranulin haploinsufficient mice. J Neurosci 33(12):5352–5361

160. Zhou X, Sun L, Brady OA, Murphy KA, Hu F (2017) Elevated TMEM106B levels exaggerate lipofuscin accumulation and lysosomal dysfunction in aged mice with progranulin deficiency. Acta Neuropathol Commun 5(1):9

161. Chang MC, Srinivasan K, Friedman BA, Suto E, Modrusan Z, Lee WP et al (2017) Progranulin deficiency causes impairment of autophagy and TDP-43 accumulation. J Exp Med 214(9):2611–2628

162. Tanaka Y, Chambers JK, Matsuwaki T, Yamanouchi K, Nishihara M (2014) Possible involvement of lysosomal dysfunction in pathological changes of the brain in aged progranulin-deficient mice. Acta Neuropathol Commun 2:78

163. Gotzl JK, Mori K, Damme M, Fellerer K, Tahirovic S, Kleinberger G et al (2014) Common pathobiochemical hallmarks of progranulin-associated frontotemporal lobar degeneration and neuronal ceroid lipofuscinosis. Acta Neuropathol 127(6):845–860

164. Gotzl JK, Colombo AV, Fellerer K, Reifschneider A, Werner G, Tahirovic S et al (2018) Early lysosomal maturation deficits in microglia triggers enhanced lysosomal activity in other brain cells of progranulin knockout mice. Mol Neurodegener 13(1):48

165. Tyynela J, Palmer DN, Baumann M, Haltia M (1993) Storage of saposins A and D in infantile neuronal ceroid-lipofuscinosis. FEBS Lett 330(1):8–12

166. Elleder M, Sokolova J, Hrebicek M (1997) Follow-up study of subunit c of mitochondrial ATP synthase (SCMAS) in Batten disease and in unrelated lysosomal disorders. Acta Neuropathol 93(4):379–390

167. Ward ME, Chen R, Huang HY, Ludwig C, Telpoukhovskaia M, Taubes A et al (2017) Individuals with progranulin haploinsufficiency exhibit features of neuronal ceroid lipofuscinosis. Sci Transl Med 9(385):eaah5642

168. Valdez C, Wong YC, Schwake M, Bu G, Wszolek ZK, Krainc D (2017) Progranulin-mediated deficiency of cathepsin D results in FTD and NCL-like phenotypes in neurons derived from FTD patients. Hum Mol Genet 26(24):4861–4872

169. Zheng Y, Brady OA, Meng PS, Mao Y, Hu F (2011) C-terminus of progranulin interacts with the beta-propeller region of sortilin to regulate progranulin trafficking. PLoS One 6(6):e21023

170. Willnow TE, Petersen CM, Nykjaer A (2008) VPS10P-domain receptors – regulators of neuronal viability and function. Nat Rev Neurosci 9(12):899–909

171. Nielsen MS, Madsen P, Christensen EI, Nykjaer A, Gliemann J, Kasper D et al (2001) The sortilin cytoplasmic tail conveys Golgi-endosome transport and binds the VHS domain of the GGA2 sorting protein. EMBO J 20(9):2180–2190

172. Lee WC, Almeida S, Prudencio M, Caulfield TR, Zhang YJ, Tay WM et al (2014) Targeted manipulation of the sortilin-progranulin axis rescues progranulin haploinsufficiency. Hum Mol Genet 23(6):1467–1478

173. Carrasquillo MM, Nicholson AM, Finch N, Gibbs JR, Baker M, Rutherford NJ et al (2010) Genome-wide screen identifies rs646776 near sortilin as a regulator of progranulin levels in human plasma. Am J Hum Genet 87(6):890–897

174. Schulze H, Sandhoff K (2014) Sphingolipids and lysosomal pathologies. Biochim Biophys Acta 1841(5):799–810

175. Zhou X, Sullivan PM, Sun L, Hu F (2017) The interaction between progranulin and prosaposin is mediated by granulins and the linker region between saposin B and C. J Neurochem 143(2):236–243

176. Nicholson AM, Finch NA, Almeida M, Perkerson RB, van Blitterswijk M, Wojtas A et al (2016) Prosaposin is a regulator of progranulin levels and oligomerization. Nat Commun 7:11992

177. Kishimoto Y, Hiraiwa M, O'Brien JS (1992) Saposins: structure, function, distribution, and molecular genetics. J Lipid Res 33(9):1255–1267

178. Cenik B, Sephton CF, Kutluk Cenik B, Herz J, Yu G (2012) Progranulin: a proteolytically processed protein at the crossroads of inflammation and neurodegeneration. J Biol Chem 287(39):32298–32306

179. Haimovich B, Tanaka JC (1995) Magainin-induced cytotoxicity in eukaryotic cells: kinetics, dose-response and channel characteristics. Biochim Biophys Acta 1240(2):149–158

180. Beel S, Moisse M, Damme M, De Muynck L, Robberecht W, Van Den Bosch L et al (2017) Progranulin functions as a cathepsin D chaperone to stimulate axonal outgrowth in vivo. Hum Mol Genet 26(15):2850–2863

181. Zhou X, Paushter DH, Feng T, Pardon CM, Mendoza CS, Hu F (2017) Regulation of cathepsin D activity by the FTLD protein progranulin. Acta Neuropathol 134(1):151–153

182. Butler VJ, Cortopassi WA, Argouarch AR, Ivry SL, Craik CS, Jacobson MP et al (2019) Progranulin stimulates the in vitro maturation of pro-Cathepsin D at acidic pH. J Mol Biol 431(5):1038–1047

183. Jian J, Zhao S, Tian QY, Liu H, Zhao Y, Chen WC et al (2016) Association between progranulin and Gaucher disease. EBioMedicine 11:127–137

184. Jian J, Tian QY, Hettinghouse A, Zhao S, Liu H, Wei J et al (2016) Progranulin recruits HSP70 to beta-Glucocerebrosidase and is therapeutic against Gaucher disease. EBioMedicine 13:212–224

185. Zhou X, Paushter DH, Pagan MD, Kim D, Nunez Santos M, Lieberman RL et al (2019) Progranulin deficiency leads to reduced glucocerebrosidase activity. PLoS One 14(7):e0212382

186. Valdez C, Ysselstein D, Young TJ, Zheng J, Krainc D (2019) Progranulin mutations result in impaired processing of prosaposin and reduced glucocerebrosidase activity. Hum Mol Genet 29(5):716–726

187. Arrant AE, Roth JR, Boyle NR, Kashyap SN, Hoffmann MQ, Murchison CF et al (2019) Impaired beta-glucocerebrosidase activity and processing in frontotemporal dementia due to progranulin mutations. Acta Neuropathol Commun 7(1):218

188. Brady RO, Kanfer JN, Shapiro D (1965) Metabolism of glucocerebrosides. Ii. Evidence of an enzymatic deficiency in Gaucher's disease. Biochem Biophys Res Commun 18:221–225

189. Do J, McKinney C, Sharma P, Sidransky E (2019) Glucocerebrosidase and its relevance to Parkinson disease. Mol Neurodegener 14(1):36

190. Sidransky E (2004) Gaucher disease: complexity in a "simple" disorder. Mol Genet Metab 83(1–2):6–15

191. Hiraiwa M, Martin BM, Kishimoto Y, Conner GE, Tsuji S, O'Brien JS (1997) Lysosomal proteolysis of prosaposin, the precursor of saposins (sphingolipid activator proteins): its mechanism and inhibition by ganglioside. Arch Biochem Biophys 341(1):17–24

192. Chen Y, Jian J, Hettinghouse A, Zhao X, Setchell KDR, Sun Y et al (2018) Progranulin associates with hexosaminidase A and ameliorates GM2 ganglioside accumulation and lysosomal storage in Tay-Sachs disease. J Mol Med (Berl) 96(12):1359–1373

193. Sandhoff K (2016) Neuronal sphingolipidoses: membrane lipids and sphingolipid activator proteins regulate lysosomal sphingolipid catabolism. Biochimie 130:146–151

194. Van Deerlin VM, Sleiman PM, Martinez-Lage M, Chen-Plotkin A, Wang LS, Graff-Radford NR et al

(2010) Common variants at 7p21 are associated with frontotemporal lobar degeneration with TDP-43 inclusions. Nat Genet 42(3):234–239

195. Finch N, Carrasquillo MM, Baker M, Rutherford NJ, Coppola G, Dejesus-Hernandez M et al (2011) TMEM106B regulates progranulin levels and the penetrance of FTLD in GRN mutation carriers. Neurology 76(5):467–474

196. Gallagher MD, Posavi M, Huang P, Unger TL, Berlyand Y, Gruenewald AL et al (2017) A dementia-associated risk variant near TMEM106B alters chromatin architecture and gene expression. Am J Hum Genet 101(5):643–663

197. Nicholson AM, Finch NA, Wojtas A, Baker MC, Perkerson RB 3rd, Castanedes-Casey M et al (2013) TMEM106B p.T185S regulates TMEM106B protein levels: implications for frontotemporal dementia. J Neurochem 126(6):781–791

198. Brady OA, Zheng Y, Murphy K, Huang M, Hu F (2013) The frontotemporal lobar degeneration risk factor, TMEM106B, regulates lysosomal morphology and function. Hum Mol Genet 22(4):685–695

199. Stagi M, Klein ZA, Gould TJ, Bewersdorf J, Strittmatter SM (2014) Lysosome size, motility and stress response regulated by fronto-temporal dementia modifier TMEM106B. Mol Cell Neurosci 61:226–240

200. Chen-Plotkin AS, Unger TL, Gallagher MD, Bill E, Kwong LK, Volpicelli-Daley L et al (2012) TMEM106B, the risk gene for frontotemporal dementia, is regulated by the microRNA-132/212 cluster and affects progranulin pathways. J Neurosci 32(33):11213–11227

201. Lang CM, Fellerer K, Schwenk BM, Kuhn PH, Kremmer E, Edbauer D et al (2012) Membrane orientation and subcellular localization of transmembrane protein 106B (TMEM106B), a major risk factor for frontotemporal lobar degeneration. J Biol Chem 287(23):19355–19365

202. Schwenk BM, Lang CM, Hogl S, Tahirovic S, Orozco D, Rentzsch K et al (2014) The FTLD risk factor TMEM106B and MAP6 control dendritic trafficking of lysosomes. EMBO J 33(5):450–467

203. Suzuki H, Matsuoka M (2016) The lysosomal trafficking transmembrane protein 106B is linked to cell death. J Biol Chem 291(41):21448–21460

204. Busch JI, Martinez-Lage M, Ashbridge E, Grossman M, Van Deerlin VM, Hu F et al (2013) Expression of TMEM106B, the frontotemporal lobar degeneration-associated protein, in normal and diseased human brain. Acta Neuropathol Commun 1:36

205. Klein ZA, Takahashi H, Ma M, Stagi M, Zhou M, Lam TT et al (2017) Loss of TMEM106B ameliorates lysosomal and frontotemporal dementia-related phenotypes in progranulin-deficient mice. Neuron 95(2):281–96 e6

206. Arrant AE, Nicholson AM, Zhou X, Rademakers R, Roberson ED (2018) Partial Tmem106b reduction does not correct abnormalities due to progranulin haploinsufficiency. Mol Neurodegener 13(1):32

207. Nicholson AM, Zhou X, Perkerson RB, Parsons TM, Chew J, Brooks M et al (2018) Loss of Tmem106b is unable to ameliorate frontotemporal dementia-like phenotypes in an AAV mouse model of C9ORF72-repeat induced toxicity. Acta Neuropathol Commun 6(1):42

208. Rohrer JD, Warren JD, Fox NC, Rossor MN (2013) Presymptomatic studies in genetic frontotemporal dementia. Rev Neurol (Paris) 169(10):820–824

209. Boeve B, Bove J, Brannelly P, Brushaber D, Coppola G, Dever R et al (2020) The longitudinal evaluation of familial frontotemporal dementia subjects protocol: framework and methodology. Alzheimers Dement 16(1):22–36

Trends in Understanding the Pathological Roles of TDP-43 and FUS Proteins

Emanuele Buratti

Abbreviations

ALS	Amyotrophic lateral sclerosis
FTLD	Frontotemporal lobar dementia
FUS/TLS	Fused in sarcoma/translocated in liposarcoma
hnRNP	Heterogeneous ribonucleoproteins
lncRNA	Long noncoding RNA
mRNA	Messenger RNA
miRNA	MicroRNA
TDP-43	TAR DNA binding protein 43 kDa

Introduction

The involvement of TAR DNA binding protein-43 (TDP-43) in neurodegenerative diseases was first described in 2006 when this protein was shown to be the main component of the characteristic aggregates found in the brains in patients with amyotrophic lateral sclerosis (ALS) and frontotemporal lobar degeneration (FTLD) [1, 2]. This discovery was swiftly followed 3 years later by the identification of fused in sarcoma (FUS) as another TDP-43-related protein that was aggregating in the neurons of a subset of familial ALS and sporadic FTLD cases [3, 4]. Since then, the number of RNA binding proteins (RBPs) that have been shown to be involved in ALS/FTLD has increased considerably. It now includes several other RNA binding proteins such as EWS (Ewing sarcoma breakpoint 1, also called EWSR1) and TAF15 (TATA box binding protein-associated factor 68 kDa) [5], heterogeneous ribonucleoproteins (hnRNP A1 and hnRNP A2/B1) [6], Matrin-3 (MATR3) [7], ataxin-2 (ATXN2) [8], and T-cell intracellular antigen 1 (TIA1) [9–11]. More recently, the identification of DNA and as a consequence RNA repeat expansions in the C9orf72 gene [12, 13] has allowed to greatly expand the crucial role of RNA alterations in the ALS/FTLD phenotype and has extended the number of RNA-mediated pathways that can lead to disease [14–18].

Taken together, all these findings have firmly established RNA metabolism as a major contributor of ALS/FTLD processes in humans [19–23], and the emerging picture is that a combination of RNA processing alterations might represent the principal contributor to the occurrence of both ALS and FTLD in patients [17, 24, 25]. This conclusion does not really simplify matters in terms of knowing exactly why neurons die because RNA processing basically regulates all the processes within a eukaryotic cell. Therefore, the number of pathways that could eventually become disrupted following TDP-43 and FUS aggregation is steadily growing and ranges from such diverse extremes as DNA plasticity and

E. Buratti (✉)
International Centre for Genetic Engineering and Biotechnology (ICGEB), Trieste, Italy
e-mail: buratti@icgeb.org

© Springer Nature Switzerland AG 2021
B. Ghetti et al. (eds.), *Frontotemporal Dementias*, Advances in Experimental Medicine and Biology 1281, https://doi.org/10.1007/978-3-030-51140-1_15

damage [26], pre-mRNA splicing [27], nucleocy-toplasmic transport [28, 29], polyadenylation [30], or RNA translation [31–34]. Once impaired, these basic mechanisms can then induce misregulation of more complex processes such as endocytosis [35], neuroinflammation [36, 37], autoimmunity [38], mitochondrial functions [39], stress granule formation [40–43], epigenetics mechanisms [44], and even alterations at the general metabolic profiles of patients [45, 46]. Therefore, the aim of this chapter will be to highlight promising future trends in TDP-43/FUS research that will hopefully lead to the identification of pathways that play an important role in disease and, most importantly, that can be considered "druggable" with the technical means at our disposal.

TDP-43 and FUS Protein Structure

TDP-43 protein structure, mutations, and its posttranslational modifications have recently been described in several recent reviews [47–50] (Fig. 1). For this reason, just a brief summary will be presented in this chapter. Basically, at the structural level, TDP-43 possesses a well-structured N-terminus region [51, 52] that carries a nuclear localization signal [53] and is involved in protein dimerization/oligomerization [51, 54]. This is important for TDP-43 splicing functions [55] that are mainly regulated by two RRM (RNA recognition motif) domains that closely follow the N-terminus of the protein. These domains are required to bind target RNA mostly in a sequence-specific manner [56, 57] but are also participating in the aggregation process of this protein through being prone to self-assembly [58]. Finally, the sequence of TDP-43 is completed by a mostly unstructured C-terminus region that has prion-like properties (PrLD), is mainly used to interact with other proteins, and plays a fundamental role in phase separation and aggregation of this protein [59].

Similar to TDP-43, FUS is an hnRNP protein originally found translocated in human liposarcomas and for this reason was also denoted as TLS [60, 61]. Originally, FUS was also called hnRNP P2 [62], and it belongs to the FET protein family that includes two other RBPs, EWS and TAF15, that have also been found to be involved in FTLD [63, 64]. The FUS structure consists of several domains that have been reviewed in detail elsewhere [65] (Fig. 1), and it consists of an N-terminal

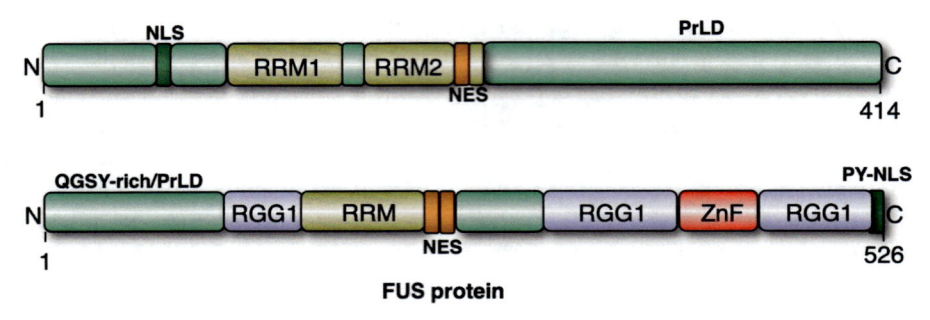

Fig. 1 This figure shows a schematic domain structure of TDP-43 and FUS. TDP-43 is a 414-long protein characterized by two RNA recognition motifs (RRM), RRM1 and RRM2 (which contains a putative nuclear export sequence, NES) that are the main regulators of RNA binding. At the N-terminus, there is a highly structured region that regulates oligomerization of this protein and contains a nuclear localization signal (NLS), while at the C-terminus there is a mostly unstructured region that is responsible for protein-protein interactions and has char-acteristics of a prion-like domain. On the other hand, the 526-residue-long FUS protein contains a single RRM domain, and its unstructured, prion-like region is localized at the N-terminus of the protein. FUS further contains two adjacent putative NES sequences, three arginine-glycine-glycine-rich domains (RGG1-3), and a zinc-finger motif (ZnF) that contribute to stabilize RNA binding. The nuclear localization signal of FUS is located at the very C-terminal of this protein (PY-NLS)

domain with a QGSY-rich region, also described as a prion-like domain (PrLD) that is responsible for FUS dimerization and binding to chromatin for regulation of transcription initiation [66]. This region is followed by a highly conserved RRM domain whose structure has been solved in solution alone [67] or bound to RNA [68]. In general, the presence of unstructured prion-like domains and the RNA binding ability are probably the major unifying factors of TDP-43 and FUS. However, they do not bind in exactly the same positions because the FUS consensus binding sequence could never be as clearly defined as it was for TDP-43. In particular, CLIP analyses have shown that FUS binds preferentially in a sawtooth manner within long introns in neuronal cells [69], while TDP-43 prefers UG-rich and a few other selected motifs [70–72].

Interestingly, and similarly to TDP-43, the FUS RRM has also been shown to be prone to self-assembly to form amyloid fibrils [73] although for the FUS RRM domain, there does not seem to be a linear consensus motif [74]. This RRM region is then followed by a zinc finger motif and multiple RGG repeats that are located at the C-terminal end and can also participate in RNA binding [68]. All these domains act together to mediate both protein-RNA and protein-protein interactions in the multiple activities of this protein.

Disease-Associated Mutations in TDP-43 and FUS

In the early days of TDP-43 and FUS research, and in the absence of robust functional studies (which later became available), the presence of TDP-43 aggregates in patients did not represent by itself a sufficient condition to establish TDP-43 as a disease-causing gene [75]. In this early context, therefore, the first direct link of TDP-43 with disease was provided by the discovery of mutations within its encoding gene *TARDBP*, which were shown to segregate with disease among family members [76–78]. To this date, more than 50 missense mutations together with few truncation and insertions/deletions have been

reported in the literature. These mutations account for about 5% of familial ALS cases and have also been found in a few FTLD cases [49]. The functional significance of these mutations is for the most part still unknown, although several have been described to affect the liquid-liquid separation properties of TDP-43 [79], RNA binding [80], or posttranslational modifications such as phosphorylation [81] and acetylation [82, 83]. In general, mutations are likely to be associated with features that can be directly connected with basic TDP-43 properties within cells, such as altered subcellular localization, protein half-life, or protein-protein interactions [49]. In some cases, mutations have been shown to affect directly important neuronal functions. For example, a disease-associated mutation of TDP-43 (A315T) has been recently described to affect dendritic spine assembly in an ALS mouse model [84], and other mutations have been shown to impair RNA axonal transport [85].

Similarly to TDP-43, mutations in FUS can be found in about 4% of familial ALS cases and less than 1% of sporadic ALS cases [65, 86]. Unlike TDP-43, however, these mutations can be found in almost all regions of the protein, although the most severe ones in terms of early occurrence of the pathology primarily affect the C-terminal domain where the nuclear localization sequence resides [3, 4]. A detailed overview of *FUS/TLS* mutations can be found in recent review articles [87]. Most importantly, mutations altering the cytoplasmic mislocalization of FUS/TLS can compromise the autoregulation process of this protein and promote its increase and abnormal accumulation in the cytoplasm [88]. At a more general level, FUS mutations have also been associated with a drastic reduction in nuclear GEM structures and decreased binding to the essential U1-snRNP splicing factor, causing a general impairment of the splicing process [89–92]. Regarding this issue, mutant FUS proteins have been shown to change their binding affinity consistently as a consequence of their abnormal localization and redistribution in the cytoplasm [93]. Alternatively, some FUS mutations have also been shown to act in a similar manner to TDP-43 mutations, for example, in their ability to sequester paraspeckle components in the cytoplasm [94, 95] or to affect the interaction with other RNA binding proteins, such as ELAVL4, that will eventually be included in the cytoplasmic inclusions [96].

In conclusion, although mutations in TDP-43 and FUS are quite rare in ALS/FTLD patients, they certainly seem to play a role in the pathology and could theoretically be used as a target in therapeutic intervention [97]. Most importantly, however, their study has also helped to uncover novel molecular aspects of the disease such as DNA damage [98, 99] or the importance of autoregulatory processes, as recently observed in a mouse expressing the TDP-43 disease-associated mutant Q331K [100].

Therefore, the functional study of TDP-43 and FUS mutations plays a very important role in our better understanding of the disease. Interestingly, it should be noted that this usefulness is not unique with regard to natural mutations but is true also for artificial mutations obtained through ENU mutagenesis. For example, a gain-of-function artificial mutation in TDP-43 (M323K) has recently allowed to reveal a novel category of splicing events controlled by TDP-43 that consists in the skipping of constitutive exons from several cellular genes that play an important role in proteostasis, and this loss was associated with adult-onset motor neuron loss and neurodegenerative changes in a mouse model of ALS [101].

Major Pathological Features of ALS/FTLD Associated with TDP-43 and FUS

Although the study of disease-associated mutations is important to better understand several aspects of disease, it is now also clear that TDP-43 and FUS mutations are very rare in patients. Therefore, they may not necessarily recapitulate the most common pathological features of TDP-43 and FUS aggregation in neurons. In most patients, in fact, the most common pathological feature shown by these proteins is represented by the aberrant aggregation of the wild-type proteins in the body of affected neurons.

In ALS, almost all cases present TDP-43 inclusions with the exception of patients with mutations in FUS or SOD1 [102, 103]. On the other hand, in FTLD, TDP-43 inclusions are present in almost half of the cases (45%), and wild-type FUS has been shown to abnormally aggregate in a subset of FTLD cases (9%) with the rest characterized by Tau pathology [104]. For reasons that are still not clear, TDP-43 and FUS proteins do not seem to co-localize in the same pathological aggregates, although wild-type TDP-43 has been observed to bind with low affinity to FUS [105], and in yeast models both proteins have been reported to co-aggregate very efficiently [106].

A second important feature of the pathology is that these proteins are variably modified at the posttranslational level. In fact, within the pathological aggregates, TDP-43 is aberrantly ubiquitinated, phosphorylated, acetylated, sumoylated, and cleaved to generate C-terminal fragments [1, 2, 82, 107, 108]. Interestingly, aberrant phosphorylation of TDP-43 has also been observed in other diseases such as inclusion body myopathy [109] or Niemann-Pick C [110]. Compared to TDP-43, FUS is methylated in its C-terminal arginine residues, and this modification seems to be specifically associated with the formation of cytoplasmic FUS inclusions in FTLD-FUS patients [111]. More recently, FUS phosphorylation has been described to occur in its low complexity domain, and this can affect its ability to phase separate and aggregation propensity [112]. How and to what extent these posttranslational modifications may contribute to the formation of pathological TDP-43 aggregates in patients still remains an open issue that deserves further investigation [48].

An important issue that is also still open regards their potential amyloid composition of the TDP-43 and FUS aggregates. Initially, aggregates that were described in patients did not seem to possess an amyloid nature [1, 2], and TDP-43 inclusion bodies are of an amorphous nature [113]; under certain conditions, TDP-43 and FUS have also been observed to adopt amyloid conformations [114, 115]. In particular, selected fragments of the TDP-43 C-terminus have a very high propensity to form amyloid-like fibrils in vitro [116–122]. Similar to TDP-43, FUS inclusions in patients could not be stained by amyloid-detecting dyes such as Congo red and thioflavin B [123]. However, a segment with a strong amyloid-forming tendency that could induce the seeded aggregation of FUS has been recently isolated [124], and FUS RRM domains have been reported to undergo irreversible unfolding to self-assemble in amyloid fibrils [73]. Taken together, all these evidences point toward a condition where, although late-stage TDP-43 and FUS aggregates do not display typical features of amyloid aggregates, a few of the steps that lead to their aggregation may include amyloid formation and could thus be inhibited by drugs that are designed against this process.

From a therapeutic point of view, however, a priority target is to better understand the mechanisms that lead to protein aggregation, and several factors have been described which promote this event, especially for TDP-43 [125]. Regarding this issue, it should be noted that TDP-43 and FUS are aggregation-prone proteins that have a tendency to aggregate even following small increases in their endogenous expression levels [125, 126]. As a result, many different types of stimuli can trigger their aggregation and include already described mutations or lower efficiency of the autophagic/proteasome pathways (discussed below).

In physiological conditions, one of the connections that can probably play an important but not exclusive role [127] in the pathological aggregation of these proteins is represented by their recruitment in stress granules (SG) [40, 128, 129]. It is now well accepted that both TDP-43 and FUS are recruited to SGs in condition of different environmental insults [130, 131], and several reports have strengthened their connection with ALS/FTLD disease [132, 133]. In stressful conditions, the purpose of SGs is to arrest translation of housekeeping proteins by transiently sequestering cellular mRNAs. In this way, SGs promote the selective translation of stress-response proteins to help cellular recovery. Following stress removal, SGs normally dissolve quickly, and mRNA translation goes back to normal. However, using advanced optogenetic techniques, it has been shown that following the persistence of a stressful condition within cells, the SGs eventually evolve to form aberrant aggregates that could then lead to the pathological cascade [43]. At present, there are no therapeutic strategies that target specifically stress granules in disease. Importantly, recent evidence has shown that a class of small planar molecules can reduce the association of SGs with TDP-43 in iPSC-derived patient motor neurons and prevent accumulation of this protein in the cytoplasm [134]. Of course, although promising, the therapeutic potential of all these approaches will have to be tested in more complex animal models.

Categorizing TDP-43 and FUS Pathological Functions Within Cells

Since their identification, many studies have targeted the issue of clarifying the pathological role of aggregates once they have sequestered the soluble pool of TDP-43 and FUS in the nuclear and cytoplasmic compartments. Although a lot of progress has taken place in this area, it is far from being understood [135]. In general, protein aggregates within neurons can be directly toxic or they can become so, by acting as "protein sinks," thus depleting the cell of active proteins. Alternatively, aggregates may also be considered protective when they serve to remove mutated proteins that might be toxic when produced in excess or whose degradation becomes impaired. Finally, a third possibility is that aggregates may represent mere epiphenomena of the disease and not directly connected with the pathology.

With regard to TDP-43, some lines of evidence support the possibility that the aggregates may have a direct toxic role [113, 136, 137] as they may interfere with the nucleocytoplasmic transport of both proteins and RNA in the cell [138] or an indirect toxic role that could be induced by the sequestration in the aggregates of other proteins with which TDP-43 and FUS are normally in close contact within the cellular environment [64, 139–141]. However, there is also evidence that aggregates may be protective, at least during the early stages of the disease. This hypothesis has been supported by studies in TDP-43 *Drosophila* models [142] and somewhat reminds a situation that has been observed for Huntington's disease [143]. More recently, random mutagenesis of the TDP-43 prion-like domain to express more than 50.000 mutants has shown that mutations that increase hydrophobicity and aggregation can decrease toxicity [144]. Nonetheless, a consensus is still lacking, and readers are referred to several reviews dealing with this specific issue [145–149].

In keeping with these views, it is very likely that a combination of both gain- and loss-of-function effects, not necessarily linked in a temporal manner, may result in the alteration of the many nuclear and cytoplasmic functions of these proteins within neurons [107, 150–152] (Fig. 2) that will be described in the next section. As can be expected, the very high evolutionary conservation of both TDP-43 and FUS means that their alteration in the nucleus and cytoplasm will result in harmful consequences with regard to many cellular processes and pathways, and they could be most pronounced in neurons given their extremely complex architecture and metabolism.

Altered Cellular Pathways Mediated by TDP-43 and FUS Within Cells

As already mentioned, following aberrant aggregation and concurrent nuclear depletion of TDP-43 and FUS, there are several nuclear and cytoplasmic pathways that are likely to become disrupted (Fig. 2). The following list aims to briefly describe all the major ones that have been identified so far:

- *Response to DNA damage.* Maintaining the integrity of DNA in postmitotic neurons that do not divide during the entire life of an individual is a particularly critical issue with regard to their survival. At the proteomic level, both TDP-43 and FUS have been described to bind several factors important for DNA repair mechanisms [153]. In the case of TDP-43, a growing body of evidence has shown that

alterations in TDP-43 expression may affect this process. For example, it has been demonstrated that this protein can induce neurodegeneration by compromising the functionality of chromatin remodeler Chd1/CHD2 that prevents appropriate expression of protective genes [154]. Moreover, overexpression of TDP-43 in *Drosophila* has been shown to induce cell death due to many alterations, including DNA damage [155]. At the moment, experimental evidence has suggested that TDP-43 may recruit the XRCC4/DNA ligase 4 complex at sites of double-strand breaks and thus act as a key component of the nonhomologous end joining (NHEJ) pathway that represents the major repair pathway in postmitotic neurons [156, 157].

- On the other hand, much more is known about the connection between FUS and DNA repair: the first evidence that FUS is involved in DNA repair came from its ability to promote D-loop formation and homologous recombination during DNA double-strand break repair [158]. In keeping with this finding, embryonic fibroblasts and B-cells from *k*nockout mice show genomic instability and chromosome breaks [159]. Following DNA damage, FUS is phosphorylated in its N-terminal serine residues by ATM and DNA-PK [160] and directly interacts with PAR polymerase and HDAC1 protein at the site of DNA damage [161–163]. Furthermore, in conditions of DNA damage, the FUS proteins increase its binding affinity toward the two histone acetyltransferases CBP and p300, thereby repressing their transcriptional activities [164].

- Although rather premature at the moment, all these indications suggest that DNA repair-targeted therapeutic avenues might become a promising avenue in the fight against ALS and FTLD.

- *Transcription.* The TDP-43 protein was originally described to repress HIV-1 virus replication when integrated in the human genome [165]. Although this property could not be confirmed in later studies [166], there are now a few genes where TDP-43 has been described to act as an "insulator," for example, in the

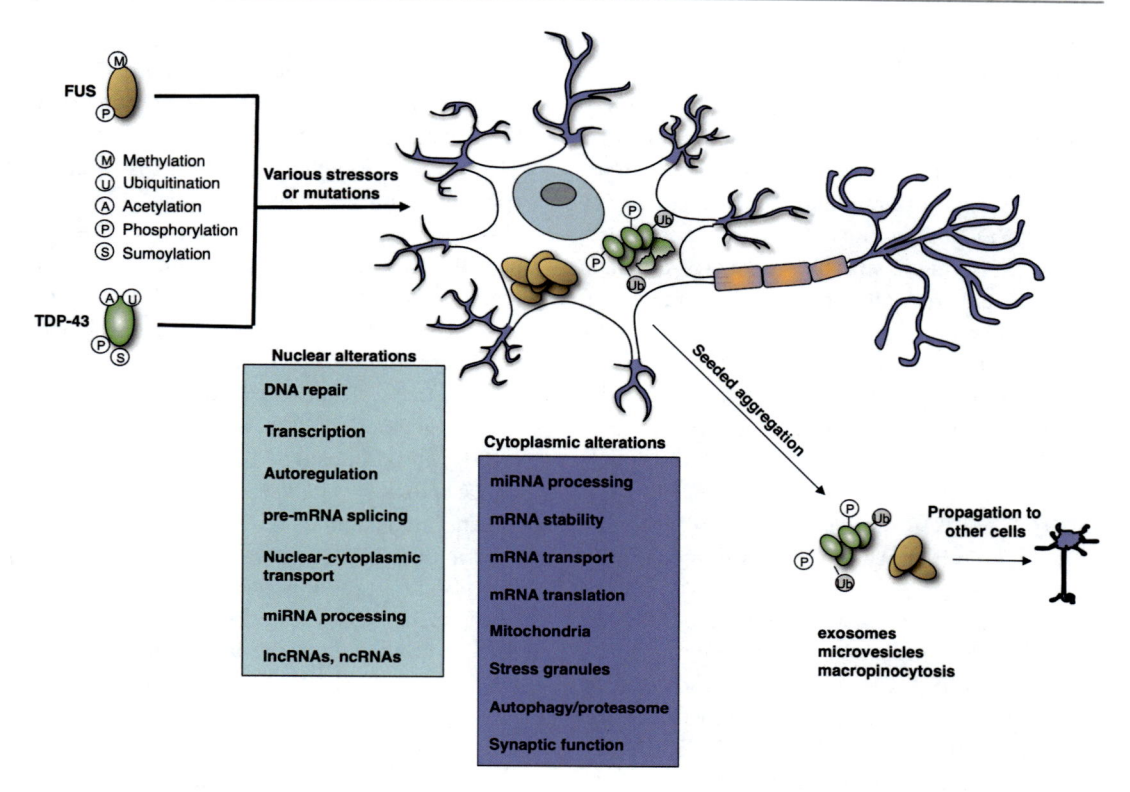

Fig. 2 This figure shows a schematic diagram of TDP-43 and FUS regulated cellular functions that can be affected following their aggregation in neurons. At the basic molecular level, the aggregation of these proteins causes a widespread dysfunction in many processes that occur both in the nucleus and the cytoplasm, from DNA damage repair mechanisms to mRNA translation. The dysfunction of these processes will then induce defects in many organ- elles (e.g., mitochondria) or complex cellular processes (such as stress granule dynamics, autophagy, and lysosomal processes). The presence of these defects, even if not immediately fatal to the neuron, will eventually induce its premature death with the possible spreading of aggregates or toxic oligomers to nearby cells through extracellular traveling mediated by vesicles (EVs) or other mechanisms

case of the *SP-10* gene in mice [167]. More conventionally, TDP-43 has also been identified as a transcriptional promoter in the case of the *TNF-alpha* gene [168] or as a transcriptional repressor of the VPS4B gene [169]. At the moment, however, it is not very clear what could be the importance of these alterations in disease and what is the molecular mechanism that mediates these effects.

– Contrary to TDP-43, the association of FUS with transcription has been much better defined. Indeed, FUS was originally identified in association with the genomic translocation of its N-terminal domain to fusion genes in a variety of liposarcomas and in myeloid leukemia to alter the transcription of the resulting chimeric genes [60, 61]. In addition to these specific events, FUS has also been shown to directly regulate the activity of RNA-pol II by controlling its phosphorylation during transcription [170], and disease-associated mutations have been shown to decrease its binding to RNA-pol II [171] and to active chromatin [66], leading to a reduced regulation of general transcription rates. Finally, the ability of FUS to bind near to alternative polyadenylation sequences has been associated with regulation also of this process [172]. At present, however, no specific therapies have been hypothesized that specifically target this characteristic feature of the FUS protein.

– *Autoregulation of their own expression.* Many RNA binding proteins regulate their own expression in cells by targeting their own pre-mRNA and inducing changes in their processing to allow proper translation or induce degradation [173]. An important pre-mRNA target of TDP-43 is its own transcript that has been shown to undergo a splicing event in its 3'UTR region to regulate the differential use of *TARDBP* alternative polyadenylation sites [174]. Although still to be exactly defined, this mechanism allows the autoregulation of protein expression within cells in order to keep TDP-43 protein levels within a physiological range [174]. At present, there are indications that perturbations in this system, either due to mutations in TDP-43 or through artificial inhibition of the splicing event in the 3'UTR of TDP-43 that regulates its recognition, could be linked to the pathological phenotype [100, 175].

- In a manner similar to TDP-43, FUS can autoregulate its expression by binding to its own pre-mRNA [88]. The autoregulation mechanism is controlled by a process called nonsense-mediated decay [176]. This autoregulation can also be mediated by specific miRNAs that bind to FUS 3'UTR sequence and whose expression is affected by FUS itself [177].

- From the point of view of the pathology, it is easy to understand how aggregation of both TDP-43 and FUS may lead to a dysfunction in autoregulation. In particular, sequestration of TDP-43 or FUS in the aggregates will result in starting a vicious cycle where lack of these proteins would result in increased expression that would then lead to even more aggregation, and this will eventually result in increasingly harmful gain- or loss-of-function effects on the RNA metabolism [178]. In summary, it is very likely that future therapeutic strategies will be aimed at modulating this specific mechanism (although it might prove difficult to avoid an excessive stimulation or degradation of TDP-43 and FUS mRNAs).

- *Pre-mRNA splicing processes.* TDP-43 initial involvement in the regulation of pre-mRNA splicing was first identified for exon 9 in the *CFTR* gene [179]. Recently, several high-throughput studies have addressed in detail all the splicing alterations that occur in human ALS/FTLD patients or TDP-43 mice disease models [71, 180–182]. However, many of these events are probably not direct targets of TDP-43 and originate from changes in other splicing factors controlled by TDP-43 [183]. Nonetheless, several direct targets of TDP-43 have been identified in recent studies, and they include *POLDIP3/SKAR* [34, 184], *SORT1* [185, 186], *STAG2, MADD* [187], and *TNIK* [188] pre-mRNAs. In addition, among these targets, there are also important proteins for neurodegeneration that include hnRNP A1 [189], Tau [190], and SMN [191]. In addition, transcriptomic analyses from ALS-FTD patients and animal models of disease have shown that TDP-43 has the very important function of repressing the inclusion of cryptic exons [101, 192, 193]. These exons are normally excluded from the mature mRNA and, when inserted, will often cause a change in the reading frame and thus the introduction of premature translational stop codons. At present, their contribution to disease has not been clearly established, although TDP-43 splicing repression seems to be a key general feature for maintaining motor neurons in good health [194]. Nonetheless, once identified, splicing events that might play a critical role in ALS pathology would represent ideal therapeutic targets considering that their inclusion is strongly repressed in normal conditions.

- Like TDP-43, FUS involvement in splicing has been well studied. At the general level, the link between FUS and splicing is supported by reports describing its binding to the SMN protein, U1 small nuclear ribonucleoprotein (snRNP), and Sm-snRNP complex [89, 195]. In particular, FUS has been shown to bind to nascent pre-mRNAs and acts as a molecular mediator between RNA-pol II and U1-snRNP [196]. At the RNA level, FUS can also control histone transcript 3' end processing during the S replication phase of the cell cycle by interacting with the U7 snRNP complex and the transcriptional apparatus [197]. As with TDP-43, several studies have tried to define the splicing targets of FUS/TLS by high-throughput assays in several disease models [69, 93, 188, 198–202]. However, like with TDP-43, these comparisons have resulted in little overlap among all published datasets [203], and more efforts will be required to determine exactly what are the splicing targets of FUS depending on individual cellular contexts. Finally, an interesting difference between TDP-43 and FUS is the observation that FUS can also affect minor intron splicing by interacting specifically with the key minor intron component U11snRNP and trapping it in the aggregates [204].

- *miRNA processing*. The dysregulation of miRNA expression following TDP-43 and FUS aggregation may have potentially very harmful consequences on neuronal cell survival in ALS [205] and other common diseases such as Alzheimer's, Parkinson's, and Huntington's [206]. Even before TDP-43 and FUS were found to be involved in neurodegenerative diseases, it was described that these proteins are present in the Drosha complex [207] and that TDP-43 is associated with perichromatin fibrils [208] that correspond to the region where miRNA processing occurs. Indeed, during neuronal differentiation, TDP-43 has also been shown to control Drosha protein stability, thus potentially affecting the biogenesis of the entire cellular miRNA population [209]. Finally, in human neuronal cell lines, TDP-43 depletion has also been associated with a consistent increase of DICER mRNA and protein levels, further supporting the connection between TDP-43 and miRNA biogenesis [188]. More specifically, follow-up studies have confirmed that TDP-43 depletion can lead to altered expression of various miRNAs, such as let-7b, miR-663, miR-9, miR1/miR206, miR-520, miR-132, miR-143, miR-574, and miR-NID1 [210–215].

- As already mentioned, FUS was also found to localize at the Drosha complex together with TDP-43 [207], and further studies have shown that this protein is able to recruit Drosha at chromatin sites of active transcription to promote pri-miRNA processing [216]. Like TDP-43, FUS depletion in human neuroblastoma

cells can alter the expression of a consistent number of analyzed miRNAs that include miR-9, miR-125b, and miR-132, which have important roles in neuronal metabolism and differentiation [216]. In the future, the challenge will be to better characterize the extent to which the TDP-43- and FUS-mediated control of miRNA expression contributes to the pathology and whether this may be targeted by specific therapeutic approaches.

- *lncRNA expression*. It is now clear that lncRNA expression is associated with the occurrence of age-related diseases and neurodegenerative disorders [217, 218]. Just like protein-coding RNAs, TDP-43 has been shown to bind and affect the expression of a variety of lncRNAs, such as *gadd7* [219], *SPA* [220], *MALAT1* [181, 221], *NEAT1_2* [181], *Xist* [222], Myolinc [223], and *lncLSTR* [224]. Alterations in the expression for some of these transcripts were detected in human FTLD brains compared to healthy controls [181] or in the spinal motoneurons of sporadic ALS patients (Nishimoto et al., 2013). Interestingly, the regulation of lncRNAs by TDP-43 is not unidirectional, because a recent study has shown that the lncRNAs known as *MIAT* can regulate TDP-43 expression [225].

- Compared to TDP-43, FUS was found to bind to a consistent fraction (30%) of all literature annotated lncRNAs, including NEAT1_2, although not in the same place as TDP-43 [198, 226]. In keeping with this finding, transcriptomic analyses of mouse embryonic stem cells derived from a FUS-ALS model showed that several lncRNAs were misregulated and potentially connected with disease [227]. Finally, as with TDP-43, it has also been reported that lncRNAs such as *hsrw* can rescue FUS toxicity in a *Drosophila* model [228], thus showing that TDP-43 and FUS influence on lncRNAs goes both ways.

- In conclusion, although a direct interaction of FUS and TDP-43 with several lncRNAs has been established, there are still many open questions that remain, such as whether these factors can affect their transcription or stability, how they can act to affect lncRNA biological properties, and what is their importance in disease [226, 229]. All these questions will need to be addressed before attempting any therapeutic strategy.

- *ncRNA expression.* In addition to binding with miRNAs and lncRNAs, TDP-43 has also been shown to bind other members of the noncoding RNA family (ncRNAs) [230]. Indeed, in-depth analysis of RNA sequencing data has shown that TDP-43 can bind to several kinds of transcripts such as SINE, LINE, and LTRs [231], and more recent evidence has highlighted in a *Drosophila* model that TDP-43 pathology leads to inhibition of all those mechanisms that are responsible for retrotransposon repression [155]. At the moment, however, the importance of these interactions in disease is not known. As a consequence, their therapeutic potential is also uncertain in the absence of further investigations.

- *Nucleocytoplasmic transport.*Defects in this process have been investigated as potentially responsible for inducing the aggregation process because they might lead directly to abnormal accumulation of TDP-43 and FUS in a specific cellular compartment [232]. Indeed, both TDP-43 and FUS continuously shuttle between the nucleus and the cytoplasm through several receptors such as Transportin-1 (Importin-β2) for FUS and Importin-β1 for TDP-43 [233, 234]. At the molecular level, transport and aggregation of FUS is modulated by arginine methylation of the RGG domain by PRMT1 which reduces binding to Transportin-1 [235–238].

- More recently, nucleocytoplasmic transport has also been linked with ALS/FTLD as part of the possible pathological role played by poly(GA) dipeptides produced from the expanded C9orf72 repeats [239]. Likewise, aggregated and disease-linked mutant TDP-43 have been recently associated with the direct sequestration and/or mislocalization of nucleoporins and transport factors in a variety of disease models, suggesting that disruption of RNA and protein import/export might represent a common feature of ALS disease

[240]. Taken together, all this emerging evidence suggests that small molecules able to rescue or prevent these transport defects may be effective for disease treatment.

- *mRNA stability.* Among the genes that play important roles in neuronal viability, such as microtubule dynamics and protein aggregation turnover, TDP-43 has been shown to affect the mRNA stability of the human low molecular weight neurofilament (*hNFL*) [241] and the histone deacetylase *HDAC6* transcripts [242, 243]. However, the list of mRNAs whose stability is controlled by TDP-43 is probably much longer if we take into account that TDP-43 binding regions are particularly abundant in the 3'UTR region of mature mRNAs [72, 244]. As a result, several targets have been described so far, such as *Add2* [245], *VEGFA* and *GRN* [72], and *IL-6* [246], Tbc1d1 [247], and G3BP [248]. In particular, the regulation of G3BP may be very important for neurodegeneration because this protein factor is a component of stress granules (SG) that play a key role in the TDP-43 protein aggregation process [41, 132, 249]. Another direct connection between mRNA stability and disease has come from the recent observation that TDP-43 can suppress Tau expression by promoting mRNA instability, thus suggesting that downregulation of TDP-43 may affect pathology in Alzheimer's disease patients and related Tau pathologies [250].

- Like TDP-43, several studies have reported the binding of FUS to the 3'UTR sequence of many target mRNAs [69, 72, 93, 198]. In particular, FUS depletion in primary cortical neurons has been shown to downregulate the AMPA receptor GluA1 protein subunit by acting on its mRNA stability [251].

- As with pre-mRNA splicing events, the importance of these studies for therapy will depend on the identification of key transcript alterations that could be corrected by RNA therapy.

- *mRNA transport.* As expected, mRNA transport into axons and dendrites is very important to maintain neuronal activity and synaptic plasticity [252]. The first experimental evidence that TDP-43 is important for mRNA transport was first provided by Fallini et al., who showed that in motor neurons TDP-43 co-localizes with other well-known transport RBPs, including SMN and FMRP, and is actively transported along axons [253] in a bidirectional movement [85]. The fact that in the adult mouse brain TDP-43 has been found to bind many mRNAs from synaptic genes has further strengthened a role of TDP-43 in regulating the transport of synaptic mRNAs into distal processes [254].

- Regarding FUS, this protein has also been shown to participate in RNA granule transport into dendrites and in the regulation of local translation at the synapse [255, 256]. In primary hippocampal neurons, FUS can regulate spine remodeling following mGluR5 activation [257] and is transported into dendrites with actin filaments myosin-V whose role is to specifically sort RNA granules into dendrites [255, 258]. More recently, FUS/TLS, TDP-43, and SOD1 have also been shown to be transported to neurite terminals by a mechanism that involves endoplasmic reticulum tubules and the neurofilament cytoskeleton [259].

- *mRNA translation.* With regard to neurodegeneration, mRNA translation is strictly coupled with the process of mRNA transport. The importance of TDP-43 in regulating local translation was first demonstrated in rat hippocampal neurons where TDP-43 was found to act as a translational repressor [260]. This observation was in keeping with several proteomic analyses which identified TDP-43 as mostly associated with the RNA splicing and translation machineries [261]. Subsequently, TDP-43 was also found associated with the heavy polysome fractions [31], and several specific targets regulated by TDP-43 at the translational level have been described: *futsch*/Map 1b, Rac1, MTHFSD, and DDX58 [32, 33, 262–264]. Finally, at the protein interaction level, TDP-43 has been shown to interact with the protein RACK1 that is a known regulator for activity-dependent translation [265] and by regulating the splicing of ribosomal S6

kinase 1 (S6K1) Aly/REF-like target (SKAR) that plays a role in the pioneering round of translation [34].

- Similarly to TDP-43, FUS involvement in controlling local translation was suggested when it was found that this protein can co-localize with APC protein in ribonucleoprotein complexes and promote translation of associated mRNAs, such as *Kank2*, *Pkp4*, and *Ddr2* [266]. Moreover, it has been recently established that mutant FUS proteins can suppress translation by sequestering components of the cellular translational machinery in inclusions [267].

- *Autophagy/lysosome system.* The altered clearance of misfolded or aggregated proteins through the autophagy-lysosome system or the ubiquitin-lysosome system probably represents one of the major causative pathways for ALS/FTLD [268, 269]. First of all, this conclusion is based on the occurrence of mutations in several genes encoding for proteins controlling these pathways, such as *VCP* (autophagosome-autolysosome maturation), *CHMP2B* (late-stage endosome-lysosome fusion), *TBK1* (phosphorylation of autophagy adaptors p62 and OPTN), *OPTN* (autophagosome formation and maturation), and *p62/SQSTM1* (autophagy receptor). In particular, the gene that seems to play a key role is *UBQLN2* (recruitment of autophagosomes to polyubiquitinated aggregates) that may represent a promising therapeutic target [270]. Their detection in patients means that their correct function is closely associated with disease development [268, 271, 272]. Unsurprisingly, impairment of degraded or misfolded proteins is likely to induce pathological aggregation [187, 273, 274]. Moreover, both TDP-43 and FUS have been shown to affect directly the expression of key autophagic/lysosome machinery components either when knocked-down, overexpressed, or in the presence of disease-associated mutations [275–278].

- At the moment, therefore, autophagy and the lysosomal system is considered a primary therapeutic target to preserve neuronal functionality, and several trials are currently under way with autophagy-inducing molecules such as rapamycin, trehalose, and other compounds that have proved to be effective in improving various aspects of TDP-43 and FUS pathology in mouse and cellular models [279–283]. However, it should be noted that many of the approaches tried until now have mainly focused on the role of autophagy in neurons and there is still considerable room for improvement, for example, in better understanding the role played by autophagy dysfunctions in glia cells that make up a considerable amount of nonneuronal cells in our brains [284].

- *Synaptic functions.* One conclusion of all the high-throughput studies that have been performed on identifying the RNA targets of TDP-43 and FUS has shown that many could be affecting synaptic transmission and plasticity [285], an event that could be closely linked with the early motor and cognitive deficits in ALS and FTD [286]. This is supported by several lines of evidence that indicate how overexpression of TDP-43 can impair presynaptic integrity [287, 288], both following disease onset and also at the presymptomatic stage [289, 290]. The connection between synapses and TDP-43 is probably a very conserved evolutionary feature because synaptic control by TDP-43 has been shown to be present also in *Drosophila* and zebrafish disease models [291, 292]. Very similarly to TDP-43, also for FUS, it has been described that missense mutations can profoundly disrupt synaptic homeostasis in a mouse model of disease [293] possibly by affecting mRNA stability of molecules such as SynGAPα2 that promotes maturation of dendritic spines [294]. At present, no therapeutic approaches have specifically targeted this particular aspect of the pathology. However, if we consider the increasing evidence that both TDP-43 and FUS may control synaptic integrity and function, it is likely that this possibility will draw more attention in the future.

TDP-43 and FUS Spreading in ALS/FTLD Diseases

Another important question about TDP-43 and FUS pathology is represented by understanding the mechanisms at the basis of the spread of the disease between different neurons and brain regions [295–297]. In recent years, the hypothesis that has gained most attention is represented by the possibility that TDP-43, FUS, and some polypeptides derived from C9orf72 may spread in a manner that resembles the prion protein [298, 299]. The importance of better understanding if and how these aggregates spread is quite self-evident, because giving an answer to this mechanism may represent a very good therapeutic target. At the structural level, the prion-like spreading hypothesis is supported by the presence of a prion-like domain in the C-terminus of TDP-43 (residues 274–414) and in the FUS N-terminus (residues 1–239) [300]. At the experimental level, support has also come from the observation that in vitro TDP-43 aggregates can induce endogenous TDP-43 aggregation when transduced in HEK293T cells [301] and SH-SY5Y neuronal cells [302]. In parallel, it has also been reported that TDP-43 oligomers can spread from cell to cell by microvesicle/exosome pathways [303, 304] and that TDP-43 aggregates are able to gain entry into cells by stimulating "membrane ruffling" and consequent macropinocytosis [305].

At the moment, the research on FUS is not as advanced as with TDP-43. Nonetheless, a study using a mutant FUS protein that is prone to form fibrils has shown that also this protein has the capability of seeding wild-type FUS [306]. These preliminary results, therefore, suggest that FUS protein may therefore spread between cells using a similar mechanism.

Clearly, TDP-43 and FUS spreading is a very promising area of research because being able to inhibit spreading could obviously represent a very effective therapy. However, there are still many open questions with regard to what are the specific propagating protein assemblies and/or conformations that make this spreading possible. For obvious reasons, finding answers to these questions is an absolute requirement to develop effective therapies.

Modifiers of TDP-43 and FUS Toxicity

Although both TDP-43 and FUS are able to act together to enhance neurodegenerative phenotypes [307], comparative analyses in *Drosophila* and zebrafish models indicate that FUS acts downstream with respect to TDP-43 [308, 309]. This property, however, does not seem to be linked with toxicity as high-throughput approaches to find yeast modifiers of TDP-43 and FUS/TLS toxicity have uncovered that they are quite different from each other [310]. This finding has not discouraged research in this area, and, at the moment, there are several modifiers of TDP-43 pathology that include hnRNP U and hnRNP A1/A2 [311], hnRNP K [312], DAZAP1 [313], ataxin-2 [314], and hUPF1 [315]. At a more general level, it has also been recently reported that upregulation of glycolysis can be induced by overexpression of the GLUT-3 protein in neurons and that this event can be neuroprotective against defects induced by TDP-43 [316]. Likewise, it has also been recently reported that upregulation of the Atg7 gene (a key regulator of macroautophagy/autophagy) can improve motor function and life span in flies that lack *TBPH*, the homologue gene of human TDP-43 [317].

In the case of FUS, it has also been shown that downregulation of several nuclear transport proteins such as Nup154 and XPO1 can prevent FUS-induced neurotoxicity [318] and that muscleblind protein can also rescue FUS-induced motor dysfunctions although in this case the molecular mechanism is still unknown [319].

Finally, it is also interesting to note that some modifiers such as HuR have been reported to act on both TDP-43 and FUS [320], thus showing that contrary to previous expectations, some modifier overlap can presumably exist.

The presence of all these modifiers is quite interesting from several points of view. First of all, many of them could be useful to explain the huge variability that is observed in the age of onset and disease course of TDP-43 and FUS proteinopathies. Secondly, depending on their identity, these modifiers may represent more viable targets for therapeutic action than TDP-43 and FUS. In keeping with this view, it was shown that reduction of ataxin-2 using antisense oligos in a TDP-43 mouse model was able to extend life span and improve the motor phenotype [321].

Conclusions and Future Perspectives

As described in this chapter, the occurrence of TDP-43 and FUS mutations, overexpression, or aggregation can have a profound impact on several important cellular pathways. Although both proteins share many similarities and act on similar pathways [322], they are also quite different according to several lines of neuropathological and experimental evidence [323, 324]. From a therapeutic point of view, therefore, the most important research priority in the future will be to obtain a full understanding of TDP-43 and FUS-controlled pathways to identify those that are mostly responsible for neuronal death (especially at the beginning of the pathology). These targets should then be used to prioritize various RNA-based therapeutic actions that modern technology is currently developing at a very fast pace [23, 325]. Unfortunately, however, both TDP-43 and FUS may not represent ideal "druggable" targets because, as described in this chapter, each of them plays many important and diverse roles within cells. Therefore, altering their general expression within neurons will probably not be very feasible in vivo, as overexpression or downregulation is likely to be considerably toxic. For this reason, a more refined approach would be that of identifying modifying factors, transcripts, or cellular conditions that might act as suppressors or enhancers of TDP-43 and FUS pathology. An advanced knowledge of these factors/events/conditions will then be useful to identify molecular targets that can be potentially addressed using modern therapeutic strategies. It is only after we have obtained a clear view of many of these still unknown issues that we will probably be able to develop novel, hypothesis-based, therapeutic approaches that could be of clinical benefit.

Acknowledgments This work was supported by AriSLA (Italy) grant (PathensTDP) and Beneficentia Stiftung (Liechtenstein).

References

1. Neumann M, Sampathu DM, Kwong LK, Truax AC, Micsenyi MC, Chou TT et al (2006) Ubiquitinated TDP-43 in frontotemporal lobar degeneration and amyotrophic lateral sclerosis. Science 314(5796):130–133
2. Arai T, Hasegawa M, Akiyama H, Ikeda K, Nonaka T, Mori H et al (2006) TDP-43 is a component of ubiquitin-positive tau-negative inclusions in frontotemporal lobar degeneration and amyotrophic lateral sclerosis. Biochem Biophys Res Commun 351(3):602–611
3. Kwiatkowski TJ Jr, Bosco DA, Leclerc AL, Tamrazian E, Vanderburg CR, Russ C et al (2009) Mutations in the FUS/TLS gene on chromosome 16 cause familial amyotrophic lateral sclerosis. Science 323(5918):1205–1208
4. Vance C, Rogelj B, Hortobagyi T, De Vos KJ, Nishimura AL, Sreedharan J et al (2009) Mutations in FUS, an RNA processing protein, cause familial amyotrophic lateral sclerosis type 6. Science 323(5918):1208–1211
5. Neumann M, Bentmann E, Dormann D, Jawaid A, DeJesus-Hernandez M, Ansorge O et al (2011) FET proteins TAF15 and EWS are selective markers that distinguish FTLD with FUS pathology from amyotrophic lateral sclerosis with FUS mutations. Brain 134(Pt 9):2595–2609
6. Kim HJ, Kim NC, Wang YD, Scarborough EA, Moore J, Diaz Z et al (2013) Mutations in prion-like domains in hnRNPA2B1 and hnRNPA1 cause multisystem proteinopathy and ALS. Nature 495(7442):467–473
7. Johnson JO, Pioro EP, Boehringer A, Chia R, Feit H, Renton AE et al (2014) Mutations in the Matrin 3 gene cause familial amyotrophic lateral sclerosis. Nat Neurosci 17(5):664–666
8. Elden AC, Kim HJ, Hart MP, Chen-Plotkin AS, Johnson BS, Fang X et al (2010) Ataxin-2 intermediate-length polyglutamine expansions are associated with increased risk for ALS. Nature 466(7310):1069–1075
9. Mackenzie IR, Nicholson AM, Sarkar M, Messing J, Purice MD, Pottier C et al (2017) TIA1 mutations in amyotrophic lateral sclerosis and frontotemporal dementia promote phase separation and alter stress granule dynamics. Neuron 95(4):808–16 e9
10. Yuan Z, Jiao B, Hou L, Xiao T, Liu X, Wang J et al (2018) Mutation analysis of the TIA1 gene in Chinese patients with amyotrophic lateral sclerosis and frontotemporal dementia. Neurobiol Aging 64:160 e9–160e12
11. Zhang K, Liu Q, Shen D, Tai H, Fu H, Liu S et al (2018) Genetic analysis of TIA1 gene in Chinese patients with amyotrophic lateral sclerosis. Neurobiol Aging 67:201 e9–201e10
12. DeJesus-Hernandez M, Mackenzie IR, Boeve BF, Boxer AL, Baker M, Rutherford NJ et al (2011) Expanded GGGGCC hexanucleotide repeat in non-coding region of C9ORF72 causes chromosome 9p-linked FTD and ALS. Neuron 72(2):245–256
13. Renton AE, Majounie E, Waite A, Simon-Sanchez J, Rollinson S, Gibbs JR et al (2011) A hexanucleotide repeat expansion in C9ORF72 is the cause of chromosome 9p21-linked ALS-FTD. Neuron 72(2):257–268
14. Mori K, Lammich S, Mackenzie IR, Forne I, Zilow S, Kretzschmar H et al (2013) hnRNP A3 binds to GGGGCC repeats and is a constituent of p62-positive/TDP43-negative inclusions in the hippocampus of patients with C9orf72 mutations. Acta Neuropathol 125(3):413–423
15. Cooper-Knock J, Walsh MJ, Higginbottom A, Robin Highley J, Dickman MJ, Edbauer D et al (2014) Sequestration of multiple RNA recognition motif-containing proteins by C9orf72 repeat expansions. Brain 137(Pt 7):2040–2051
16. Lee YB, Chen HJ, Peres JN, Gomez-Deza J, Attig J, Stalekar M et al (2013) Hexanucleotide repeats in ALS/FTD form length-dependent RNA foci, sequester RNA binding proteins, and are neurotoxic. Cell Rep 5(5):1178–1186
17. Balendra R, Isaacs AM (2018) C9orf72-mediated ALS and FTD: multiple pathways to disease. Nat Rev Neurol 14(9):544–558
18. Vatsavayai SC, Nana AL, Yokoyama JS, Seeley WW (2019) C9orf72-FTD/ALS pathogenesis: evidence from human neuropathological studies. Acta Neuropathol 137(1):1–26
19. Renoux AJ, Todd PK (2012) Neurodegeneration the RNA way. Prog Neurobiol 97(2):173–189
20. Sephton CF, Yu G (2015) The function of RNA-binding proteins at the synapse: implications for neurodegeneration. Cell Mol Life Sci 72(19):3621–3635
21. Zhao M, Kim JR, van Bruggen R, Park J (2018) RNA-binding proteins in amyotrophic lateral sclerosis. Mol Cells 41(9):818–829

22. Butti Z, Patten SA (2018) RNA dysregulation in amyotrophic lateral sclerosis. Front Genet 9:712

23. Nussbacher JK, Tabet R, Yeo GW, Lagier-Tourenne C (2019) Disruption of RNA metabolism in neurological diseases and emerging therapeutic interventions. Neuron 102(2):294–320

24. Hardy J, Rogaeva E (2014) Motor neuron disease and frontotemporal dementia: sometimes related, sometimes not. Exp Neurol 262(Pt B):75–83

25. Solomon DA, Mitchell JC, Salcher-Konrad MT, Vance CA, Mizielinska S (2019) Review: modelling the pathology and behaviour of frontotemporal dementia. Neuropathol Appl Neurobiol 45(1):58–80

26. Penndorf D, Witte OW, Kretz A (2018) DNA plasticity and damage in amyotrophic lateral sclerosis. Neural Regen Res 13(2):173–180

27. Chabot B, Shkreta L (2016) Defective control of pre-messenger RNA splicing in human disease. J Cell Biol 212(1):13–27

28. Fahrenkrog B, Harel A (2018) Perturbations in traffic: aberrant nucleocytoplasmic transport at the heart of neurodegeneration. Cell 7(12):232

29. Burk K, Pasterkamp RJ (2019) Disrupted neuronal trafficking in amyotrophic lateral sclerosis. Acta Neuropathol 137(6):859–877

30. Rot G, Wang Z, Huppertz I, Modic M, Lence T, Hallegger M et al (2017) High-resolution RNA maps suggest common principles of splicing and polyadenylation regulation by TDP-43. Cell Rep 19(5):1056–1067

31. Higashi S, Kabuta T, Nagai Y, Tsuchiya Y, Akiyama H, Wada K (2013) TDP-43 associates with stalled ribosomes and contributes to cell survival during cellular stress. J Neurochem 126(2):288–300

32. MacNair L, Xiao S, Miletic D, Ghani M, Julien JP, Keith J et al (2016) MTHFSD and DDX58 are novel RNA-binding proteins abnormally regulated in amyotrophic lateral sclerosis. Brain 139:86–100

33. Majumder P, Chen YT, Bose JK, Wu CC, Cheng WC, Cheng SJ et al (2012) TDP-43 regulates the mammalian spinogenesis through translational repression of Rac1. Acta Neuropathol 124(2):231–245

34. Fiesel FC, Weber SS, Supper J, Zell A, Kahle PJ (2012) TDP-43 regulates global translational yield by splicing of exon junction complex component SKAR. Nucleic Acids Res 40(6):2668–2682

35. Liu G, Coyne AN, Pei F, Vaughan S, Chaung M, Zarnescu DC et al (2017) Endocytosis regulates TDP-43 toxicity and turnover. Nat Commun 8(1):2092

36. Beers DR, Appel SH (2019) Immune dysregulation in amyotrophic lateral sclerosis: mechanisms and emerging therapies. Lancet Neurol 18(2):211–220

37. Trageser KJ, Smith C, Herman FJ, Ono K, Pasinetti GM (2019) Mechanisms of immune activation by c9orf72-expansions in amyotrophic lateral sclerosis and frontotemporal dementia. Front Neurosci 13:1298

38. Borroni B, Alberici A, Buratti E (2019) Review: molecular pathology of frontotemporal lobar degenerations. Neuropathol Appl Neurobiol 45(1):41–57

39. Briston T, Hicks AR (2018) Mitochondrial dysfunction and neurodegenerative proteinopathies: mechanisms and prospects for therapeutic intervention. Biochem Soc Trans 46(4):829–842

40. Aulas A, Vande VC (2015) Alterations in stress granule dynamics driven by TDP-43 and FUS: a link to pathological inclusions in ALS? Front Cell Neurosci 9:423

41. Bentmann E, Haass C, Dormann D (2013) Stress granules in neurodegeneration – lessons learnt from TAR DNA binding protein of 43 kDa and fused in sarcoma. FEBS J 280(18):4348–4370

42. Chew J, Cook C, Gendron TF, Jansen-West K, Del Rosso G, Daughrity LM et al (2019) Aberrant deposition of stress granule-resident proteins linked to C9orf72-associated TDP-43 proteinopathy. Mol Neurodegener 14(1):9

43. Zhang P, Fan B, Yang P, Temirov J, Messing J, Kim HJ et al (2020) Chronic optogenetic induction of stress granules is cytotoxic and reveals the evolution of ALS-FTD pathology. Elife 8:e39578

44. Bennett SA, Tanaz R, Cobos SN, Torrente MP (2019) Epigenetics in amyotrophic lateral sclerosis: a role for histone post-translational modifications in neurodegenerative disease. Transl Res 204:19–30

45. Jawaid A, Khan R, Polymenidou M, Schulz PE (2018) Disease-modifying effects of metabolic perturbations in ALS/FTLD. Mol Neurodegener 13(1):63

46. Joppe K, Roser AE, Maass F, Lingor P (2019) The contribution of iron to protein aggregation disorders in the central nervous system. Front Neurosci 13:15

47. Francois-Moutal L, Perez-Miller S, Scott DD, Miranda VG, Mollasalehi N, Khanna M (2019) Structural insights Into TDP-43 and effects of post-translational modifications. Front Mol Neurosci 12:301

48. Buratti E (2018) TDP-43 post-translational modifications in health and disease. Expert Opin Ther Targets 22(3):279–293

49. Buratti E (2015) Functional significance of TDP-43 mutations in disease. Adv Genet 91:1–53

50. Afroz T, Perez-Berlanga M, Polymenidou M (2019) Structural transition, function and dysfunction of TDP-43 in neurodegenerative diseases. Chimia (Aarau) 73(6):380–390

51. Mompean M, Romano V, Pantoja-Uceda D, Stuani C, Baralle FE, Buratti E et al (2016) The TDP-43 N-terminal domain structure at high resolution. FEBS J 283(7):1242–1260

52. Qin H, Lim LZ, Wei Y, Song J (2014) TDP-43 N terminus encodes a novel ubiquitin-like fold and its unfolded form in equilibrium that can be shifted by binding to ssDNA. Proc Natl Acad Sci U S A 111(52):18619–18624

53. Winton MJ, Igaz LM, Wong MM, Kwong LK, Trojanowski JQ, Lee VM (2008) Disturbance of nuclear and cytoplasmic TAR DNA-binding protein (TDP-43) induces disease-like redistribution,

sequestration, and aggregate formation. J Biol Chem 283(19):13302–13309

54. Afroz T, Hock EM, Ernst P, Foglieni C, Jambeau M, Gilhespy LAB et al (2017) Functional and dynamic polymerization of the ALS-linked protein TDP-43 antagonizes its pathologic aggregation. Nat Commun 8(1):45

55. Jiang LL, Xue W, Hong JY, Zhang JT, Li MJ, Yu SN et al (2017) The N-terminal dimerization is required for TDP-43 splicing activity. Sci Rep 7(1):6196

56. Buratti E, Baralle FE (2001) Characterization and functional implications of the RNA binding properties of nuclear factor TDP-43, a novel splicing regulator of CFTR exon 9. J Biol Chem 276(39):36337–36343

57. Lukavsky PJ, Daujotyte D, Tollervey JR, Ule J, Stuani C, Buratti E et al (2013) Molecular basis of UG-rich RNA recognition by the human splicing factor TDP-43. Nat Struct Mol Biol 20(12):1443–1449

58. Agrawal S, Kuo PH, Chu LY, Golzarroshan B, Jain M, Yuan HS (2019) RNA recognition motifs of disease-linked RNA-binding proteins contribute to amyloid formation. Sci Rep 9(1):6171

59. Babinchak WM, Haider R, Dumm BK, Sarkar P, Surewicz K, Choi JK et al (2019) The role of liquid-liquid phase separation in aggregation of the TDP-43 low-complexity domain. J Biol Chem 294(16):6306–6317

60. Crozat A, Aman P, Mandahl N, Ron D (1993) Fusion of CHOP to a novel RNA-binding protein in human myxoid liposarcoma. Nature 363(6430):640–644

61. Ichikawa H, Shimizu K, Hayashi Y, Ohki M (1994) An RNA-binding protein gene, TLS/FUS, is fused to ERG in human myeloid leukemia with t(16;21) chromosomal translocation. Cancer Res 54(11):2865–2868

62. Calvio C, Neubauer G, Mann M, Lamond AI (1995) Identification of hnRNP P2 as TLS/FUS using electrospray mass spectrometry. RNA 1(7):724–733

63. Morohoshi F, Ootsuka Y, Arai K, Ichikawa H, Mitani S, Munakata N et al (1998) Genomic structure of the human RBP56/hTAFII68 and FUS/TLS genes. Gene 221(2):191–198

64. Mackenzie IR, Neumann M (2012) FET proteins in frontotemporal dementia and amyotrophic lateral sclerosis. Brain Res 1462:40–43

65. Deng H, Gao K, Jankovic J (2014) The role of FUS gene variants in neurodegenerative diseases. Nat Rev Neurol 10(6):337–348

66. Yang L, Gal J, Chen J, Zhu H (2014) Self-assembled FUS binds active chromatin and regulates gene transcription. Proc Natl Acad Sci U S A 111(50):17809–17814

67. Liu X, Niu C, Ren J, Zhang J, Xie X, Zhu H et al (2013) The RRM domain of human fused in sarcoma protein reveals a non-canonical nucleic acid binding site. Biochim Biophys Acta 1832(2):375–385

68. Loughlin FE, Lukavsky PJ, Kazeeva T, Reber S, Hock EM, Colombo M et al (2019) The solution structure of FUS bound to RNA reveals a bipartite

mode of RNA recognition with both sequence and shape specificity. Mol Cell 73(3):490–504 e6

69. Rogelj B, Easton LE, Bogu GK, Stanton LW, Rot G, Curk T et al (2012) Widespread binding of FUS along nascent RNA regulates alternative splicing in the brain. Sci Rep 2:603

70. Bhardwaj A, Myers MP, Buratti E, Baralle FE (2013) Characterizing TDP-43 interaction with its RNA targets. Nucleic Acids Res 41(9):5062–5074

71. Polymenidou M, Lagier-Tourenne C, Hutt KR, Huelga SC, Moran J, Liang TY et al (2011) Long pre-mRNA depletion and RNA missplicing contribute to neuronal vulnerability from loss of TDP-43. Nat Neurosci 14(4):459–468

72. Colombrita C, Onesto E, Megiorni F, Pizzuti A, Baralle FE, Buratti E et al (2012) TDP-43 and FUS RNA-binding proteins bind distinct sets of cytoplasmic messenger RNAs and differently regulate their post-transcriptional fate in motoneuron-like cells. J Biol Chem 287(19):15635–15647

73. Lu Y, Lim L, Song J (2017) RRM domain of ALS/FTD-causing FUS characteristic of irreversible unfolding spontaneously self-assembles into amyloid fibrils. Sci Rep 7(1):1043

74. Wang X, Schwartz JC, Cech TR (2015) Nucleic acid-binding specificity of human FUS protein. Nucleic Acids Res 43(15):7535–7543

75. Rothstein JD (2007) TDP-43 in amyotrophic lateral sclerosis: pathophysiology or patho-babel? Ann Neurol 61(5):382–384

76. Kabashi E, Valdmanis PN, Dion P, Spiegelman D, McConkey BJ, Vande Velde C et al (2008) TARDBP mutations in individuals with sporadic and familial amyotrophic lateral sclerosis. Nat Genet 40(5):572–574

77. Sreedharan J, Blair IP, Tripathi VB, Hu X, Vance C, Rogelj B et al (2008) TDP-43 mutations in familial and sporadic amyotrophic lateral sclerosis. Science 319(5870):1668–1672

78. Gitcho MA, Baloh RH, Chakraverty S, Mayo K, Norton JB, Levitch D et al (2008) TDP-43 A315T mutation in familial motor neuron disease. Ann Neurol 63(4):535–538

79. Patel A, Lee HO, Jawerth L, Maharana S, Jahnel M, Hein MY et al (2015) A liquid-to-solid phase transition of the ALS protein FUS accelerated by disease mutation. Cell 162(5):1066–1077

80. Chen HJ, Topp SD, Hui HS, Zacco E, Katarya M, McLoughlin C et al (2019) RRM adjacent TARDBP mutations disrupt RNA binding and enhance TDP-43 proteinopathy. Brain 142(12):3753–3770

81. Newell K, Paron F, Mompean M, Murrell J, Salis E, Stuani C et al (2019) Dysregulation of TDP-43 intracellular localization and early onset ALS are associated with a TARDBP S375G variant. Brain Pathol 29(3):397–413

82. Cohen TJ, Hwang AW, Restrepo CR, Yuan CX, Trojanowski JQ, Lee VM (2015) An acetylation switch controls TDP-43 function and aggregation propensity. Nat Commun 6:5845

83. Wang P, Wander CM, Yuan CX, Bereman MS, Cohen TJ (2017) Acetylation-induced TDP-43 pathology is suppressed by an HSF1-dependent chaperone program. Nat Commun 8(1):82

84. Jiang T, Handley E, Brizuela M, Dawkins E, Lewis KEA, Clark RM et al (2019) Amyotrophic lateral sclerosis mutant TDP-43 may cause synaptic dysfunction through altered dendritic spine function. Dis Model Mech 12(5):dmm038109

85. Alami NH, Smith RB, Carrasco MA, Williams LA, Winborn CS, Han SS et al (2014) Axonal transport of TDP-43 mRNA granules is impaired by ALS-causing mutations. Neuron 81(3):536–543

86. Naumann M, Peikert K, Gunther R, van der Kooi AJ, Aronica E, Hubers A et al (2019) Phenotypes and malignancy risk of different FUS mutations in genetic amyotrophic lateral sclerosis. Ann Clin Transl Neurol 6(12):2384–2394

87. Lattante S, Rouleau GA, Kabashi E (2013) TARDBP and FUS mutations associated with amyotrophic lateral sclerosis: summary and update. Hum Mutat 34(6):812–826

88. Zhou Y, Liu S, Liu G, Ozturk A, Hicks GG (2013) ALS-associated FUS mutations result in compromised FUS alternative splicing and autoregulation. PLoS Genet 9(10):e1003895

89. Yamazaki T, Chen S, Yu Y, Yan B, Haertlein TC, Carrasco MA et al (2012) FUS-SMN protein interactions link the motor neuron diseases ALS and SMA. Cell Rep 2(4):799–806

90. Tsuiji H, Iguchi Y, Furuya A, Kataoka A, Hatsuta H, Atsuta N et al (2013) Spliceosome integrity is defective in the motor neuron diseases ALS and SMA. EMBO Mol Med 5(2):221–234

91. Sun S, Ling SC, Qiu J, Albuquerque CP, Zhou Y, Tokunaga S et al (2015) ALS-causative mutations in FUS/TLS confer gain and loss of function by altered association with SMN and U1-snRNP. Nat Commun 6:6171

92. Yu Y, Chi B, Xia W, Gangopadhyay J, Yamazaki T, Winkelbauer-Hurt ME et al (2015) U1 snRNP is mislocalized in ALS patient fibroblasts bearing NLS mutations in FUS and is required for motor neuron outgrowth in zebrafish. Nucleic Acids Res 43(6):3208–3218

93. Hoell JI, Larsson E, Runge S, Nusbaum JD, Duggimpudi S, Farazi TA et al (2011) RNA targets of wild-type and mutant FET family proteins. Nat Struct Mol Biol 18(12):1428–1431

94. Nishimoto Y, Nakagawa S, Hirose T, Okano HJ, Takao M, Shibata S et al (2013) The long noncoding RNA nuclear-enriched abundant transcript 1_2 induces paraspeckle formation in the motor neuron during the early phase of amyotrophic lateral sclerosis. Mol Brain 6:31

95. Shelkovnikova TA, Robinson HK, Connor-Robson N, Buchman VL (2013) Recruitment into stress granules prevents irreversible aggregation of FUS protein mislocalized to the cytoplasm. Cell Cycle 12(19):3194–3202

96. De Santis R, Alfano V, de Turris V, Colantoni A, Santini L, Garone MG et al (2019) Mutant FUS and ELAVL4 (HuD) aberrant crosstalk in amyotrophic lateral sclerosis. Cell Rep 27(13):3818–31 e5

97. Nishimura AL, Shum C, Scotter EL, Abdelgany A, Sardone V, Wright J et al (2014) Allele-specific knockdown of ALS-associated mutant TDP-43 in neural stem cells derived from induced pluripotent stem cells. PLoS One 9(3):e91269

98. Guerrero EN, Mitra J, Wang H, Rangaswamy S, Hegde PM, Basu P et al (2019) Amyotrophic lateral sclerosis-associated TDP-43 mutation Q331K prevents nuclear translocation of XRCC4-DNA ligase 4 complex and is linked to genome damage-mediated neuronal apoptosis. Hum Mol Genet 28(18):3161–3162

99. Naumann M, Pal A, Goswami A, Lojewski X, Japtok J, Vehlow A et al (2018) Impaired DNA damage response signaling by FUS-NLS mutations leads to neurodegeneration and FUS aggregate formation. Nat Commun 9(1):335

100. White MA, Kim E, Duffy A, Adalbert R, Phillips BU, Peters OM et al (2018) TDP-43 gains function due to perturbed autoregulation in a Tardbp knock-in mouse model of ALS-FTD. Nat Neurosci 21(4):552–563

101. Fratta P, Sivakumar P, Humphrey J, Lo K, Ricketts T, Oliveira H et al (2018) Mice with endogenous TDP-43 mutations exhibit gain of splicing function and characteristics of amyotrophic lateral sclerosis. EMBO J 37(11):e98684

102. Mackenzie IR, Bigio EH, Ince PG, Geser F, Neumann M, Cairns NJ et al (2007) Pathological TDP-43 distinguishes sporadic amyotrophic lateral sclerosis from amyotrophic lateral sclerosis with SOD1 mutations. Ann Neurol 61(5):427–434

103. Tan CF, Eguchi H, Tagawa A, Onodera O, Iwasaki T, Tsujino A et al (2007) TDP-43 immunoreactivity in neuronal inclusions in familial amyotrophic lateral sclerosis with or without SOD1 gene mutation. Acta Neuropathol (Berl) 113(5):535–542

104. Ling SC, Polymenidou M, Cleveland DW (2013) Converging mechanisms in ALS and FTD: disrupted RNA and protein homeostasis. Neuron 79(3):416–438

105. Ling SC, Albuquerque CP, Han JS, Lagier-Tourenne C, Tokunaga S, Zhou H et al (2010) ALS-associated mutations in TDP-43 increase its stability and promote TDP-43 complexes with FUS/TLS. Proc Natl Acad Sci U S A 107(30):13318–13323

106. Kryndushkin D, Wickner RB, Shewmaker F (2011) FUS/TLS forms cytoplasmic aggregates, inhibits cell growth and interacts with TDP-43 in a yeast model of amyotrophic lateral sclerosis. Protein Cell 2(3):223–236

107. Lee EB, Lee VM, Trojanowski JQ (2012) Gains or losses: molecular mechanisms of TDP43-mediated neurodegeneration. Nat Rev Neurosci 13(1):38–50

108. Seyfried NT, Gozal YM, Dammer EB, Xia Q, Duong DM, Cheng D et al (2010) Multiplex SILAC analysis

of a cellular TDP-43 proteinopathy model reveals protein inclusions associated with SUMOylation and diverse polyubiquitin chains. Mol Cell Proteomics 9(4):705–718

109. Murnyak B, Bodoki L, Vincze M, Griger Z, Csonka T, Szepesi R et al (2015) Inclusion body myositis – pathomechanism and lessons from genetics. Open Med (Wars) 10(1):188–193

110. Dardis A, Zampieri S, Canterini S, Newell KL, Stuani C, Murrell JR et al (2016) Altered localization and functionality of TAR DNA Binding Protein 43 (TDP-43) in niemann-pick disease type C. Acta Neuropathol Commun 4(1):52

111. Neumann M (2013) Frontotemporal lobar degeneration and amyotrophic lateral sclerosis: molecular similarities and differences. Rev Neurol (Paris) 169(10):793–798

112. Monahan Z, Ryan VH, Janke AM, Burke KA, Rhoads SN, Zerze GH et al (2017) Phosphorylation of the FUS low-complexity domain disrupts phase separation, aggregation, and toxicity. EMBO J 36(20):2951–2967

113. Capitini C, Conti S, Perni M, Guidi F, Cascella R, De Poli A et al (2014) TDP-43 inclusion bodies formed in bacteria are structurally amorphous, non-amyloid and inherently toxic to neuroblastoma cells. PLoS One 9(1):e86720

114. Robinson JL, Geser F, Stieber A, Umoh M, Kwong LK, Van Deerlin VM et al (2013) TDP-43 skeins show properties of amyloid in a subset of ALS cases. Acta Neuropathol 125(1):121–131

115. Bigio EH, Wu JY, Deng HX, Bit-Ivan EN, Mao Q, Ganti R et al (2013) Inclusions in frontotemporal lobar degeneration with TDP-43 proteinopathy (FTLD-TDP) and amyotrophic lateral sclerosis (ALS), but not FTLD with FUS proteinopathy (FTLD-FUS), have properties of amyloid. Acta Neuropathol 125(3):463–465

116. Chen AK, Lin RY, Hsieh EZ, Tu PH, Chen RP, Liao TY et al (2010) Induction of amyloid fibrils by the C-terminal fragments of TDP-43 in amyotrophic lateral sclerosis. J Am Chem Soc 132(4):1186–1187

117. Saini A, Chauhan VS (2011) Delineation of the core aggregation sequences of TDP-43 C-terminal fragment. Chembiochem 12(16):2495–2501

118. Saini A, Chauhan VS (2014) Self-assembling properties of peptides derived from TDP-43 C-terminal fragment. Langmuir 30(13):3845–3856

119. Guo W, Chen Y, Zhou X, Kar A, Ray P, Chen X et al (2011) An ALS-associated mutation affecting TDP-43 enhances protein aggregation, fibril formation and neurotoxicity Nat Struct Mol Biol 18(7):822–830

120. Sun CS, Wang CY, Chen BP, He RY, Liu GC, Wang CH et al (2014) The influence of pathological mutations and proline substitutions in TDP-43 glycine-rich peptides on its amyloid properties and cellular toxicity. PLoS One 9(8):e103644

121. Mompean M, Buratti E, Guarnaccia C, Brito RM, Chakrabartty A, Baralle FE et al (2014) Structural characterization of the minimal segment of TDP-43

competent for aggregation. Arch Biochem Biophys 545C:53–62

122. Mompean M, Hervas R, Xu Y, Tran TH, Guarnaccia C, Buratti E et al (2015) Structural evidence of amyloid fibril formation in the putative aggregation domain of TDP-43. J Phys Chem Lett 6(13):2608–2615

123. Shelkovnikova TA, Peters OM, Deykin AV, Connor-Robson N, Robinson H, Ustyugov AA et al (2013) Fused in sarcoma (FUS) protein lacking nuclear localization signal (NLS) and major RNA binding motifs triggers proteinopathy and severe motor phenotype in transgenic mice. J Biol Chem 288(35):25266–25274

124. Ding X, Sun F, Chen J, Chen L, Tobin-Miyaji Y, Xue S et al (2020) Amyloid-forming segment induces aggregation of FUS-LC domain from phase separation modulated by site-specific phosphorylation. J Mol Biol 432(2):467–483

125. Buratti E, Baralle FE (2012) TDP-43: gumming up neurons through protein-protein and protein-RNA interactions. Trends Biochem Sci 37(6):237–247

126. Shelkovnikova TA (2013) Modelling FUSopathies: focus on protein aggregation. Biochem Soc Trans 41(6):1613–1617

127. Hans F, Glasebach H, Kahle PJ (2020) Multiple distinct pathways lead to hyperubiquitylated insoluble TDP-43 protein independent of its translocation into stress granules. J Biol Chem 295(3):673–689

128. Wolozin B, Ivanov P (2019) Stress granules and neurodegeneration. Nat Rev Neurosci 20(11):649–666

129. Baradaran-Heravi Y, Van Broeckhoven C, van der Zee J (2020) Stress granule mediated protein aggregation and underlying gene defects in the FTD-ALS spectrum. Neurobiol Dis 134:104639

130. Colombrita C, Zennaro E, Fallini C, Weber M, Sommacal A, Buratti E et al (2009) TDP-43 is recruited to stress granules in conditions of oxidative insult. J Neurochem 111(4):1051–1061

131. Bosco DA, Lemay N, Ko HK, Zhou H, Burke C, Kwiatkowski TJ Jr et al (2010) Mutant FUS proteins that cause amyotrophic lateral sclerosis incorporate into stress granules. Hum Mol Genet 19(21):4160–4175

132. Li YR, King OD, Shorter J, Gitler AD (2013) Stress granules as crucibles of ALS pathogenesis. J Cell Biol 201(3):361–372

133. Dormann D, Haass C (2013) Fused in sarcoma (FUS): an oncogene goes awry in neurodegeneration. Mol Cell Neurosci 56:475–486

134. Fang MY, Markmiller S, Vu AQ, Javaherian A, Dowdle WE, Jolivet P et al (2019) Small-molecule modulation of TDP-43 recruitment to stress granules prevents persistent TDP-43 accumulation in ALS/FTD. Neuron 103(5):802–19 e11

135. Blokhuis AM, Groen EJ, Koppers M, van den Berg LH, Pasterkamp RJ (2013) Protein aggregation in amyotrophic lateral sclerosis. Acta Neuropathol 125(6):777–794

136. Fang YS, Tsai KJ, Chang YJ, Kao P, Woods R, Kuo PH et al (2014) Full-length TDP-43 forms toxic

amyloid oligomers that are present in frontotemporal lobar dementia-TDP patients. Nat Commun 5:4824

137. Johnson BS, Snead D, Lee JJ, McCaffery JM, Shorter J, Gitler AD (2009) TDP-43 is intrinsically aggregation-prone, and amyotrophic lateral sclerosis-linked mutations accelerate aggregation and increase toxicity. J Biol Chem 284(30):20329–20339

138. Woerner AC, Frottin F, Hornburg D, Feng LR, Meissner F, Patra M et al (2016) Cytoplasmic protein aggregates interfere with nucleo-cytoplasmic transport of protein and RNA. Science 351(6269):173–176

139. Dammer EB, Fallini C, Gozal YM, Duong DM, Rossoll W, Xu P et al (2012) Coaggregation of RNA-binding proteins in a model of TDP-43 proteinopathy with selective RGG motif methylation and a role for RRM1 ubiquitination. PLoS One 7(6):e38658

140. Collins M, Riascos D, Kovalik T, An J, Krupa K, Hood BL et al (2012) The RNA-binding motif 45 (RBM45) protein accumulates in inclusion bodies in amyotrophic lateral sclerosis (ALS) and frontotemporal lobar degeneration with TDP-43 inclusions (FTLD-TDP) patients. Acta Neuropathol 124(5):717–732

141. Shelkovnikova TA, Robinson HK, Troakes C, Ninkina N, Buchman VL (2014) Compromised paraspeckle formation as a pathogenic factor in FUSopathies. Hum Mol Genet 23(9):2298–2312

142. Cragnaz L, Klima R, Skoko N, Budini M, Feiguin F, Baralle FE (2014) Aggregate formation prevents dTDP-43 neurotoxicity in the Drosophila melanogaster eye. Neurobiol Dis 71:74–80

143. Leitman J, Ulrich Hartl F, Lederkremer GZ (2013) Soluble forms of polyQ-expanded huntingtin rather than large aggregates cause endoplasmic reticulum stress. Nat Commun 4:2753

144. Bolognesi B, Faure AJ, Seuma M, Schmiedel JM, Tartaglia GG, Lehner B (2019) The mutational landscape of a prion-like domain. Nat Commun 10(1):4162

145. Wegorzewska I, Baloh RH (2011) TDP-43-based animal models of neurodegeneration: new insights into ALS pathology and pathophysiology. Neurodegener Dis 8(4):262–274

146. Armstrong GA, Drapeau P (2013) Loss and gain of FUS function impair neuromuscular synaptic transmission in a genetic model of ALS. Hum Mol Genet 22(21):4282–4292

147. Xu ZS (2012) Does a loss of TDP-43 function cause neurodegeneration? Mol Neurodegener 7:27

148. Vanden Broeck L, Callaerts P, Dermaut B (2014) TDP-43-mediated neurodegeneration: towards a loss-of-function hypothesis? Trends Mol Med 20(2):66–71

149. Kino Y, Washizu C, Kurosawa M, Yamada M, Miyazaki H, Akagi T et al (2015) FUS/TLS deficiency causes behavioral and pathological abnormalities distinct from amyotrophic lateral sclerosis. Acta Neuropathol Commun 3:24

150. Buratti E, Baralle FE (2009) The molecular links between TDP-43 dysfunction and neurodegeneration. Adv Genet 66:1–34

151. Halliday G, Bigio EH, Cairns NJ, Neumann M, Mackenzie IR, Mann DM (2012) Mechanisms of disease in frontotemporal lobar degeneration: gain of function versus loss of function effects. Acta Neuropathol 124(3):373–382

152. Birsa N, Bentham MP, Fratta P (2020) Cytoplasmic functions of TDP-43 and FUS and their role in ALS. Semin Cell Dev Biol 99:193–201

153. Kawaguchi T, Rollins MG, Moinpour M, Morera AA, Ebmeier CC, Old WM et al (2020) Changes to the TDP-43 and FUS interactomes induced by DNA damage. J Proteome Res 19(1):360–370

154. Berson A, Sartoris A, Nativio R, Van Deerlin V, Toledo JB, Porta S et al (2017) TDP-43 promotes neurodegeneration by impairing chromatin remodeling. Curr Biol 27(23):3579–90 e6

155. Krug L, Chatterjee N, Borges-Monroy R, Hearn S, Liao WW, Morrill K et al (2017) Retrotransposon activation contributes to neurodegeneration in a Drosophila TDP-43 model of ALS. PLoS Genet 13(3):e1006635

156. Guerrero EN, Mitra J, Wang H, Rangaswamy S, Hegde PM, Basu P et al (2019) Amyotrophic lateral sclerosis-associated TDP-43 mutation Q331K prevents nuclear translocation of XRCC4-DNA ligase 4 complex and is linked to genome damage-mediated neuronal apoptosis. Hum Mol Genet 28(5):2459–2476

157. Mitra J, Guerrero EN, Hegde PM, Liachko NF, Wang H, Vasquez V et al (2019) Motor neuron disease-associated loss of nuclear TDP-43 is linked to DNA double-strand break repair defects. Proc Natl Acad Sci U S A 116(10):4696–4705

158. Baechtold H, Kuroda M, Sok J, Ron D, Lopez BS, Akhmedov AT (1999) Human 75-kDa DNA-pairing protein is identical to the pro-oncoprotein TLS/FUS and is able to promote D-loop formation. J Biol Chem 274(48):34337–34342

159. Hicks GG, Singh N, Nashabi A, Mai S, Bozek G, Klewes L et al (2000) Fus deficiency in mice results in defective B-lymphocyte development and activation, high levels of chromosomal instability and perinatal death. Nat Genet 24(2):175–179

160. Gardiner M, Toth R, Vandermoere F, Morrice NA, Rouse J (2008) Identification and characterization of FUS/TLS as a new target of ATM. Biochem J 415(2):297–307

161. Rulten SL, Rotheray A, Green RL, Grundy GJ, Moore DA, Gomez-Herreros F et al (2014) PARP-1 dependent recruitment of the amyotrophic lateral sclerosis-associated protein FUS/TLS to sites of oxidative DNA damage. Nucleic Acids Res 42(1):307–314

162. Mastrocola AS, Kim SH, Trinh AT, Rodenkirch LA, Tibbetts RS (2013) The RNA-binding protein fused in sarcoma (FUS) functions downstream of poly(ADP-ribose) polymerase (PARP) in response to DNA damage. J Biol Chem 288(34):24731–24741

163. Wang WY, Pan L, Su SC, Quinn EJ, Sasaki M, Jimenez JC et al (2013) Interaction of FUS and

HDAC1 regulates DNA damage response and repair in neurons. Nat Neurosci 16(10):1383–1391

164. Wang X, Arai S, Song X, Reichart D, Du K, Pascual G et al (2008) Induced ncRNAs allosterically modify RNA-binding proteins in cis to inhibit transcription. Nature 454(7200):126–130

165. Ou SH, Wu F, Harrich D, Garcia-Martinez LF, Gaynor RB (1995) Cloning and characterization of a novel cellular protein, TDP-43, that binds to human immunodeficiency virus type 1 TAR DNA sequence motifs. J Virol 69(6):3584–3596

166. Nehls J, Koppensteiner H, Brack-Werner R, Floss T, Schindler M (2014) HIV-1 replication in human immune cells is independent of TAR DNA binding protein 43 (TDP-43) expression. PLoS One 9(8):e105478

167. Abhyankar MM, Urekar C, Reddi PP (2007) A novel CpG-free vertebrate insulator silences the testis-specific SP-10 gene in somatic tissues: role for TDP-43 in insulator function. J Biol Chem 282(50):36143–36154

168. Murata H, Hattori T, Maeda H, Takashiba S, Takigawa M, Kido J et al (2015) Identification of transactivation-responsive DNA-binding protein 43 (TARDBP43; TDP-43) as a novel factor for TNF-alpha expression upon lipopolysaccharide stimulation in human monocytes. J Periodontal Res 50(4):452–460

169. Schwenk BM, Hartmann H, Serdaroglu A, Schludi MH, Hornburg D, Meissner F et al (2016) TDP-43 loss of function inhibits endosomal trafficking and alters trophic signaling in neurons. EMBO J 35(21):2350–2370

170. Schwartz JC, Ebmeier CC, Podell ER, Heimiller J, Taatjes DJ, Cech TR (2012) FUS binds the CTD of RNA polymerase II and regulates its phosphorylation at Ser2. Genes Dev 26(24):2690–2695

171. Schwartz JC, Podell ER, Han SS, Berry JD, Eggan KC, Cech TR (2014) FUS is sequestered in nuclear aggregates in ALS patient fibroblasts. Mol Biol Cell 25(17):2571–2578

172. Masuda A, Takeda J, Okuno T, Okamoto T, Ohkawara B, Ito M et al (2015) Position-specific binding of FUS to nascent RNA regulates mRNA length. Genes Dev 29(10):1045–1057

173. Buratti E, Baralle FE (2011) TDP-43: new aspects of autoregulation mechanisms in RNA binding proteins and their connection with human disease. FEBS J 278(19):3530–3538

174. Ayala YM, De Conti L, Avendano-Vazquez SE, Dhir A, Romano M, D'Ambrogio A et al (2011) TDP-43 regulates its mRNA levels through a negative feedback loop. EMBO J 30(2):277–288

175. Sugai A, Kato T, Koyama A, Koike Y, Konno T, Ishihara T et al (2019) Non-genetically modified models exhibit TARDBP mRNA increase due to perturbed TDP-43 autoregulation. Neurobiol Dis 130:104534

176. Maquat LE (2005) Nonsense-mediated mRNA decay in mammals. J Cell Sci 118(Pt 9):1773–1776

177. Dini Modigliani S, Morlando M, Errichelli L, Sabatelli M, Bozzoni I (2014) An ALS-associated mutation in the FUS 3'-UTR disrupts a microRNA-FUS regulatory circuitry. Nat Commun 5:4335

178. Budini M, Buratti E (2011) TDP-43 autoregulation: implications for disease. J Mol Neurosci 45(3):473–479

179. Buratti E, Dork T, Zuccato E, Pagani F, Romano M, Baralle FE (2001) Nuclear factor TDP-43 and SR proteins promote in vitro and in vivo CFTR exon 9 skipping. EMBO J 20(7):1774–1784

180. Sephton CF, Good SK, Atkin S, Dewey CM, Mayer P 3rd, Herz J et al (2010) TDP-43 is a developmentally regulated protein essential for early embryonic development. J Biol Chem 285(9):6826–6834

181. Tollervey JR, Curk T, Rogelj B, Briese M, Cereda M, Kayikci M et al (2011) Characterizing the RNA targets and position-dependent splicing regulation by TDP-43. Nat Neurosci 14(4):452–458

182. Xiao S, Sanelli T, Dib S, Sheps D, Findlater J, Bilbao J et al (2011) RNA targets of TDP-43 identified by UV-CLIP are deregulated in ALS. Mol Cell Neurosci 47(3):167–180

183. Buratti E, Romano M, Baralle FE (2013) TDP-43 high throughput screening analyses in neurodegeneration: advantages and pitfalls. Mol Cell Neurosci 56C:465–474

184. Shiga A, Ishihara T, Miyashita A, Kuwabara M, Kato T, Watanabe N et al (2012) Alteration of POLDIP3 splicing associated with loss of function of TDP-43 in tissues affected with ALS. PLoS One 7(8):e43120

185. Prudencio M, Jansen-West KR, Lee WC, Gendron TF, Zhang YJ, Xu YF et al (2012) Misregulation of human sortilin splicing leads to the generation of a nonfunctional progranulin receptor. Proc Natl Acad Sci U S A 109(52):21510–21515

186. Mohagheghi F, Prudencio M, Stuani C, Cook C, Jansen-West K, Dickson DW et al (2016) TDP-43 functions within a network of hnRNP proteins to inhibit the production of a truncated human SORT1 receptor. Hum Mol Genet 25(3):534–545

187. De Conti L, Akinyi MV, Mendoza-Maldonado R, Romano M, Baralle M, Buratti E (2015) TDP-43 affects splicing profiles and isoform production of genes involved in the apoptotic and mitotic cellular pathways. Nucleic Acids Res 43(18):8990–9005

188. Colombrita C, Onesto E, Buratti E, de la Grange P, Gumina V, Baralle FE et al (2015) From transcriptomic to protein level changes in TDP-43 and FUS loss-of-function cell models. Biochim Biophys Acta 1849(12):1398–1410

189. Deshaies JE, Shkreta L, Moszczynski AJ, Sidibe H, Semmler S, Fouillen A et al (2018) TDP-43 regulates the alternative splicing of hnRNP A1 to yield an aggregation-prone variant in amyotrophic lateral sclerosis. Brain 141(5):1320–1333

190. Gu J, Chen F, Iqbal K, Gong CX, Wang X, Liu F (2017) Transactive response DNA-binding protein 43 (TDP-43) regulates alternative splicing of tau

exon 10: implications for the pathogenesis of tauopathies. J Biol Chem 292(25):10600–10612

191. Bose JK, Wang IF, Hung L, Tarn WY, Shen CK (2008) TDP-43 overexpression enhances exon 7 inclusion during the survival of motor neuron pre-mRNA splicing. J Biol Chem 283(43):28852–28859

192. Ling JP, Pletnikova O, Troncoso JC, Wong PC (2015) TDP-43 repression of nonconserved cryptic exons is compromised in ALS-FTD. Science 349(6248):650–655

193. Tan Q, Yalamanchili HK, Park J, De Maio A, Lu HC, Wan YW et al (2016) Extensive cryptic splicing upon loss of RBM17 and TDP43 in neurodegeneration models. Hum Mol Genet 25(23):5083–5093

194. Donde A, Sun M, Ling JP, Braunstein KE, Pang B, Wen X et al (2019) Splicing repression is a major function of TDP-43 in motor neurons. Acta Neuropathol 138(5):813–826

195. Gerbino V, Carri MT, Cozzolino M, Achsel T (2013) Mislocalised FUS mutants stall spliceosomal snRNPs in the cytoplasm. Neurobiol Dis 55:120–128

196. Yu Y, Reed R (2015) FUS functions in coupling transcription to splicing by mediating an interaction between RNAP II and U1 snRNP. Proc Natl Acad Sci U S A 112(28):8608–8613

197. Raczynska KD, Ruepp MD, Brzek A, Reber S, Romeo V, Rindlisbacher B et al (2015) FUS/TLS contributes to replication-dependent histone gene expression by interaction with U7 snRNPs and histone-specific transcription factors. Nucleic Acids Res 43(20):9711–9728

198. Lagier-Tourenne C, Polymenidou M, Hutt KR, Vu AQ, Baughn M, Huelga SC et al (2012) Divergent roles of ALS-linked proteins FUS/TLS and TDP-43 intersect in processing long pre-mRNAs. Nat Neurosci 15(11):1488–1497

199. Ishigaki S, Masuda A, Fujioka Y, Iguchi Y, Katsuno M, Shibata A et al (2012) Position-dependent FUS-RNA interactions regulate alternative splicing events and transcriptions. Sci Rep 2:529

200. van Blitterswijk M, Wang ET, Friedman BA, Keagle PJ, Lowe P, Leclerc AL et al (2013) Characterization of FUS mutations in amyotrophic lateral sclerosis using RNA-Seq. PLoS One 8(4):e60788

201. Nakaya T, Alexiou P, Maragkakis M, Chang A, Mourelatos Z (2013) FUS regulates genes coding for RNA-binding proteins in neurons by binding to their highly conserved introns. RNA 19(4):498–509

202. Honda D, Ishigaki S, Iguchi Y, Fujioka Y, Udagawa T, Masuda A et al (2013) The ALS/FTLD-related RNA-binding proteins TDP-43 and FUS have common downstream RNA targets in cortical neurons. FEBS Open Bio 4:1–10

203. Orozco D, Edbauer D (2013) FUS-mediated alternative splicing in the nervous system: consequences for ALS and FTLD. J Mol Med (Berl) 91(12):1343–1354

204. Reber S, Stettler J, Filosa G, Colombo M, Jutzi D, Lenzken SC et al (2016) Minor intron splicing is regulated by FUS and affected by ALS-associated FUS mutants. EMBO J 35(14):1504–1521

205. Volonte C, Apolloni S, Parisi C (2015) MicroRNAs: newcomers into the ALS picture. CNS Neurol Disord Drug Targets 14(2):194–207

206. Goodall EF, Heath PR, Bandmann O, Kirby J, Shaw PJ (2013) Neuronal dark matter: the emerging role of microRNAs in neurodegeneration. Front Cell Neurosci 7:178

207. Gregory RI, Yan KP, Amuthan G, Chendrimada T, Doratotaj B, Cooch N et al (2004) The microprocessor complex mediates the genesis of microRNAs. Nature 432(7014):235–240

208. Casafont I, Bengoechea R, Tapia O, Berciano MT, Lafarga M (2009) TDP-43 localizes in mRNA transcription and processing sites in mammalian neurons. J Struct Biol 167(3):235–241

209. Di Carlo V, Grossi E, Laneve P, Morlando M, Dini Modigliani S, Ballarino M et al (2013) TDP-43 regulates the microprocessor complex activity during in vitro neuronal differentiation. Mol Neurobiol 48(3):952–963

210. Buratti E, De Conti L, Stuani C, Romano M, Baralle M, Baralle F (2010) Nuclear factor TDP-43 can affect selected microRNA levels. FEBS J 277(10):2268–2281

211. King IN, Yartseva V, Salas D, Kumar A, Heidersbach A, Ando DM et al (2014) The RNA-binding protein TDP-43 selectively disrupts microRNA-1/206 incorporation into the RNA-induced silencing complex. J Biol Chem 289(20):14263–14271

212. Zhang Z, Almeida S, Lu Y, Nishimura AL, Peng L, Sun D et al (2013) Downregulation of microRNA-9 in iPSC-derived neurons of FTD/ALS patients with TDP-43 mutations. PLoS One 8(10):e76055

213. Park YY, Kim SB, Han HD, Sohn BH, Kim JH, Liang J et al (2013) Tat-activating regulatory DNA-binding protein regulates glycolysis in hepatocellular carcinoma by regulating the platelet isoform of phosphofructokinase through microRNA 520. Hepatology 58(1):182–191

214. Kawahara Y, Mieda-Sato A (2012) TDP-43 promotes microRNA biogenesis as a component of the Drosha and Dicer complexes. Proc Natl Acad Sci U S A 109(9):3347–3352

215. Fan Z, Chen X, Chen R (2014) Transcriptome-wide analysis of TDP-43 binding small RNAs identifies miR-NID1 (miR-8485), a novel miRNA that represses NRXN1 expression. Genomics 103(1):76–82

216. Morlando M, Dini Modigliani S, Torrelli G, Rosa A, Di Carlo V, Caffarelli E et al (2012) FUS stimulates microRNA biogenesis by facilitating co-transcriptional Drosha recruitment. EMBO J 31(24):4502–4510

217. Grammatikakis I, Panda AC, Abdelmohsen K, Gorospe M (2014) Long noncoding RNAs(lncRNAs) and the molecular hallmarks of aging. Aging (Albany NY) 6(12):992–1009

218. Roberts TC, Morris KV, Wood MJ (2014) The role of long non-coding RNAs in neurodevelopment, brain function and neurological disease. Philos Trans R Soc Lond Ser B Biol Sci 369(1652):20130507.

219. Liu X, Li D, Zhang W, Guo M, Zhan Q (2012) Long non-coding RNA gadd7 interacts with TDP-43 and regulates Cdk6 mRNA decay. EMBO J 31(23):4415–4427

220. Wu H, Yin QF, Luo Z, Yao RW, Zheng CC, Zhang J et al (2016) Unusual processing generates SPA LncRNAs that sequester multiple RNA binding proteins. Mol Cell 64(3):534–548

221. Guo F, Jiao F, Song Z, Li S, Liu B, Yang H et al (2015) Regulation of MALAT1 expression by TDP43 controls the migration and invasion of non-small cell lung cancer cells in vitro. Biochem Biophys Res Commun 465(2):293–298

222. Balas MM, Porman AM, Hansen KC, Johnson AM (2018) SILAC-MS profiling of reconstituted human chromatin platforms for the study of transcription and RNA regulation. J Proteome Res 17(10):3475–3484

223. Militello G, Hosen MR, Ponomareva Y, Gellert P, Weirick T, John D et al (2018) A novel long non-coding RNA myolinc regulates myogenesis through TDP-43 and Filip1. J Mol Cell Biol 10(2):102–117

224. Li P, Ruan X, Yang L, Kiesewetter K, Zhao Y, Luo H et al (2015) A liver-enriched long non-coding RNA, lncLSTR, regulates systemic lipid metabolism in mice. Cell Metab 21(3):455–467

225. Zhao XS, Tao N, Zhang C, Gong CM, Dong CY (2019) Long noncoding RNA MIAT acts as an oncogene in Wilms' tumor through regulation of DGCR8. Eur Rev Med Pharmacol Sci 23(23):10257–10263

226. Lourenco GF, Janitz M, Huang Y, Halliday GM (2015) Long noncoding RNAs in TDP-43 and FUS/TLS-related frontotemporal lobar degeneration (FTLD). Neurobiol Dis 82:445–454

227. Biscarini S, Capauto D, Peruzzi G, Lu L, Colantoni A, Santini T et al (2018) Characterization of the lncRNA transcriptome in mESC-derived motor neurons: implications for FUS-ALS. Stem Cell Res 27:172–179

228. Lo Piccolo L, Jantrapirom S, Nagai Y, Yamaguchi M (2017) FUS toxicity is rescued by the modulation of lncRNA hsromega expression in Drosophila melanogaster. Sci Rep 7(1):15660

229. Gagliardi S, Pandini C, Garofalo M, Bordoni M, Pansarasa O, Cereda C (2018) Long non coding RNAs and ALS: still much to do. Noncoding RNA Res 3(4):226–231

230. Cirillo D, Agostini F, Klus P, Marchese D, Rodriguez S, Bolognesi B et al (2013) Neurodegenerative diseases: quantitative predictions of protein-RNA interactions. RNA 19(2):129–140

231. Li W, Jin Y, Prazak L, Hammell M, Dubnau J (2012) Transposable elements in TDP-43-mediated neurodegenerative disorders. PLoS One 7(9):e44099

232. Dormann D, Haass C (2011) TDP-43 and FUS: a nuclear affair. Trends Neurosci 34(7):339–348

233. Dormann D, Rodde R, Edbauer D, Bentmann E, Fischer I, Hruscha A et al (2010) ALS-associated fused in sarcoma (FUS) mutations disrupt Transportin-mediated nuclear import. EMBO J 29(16):2841–2857

234. Nishimura AL, Zupunski V, Troakes C, Kathe C, Fratta P, Howell M et al (2010) Nuclear import impairment causes cytoplasmic trans-activation response DNA-binding protein accumulation and is associated with frontotemporal lobar degeneration. Brain 133(Pt 6):1763–1771

235. Tradewell ML, Yu Z, Tibshirani M, Boulanger MC, Durham HD, Richard S (2012) Arginine methylation by PRMT1 regulates nuclear-cytoplasmic localization and toxicity of FUS/TLS harbouring ALS-linked mutations. Hum Mol Genet 21(1):136–149

236. Dormann D, Madl T, Valori CF, Bentmann E, Tahirovic S, Abou-Ajram C et al (2012) Arginine methylation next to the PY-NLS modulates Transportin binding and nuclear import of FUS. EMBO J 31(22):4258–4275

237. Scaramuzzino C, Monaghan J, Milioto C, Lanson NA Jr, Maltare A, Aggarwal T et al (2013) Protein arginine methyltransferase 1 and 8 interact with FUS to modify its sub-cellular distribution and toxicity in vitro and in vivo. PLoS One 8(4):e61576

238. Yamaguchi A, Kitajo K (2012) The effect of PRMT1-mediated arginine methylation on the subcellular localization, stress granules, and detergent-insoluble aggregates of FUS/TLS. PLoS One 7(11):e49267

239. Zhang YJ, Gendron TF, Grima JC, Sasaguri H, Jansen-West K, Xu YF et al (2016) C9ORF72 poly(GA) aggregates sequester and impair HR23 and nucleocytoplasmic transport proteins. Nat Neurosci 19(5):668–677

240. Chou CC, Zhang Y, Umoh ME, Vaughan SW, Lorenzini I, Liu F et al (2018) TDP-43 pathology disrupts nuclear pore complexes and nucleocytoplasmic transport in ALS/FTD. Nat Neurosci 21(2):228–239

241. Strong MJ, Volkening K, Hammond R, Yang W, Strong W, Leystra-Lantz C et al (2007) TDP43 is a human low molecular weight neurofilament (hNFL) mRNA-binding protein. Mol Cell Neurosci 35(2):320–327

242. Kim SH, Shanware NP, Bowler MJ, Tibbetts RS (2010) Amyotrophic lateral sclerosis-associated proteins TDP-43 and FUS/TLS function in a common biochemical complex to co-regulate HDAC6 mRNA. J Biol Chem 285(44):34097–34105

243. Fiesel FC, Schurr C, Weber SS, Kahle PJ (2011) TDP-43 knockdown impairs neurite outgrowth dependent on its target histone deacetylase 6. Mol Neurodegener 6:64

244. Sephton CF, Cenik C, Kucukural A, Dammer EB, Cenik B, Han Y et al (2011) Identification of neuronal RNA targets of TDP-43-containing ribonucleoprotein complexes. J Biol Chem 286(2):1204–1215

245. Costessi L, Porro F, Iaconcig A, Muro AF (2014) TDP-43 regulates beta-adducin (Add2) transcript stability. RNA Biol 11(10):1280–1290

246. Lee S, Lee TA, Lee E, Kang S, Park A, Kim SW et al (2015) Identification of a subnuclear body involved in sequence-specific cytokine RNA processing. Nat Commun 6:5791

247. Stallings NR, Puttaparthi K, Dowling KJ, Luther CM, Burns DK, Davis K et al (2013) TDP-43, an ALS linked protein, regulates fat deposition and glucose homeostasis. PLoS One 8(8):e71793

248. McDonald KK, Aulas A, Destroismaisons L, Pickles S, Beleac E, Camu W et al (2011) TAR DNA-binding protein 43 (TDP-43) regulates stress granule dynamics via differential regulation of G3BP and TIA-1. Hum Mol Genet 20(7):1400–1410

249. Vanderweyde T, Youmans K, Liu-Yesucevitz L, Wolozin B (2013) Role of stress granules and RNA-binding proteins in neurodegeneration: a mini-review. Gerontology 59(6):524–533

250. Gu J, Wu F, Xu W, Shi J, Hu W, Jin N et al (2017) TDP-43 suppresses tau expression via promoting its mRNA instability. Nucleic Acids Res 45(10):6177–6193

251. Udagawa T, Fujioka Y, Tanaka M, Honda D, Yokoi S, Riku Y et al (2015) FUS regulates AMPA receptor function and FTLD/ALS-associated behaviour via GluA1 mRNA stabilization. Nat Commun 6:7098

252. Swanger SA, Bassell GJ (2011) Making and breaking synapses through local mRNA regulation. Curr Opin Genet Dev 21(4):414–421

253. Fallini C, Bassell GJ, Rossoll W (2012) The ALS disease protein TDP-43 is actively transported in motor neuron axons and regulates axon outgrowth. Hum Mol Genet 21(16):3703–3718

254. Narayanan RK, Mangelsdorf M, Panwar A, Butler TJ, Noakes PG, Wallace RH (2013) Identification of RNA bound to the TDP-43 ribonucleoprotein complex in the adult mouse brain. Amyotroph Lateral Scler Frontotemporal Degener 14(4):252–260

255. Fujii R, Okabe S, Urushido T, Inoue K, Yoshimura A, Tachibana T et al (2005) The RNA binding protein TLS is translocated to dendritic spines by mGluR5 activation and regulates spine morphology. Curr Biol 15(6):587–593

256. Belly A, Moreau-Gachelin F, Sadoul R, Goldberg Y (2005) Delocalization of the multifunctional RNA splicing factor TLS/FUS in hippocampal neurones: exclusion from the nucleus and accumulation in dendritic granules and spine heads. Neurosci Lett 379(3):152–157

257. Fujii R, Takumi T (2005) TLS facilitates transport of mRNA encoding an actin-stabilizing protein to dendritic spines. J Cell Sci 118(Pt 24):5755–5765

258. Yoshimura A, Fujii R, Watanabe Y, Okabe S, Fukui K, Takumi T (2006) Myosin-Va facilitates the accumulation of mRNA/protein complex in dendritic spines. Curr Biol 16(23):2345–2351

259. Muresan V, Ladescu MZ (2016) Shared molecular mechanisms in Alzheimer's disease and amyotrophic lateral sclerosis: neurofilament-dependent transport of sAPP, FUS, TDP-43 and SOD1, with endoplasmic reticulum-like tubules. Neurodegener Dis 16(1–2):55–61

260. Wang IF, Wu LS, Chang HY, Shen CK (2008) TDP-43, the signature protein of FTLD-U, is a neuronal activity-responsive factor. J Neurochem 105(3):797–806

261. Freibaum BD, Chitta RK, High AA, Taylor JP (2010) Global analysis of TDP-43 interacting proteins reveals strong association with RNA splicing and translation machinery. J Proteome Res 9(2):1104–1120

262. Godena VK, Romano G, Romano M, Appocher C, Klima R, Buratti E et al (2011) TDP-43 regulates drosophila neuromuscular junctions growth by modulating Futsch/MAP1B levels and synaptic microtubules organization. PLoS One 6(3):e17808

263. Coyne AN, Siddegowda BB, Estes PS, Johannesmeyer J, Kovalik T, Daniel SG et al (2014) Futsch/MAP1B mRNA is a translational target of TDP-43 and is neuroprotective in a Drosophila model of amyotrophic lateral sclerosis. J Neurosci 34(48):15962–15974

264. Romano M, Feiguin F, Buratti E (2016) TBPH/TDP-43 modulates translation of Drosophila futsch mRNA through an UG-rich sequence within its 5'UTR. Brain Res 1647:50–56

265. Russo A, Scardigli R, La Regina F, Murray ME, Romano N, Dickson DW et al (2017) Increased cytoplasmic TDP-43 reduces global protein synthesis by interacting with RACK1 on polyribosomes. Hum Mol Genet 26(8):1407–1418

266. Yasuda K, Zhang H, Loiselle D, Haystead T, Macara IG, Mili S (2013) The RNA-binding protein Fus directs translation of localized mRNAs in APC-RNP granules. J Cell Biol 203(5):737–746

267. Kamelgarn M, Chen J, Kuang L, Jin H, Kasarskis EJ, Zhu H (2018) ALS mutations of FUS suppress protein translation and disrupt the regulation of nonsense-mediated decay. Proc Natl Acad Sci U S A 115(51):E11904–E11E13

268. Budini M, Buratti E, Morselli E, Criollo A (2017) Autophagy and Its impact on neurodegenerative diseases: new roles for TDP-43 and C9orf72. Front Mol Neurosci 10:170

269. Gotzl JK, Lang CM, Haass C, Capell A (2016) Impaired protein degradation in FTLD and related disorders. Ageing Res Rev 32:122–139

270. Renaud L, Picher-Martel V, Codron P, Julien JP (2019) Key role of UBQLN2 in pathogenesis of amyotrophic lateral sclerosis and frontotemporal dementia. Acta Neuropathol Commun 7(1):103

271. Rainero I, Rubino E, Michelerio A, D'Agata F, Gentile S, Pinessi L (2017) Recent advances in the molecular genetics of frontotemporal lobar degeneration. Funct Neurol 32(1):7–16

272. Pensato V, Magri S, Bella ED, Tannorella P, Bersano E, Soraru G et al (2020) Sorting rare als genetic variants by targeted re-sequencing panel in Italian patients: OPTN, VCP, and SQSTM1 variants account for 3% of rare genetic forms. J Clin Med 9(2):412

273. Polymenidou M, Lagier-Tourenne C, Hutt KR, Bennett CF, Cleveland DW, Yeo GW (2012) Misregulated RNA processing in amyotrophic lateral sclerosis. Brain Res 1462:3–15

274. Tank EM, Figueroa-Romero C, Hinder LM, Bedi K, Archbold HC, Li X et al (2018) Abnormal RNA stability in amyotrophic lateral sclerosis. Nat Commun 9(1):2845

275. Soo KY, Sultana J, King AE, Atkinson R, Warraich ST, Sundaramoorthy V et al (2015) ALS-associated mutant FUS inhibits macroautophagy which is restored by overexpression of Rab1. Cell Death Dis 1:15030

276. Lin TW, Chen MT, Lin LT, Huang PI, Lo WL, Yang YP et al (2017) TDP-43/HDAC6 axis promoted tumor progression and regulated nutrient deprivation-induced autophagy in glioblastoma. Oncotarget 8(34):56612–56625

277. Ling SC, Dastidar SG, Tokunaga S, Ho WY, Lim K, Ilieva H et al (2019) Overriding FUS autoregulation in mice triggers gain-of-toxic dysfunctions in RNA metabolism and autophagy-lysosome axis. Elife 8:e40811

278. Xia Q, Wang H, Hao Z, Fu C, Hu Q, Gao F et al (2016) TDP-43 loss of function increases TFEB activity and blocks autophagosome-lysosome fusion. EMBO J 35(2):121–142

279. Wang Y, Liu FT, Wang YX, Guan RY, Chen C, Li DK et al (2018) Autophagic modulation by trehalose reduces accumulation of TDP-43 in a cell model of amyotrophic lateral sclerosis via TFEB activation. Neurotox Res 34(1):109–120

280. Wang IF, Guo BS, Liu YC, Wu CC, Yang CH, Tsai KJ et al (2012) Autophagy activators rescue and alleviate pathogenesis of a mouse model with proteinopathies of the TAR DNA-binding protein 43. Proc Natl Acad Sci U S A 109(37):15024–15029

281. Barmada SJ, Serio A, Arjun A, Bilican B, Daub A, Ando DM et al (2014) Autophagy induction enhances TDP43 turnover and survival in neuronal ALS models. Nat Chem Biol 10(8):677–685

282. Zhou F, Dong H, Liu Y, Yan L, Sun C, Hao P et al (2018) Raloxifene, a promising estrogen replacement, limits TDP-25 cell death by enhancing autophagy and suppressing apoptosis. Brain Res Bull 140:281–290

283. Marrone L, Drexler HCA, Wang J, Tripathi P, Distler T, Heisterkamp P et al (2019) FUS pathology in ALS is linked to alterations in multiple ALS-associated proteins and rescued by drugs stimulating autophagy. Acta Neuropathol 138(1):67–84

284. Strohm L, Behrends C (2020) Glia-specific autophagy dysfunction in ALS. Semin Cell Dev Biol 99:172–182

285. Koza P, Beroun A, Konopka A, Gorkiewicz T, Bijoch L, Torres JC et al (2019) Neuronal TDP-43 depletion affects activity-dependent plasticity. Neurobiol Dis 130:104499

286. Ling SC (2018) Synaptic paths to neurodegeneration: the emerging role of TDP-43 and FUS in synaptic functions. Neural Plast 2018:8413496

287. Heyburn L, Moussa CE (2016) TDP-43 overexpression impairs presynaptic integrity. Neural Regen Res 11(12):1910–1911

288. Gulino R, Forte S, Parenti R, Gulisano M (2015) TDP-43 as a modulator of synaptic plasticity in a mouse model of spinal motoneuron degeneration. CNS Neurol Disord Drug Targets 14(1):55–60

289. Fogarty MJ, Klenowski PM, Lee JD, Drieberg-Thompson JR, Bartlett SE, Ngo ST et al (2016) Cortical synaptic and dendritic spine abnormalities in a presymptomatic TDP-43 model of amyotrophic lateral sclerosis. Sci Rep 6:37968

290. Chand KK, Lee KM, Lee JD, Qiu H, Willis EF, Lavidis NA et al (2018) Defects in synaptic transmission at the neuromuscular junction precede motor deficits in a TDP-43(Q331K) transgenic mouse model of amyotrophic lateral sclerosis. FASEB J 32(5):2676–2689

291. Romano G, Holodkov N, Klima R, Grilli F, Guarnaccia C, Nizzardo M et al (2018) Downregulation of glutamic acid decarboxylase in Drosophila TDP-43-null brains provokes paralysis by affecting the organization of the neuromuscular synapses. Sci Rep 8(1):1809

292. Petel Legare V, Harji ZA, Rampal CJ, Allard-Chamard X, Rodriguez EC, Armstrong GAB (2019) Augmentation of spinal cord glutamatergic synaptic currents in zebrafish primary motoneurons expressing mutant human TARDBP (TDP-43). Sci Rep 9(1):9122

293. Sephton CF, Tang AA, Kulkarni A, West J, Brooks M, Stubblefield JJ et al (2014) Activity-dependent FUS dysregulation disrupts synaptic homeostasis. Proc Natl Acad Sci U S A 111(44):E4769–E4778

294. Yokoi S, Udagawa T, Fujioka Y, Honda D, Okado H, Watanabe H et al (2017) 3'UTR length-dependent control of SynGAP isoform alpha2 mrna by fus and elav-like proteins promotes dendritic spine maturation and cognitive function. Cell Rep 20(13):3071–3084

295. Polymenidou M, Cleveland DW (2012) Prion-like spread of protein aggregates in neurodegeneration. J Exp Med 209(5):889–893

296. Lee S, Kim HJ (2015) Prion-like mechanism in amyotrophic lateral sclerosis: are protein aggregates the key? Exp Neurobiol 24(1):1–7

297. Pradat PF, Kabashi E, Desnuelle C (2015) Deciphering spreading mechanisms in amyotrophic lateral sclerosis: clinical evidence and potential molecular processes. Curr Opin Neurol 28(5):455–461

298. Ayers JI, Cashman NR (2018) Prion-like mechanisms in amyotrophic lateral sclerosis. Handb Clin Neurol 153:337–354

299. Harrison AF, Shorter J (2017) RNA-binding proteins with prion-like domains in health and disease. Biochem J 474(8):1417–1438

300. Cushman M, Johnson BS, King OD, Gitler AD, Shorter J (2010) Prion-like disorders: blurring the divide between transmissibility and infectivity. J Cell Sci 123(Pt 8):1191–1201

301. Furukawa Y, Kaneko K, Watanabe S, Yamanaka K, Nukina N (2011) A seeding reaction reca-

pitulates intracellular formation of Sarkosyl-insoluble transactivation response element (TAR) DNA-binding protein-43 inclusions. J Biol Chem 286(21):18664–18672

302. Nonaka T, Masuda-Suzukake M, Arai T, Hasegawa Y, Akatsu H, Obi T et al (2013) Prion-like properties of pathological TDP-43 aggregates from diseased brains. Cell Rep 4(1):124–134

303. Feiler MS, Strobel B, Freischmidt A, Helferich AM, Kappel J, Brewer BM et al (2015) TDP-43 is intercellularly transmitted across axon terminals. J Cell Biol 211(4):897–911

304. Maggio S, Ceccaroli P, Polidori E, Cioccoloni A, Stocchi V, Guescini M (2019) Signal Exchange through Extracellular Vesicles in Neuromuscular Junction Establishment and Maintenance: From Physiology to Pathology. Int J Mol Sci 20(11):2804

305. Zeineddine R, Pundavela JF, Corcoran L, Stewart EM, Do-Ha D, Bax M et al (2015) SOD1 protein aggregates stimulate macropinocytosis in neurons to facilitate their propagation. Mol Neurodegener 10(1):57

306. Nomura T, Watanabe S, Kaneko K, Yamanaka K, Nukina N, Furukawa Y (2014) Intranuclear aggregation of mutant FUS/TLS as a molecular pathomechanism of amyotrophic lateral sclerosis. J Biol Chem 289(2):1192–1202

307. Lanson NA Jr, Maltare A, King H, Smith R, Kim JH, Taylor JP et al (2011) A Drosophila model of FUS-related neurodegeneration reveals genetic interaction between FUS and TDP-43. Hum Mol Genet 20(13):2510–2523

308. Wang JW, Brent JR, Tomlinson A, Shneider NA, McCabe BD (2011) The ALS-associated proteins FUS and TDP-43 function together to affect Drosophila locomotion and life span. J Clin Invest 121(10):4118–4126

309. Kabashi E, Bercier V, Lissouba A, Liao M, Brustein E, Rouleau GA et al (2011) FUS and TARDBP but not SOD1 interact in genetic models of amyotrophic lateral sclerosis. PLoS Genet 7(8):e1002214

310. Sun Z, Diaz Z, Fang X, Hart MP, Chesi A, Shorter J et al (2011) Molecular determinants and genetic modifiers of aggregation and toxicity for the ALS disease protein FUS/TLS. PLoS Biol 9(4):e1000614

311. Suzuki H, Shibagaki Y, Hattori S, Matsuoka M (2015) Nuclear TDP-43 causes neuronal toxicity by escaping from the inhibitory regulation by hnRNPs. Hum Mol Genet 24(6):1513–1527

312. Moujalled D, James JL, Yang S, Zhang K, Duncan C, Moujalled DM et al (2015) Phosphorylation of hnRNP K by cyclin-dependent kinase 2 controls cytosolic accumulation of TDP-43. Hum Mol Genet 24(6):1655–1669

313. Appocher C, Mohagheghi F, Cappelli S, Stuani C, Romano M, Feiguin F et al (2017) Major hnRNP proteins act as general TDP-43 functional modifiers both in Drosophila and human neuronal cells. Nucleic Acids Res 45(13):8026–8045

314. Hart MP, Gitler AD (2012) ALS-associated ataxin 2 polyQ expansions enhance stress-induced caspase 3 activation and increase TDP-43 pathological modifications. J Neurosci 32(27):9133–9142

315. Barmada SJ, Ju S, Arjun A, Batarse A, Archbold HC, Peisach D et al (2015) Amelioration of toxicity in neuronal models of amyotrophic lateral sclerosis by hUPF1. Proc Natl Acad Sci U S A 112(25):7821–7826

316. Manzo E, Lorenzini I, Barrameda D, O'Conner AG, Barrows JM, Starr A et al (2019) Glycolysis upregulation is neuroprotective as a compensatory mechanism in ALS. Elife 8:e45114

317. Donde A, Sun M, Jeong YH, Wen X, Ling J, Lin S et al (2020) Upregulation of ATG7 attenuates motor neuron dysfunction associated with depletion of TARDBP/TDP-43. Autophagy 16(4):672–682

318. Steyaert J, Scheveneels W, Vanneste J, Van Damme P, Robberecht W, Callaerts P et al (2018) FUS-induced neurotoxicity in Drosophila is prevented by downregulating nucleocytoplasmic transport proteins. Hum Mol Genet 27(23):4103–4116

319. Casci I, Krishnamurthy K, Kour S, Tripathy V, Ramesh N, Anderson EN et al (2019) Muscleblind acts as a modifier of FUS toxicity by modulating stress granule dynamics and SMN localization. Nat Commun 10(1):5583

320. Lu L, Zheng L, Si Y, Luo W, Dujardin G, Kwan T et al (2014) Hu antigen R (HuR) is a positive regulator of the RNA-binding proteins TDP-43 and FUS/TLS: implications for amyotrophic lateral sclerosis. J Biol Chem 289(46):31792–31804

321. Becker LA, Huang B, Bieri G, Ma R, Knowles DA, Jafar-Nejad P et al (2017) Therapeutic reduction of ataxin-2 extends lifespan and reduces pathology in TDP-43 mice. Nature 544(7650):367–371

322. Mackenzie IR, Rademakers R, Neumann M (2010) TDP-43 and FUS in amyotrophic lateral sclerosis and frontotemporal dementia. Lancet Neurol 9(10):995–1007

323. Neumann M, Rademakers R, Roeber S, Baker M, Kretzschmar HA, Mackenzie IR (2009) A new subtype of frontotemporal lobar degeneration with FUS pathology. Brain 132(Pt 11):2922–2931

324. Baloh RH (2012) How do the RNA-binding proteins TDP-43 and FUS relate to amyotrophic lateral sclerosis and frontotemporal degeneration, and to each other? Curr Opin Neurol 25(6):701–707

325. Palomo V, Tosat-Bitrian C, Nozal V, Nagaraj S, Martin-Requero A, Martinez A (2019) TDP-43: A Key Therapeutic Target beyond Amyotrophic Lateral Sclerosis. ACS Chem Neurosci 10(3):1183–1196

A Multi-omics Data Resource for Frontotemporal Dementia Research

Peter Heutink, Kevin Menden, and Anupriya Dalmia

Introduction

Frontotemporal dementia (FTD) is a devastating early-onset dementia characterized by the deterioration of the frontal and temporal lobes, severe changes in social and personal behaviour and blunting of emotions [1]. Up to 40% of cases have a positive family history, and mutations in at least ten genes explain almost 50% of familial cases, and this has been the key to the remarkable progress in our understanding of the molecular basis of FTD. Among the familial cases, mutations in the microtubule-associated protein tau (*MAPT*), granulin (*GRN*) and *C9orf72* are responsible for the majority of cases [2]. Neuropathologically, mutations in *MAPT* are associated with neurofibrillary tangles consisting of hyperphosphorylated tau protein, and mutations in *GRN* and *C9orf72* lead to accumulation of the transactive response DNA-binding protein 43 kDa (TDP-43). Although all three genes are associated with a clinical FTD phenotype, their cellular functions are quite diverse, and how these different genes lead to a similar clinical phenotype is still an unanswered question. Currently, there is no cure for FTD, and for the development of successful therapies, it is essential to understand the role of all genetic and environmental risk factors in the disease process, and to investigate which factors are important in the progression of the disease in all patients and which are specific for subgroups of patients.

It is therefore of utmost importance to identify the regulatory mechanisms that lead to neurodegeneration as a consequence of the already identified mutations and novel genes that are being identified by whole-genome sequencing (WGS) and whole-exome sequencing (WES) studies and genome-wide association studies (GWAS).

Publicly available data resources such as Genotype-Tissue Expression (GTEx) (https://gtexportal.org/home/), Encyclopedia of DNA Elements (ENCODE) [3, 4] and the Functional Annotation of the Mammalian Genome (FANTOM) [5] provide excellent tools to investigate the molecular processes in which identified genes and candidate genes for FTD are involved and can help to determine the processes that regulate the expression of these genes, but an

P. Heutink (✉)
Department of Genome Biology of Neurodegenerative Diseases, German Center for Neurodegenerative Diseases (DZNE), Tübingen, Germany
e-mail: peter.heutink@dzne.de

K. Menden
Department of Genome Biology of Neurodegenerative Diseases, Deutsches Zentrum für Neurodegenerative Erkrankungen, Tübingen, Baden-Württemberg, Germany

A. Dalmia
Department of Genome Biology, Deutsches Zentrum für Neurodegenerative Erkrankungen, Tübingen, Baden-Württemberg, Germany
e-mail: anupriya.dalmia@dzne.de

© Springer Nature Switzerland AG 2021
B. Ghetti et al. (eds.), *Frontotemporal Dementias*, Advances in Experimental Medicine and Biology 1281, https://doi.org/10.1007/978-3-030-51140-1_16

important limitation is that all these resources have been generated from human tissues and cellular models of unaffected controls. To understand the role of identified genes in the disease situation, there is a need to generate a publicly available resource from affected cells and tissues obtained from patients and animal models. As part of the European Union (EU) Joint Programme – Neurodegenerative Diseases Research (JPND), we formed the Risk and modifying factors in FTD (RiMod-FTD) consortium with the aim to investigate common and distinctly affected processes in different groups of FTD patients, using a combination of genomic and cell biological approaches on tissues of selected patient groups and corresponding animal and cellular model systems. Our integrative approach allows an unbiased selection of the most suitable targets that can improve our understanding of disease progression and, in addition, will help identify the key genes in the disease process that are the most suitable targets to modify the disease phenotype, and thus provide better choices for therapy development. Here, we describe the current state of our resource and provide examples of how the data can be mined to understand the molecular processes associated with identified genes for FTD and help to prioritize candidate genes identified through WGS/WES and GWAS studies.

The Risk and Modifying Factors in Frontotemporal Dementia Resource

In order to generate a comprehensive multi-omics data resource, we collected frozen post-mortem brain tissue from seven regions (frontal, temporal and occipital lobes, hippocampus, cerebellum, putamen, caudate) of patients carrying mutations in the three most commonly mutated genes in FTD—*MAPT, GRN* and *C9orf72*—and controls without neurological disease for multi-omics characterization. Extensive quality control measures ensured we only included samples that provided us with high-quality ribonucleic acid (RNA), epigenetic and protein data. Because

human post-mortem brain represents the disease end stage, we have also collected tissue at different time points of the development of pathology from the frontal lobes of established mouse models for the same three genes. In addition, we have used human immune pluripotent stem (iPS) lines carrying the same mutations, differentiated them into neurons and performed similar analyses. In this way, we have created a resource that can be used to mine molecular data at the end stage of disease but also during life and early differentiation. The inclusion of iPS lines provides us with the additional possibility to investigate and validate identified pathways by targeted perturbation studies with, for example, RNAi and CRISPR-Cas9 (Table 1).

To thoroughly characterize the molecular mechanisms in post-mortem human brain tissue, mouse models and induced pluripotent stem cell (iPSC)-derived neurons, we generated various omics-datasets. RNA-sequencing (RNA-seq), the most widely used omics-technology [6], allows to measure the gene expression of the entire transcriptome, and it thus represents a central dataset in the resource. Additionally, we generated Cap Analysis of Gene Expression sequencing (CAGE-seq) [7] data, which captures the 5′-end of transcripts and can thus be used to profile the transcription start site (TSS) of genes. The CAGE-seq data thus represents a complementary dataset to the RNA-seq data, as it can not only be used to measure gene expression but also to identify different TSS or promoter usage as well as enhancers [8]. The transcriptome is heavily influenced by the epigenome, for instance, by CpG methylation [9]. To assess potential epigenomic changes in FTD, and to help explain observed transcriptomic aberrations, we profiled over 800,000 CpG sites for methylation. Since for all protein-coding genes, the end-product of gene expression is a protein, we used proteomics technology to quantify the expression of thousands of proteins as an important complementary readout to the transcriptome. As both gene expression and translation are regulated, in part, by micro RNAs (miRNAs), we performed small RNA-sequencing (smRNA-seq) to identify important regulator miRNAs and potentially explain

Table 1 List of datasets that have already been generated and processed for RiMod-FTD

Post-mortem human brain tissue

Data type	Brain region	Samples (control, MAPT, GRN, C9orf72, sporadic)
RNA-seq	Frontal	47 (16, 11, 7, 13, 0)
CAGE-seq	Frontal, temporal, caudate, hippocampus, occipital, cerebellum, and putamen	248 (66, 61, 42, 53, 24)
smRNA-seq	Frontal and temporal	87 (27, 25, 14, 21, 0)
Proteomics	Frontal and temporal	69 (16, 24, 12, 17, 0)
Methylation	Frontal	48 (14, 13, 7, 14, 0)
ChIP-seq H3K4me3	Frontal	16 (4, 4, 4, 4, 0)
ChIP-seq H3K4me3	Sorted neurons (frontal)	25 (8, 8, 3, 6, 0)

Mouse models

Data type	Model Mouse line	Samples
CAGE-seq	MAPT-P301L rTg(TauP301L)4510	32 (control: 16, transgenic: 16)
CAGE-seq	GRN knockout Grm$^{tm1.1Pvd}$	33 (control: 17, knockout: 16)
CAGE-seq	C9orf72 knockdown C57BL/6j-Tg(C9orf72_i3)112Lutzy/J	29 (WT: 12, scramble: 9, knockdown: 8)
Proteomics	MAPT-P301L rTg(TauP301L)4510	33 (control: 16, transgenic: 17)
Proteomics	GRN knockout Grm$^{tm1.1Pvd}$	33 (control: 17, knockout: 16)
Proteomics	C9orf72 knockdown C57BL/6j-Tg(C9orf72_i3)112Lutzy/J	31 (WT: 12, scramble: 9, knockdown: 10)

iPSC-derived cells

Data type	Cell type	Samples (control, MAPT, GRN, C9orf72)
smRNA-seq	Neurons	21 (9, 7, 4, 6)

changes observed in the transcriptome or proteome. Finally, Chromatin Immuno-Precipitation sequencing (ChIP-seq) was performed for the H3K4me3 protein to identify active promoters. All the above-mentioned genomics data types that have been generated for the RiMod-FTD resource focus on different parts of the cellular transcriptional machinery. By combining these different datasets, it is possible to generate better hypotheses about the disease-causing regulatory mechanisms or to validate existing hypotheses using multiple data modalities. A graphical overview of the datasets already generated and planned for future releases is depicted in Fig. 1.

Analysing Multi-omics Datasets

Generating a multi-omics data resource is, of course, only the first step on the path to gain new knowledge about the condition of interest. The next step is to rigorously analyse the data and/or integrate it with genetic data to generate new hypotheses about disease mechanisms. For large and complex datasets such as those found in a multi-omics data resource, there exists a plethora of bioinformatics methods that can be applied to gather new information. For conventional techniques like RNA-seq, there are several accessible and established tools. For others, the researchers might have to write new algorithms themselves. In recent years, specialized algorithms have been developed that allow the integration of multiple experiments from different technologies [10]. Combining the different datasets with the possibilities of modern bioinformatics can then lead to new insights. Moreover, having a central disease-specific data resource available is beneficial in more ways than just to create new insights based on the resource datasets alone. It depicts a valuable asset that FTD-researchers can use to better interpret their own experiments or test their hypotheses. For instance, a clinician or biologist may state a hypothesis about the involvement of a new gene in FTD pathology based on results from an experiment. Before investing more resources in further investigating the role of this gene, the researcher would like to see some more

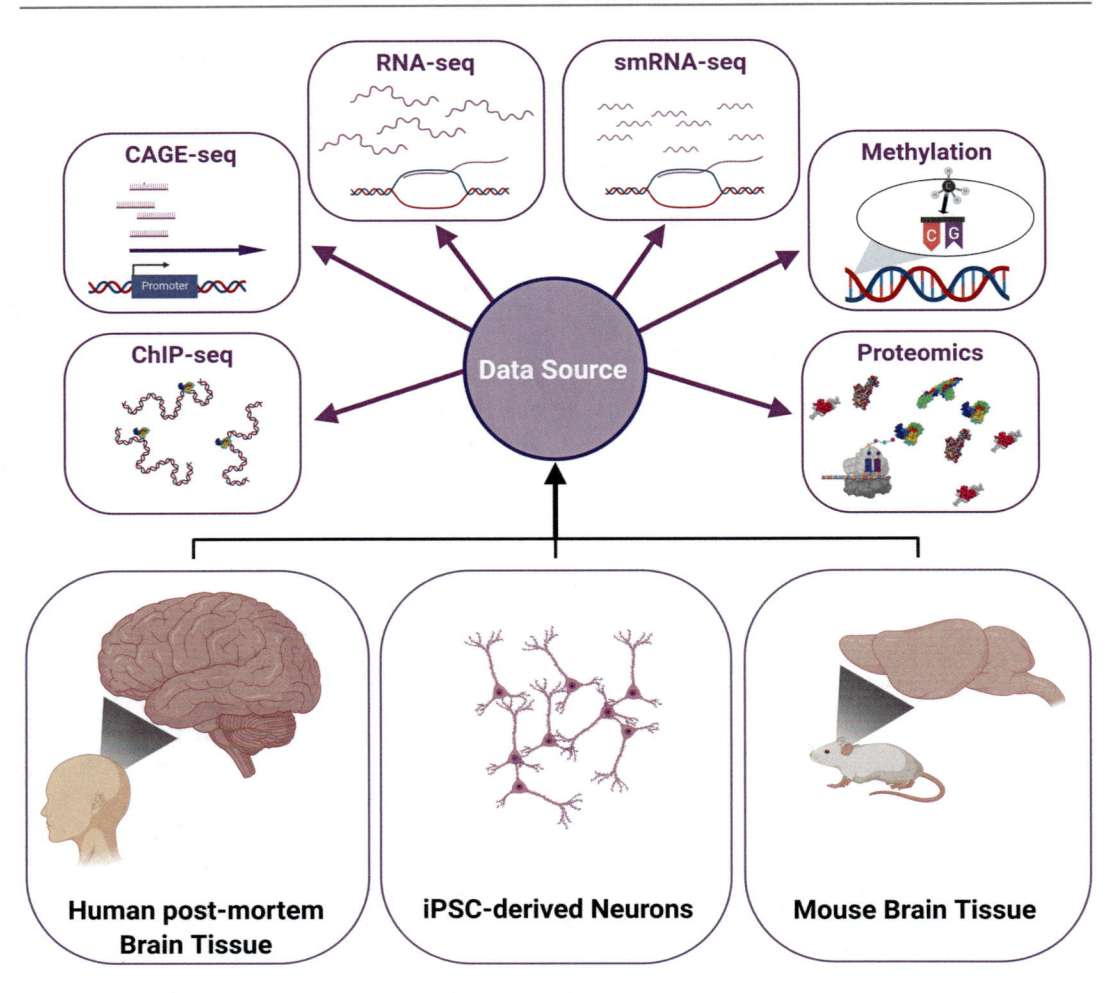

Fig. 1 The RiMod-FTD data resource consists of datasets generated from post-mortem human brain tissue, iPSC-derived neurons and brain tissue from mouse models covering FTD caused by MAPT, GRN and C9orf72. The multi-omics technologies used to generate the data cover ChIP-seq, CAGE-seq, RNA-seq, smRNA-seq, epigenetic arrays and proteomics

evidence. In such a case, RiMod-FTD allows to quickly check the transcriptional state of this gene in several FTD subtypes or whether the quantities of the protein product are changed in the disease. Additionally, the researcher could examine whether the gene is differentially methylated and, finally, check whether aberrant regulation of the gene can be observed in multiple model systems. With more datasets added to the resource in the future, the possibilities for validating experimental results will further increase. Being able to validate scientific findings from own experiments in public data is obviously of

great value and helps to identify the best research paths to pursue and thus to accelerate the scientific progress. In the following, we cover the different technologies used to generate the datasets found in the resource, how these data can be analysed and, where suitable, we present some examples related to FTD.

Pre-processing

Before any dataset generated in the wet lab can be mined for interesting results, it first has to be processed and brought into a format suitable for analysis. While great efforts have been undertaken to simplify this part of the analysis, it remains a very crucial and important step in bioinformatics. The process of converting the raw data that come, for instance, from a sequencing machine, into interpretable and biologically meaningful data points usually requires several steps, each of which is executed with a specialized algorithm. This sequence of steps is commonly called a processing or analysis pipeline. Writing such a pipeline for any omics-data type requires extensive technical knowledge about the data-generating process as well as a good understanding of bioinformatics algorithms capable of handling the respective data. All datasets in RiMod-FTD have been processed and analysed carefully and are available in raw data as well as processed data format. This makes the data more accessible for scientists without extensive domain knowledge, while preserving the raw data for any scientist who wants to process the data with a different pipeline.

Analysing the Transcriptome with Ribonucleic Acid Sequencing

The transcriptome is probably the most commonly studied 'ome' and plays a central role in many studies. Rightfully so, as regulation of gene expression underlies most cellular processes, it is aberrant in many diseases and depicts the closest readout for effects from genetic and epigenetic variation. While multiple technologies exist that can measure gene expression, RNA-seq is the most common one nowadays. Because of this, and because of the importance of the transcriptome, excellent tools exist that help to analyse RNA-seq data. Usually analysis of transcriptomic starts with identifying differentially expressed genes (DEGs) between different groups of samples. Several software packages for this purpose, called differential expression (DE) analysis, exist, such as DESeq [11] or edgeR [12], which allow to apply carefully developed statistical models to calculate fold-changes and p-values for every gene. Although DE analysis is a very standard approach and the above-mentioned software packages are easy to use, care must be taken by the user to specify the design matrix correctly and to account for confounding variables such as

age, gender or experiment batches. The results of DE analysis constitute the basics of many downstream methods and help the experimenter to identify pathways that are most affected by a condition. Along with raw RNA-seq data, the RiMod-FTD resource contains pre-calculated fold-changes and p-values for the most important comparisons of the contained transcriptomic datasets. This makes it easy to quickly check the status of a specific gene in multiple FTD subgroups or model systems, without the need to first process and analyse the data.

The entire set of DEGs defined by DE analysis can be used in combination with public databases of pathways and gene sets that have been curated by experts to test for enrichment of DEGs in some of these pathways. Results from such analyses can be of great value, as they, if done correctly, immediately highlight the cellular processes different between conditions. In a recent study, Dickson et al. [13] performed RNA-sequencing on human brain samples of patients with $C9orf72$ repeat expansion, patients without this mutation and control subjects. Using pathway analysis in combination with weighted gene co-expression network analysis (WGCNA), they found that vesicular transport pathways are especially affected by $C9orf72$ repeat expansions. Using only transcriptomic data, the authors could highlight several affected pathways in $C9orf72$ mutation carriers and identified biomarker candidate genes by applying LASSO regression. Importantly, RiMod-FTD contains datasets from patients not only with $C9orf72$ but also with GRN and $MAPT$ mutations, and it thus allows to test for commonalities between the disease subgroups in terms of affected pathways or WGCNA modules. For example, analysing the RNA-seq data from the RiMod-FTD resource, we have found that oxidative phosphorylation is impaired in both FTD-GRN and FTD-$MAPT$. However, membrane-trafficking-associated pathways appear to be strongly down-regulated in FTD-$MAPT$, while FTD-GRN shows a stronger enrichment for immune system–related pathways. Moreover, as lists of affected pathways are available in the resource, a scientist with an interest in a specific pathway can quickly investigate whether this pathway is affected in some FTD subtype or model system.

Complex tissue, like post-mortem brain tissue, consists of several transcriptionally different cell types. When interpreting RNA-seq experiments on such tissues, it is important to keep in mind that systematic differences in cell-type compositions between sample groups can lead to false-positive DEGs in the analysis. To account for this problem, several cell deconvolution methods have been developed that allow to estimate the cellular composition of each sample from RNA-seq data. Not only does this help to control for false positives, but it can also uncover unknown cellular composition changes in a disease. Examples for cell deconvolution algorithms are MuSiC [14] and Scaden [15]. The latter has been developed for the analysis of data from the RiMod-FTD project and showed best performance on post-mortem brain tissue when compared to other algorithms.

Co-expression Analysis

If an expression dataset is sufficiently large, gene co-expression analysis can be used to obtain dataset-specific expression modules that are relevant to the disease. WGCNA, which was mentioned earlier, is the most popular algorithm for this task [16]. Briefly, WGCNA calculates co-expression values of genes across a dataset, which can then be used to cluster genes into co-expression modules. The underlying assumption is that genes with similar expression patterns tend to have similar functions or are involved in overlapping regulatory mechanisms. A module eigengene, which is the first principal component of the expression matrix, can be used to associate traits with modules—which allows to identify disease-associated modules. Other, module-internal metrics calculated by WGCNA help to identify module hub-genes that might be of special importance. In the study mentioned earlier by Dickson et al., WGCNA was used to identify co-expression modules that are associated with the *C9orf72* repeat expansion. Through module analysis, they identified a module that contained the gene *C9orf72* and was enriched for metabolic pathways, indicating that *C9orf72* might have a

similar function or affect these pathways. Another study from Swarup and colleagues [17] performed WGCNA on RNA-seq data from brain tissue of mouse models for *MAPT* and *GRN* mutations. The authors identified two modules that are significantly correlated with tau hyperphosphorylation, a marker of disease progression in FTD and Alzheimer's disease (AD) [18]. By further analysing these modules, they were able to highlight multiple genes with potentially important roles in the pathways represented by the modules. These studies show how valuable information can be extracted from transcriptomic data alone using pathway- and module-based approaches. A great advantage of RiMod-FTD is the availability of transcriptomics datasets from several tissues and model systems. This allows us to evaluate the robustness of co-expression modules—which are often to some extent dataset-specific—longitudinally and across different model systems. Furthermore, modules or pathways that a researcher has identified in their own dataset can be tested for reproducibility in the various FTD-related datasets of RiMod-FTD. We believe that lacking reproducibility of results generated with genomics technologies is a major hurdle to the scientific progress, and public resources with easily accessible datasets like RiMod-FTD are one way of addressing this problem.

Alternative Splicing of Transcripts

While it is common to perform most analyses with RNA-seq data on the gene level, it is possible to infer transcript-level information from this data as well. However, estimating transcript abundances from RNA-seq data is substantially more challenging, as the sequence of isoforms overlaps to a large part, and, consequently, most reads could be assigned to multiple transcripts. Furthermore, the downstream analysis options are currently not as rich for transcripts as for genes, since many tools (e.g. pathway databases) operate mainly on the gene level. Nevertheless, various tools for the quantification of transcripts and the detection of alternative splicing have been developed. For instance, Leafcutter and MAJIQ are two modern examples of algorithms that can identify alternative splicing events from RNA-seq data [19, 20]. Both tools circumvent the problem of transcript quantification by focusing on exon splice junctions, and thus the exclusion of introns, instead of the inclusion of exons [19]. Although differential splicing analysis is still not routinely done with RNA-seq data, it has long been known that aberrant splicing can have devastating effects and lead to disease. For instance, the authors of MAJIQ reported differential splicing of the *CAM2K* gene in Alzheimer's disease (AD) [20]. The gene *MAPT* is another prominent example. Mutations in *MAPT* lead to a ratio change of tau isoforms, the protein product of the gene. The isoforms have different chemical properties, and the disrupted balance between them can cause disease [21]. Mutations in the genes for TDP-43 and FUS have been associated with alternative splicing in amyotrophic lateral sclerosis (ALS) [22, 23], and a mutation in the gene *PINK1* was shown to activate a cryptic splice-site in Parkinson's disease [24]. Many other mutations can cause alterations in splicing and cause disease, showing that the interrogation of differential splicing represents an important aspect of RNA-seq data analysis. The RNA-seq datasets in the RiMod-FTD resource have been analysed for alternative splicing and can be easily queried for evidence of alternative splicing of a gene of interest in a specific FTD subgroup. Transcriptomic regulation via alternative splicing is a complex mechanism that certainly has not been fully interrogated, and we hope that the diverse RNA-seq data available in RiMod-FTD can help to elucidate the role of gene isoforms in FTD.

Detecting Regulatory Mechanisms

Once deregulated cellular pathways in a disease have been identified using methods such as DE analysis, pathway enrichment or WGCNA, it is often of great interest to identify the regulatory mechanisms that drive these changes. Indeed, this depicts the major goal of many studies. Understanding the regulatory mechanisms that underlie a disease greatly helps to identify druggable targets that can be further interrogated and potentially help to develop treatments. However,

the regulation of the transcriptome involves numerous players that work with and against each other, and no single assay can capture all of them. Therefore, a multi-omics approach is essential. The great advantage of RiMod-FTD is that it contains multi-omics datasets from matching samples, which measure different aspects of transcriptomic regulation. This makes it possible to identify potential regulatory mechanisms or confirm or deny hypotheses about transcriptomic regulation. In the following, we cover different modes of regulation, assays available in RiMod-FTD that can be used to understand them and bioinformatics algorithms that help to extract the desired information.

Regulation by Transcription Factors

The most well-known players in the regulation of gene expression are transcription factors (TFs), which bind to promoters and can increase or repress the expression of one or several genes. Multiple bioinformatics tools have been developed to identify candidate TFs responsible for observed expression patterns. They differ in the data that they require as input and the information they use to generate TF rankings. One method to identify active TFs is to look for enrichment of transcription factor binding sites (TFBS) in the promotor region of a set of genes compared to a background. CAGEd-oPOSSUM [25] uses user-provided CAGE-seq data to generate promoter-proximal regions, which are then scanned for TFBS enrichment. Promoters, which are often in the vicinity of the TSS, are thus frequently enriched in the region around CAGE-peaks. A different approach is taken by ChEA3, which only needs a list of genes as input [26]. The algorithm then integrates information gathered from various sources to rank TFs according to consistent evidence across information sources. As this approach only relies on a list of, for example, up-regulated genes, which can be readily inferred from RNA-seq data, it is widely applicable. Because RiMod-FTD contains both CAGE-seq and RNA-seq data, both above-discussed methods can be applied, in complementary fashion, to the data. Chromatin Immunoprecipitation sequencing (ChIP-seq) is another technology that can be used to study regulation by TFs [27]. With ChIP-seq, the experimenter can identify DNA elements to which a protein of interest binds. As TFs bind to DNA, a ChIP-seq experiment for a particular TF will identify promoters and enhancers that are bound by the TF of interest, which can be used to identify genes regulated by these promoters. The analysis of ChIP-seq data requires specialized algorithms that discriminate between real binding sites and background signal. A very popular tool for this purpose is MACS2 [28]. Although RiMod-FTD currently does not contain ChIP-seq data for specific transcription factors, it contains H3K4me3 ChIP-seq data. H3K4me3 is associated with active promoters and can thus be used to identify active genes and TFs that potentially drive the expression (similar to CAGE-seq). In addition to RNA-seq, CAGE-seq and ChIP-seq, RiMod-FTD also contains proteomic data that can be assessed for TF quantities, which give a more direct readout than using mRNA levels as proxy. However, on a more cautious note, we want to mention that TFs are usually of low abundance in the cell and are thus not always caught by proteomics experiments [29]. It is thus important to use all available datasets for inferring relevant TFs.

Regulation by Micro-RNAs

Micro-RNAs (miRNAs) are another type of important transcriptional regulator that mainly works by binding to the 3′-end of messenger RNAs (mRNAs) to decrease the mRNA stability or to repress the rate of translation [30]. Hence, they affect both the abundance of mRNA and the rate of protein production. Because miRNAs are very short (21–25 nucleotides), specialized protocols must be used for miRNA expression profiling, which is why their activity cannot reliably be inferred from a typical RNA-seq experiment, which measures mRNA or total RNA expression. RiMod-FTD contains smRNA-seq and RNA-seq data from matched samples. This is of great value, as it allows to identify potential miRNA-target pairings with greater confidence. First, candidate targets for each miRNA are predicted, a task for which several computational tools have

been developed. These algorithms incorporate knowledge about miRNA-biology, such as the seed sequence of miRNAs—which must be complementary to a region in the target gene—or evolutionary information. However, as the seed regions used for binding to targets are very small, computationally predicted targets contain high numbers of false positives [31]. Paired information of gene and miRNA expression can be used to perform correlation analysis of miRNA-target pairs [32]. The assumption here is that a negative correlation should be observed when the miRNA regulates a target candidate. If no negative correlation is observed, then either the target prediction is wrong or the regulation by the miRNA is overshadowed by other regulatory effects.

As an example for this approach, we want to highlight a study by Swarup and colleagues, where the authors used protein coding gene and miRNA expression data to identify the miRNA—miR-203—as a potential regulator for a disease-associated co-expression module in mouse models of FTD [17]. After highlighting this miRNA as a potential regulator, the authors went further and overexpressed this miRNA in mouse neuronal cell cultures, where they could observe down-regulation of the predicted targets along with increased apoptosis, thus validating their findings from the transcriptomic data. Replication of such candidate miRNAs in other datasets is important. The RiMod-FTD resource contains several datasets of matched gene- and miRNA-expression, which can be used to infer potentially important regulator miRNAs or to validate findings from other studies, such as those from Swarup et al.

Regulation by Deoxyribonucleic Acid Methylation

The methylation of DNA residues can have strong regulatory effects on gene expression. Cytosine residues can be methylated at their fifth carbon molecule, usually in the context of CpG dinucleotides [9]. CpG methylation at the promoter of genes causes transcriptional repression of that gene. Aberrant methylation can therefore directly affect the transcriptome, and many human diseases have now been associated with methylation [33]. Many technologies for measuring DNA methylation exist, of which methylation array chips are a popular method that nowadays cover over 850,000 different CpG sites across the genome. Specialized software packages have been developed to analyse this data. Like DE analysis, differentially methylated CpG sites between two conditions can be inferred. RiMod-FTD contains methylation data of the newest technology, covering over 850,000 different CpG sites. These data serve as an additional resource for identifying underlying regulatory mechanisms and can help to elucidate disease-related changes in the epigenome. As an example for the relevance of DNA methylation in FTD, repeat expansions in the *C9orf72* gene—a common cause of FTD and ALS—are associated with hypermethylation of the repeat itself and *C9orf72*-flanking CpG island [34]. Gijselinck and colleagues reported that the repeat size correlates with the degree of hypermethylation, with longer repeats leading to more methylation of the flanking CpG island [35]. Repeat size and methylation state are also correlated with age at onset, and the authors suggested that the increased methylation might be a factor explaining the differences in age at onset of the disease.

Proteomics

Being the end-product of gene expression, splicing and translation, proteins constitute the major functional molecules in the cell. Although higher gene expression generally leads to higher quantities of the protein product, the correlation of these two quantities varies significantly [36]. Measuring mRNA concentration is hence not enough to infer protein concentrations [37]. It is obvious that the interrogation of the proteome is a fundamentally important step on the path to understanding cellular pathways and diseases that complement transcriptomic and epigenomic profiling. While the mature RNA-seq technology can be readily used to measure the expression of the entire transcriptome, quantification of the proteome depicts a more difficult challenge. The current technology works by digesting proteins into smaller peptides, which are subsequently measured by lipid chromatography (LC) and

mass spectrography (MS). Bioinformatic algorithms are then employed, in combination with databases, to translate the quantified peptides into protein-level information [38]. Like gene expression, differences of protein quantities between conditions can then be assessed. In addition to the transcriptomic and regulatory assays, RiMod-FTD contains several proteomics datasets from diverse resources, such as multiple brain tissues, patients with different causal mutations or different mouse models. While these datasets cannot cover the entire transcriptome, they represent valuable complementary measurements that help to examine how transcriptional aberrances translate into the proteome. As proteomics experiments are less often conducted than RNA-seq experiments, we believe that the proteomics datasets of RiMod-FTD will be of especially high value for scientists working in the field.

Advantages of Multi-Model Approaches

As shown earlier, the use of multiple omics technologies to profile a biological system and to understand a disease is of great value. It allows us to study several, albeit not all, parts of the highly interconnected regulatory machine that is the cell and is therefore indispensable for widening the systems-level understanding. However, most diseases, especially neurodegenerative diseases such as FTD, arise through complex mechanisms that lead from disease onset to the final disease stages. Understanding these temporal pathway activity patterns and interactions is essential for a complete understanding of a disease, and most probably necessary to eventually develop remedies. To study neurodegeneration, brain tissue is often used—which is only available post-mortem (with some exceptions) and therefore represents the very end stage of the disease. Especially for diseases that develop over many years, only examining the end stage will not allow us to fully understand how the disease develops. It is therefore crucial to use a multi-model approach to study a complex disease like FTD. For instance, mouse models of neurodegeneration allow to profile the disease development over different temporal stages [39]. Of course, other ramifications exist for these models, as findings in mice rarely entirely translate to humans, and a mouse disease model never completely recapitulates the actual disease [40]. Nevertheless, they depict a valuable complementary model to human post-mortem brain tissue. To increase the value of using mouse models, modern machine learning–based approaches have been developed that help to translate the findings from mice to humans [41].

A further level of complexity arises when considering the complex multicellular nature of both human and mouse brain tissue. While many cell types are typically affected in neurodegenerative diseases, the dysregulated pathways likely differ from type to type. This has been increasingly recognized in recent years. As an example, microglia have been identified as being a major factor in the development of AD [42]. In addition to tissue-level models, studying specific cell types is therefore necessary to understand the causal mechanisms behind the development of neurodegenerative diseases. In the past decade, several methods have been developed that made it possible to differentiate patient-derived induced pluripotent stem cells (iPSCs) into all the major cell types found in the brain [43]. This makes it possible to study the effects of the patient-specific genetic background on specific cell types, for instance, neurons. IPSC-derived neurons thus represent a valuable approach to study cell type–specific effects under controlled conditions that cannot be examined in complex tissues. Zhang and colleagues differentiated iPSCs derived from a patient with a mutation in the FTD-causing *CHMP2B* gene into cortical neurons, which allowed them to study neuronal-specific effects of this mutation [44]. The authors identified abnormalities in endosomes and mitochondria as the most significant alterations caused by this mutation, providing insights into the causal mechanisms of *CHMP2B* mutations in neurons. The authors of a different study used iPSC-derived neurons from a patient with *MAPT* mutation and identified transcriptional changes of GABA receptor genes, which they verified in other data from mouse modes and human brain

tissue [45]. These results show how iPSC-derived neurons can be used to study neuron-specific disease mechanisms that are directly caused by a genetic alteration.

The consideration of the above-mentioned advantages and disadvantages of different model systems and tissues led to the decision to make RiMod-FTD a disease-specific data resource that contains datasets from multiple model systems. Having these multi-model datasets facilitates the discovery of mechanisms that translate from model to model, or tissue to model and enables to derive much more robust hypotheses.

Genetics Analysis

Even though almost 40% of patients with FTD have a positive family history, there exists a large gap of missing heritability to explain close to half of these cases, with the rest carrying mutations in known FTD genes such as *MAPT, GRN* and *C9orf72* [2]. With a massive influx of advancement in genetic methodologies in the past two decades, the scope to identify and study disease-causing mutations has amplified and goes beyond linkage analysis and candidate gene studies. The human genome has 100 million single-nucleotide polymorphisms (SNPs) identified to date, which can quickly and cost-effectively be genotyped using arrays. Genome-wide association studies (GWAS) are a classic example of using genotyped data to compare SNPs between healthy and diseased individuals. Strides in next-generation sequencing have also helped identify novel genetic factors and rare damaging variants implicated in FTD.

Genome-Wide Association Studies
A GWAS is based on the concept of linkage disequilibrium, which allows for a subset of SNPs to be used as proxies to genotype the entire genome. It relies on the 'common variants' theory to identify risk factors with modest effect and, in turn, risk loci in the genome that may be used to identify genes that can be clumped together to confirm pathways and processes relevant to that disease [46]. In the largest FTD-GWAS cohort, to

date, alterations in the immune system, lysosomal and autophagic pathways were identified as associated to FTD risk [47]. Since GWASs rely on finding SNPs with moderate effects, it is important to have large cohorts to be able to achieve enough statistical power to see a true biological effect. This study included a two-stage GWAS (discovery phase and replication phase) for clinical FTD, utilizing samples from 44 international research groups. The most widely used tool for GWAS is PLINK [48, 49].

As a follow-up, they performed expression and methylation quantitative loci analysis to study their effect on the associated SNPs. These types of analyses are frequently clubbed together to help discriminate causation from association as it is an important point of note that while proxy SNPs are associated with traits, they are seldom causative. The RiMod-FTD resource of multi-omic data from different brain regions of FTD patients can be useful in mining the hits found in such large-scale GWAS studies and understand the biology lying underneath the association.

For example, a recent GWAS study, shows that the rs72824905-G allele in the gene PLCG2 is associated with decreased risk in FTD as well as increased changes of longevity [50]. Following up on this finding using the RiMod-FTD RNA-seq data, we found that PLCG2 is up-regulated in patients carrying a *GRN* mutation. Loss of *GRN* function has been associated with elevated microglial neuroinflammation [51]; this finding may lend evidence to the protective effect of PLCG2 in brain immune function.

To verify this link between genes involved in brain immune function analysis and FTD and the mechanism by which they act, integrative analysis involving the results from the different omics data under the RiMod-FTD resource can help utilize the plethora of information that all of these different techniques shed a light on.

Next-Generation Sequencing

Identification of rare variants that play a role in disease progression cannot be accomplished with GWA studies that rely on the 'common variants theory'. Association of rare variants with patient status can be assessed using burden tests using the SNP-set (Sequence) Kernel Association Test (SKAT) [52]. Such tests collapse variants into genetic scores and are extremely powerful at detecting high-impact variants that are causal in the same direction. Other tests that have been used are variance tests and combined variance tests that combine burden and variance tests. These tests rely on estimating the variance of genetic effects to uncover the missing heritability. PLINK can be used to perform all of these different types of tests to elucidate the effects of rare variants in FTD, which are often of higher impact than common variants.

In the FTLD-TDP whole-genome sequencing consortium [53], WGS data from 517 unrelated patients and 838 controls were used as a discovery cohort to perform a gene-level analysis of rare variants. The authors used gene-burden analyses to prioritize 61 genes in which LOF variants were observed in at least three patients. TBK1 showed the most LOF mutation carriers, along with genes involved in the TBK1-immunity pathway. TBK1 LOF mutations are also third most frequent in the Belgian FTD cohort from the BELNEU Consortium [54], after *C9orf72* and *GRN*. While this association has been confirmed by multiple studies, the mechanisms are yet to be confirmed. Using RNA-seq and CAGE-seq data from the RiMod-FTD resource, pathway and gene-set enrichment analysis can be performed to explain the mechanism in which TBK1 mutations implicate patient status for FTD. Interestingly, TBK1, unlike PLCG2 was down-regulated in patients carrying a *GRN* mutation in the RiMod-FTD RNA-seq data. These findings offer an opportunity at a deeper understanding at the mechanism behind these correlations and the potential to uncover therapeutic targets.

Public Resource

The primary goal of RiMod-FTD is to generate a versatile data resource that can help to accelerate and support the field of FTD research. To this end, all datasets generated during the project, accompanied by useful analysis results, are made available at the European Genome-phenome

Archive (EGA) [55]. Additional to making the data available in the central and well-known database EGA, it is our plan to develop a graphical user interface that facilitates to visually inspect the data directly in the browser, without any need to download it or analyse it. This will make RiMod-FTD further accessible, especially for scientists or clinicians who only want to check the expression of a single gene or pathway.

Concluding Remarks and Outlook

An ongoing effort of RiMod-FTD is to increase the number of diverse and useful datasets over time. In addition to completing the set of currently used multi-omics experiments for all tissues and model systems available, other experiments are planned as well. We aim to extend human post-mortem brain samples and mouse models to additional mutations, sporadic cases and spectrum disorders such as progressive supranuclear palsy (PSP) and amyotrophic lateral sclerosis (ALS). We also aim to extend over brain regions to be able to compare strongly affected regions with relatively preserved regions. The development of single-cell approaches and spatial transcriptomics has enabled us to examine changes at single-cell resolution, which is necessary to disentangle the cell-type-specific transcriptomic changes. Adding single-cell experiments to RiMod-FTD will therefore increase the value of the resource. Complementary to single-cell RNA-sequencing (scRNA-seq) approaches, we aim to differentiate patient-derived iPSCs into different relevant cell types, such as microglia and co-cultures. This will be done for additional mutations as well.

With these planned efforts and the already existing data, we hope to further untangle the cellular mechanisms behind the complex disease FTD and believe that the RiMod-FTD resource constitutes a significant contribution to the field of FTD research that will help to accelerate the scientific progress towards better disease understanding, diagnosis and eventually treatment.

Acknowledgements This study was supported, in part, by RiMod-FTD an EU Joint Programme – Neurodegenerative Disease Research (JPND) to PH, KM; BMBF Integrative Data Semantics for Neurodegenerative research (IDSN) to PH and the DZNE and NOMIS Foundation to PH, KM.

References

1. Rohrer JD, Guerreiro R, Vandrovcova J et al (2009) The heritability and genetics of frontotemporal lobar degeneration. Neurology 73(18):1451–1456
2. Bang J, Spina S, Miller BL (2015) Frontotemporal dementia. Lancet 386(10004):1672–1682
3. Davis CA, Hitz BC, Sloan CA et al (2018) The Encyclopedia of DNA elements (ENCODE): Data portal update. Nucleic Acids Res 46(D1):D794–D801
4. Dunham I, Kundaje A, Aldred SF et al (2012) An integrated encyclopedia of DNA elements in the human genome. Nature 489(7414):57–74
5. Lizio M, Harshbarger J, Shimoji H et al (2015) Gateways to the FANTOM5 promoter level mammalian expression atlas. Genome Biol 16(1):22
6. Wang Z, Gerstein M, Snyder M (2009) RNA-Seq: a revolutionary tool for transcriptomics. Nat Rev Genet
7. Shiraki T, Kondo S, Katayama S et al (2003) Cap analysis gene expression for high-throughput analysis of transcriptional starting point and identification of promoter usage. Proc Natl Acad Sci USA 100(26):15776–15781
8. Andersson R, Gebhard C, Miguel-Escalada I et al (2014) An atlas of active enhancers across human cell types and tissues. Nature 507(7493):455–461
9. Greenberg MVC, Bourc'his D (2019) The diverse roles of DNA methylation in mammalian development and disease. Nat Rev Mol Cell Biol. Nature Publishing Group 20:590–607
10. Argelaguet R, Velten B, Arnol D et al (2018) Multi-omics factor analysis-a framework for unsupervised integration of multi-omics data sets. Mol Sys Biol 14(6):e8124
11. Love MI, Huber W, Anders S (2014) Moderated estimation of fold change and dispersion for RNA-seq data with DESeq2. Genome Biol 15(12):550
12. Robinson, M. D., McCarthy, D. J., & Smyth, G. K. (2010). edgeR: a Bioconductor package for differential expression analysis of digital gene expression data. Bioinformatics (Oxford, England), 26(1):139–140
13. Dickson DW, Baker MC, Jackson JL et al (2019) Extensive transcriptomic study emphasizes importance of vesicular transport in C9orf72 expansion carriers. Acta Neuropathol Commun 7(1):150
14. Wang X, Park J, Susztak K et al (2019) Bulk tissue cell type deconvolution with multi-subject single-cell expression reference. Nat Commun 10:1):1–1):9

15. Menden K, Marouf M, Dalmia A et al (2019) Deep-learning-based cell composition analysis from tissue expression profiles. bioRxiv 659227

16. Langfelder P, Horvath S (2008) WGCNA: an R package for weighted correlation network analysis. BMC Bioinformatics 9:1

17. Swarup V, Hinz FI, Rexach JE et al (2019) Identification of evolutionarily conserved gene networks mediating neurodegenerative dementia. Nat Med 25(1):152–164

18. Rademakers R, Cruts M, Van Broeckhoven C (2004) The role of tau (MAPT) in frontotemporal dementia and related tauopathies. Hum Mutat 24:277–295

19. Li YI, Knowles DA, Humphrey J et al (2018) Annotation-free quantification of RNA splicing using LeafCutter. Nat Genet 50(1):151–158

20. Vaquero-Garcia J, Barrera A, Gazzara MR et al (2016) A new view of transcriptome complexity and regulation through the lens of local splicing variations. elife 5(February):e11752

21. Buée L, Bussière T, Buée-Scherrer V et al (2000) Tau protein isoforms, phosphorylation and role in neurodegenerative disorders. Brain Res Rev. Elsevier B.V 33:95–130

22. Arnold ES, Ling SC, Huelga SC et al (2013) ALS-linked TDP-43 mutations produce aberrant RNA splicing and adult-onset motor neuron disease without aggregation or loss of nuclear TDP-43. Proc Natl Acad Sci USA 110(8):E736–E745

23. Sun S, Ling SC, Qiu J et al (2015) ALS-causative mutations in FUS/TLS confer gain and loss of function by altered association with SMN and U1-snRNP. Nat Commun 6:6171

24. Samaranch L, Lorenzo-Betancor O, Arbelo JM et al (2010) PINK1-linked parkinsonism is associated with Lewy body pathology. Brain 133(4):1128–1142

25. Arenillas DJ, Forrest ARR, Kawaji H et al (2016) CAGEd-oPOSSUM: motif enrichment analysis from CAGE-derived TSSs. Bioinformatics 32(18):2858–2860

26. Keenan AB, Torre D, Lachmann A et al (2019) ChEA3: transcription factor enrichment analysis by orthogonal omics integration. Nucleic Acids Res 47(W1):W212–W224

27. Johnson DS, Mortazavi A, Myers RM et al (2007) Genome-wide mapping of in vivo protein-DNA interactions. Science (80-) 316(5830):1497–1502

28. Gaspar JM 2018 Improved peak-calling with MACS2. bioRxiv 496521

29. Ding C, Chan DW, Liu W et al (2013) Proteome-wide profiling of activated transcription factors with a concatenated tandem array of transcription factor response elements. Proc Natl Acad Sci USA 110(17):6771–6776

30. Hausser J, Zavolan M (2014) Identification and consequences of miRNA–target interactions — beyond repression of gene expression. Nat Rev Genet 15(9):599–612

31. Sethupathy P, Megraw M, Hatzigeorgiou AG (2006) A guide through present computational approaches for the identification of mammalian microRNA targets. Nat Methods 3(11):881–886

32. Borgmästars E, de Weerd HA, Lubovac-Pilav Z et al (2019) miRFA: an automated pipeline for microRNA functional analysis with correlation support from TCGA and TCPA expression data in pancreatic cancer. BMC Bioinformatics 20(1):393

33. Jin Z, Liu Y (2018) DNA methylation in human diseases. Genes Dis. Chongqing yi ke da xue, di 2 lin chuang xue yuan Bing du xing gan yan yan jiu suo 5:1–8

34. Xi Z, Zhang M, Bruni AC et al (2015) The C9orf72 repeat expansion itself is methylated in ALS and FTLD patients. Acta Neuropathol 129(5):715–727

35. Gijselinck I, Van Mossevelde S, Van Der Zee J et al (2016) The C9orf72 repeat size correlates with onset age of disease, DNA methylation and transcriptional downregulation of the promoter. Mol Psychiatry 21(8):1112–1124

36. Schwanhüusser B, Busse D, Li N et al (2011) Global quantification of mammalian gene expression control. Nature 473(7347):337–342

37. Liu Y, Beyer A, Aebersold R (2016) On the dependency of cellular protein levels on mRNA abundance. Cell. Cell Press 165:535–550

38. Altelaar AFM, Munoz J, Heck AJR (2013) Next-generation proteomics: towards an integrative view of proteome dynamics. Nat Rev Genet 14:35–48

39. Trancikova A, Ramonet D, Moore DJ (2011) Genetic mouse models of neurodegenerative diseases. In: Progress in molecular biology and translational science. Elsevier B.V, Amsterdam, pp 419–482

40. Seok J, Shaw Warren H, Alex GC et al (2013) Genomic responses in mouse models poorly mimic human inflammatory diseases. Proc Natl Acad Sci USA 110(9):3507–3512

41. Normand R, Du W, Briller M et al (2018) Found in translation: a machine learning model for mouse-to-human inference. Nat Methods 15(12):1067–1073

42. McQuade A, Blurton-Jones M (2019) Microglia in Alzheimer's disease: exploring how genetics and phenotype influence risk. J Mol Biol. Academic Press 431:1805–1817

43. Penney J, Ralveniu, WT, Tsai, L (2020) Modeling Alzheimer's disease with iPSC-derived brain cells. Mol Psychiatry 25:148–167

44. Zhang Y, Schmid B, Nikolaisen NK et al (2017) Patient iPSC-derived neurons for disease modeling of frontotemporal dementia with mutation in CHMP2B. Stem Cell Rep 8(3):648–658

45. Jiang S, Wen N, Li Z et al. (2018) Integrative system biology analyses of CRISPR-edited iPSC-derived neurons and human brains reveal deficiencies of presynaptic signaling in FTLD and PSP. Transl Psychiatry 8:265

46. Ferrari R, Grassi M, Salvi E et al (2015) A genome-wide screening and SNPs-to-genes approach to identify novel genetic risk factors associated

with frontotemporal dementia. Neurobiol Aging 36(10):2904.e13–2904.e26

47. Ferrari R, Hernandez DG, Nalls MA et al (2014) Frontotemporal dementia and its subtypes: a genome-wide association study. Lancet Neurol 13(7):686–699

48. Shaun Purcell. PLINK. 2017

49. Purcell S, Neale B, Todd-Brown K et al (2007) PLINK: a tool set for whole-genome association and population-based linkage analyses. Am J Hum Genet 81(3):559–575

50. van der Lee SJ, Conway OJ, Jansen I et al (2019) A nonsynonymous mutation in PLCG2 reduces the risk of Alzheimer's disease, dementia with Lewy bodies and frontotemporal dementia, and increases the likelihood of longevity. Acta Neuropathol 138(2):237–250

51. Martens LH, Zhang J, Barmada SJ et al (2012) Progranulin deficiency promotes neuroinflammation and neuron loss following toxin-induced injury. J Clin Invest 122(11):3955–3959

52. Wu MC, Lee S, Cai T et al (2011) Rare-variant association testing for sequencing data with the sequence kernel association test. Am J Hum Genet 89(1):82–93

53. Pottier C, Ren Y, Perkerson RB et al (2019) Genome-wide analyses as part of the international FTLD-TDP whole-genome sequencing consortium reveals novel disease risk factors and increases support for immune dysfunction in FTLD. Acta Neuropathol 137(6):879–899

54. Gijselinck I, Van Mossevelde S, Van Der Zee J et al (2015) Loss of TBK1 is a frequent cause of frontotemporal dementia in a Belgian cohort. Neurology 85(24):2116–2125

55. Lappalainen I, Almeida-King J, Kumanduri V et al (2015) The European genome-phenome archive of human data consented for biomedical research. Nat Genet. Nature Publishing Group 47:692–695

Mendelian and Sporadic FTD: Disease Risk and Avenues from Genetics to Disease Pathways Through In Silico Modelling

Claudia Manzoni and Raffaele Ferrari

Introduction

Complex disorders are by definition non-linear conditions where environmental and genetic factors play an intertwined role in contributing to disease pathogenesis and progression. Environmental factors are challenging in that it is difficult to identify and measure those that specifically impact disease [1]. Conversely, the dissection of genetic factors has benefitted from constant improvements in the technologies for generating high-resolution data and analytical tools (Wetterstrand KA. 2019. https://www.genome.gov/about-genomics/fact-sheets/Sequencing-Human-Genome-cost).

We have come to appreciate that, on the basis of genetics, there are two broad categories of patients: (i) a minority of so-called familial cases where pathogenic (Mendelian) mutations in single candidate genes (i.e. Mendelian genes) co-segregate with disease and (ii) a majority of so-called sporadic cases where, in the absence of Mendelian mutations, multiple genetic variants with small effect size increase the risk for developing disease.

Mendelian genes have been classically isolated via linkage analysis and/or whole-exome/genome sequencing of trios, first-degree relatives or well-phenotyped pedigrees [2]. Sporadic forms of disease are conveniently investigated through case/control association studies, e.g. genome-wide association studies (GWAS) [3]. The idea that genetic investigation of *familial cases* is straightforward is only apparent. It is, in fact, worth noting that there are uncharacterised *familial cases* where Mendelian mutations have not been isolated [4]. Also, functional investigation of Mendelian genotype-phenotype correlation has proven neither time- nor cost-effective, to date. Moreover, the genetic architecture of risk for *sporadic cases* is challenging to assess and even harder to model, especially considering that multiple variants with small effect size are to be taken into account, simultaneously.

In this chapter, we focus on the heterogeneous features of frontotemporal dementia (FTD) touching upon its complex genetic landscape and discuss how novel approaches (e.g. in silico systems biology) promise to revolutionise the translation of genetic information into functional understanding of disease. These approaches represent a stepping-stone towards functional validation of risk pathways and, possibly, drug target identification. All this holds relevance as the

C. Manzoni
Department of Pharmacology, UCL School of Pharmacy, London, UK

School of Pharmacy, University of Reading, Reading, UK

R. Ferrari (✉)
Department of Neurodegenerative Disease, University College London, Institute of Neurology, London, UK
e-mail: r.ferrari@ucl.ac.uk

B. Ghetti et al. (eds.), *Frontotemporal Dementias*, Advances in Experimental Medicine and Biology
1281, https://doi.org/10.1007/978-3-030-51140-1_17

field is accelerating towards effective clinical trial design and the development of measures for early diagnosis, disease prevention/monitoring and cure.

FTD and Disease Risk

Environmental Factors

The environmental exposure contributing to FTD pathogenesis is an understudied and complicated matter. It is widely accepted that complex neurodegenerative conditions, including FTD, are influenced by environmental risk factors acting in concert with the genetic risk architecture within a process referred to as gene-environment interaction [5].

No single environmental factor clearly leading to FTD has ever been indicated. Only concepts such as 'cognitive reserve' [6, 7] or 'aging' [8] have been suggested to influence disease risk and modulate age at onset. Additionally, few epidemiological studies highlighted possible links between FTD, cardiovascular disease and diabetes risk factors [9–11].

The environment is believed to influence risk for complex neurodegenerative disorders via, at least, two mechanisms. On the one hand, the environmental exposure (e.g. aging) may modulate methylation profiles in the genome or the activity of non-coding RNAs (ncRNAs) impacting gene expression and influencing disease onset and progression [12, 13]. On the other hand, the environmental exposure can represent the direct mechanistic insult triggering processes that lead to disease. For example, lessons learned from other complex neurodegenerations, such as Parkinson's disease (PD), indicate that certain toxins and pesticides can cause a cascade of effects resulting in oxidative stress that ultimately influences disease pathogenesis [14]. Also, traumatic brain concussions have been implicated in certain forms of dementia (including Alzheimer's disease [AD] and FTD) [15], and it was suggested that physical insults were linked to toxic stress resulting in mitochondria alteration, oxidative stress [16] or amyloid aggregation [17], globally impacting brain homeostasis and, subsequently, disease pathogenesis.

A better understanding of the environmental risk factors playing a role in complex neurodegenerations, such as FTD, would critically complement our dissection of disease biology (e.g. it would help highlighting impacted pathways and molecular mechanisms). A substantial caveat here is represented by the lack of efficient and reliable methods to investigate and measure the environmental exposure(s) that influence and/or contribute to the pathogenesis of complex neurodegenerations. Nevertheless, a promising approach that might aid in closing this critical gap is Mendelian randomisation (MR). MR is a statistical approach where common variants such as single nucleotide polymorphisms (SNPs) that are associated with a certain environmental exposure (e.g. SNPs which increase individual risk/ chance of smoking, drinking, developing cardiovascular disease) are used as proxies to assess association with SNPs in the disease under investigation [5]. This approach is still to be explored in FTD, yet it promises to shed light on those environmental exposures that might be relevant to FTD pathogenesis: power issues associated with GWAS performed in FTD have hampered the possibility of performing effective MR studies, to date.

Genetics

In line with its heterogeneous clinical and pathological characteristics (which can be reviewed in [18–20]), FTD's genetic features mirror its complicated global phenotypic picture [21, 22]. A positive familial history, familial (fFTD) or Mendelian, is seen in ~10–30% of cases [23–25], whilst a remainder ~70% of cases, individuals with disease but no clear familial history and/or genetic aetiology, are categorised as *sporadic* (sFTD) [21, 22].

Mendelian FTD

The vast majority (≥25%) of fFTDs associates with pathogenic mutations in *MAPT* [26], *GRN* [27] and *C9orf72* [28, 29], whilst a small minority

(<5%) associates with (very) rare mutations in *CHMP2B* [30, 31], *VCP* [32], *TBK1* [33–35], *IFT74* [36], *OPTN* [35], *SQSTM1* [37], *UBQLN2* [38], *CHCHD10* [39] and *TIA1* [40].

Mutations in *MAPT*, *GRN* and *CHMP2B* have almost exclusively been described in 'pure' FTD cases [21]. In few occasions, issues were raised on whether (all) Mendelian mutations are fully penetrant (e.g. *GRN* mutations have shown to be associated with variable age at onset or a spectrum of phenotypes within the same family [22]). Expansions in *C9orf72* have shown to be ubiquitous across neurodegenerative disease. Although they are most frequently found in cases diagnosed with FTD and amyotrophic lateral sclerosis (ALS) or within the FTD-ALS spectrum, they have also been reported in a range of phenotypes, including AD, Parkinsonian syndromes, Huntington's disease (HD), corticobasal syndrome/degeneration (CBS/D) and non-demented elderly individuals [29, 41–49]. Mutations in the remainder genes have been isolated in small numbers of (at times even single) families displaying substantial syndrome heterogeneity: a complex phenotypic signature characterised by inclusion body myopathy (IBM), Paget's disease of the bone (PDB) and FTD (IBMPFD) for *VCP* [50] and ALS and/or the FTD-ALS spectrum for *SQSTM1*, *UBQLN2*, *IFT74*, *OPTN*, *CHCHD10*, *TBK1* and *TIA1* [21, 22]. Of note, *TARDBP* and *FUS* mutations have been mainly reported in ALS whilst very rarely in FTD cases [51, 52]. It is thus still debated whether or to what extent *TARDBP* and *FUS* are to be considered 'FTD genes' [52, 53] (despite the fact that TDP-43 and FUS are clear pathological hallmarks of FTD [54]).

Regardless of complexity and heterogeneity, a key point is that Mendelian (i.e. for the most, coding) mutations, provided their large effect size, appear to be sufficient to trigger disease. Therefore, although quite rare and exclusive to a (rather small) number of families or private cases, they are indeed informative candidate genes/targets to model disease.

Sporadic FTD

Sporadic FTD cases (sFTDs) are generally screened for known candidate genes: pathogenic variants have been reported in *MAPT*, *GRN*, *C9orf72* or *TBK1* in ≤10% of cases [21, 22, 55, 56]. These might be due to de novo mutations that can (very rarely) occur in the population or (likely) to the fact that they might be cryptic Mendelian cases.

Genetics of sFTD is still poorly understood. Sporadic cases are investigated through GWAS where millions of SNPs are compared across thousands of cases and controls [3]. A GWAS assesses allele frequencies of 'common' genetic markers (SNPs) (i.e. they are present in the general population) in the two sample sets. Those markers that associate with increased risk for disease display a significantly increased frequency in cases when compared to controls. Genetic risk markers identified through GWAS are generally non-coding variants, and they are characterised by small effect sizes; thus, one single SNP is neither necessary nor sufficient to lead to disease [57]. Rather, multiple SNPs cumulatively contribute to disease pathogenesis and represent the so-called genetic architecture of disease (i.e. the genome-wide asset of genetic risk) [58].

To date, a handful of GWAS have been performed in sFTD [4]. GWAS require large cohorts of cases and controls (n = thousands), and this may sometimes represent a drawback (especially when a disease is rare or heterogeneous). In order to cope with sample collection and power issues for genetic studies of sFTD, multicentre initiatives such as the International Frontotemporal Dementia Genomics Consortium (IFGC; https://ifgcsite.wordpress.com/) and the International FTLD-TDP Whole-Genome Sequencing Consortium [56] have been established. Networks of this kind allow to share expertise and collate large numbers of samples across research centres to increase the statistical power of sFTD genetic studies.

The first FTLD-GWAS was published in 2010 by Van Deerlin et al. using a cohort of 604 cases with either pathologically confirmed frontotemporal lobar degeneration with TDP-43 pathology (FTLD-TDP) or cases carrying a *GRN* mutation (515 discovery phase; 89 replication phase). This study highlighted risk variants at a locus on chromosome 7p21 [59]. Subsequently, a larger GWAS

was published in 2014 by Ferrari et al. using a cohort of 3526 clinically diagnosed sFTD cases (2154 discovery phase; 1372 replication phase) leading to the identification of a risk locus on chromosomes 6p21.3 (for the entire cohort) and a suggestive risk locus on chromosomes 11q14 (for behavioural variant FTD [bvFTD]) [60]. A smaller GWAS was then performed by Ferrari et al. in a population-specific cohort of 530 Italian sFTDs: two suggestive signals were indicated by this study in loci mapping to chromosomes 2p16.3 and 17q25.3 [61].

Genome-wide approaches can clearly be applied in the context of multiple and different experimental designs. In FTD, this was the case of a couple of studies that analysed common variants in cohorts characterised by a genetic signature carried in two FTD genes – *GRN* and *C9orf72* – to specifically look for disease modifiers (i.e. genetic factors which influence measurable variables such as age at onset or disease progression). Both studies were published in 2018: (i) one by Pottier et al. assessing a cohort of 592 patients (382 discovery phase; 210 replication phase) carrying Mendelian mutations in *GRN* (and some being pathologically defined as FTLD-TDPs without *GRN* mutations) that led to the replication of the above-described locus on chromosome 7p21 and the identification of a new locus on chromosome 8p21.3 [62] and (ii) one by Zhang et al. assessing a cohort of 331 (144 discovery phase; 187 replication phase) *C9orf72* expansion carriers that suggested a locus on chromosome 6 acting as a modifier for age at onset [63]. Of note, a previous study by Barbier et al. conducted on a cohort of 504 patients belonging to 133 families with pathogenic mutations in both *GRN* and *C9orf72* indicated potential chromosome X-linked modifiers of age at onset (for *C9orf72* expansions carriers but not for *GRN* mutation carriers) [64]. More recently, a GWAS on 636 FTLD-TDP pathologically confirmed cases (517 discovery phase; 119 replication phase) – and not carrying mutations in any of the known FTD genes – by Pottier et al. suggested three risk loci on chromosomes 7q36, 19p13.11 and 6p21.32 [56]. Of note, provided there being different pathological subtypes within the FTLD-

TDP spectrum (i.e. subtypes 'A', 'B', 'C' and 'D'; c.f [65]), this study suggested that (i) although the 7q36 locus had been previously associated with idiopathic ALS, here the signal represented an independent association; (ii) the association with the 19p13.11 locus appeared to be the same as previously indicated in ALS studies, and it was specific to the FTLD-TDP subtype 'B'; and (iii) the rare T-allele of rs5848, located within *GRN*'s 3'-UTR, appeared to specifically (and exclusively) increase risk for cases belonging to the FTLD-TDP subtype 'A' [56].

GWAS results described in this section are summarised in Table 1.

Although one might gather from these sections that the FTD genetics arena is globally quite heterogeneous, there are reasons to suspect that homogeneous subpopulations of patients exist and can be better defined and predicted through tailored genetic (and bioinformatics) studies [21, 22].

Missing Heritability

Despite heterogeneity, it might be argued that FTD is a disorder with a robust hereditary component. However, our genetic understanding of FTD is still considerably incomplete in sporadic as well as in familial FTD (e.g. there are families where Mendelian mutations have not been isolated) [4]. It follows that missing heritability is a critical unresolved issue in FTD [66].

Recently, a number of sequencing projects in FTLD-TDP, clinical FTD and FTD-ALS cases further characterised mutations in either already established Mendelian or what could be considered as 'novel' FTD genes. For example, an excess of loss-of-function variants in FTLD-TDP cases was evident in a number of genes (i.e. *DHX58*, *IRF3*, *IRF7*, *IRF8*, *NOD2* and *TRIM21*) suggested to be in strong functional link with *TBK1* within inflammatory response pathways [56]. Further, mutations were described in *SORT1* and in a Belgian FTD cohort and subsequently confirmed in Mediterranean FTD cases [67]; *CCNF* in FTD and ALS cases [68]; *TREM2*, *CSF1R* and *AARS2* in Asian FTD cases [69, 70]; and *TYROBP* in Italian FTD-ALS pedigrees [71]. Besides many of these mutations needing addi-

Table 1 Summary of GWAS studies in FTD

FTD cohort	Total cases	Locus	markers	p-values (joint)	Affected gene	Biological meaning	Year	Reference
FTLD-TDP	604	chr 7p21	rs1020004	5.00×10^{-11}	TMEM106B	Increased expression of TMEM106B; endolysosomes	2010	59
			rs6966915	1.63×10^{-11}				
			rs1990622	1.08×10^{-11}				
Clinical FTD	2,154	chr 6p21.3	rs1980493	1.57×10^{-8}	BTNL2	Changes in methylation pattern at HLA-DRA; immune response	2014	60
			rs9268856	5.51×10^{-9}	HLA-DRA / DRB			
			rs9268877	1.05×10^{-8}				
Clinical bvFTD	1,377	chr 11q14	rs302668	2.44×10^{-7}	RAB38	Decreased expression of RAB38		
Italian FTD	530	chr 2p16.3	rs17042852	2.01×10^{-7}	NA	NA	2015	61
		chr 17q25.3	rs906175	1.22×10^{-7}	RFNG; AATK; MIR1250	Decreased expression of RFNG, AATK, MIR1250; neurogenesis; neuronal apoptosis; regulation of gene expression		
GRN mutations / C9orf72 expansion carriers	504	chr X-linked modifiers	NA	NA	NA	Effect on AAO* in C9orf72 expansion carriers	2017	64
GRN mutations carriers	592	chr 7p21	rs1990622	3.54×10^{-16}	TMEM106B	Increased expression of TMEM106B; endolysosomes	2018	62
		chr 8p21.3	rs36196656	1.58×10^{-8}	GFRA2	Decreased expression of GFRA2; GDNF signalling pathway		
C9orf72 expansion carriers	331	chr 6	rs9357140	1.0×10^{-6}	NA	Changes in methylation pattern; effect on AAO*; increased expression of HLA-DRB1; immune response	2018	63
FTLD-TDP	636	chr 7q36	rs118113626	4.8×10^{-8}	DPP6	NA	2019	56
		chr 19p13.11	rs1297319	1.27×10^{-8}	UNC13A	FTD-TDP subtype 'B' signature		
		chr 6p21.32	rs17219281	3.22×10^{-8}	HLA-DQA2 / -DQB2	Increased expression of HLA-DQA2 / -DQB2		

* AAO: age at onset

tional replication, the above studies further support the notion of population and syndrome heterogeneity characterising genetics of FTD.

Considering sFTD, the scenario is possibly even more complicated. A first issue is that GWAS in FTD have still been quite underpowered to date. This can, e.g. be appreciated by comparing numbers of cases studied across different neurodegenerative diseases such as AD (n ~ 90,000 [72]) and PD (n ~ 40,000 [73]) vs. the largest FTD-GWAS so far (n ~ 3500 [60]). A second issue is represented by the fact that underpowered GWAS in FTD have hampered appreciating the global contribution of the multiple risk markers with small effect size through, e.g. polygenic risk scoring (PRS). PRS would indeed serve the purpose of measuring how well the global genetic architecture of risk discriminates sFTD cases from controls (and/or other closely related neurodegenerations). PRS aggregates whole-genome genetic risk into a single score using a test sample to weight SNP contribution to a trait and assesses such weights in an independent target sample [74]. Since PRS has never been done in FTD, the actual genetic architecture that confers globally increased risk for developing sFTD remains elusive, even more so when considering the different FTD subtypes: (i) the clinical syndromes belonging to the core FTD

spectrum, i.e. the behavioural and language variants [18, 20], and (ii) the pathologically defined subtypes characterised by Tau and TDP-43 (FTLD-tau, FTLD-TDP) or p62 (FTLD-UPS [ubiquitin proteasome]) or FUS, EWS and TAF15 (collectively referred to as FTLD-FET) protein aggregates [54, 65].

Although a large GWAS meta-analysis for sFTD is currently (at the time this chapter is being written) ongoing within the IFGC program – including over 5000 cases – it is clear that the genetic architecture underpinning sFTD (and its various subtypes) is still poorly defined and understood; thus, more work in this area is warranted.

From Genetics to Disease Biology

Despite our poor understanding of environmental risk factor in FTD and the work ahead in further characterising the genetic architecture of risk, there is an important issue we can start addressing now: translation of our current knowledge of FTD's genetics into functional understanding of disease. This is indeed among the major topics gaining momentum in the biomedical field focusing on complex neurodegenerative disorders (including FTD) [75].

Translating GWAS Genetics into Biological Meaning

One of the biggest challenges in population genetics is the interpretation of the risk signals derived from GWAS. Whilst GWAS are instrumental in discriminating genetic risk markers and loci that associate with a trait of interest, such signals are not directly informative on the impacted gene(s) or disease mechanism(s) [76]. SNPs highlighted by GWAS are for the very vast majority non-coding (intronic or intergenic) meaning that additional investigations are required to identify the actual gene(s) and pathway(s) targeted by the risk variants within the risk locus [3, 77]. This is not a trivial issue since the understanding of impacted genes and pathways is of primary importance to untangle the functional role of the risk variants and generate more accurate disease models.

Besides increasing the resolution in prioritising genes at GWAS loci, e.g. through ad-hoc gene-burden analyses [78], other strategies involving integration of genetic and other types of data – e.g. gene expression, protein-protein interaction and pathway analyses – are being fine-tuned [76]. Indeed, a first point to clarify is whether any SNP highlighted by a GWAS exerts an effect on gene expression: this is done by assessing expression quantitative trait loci (eQTL) [79], a bioinformatics technique that evaluates expression levels (mRNA) of genes in *cis* with the risk allele(s) of the associated SNPs within the locus of interest. When the risk allele significantly associates with a change of expression of a *cis*-gene, the latter might be bona fide considered the biological target of the genetic variant. There are other types of QTL analyses, e.g. methylation (mQTL), splicing (sQTL) and protein (pQTL) [80], that focus on the identification of alterations in methylation profile, splicing or protein levels. Such quantitative traits might be used as proxies to prioritise genes and support the definition of molecular mechanisms modulated by GWAS SNPs. And, clearly, these will need to be further validated in functional assays to confirm they are truly associated with a possible disease mechanism.

The FTLD-TDP GWAS, showing association with SNPs at the locus on chromosome 7p21 [59], revealed the risk alleles to affect expression levels (increased) of the *cis*-gene *TMEM106B* [59]. Further analyses showed elevated basal levels of *TMEM106B* in FTLD brains affected by TDP-43 pathology [81]. Also, multiple follow-up studies confirmed *TMEM106B* to be functionally relevant for FTD hinting at an interplay with two known fFTD (Mendelian) genes, i.e. *GRN* and *CHMP2B*. Studies on TMEM106B protein suggested its involvement in the endolysosomal system together with CHMP2B [82]. Furthermore, overexpression of *TMEM106B* was shown to be associated with impairment of the endolysosomal system and an increase in the levels of GRN [81], whilst ablation/reduction of *TMEM106B* was able to rescue the endolysosomal phenotype observed in *Grn*-deficient mice [83] or in *CHMP2B* mutants [84]. The GWAS on *GRN* mutation carriers [62] supported the notion that *TMEM106B* is a modifier in *GRN* mutation carriers (in line with the original study [59]) and, additionally, suggested the risk allele of the top SNP at the chromosome 8p21.3 locus being a *cis*-eQTL of the GDNF family receptor alpha 2 (*GFRA2*) gene. The GFRA2 protein was shown to co-precipitate with the GRN protein possibly inferring to a potential involvement of the GDNF signalling pathway (a pathway promoting survival of neurons) in *GRN* mutation carriers. The clinical FTD-GWAS [60] indicated that both an mQTL for *HLA-DRA* (6p21.3 locus) and an eQTL for *RAB38* (11q14 locus) appeared to explain how the biological effect at those loci was possibly mediated. mQTLs at the *HLA* locus were also suggested in Zhang et al. where regulation of expression in brain cortex of pro-inflammatory elements seemed to influence age at onset in FTD patients [63]. Further support for the involvement of the immune system in FTLD-TDP pathogenesis was more recently provided by Pottier et al. who showed (i) eQTLs driven by the risk allele of the top SNP at the chromosome 6p21.32 locus leading to increased expression of *HLA-DQA2* and *HLA-DQB2* in the brain and (ii) excess of genetic burden in a number of genes acting in epistasis with *TBK1* within innate immune signalling pathways [56].

The locus characterisation described in the above paragraph are summarised in Fig. 1.

Clearly, several of the above studies strongly suggest that perturbation of multiple genes and pathways of the immune system might specifically underpin subpopulations of patients and contribute to FTD pathogenesis. This view appears to be further supported by a handful of earlier studies hinting at altered cytokine profiling in the cerebrospinal fluid (CSF) and/or serum of FTD patients [85, 86] and the identification of changes in the expression of FTD-immune pleiotropic genes (within the *HLA* region) in post-mortem brain tissue of FTD patients with an enriched microglia/macrophage signature [87].

Are Mendelian and Sporadic FTD the Same Disorder?

A relevant point in FTD research is that Mendelian genes are instrumental for disease modelling, i.e. they can be studied in in vitro/in vivo model systems (e.g. transgenic cellular and animal models or patient-derived iPS cells) to gather insights into the molecular mechanisms of disease. This is fundamental to understand the cellular functions that are compromised during disease onset and progression and to identify potential targets for therapeutic intervention.

This approach is hardly applicable to sporadic disease. Sporadic cases are associated with multiple risk factors that are very difficult to model

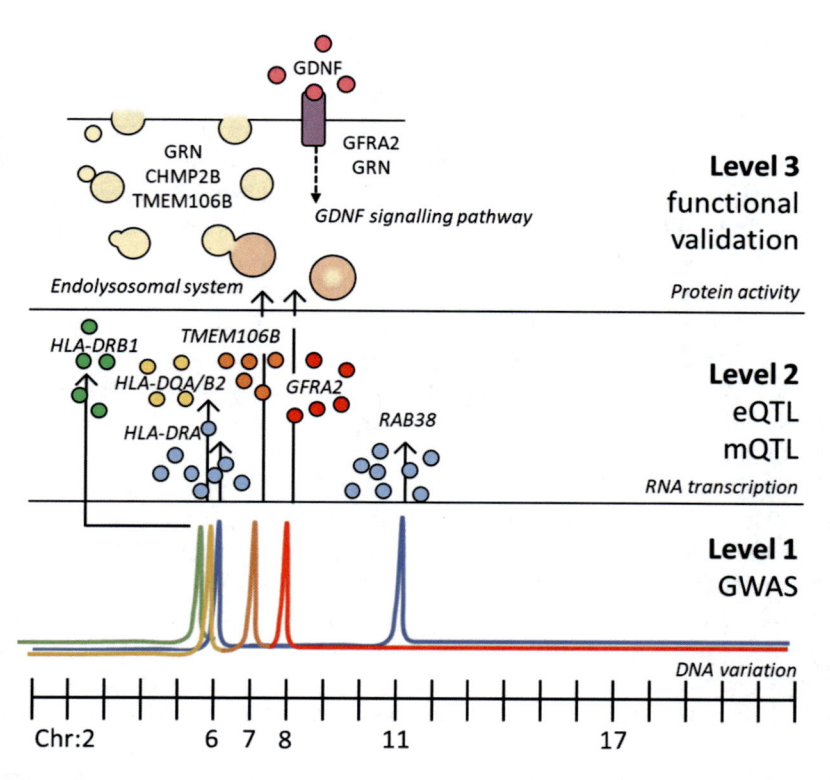

Fig. 1 Translating (sporadic) genetics into functional meaning. The pipeline for translating GWAS genetic signals into biological functions is illustrated. A GWAS is conducted to isolate 'DNA level information' on risk variants associated with FTD (*level 1*). The risk variants at the risk locus are assessed for effect(s) on gene transcription levels and/or methylation patterns (*level 2*). Validation at the protein level is pursued through functional models to characterise the impacted pathway(s) and the associated molecular mechanisms of disease (*level 3*). The original FTLD-TDP GWAS signals are depicted in orange, the International FTLD-TDP GWAS signals are depicted in red, the GRN-GWAS signals are depicted in yellow, the methylation GWAS on *C9orf72* expansion carrier signals are depicted in green and the clinical FTD-GWAS signals are depicted in blue

because they (i) feature small effect size, (ii) act as a whole (thus, the experimental system would need to model multiple risk factors at the same time) and (iii) are non-coding (thus, it is for the most unclear which gene/protein they impact). On top, the contribution of environmental exposures is, to date, impossible to model [77].

Familial models of disease do not fully capture or reflect disease complexity. In fact, by almost exclusively focusing on fFTD, FTD models are currently limited (despite a number of studies on *TMEM106B* [22, 88]) to models focused on Mendelian genes (*MAPT, GRN, C9orf72*) or models of tau pathology, a feature that is seen in FTLD-tau and beyond (e.g. AD but also progressive supranuclear palsy [PSP] or CBD). As a consequence, using the familial models as proxies for the entire disease spectrum (only because models for the sporadic forms of disease are not available) might not be entirely successful. Such *modus operandi* indirectly relies on the assumption that, since familial and sporadic FTD are clinically classified under the 'same label', the molecular mechanisms and pathways altered in familial cases might be the same or similar to those in the sporadic ones. This is, however, still an open and unexplored question. One possible example of shared mechanisms comes from the *MAPT* locus. In FTLD-tau, *MAPT* mutations (i.e. coding variants in exons 1, 9–13 [89, 90]) or heterogeneous genetic variability (e.g. intronic variants affecting expression and/or splicing of exon 10 [91, 92] or structural variants [93, 94]) cause disease and lead to tau pathology. At the same time, when considering the ~900 kb H1/H2 haplotype inversion at the *MAPT* locus [95], a yet to be identified combination of markers on this stretch may increase disease risk in a subgroup of patients with parkinsonism or broad FTD-like dementia phenotypes [96]. Further studying the genetics at the basis of tau pathology might help shedding light on communal disease mechanisms across fFTDs and sFTDs, as well as FTD and other tauopathies.

Moreover, one must not forget about a number of critical issues associated with the study of familial and/or pathologically defined cohorts: (i) they represent a minority of all FTD cases, (ii)

they might be underpowered, (iii) they might provide little or inadequate information on disease mechanism(s) underpinning the various clinical syndromes and (iv) drugs and intervention measures, currently under preclinical and clinical investigation (trials), appear tailored to fFTD or FTLD-tau only [97].

There is therefore an urgent need to expand the focus to sporadic FTD and assess disease pathways that might be communal across fFTDs and sFTDs, knowledge that will be critical and instrumental to pave the way for developing clinical trials and means for therapeutic intervention addressing all FTD cases.

Risk Pathway In Silico Modelling

Multiple genes and genetic risk variants associate with FTD. However, as in the case of other complex neurodegenerations such as PD and AD, it is difficult to portrait why and how so many different genetic elements lead to the 'same disease'.

It is well known that functional research is still not well equipped to model multiple genetic players at the same time. The classical approach relies on studying single genes (and risk factors) in isolation, collating reductionist pieces of information to recreate a global picture of disease. However, whilst this approach has been successful, e.g. the amyloid cascade hypothesis in AD based on functional work assessing mutations in *APP* and *PSEN*s [98], it appears promising, e.g. ongoing studies focusing on tau pathology [99] and the biology of *GRN, C9orf72* and *TMEM106B* [21, 22], only in a limited number of cases due to intense and costly mechanistic studies that impact the timely dissection of disease mechanisms [100].

Conversely, more recent bioinformatics and systems biology methods – incorporating notions from graph theory, network analysis and machine learning – have seen the light to model the genetic landscape associated with a complex trait and predict risk pathways to assist hypothesis-driven functional validation in the wet lab. This represents a holistic paradigm shift where risk pathway(s) are hypothesised, in silico, a priori, in a time- and cost-effective fashion, and can be subsequently tested. Systems biology approaches based on network analysis have started being applied to FTD to evaluate possible functional commonalities across FTD genes.

Weighted gene co-expression network analysis (WGCNA) – a bioinformatics method that applies mathematics, statistics and graph theory to expression (and possibly tissue-specific) level data [101] – was applied to evaluate impacted biological processes/pathways and connectivity of genes of interest within co-expression networks in knowingly impacted brain regions [102]. Specifically, FTD-relevant genes (called 'seeds' in this context) were mapped to modules representative of expression profiles in the brain and mathematically assessed for their relevance within each module, prior functionally annotating each module. Such a pipeline allows to swiftly investigate the set of functions in which each single FTD genes might be expected to be involved. At the same time, it allows to evaluate possible functional overlap(s) across several different genes in a brain regional-specific manner. The FTD-WGCNA work [102] did reduce the impacted biological processes/pathways (for both familial and sporadic forms of disease) down to (i) gene expression, DNA protection (e.g. DNA damage repair) and protein metabolism (e.g. waste disposal) processes for a majority of FTD-Mendelian genes and (ii) immune response and endolysosomal metabolism for sFTD risk factors. The intrinsic novelties of this approach can be summarised as follows: (i) the annotated modules are critical in mapping specific impacted biological processes to specific brain regions relevant to disease, and (ii) the list of genes found to be co-expressed with the FTD-relevant genes might provide informative suggestions on novel potential genetic and/or functional candidates. For example, *TBK1* mapped to a co-expression module together with *C9orf72*, *VCP*, *UBQLN2* and *OPTN* [102]. The fact that mutations in *TBK1* were isolated in the FTD and FTD-ALS spectrum reinforces the notion that members of modules including FTD-relevant genes might be (retrospectively) considered for prioritising sequencing and burden analyses aimed at the discovery of novel genes associated with disease.

Weighted protein-protein interaction network analysis (WPPINA) – another bioinformatics approach, this time taking into account protein-protein interactions (PPI) – was applied to extract physical interactors of the protein products of FTD-relevant genes [103]. This method first determined (two-layered) protein interactomes around each FTD-relevant gene (or 'seed') and then investigated communal nodes (interactors) across as many seeds as possible. Such interconnectome (made of so-called inter-interactome hubs [IIH]) was then used to perform functional annotation analysis (similarly to the case of the WGCNA modules). The FTD-WPPINA work [103] confirmed three major biological processes/pathways shared across FTD-relevant genes (previously also suggested by the FTD-WGCNA) such as gene expression, DNA damage response and waste disposal. Similarly (although slightly differently) to the WGCNA approach described above, WPPINA was instrumental in indicating, in addition to the above highlighted impacted pathways, a list of potential genetic and/or functional candidates either directly or indirectly interacting with the protein products of FTD-relevant genes. This is all the more important in that it provides protein targets within impacted pathways to be taken forwards for (i) designing ad hoc functional assays to model disease and (ii) lead to the identification of potential drug targets. Moreover, WPPINA proved promising in other contexts such as those of prioritising genes within GWAS loci and comparing/discriminating impacted biological processes across neurodegenerative diseases. Specifically, WPPINA was helpful in narrowing down potential functional candidates at PD-GWAS loci and proved useful in computationally discriminating specific subcellular pathways whilst comparing FTD and PD [104]. WPPINA suggested that, for same (or similar) impacted biological processes (e.g. biology of 'stress' and 'waste disposal'), it was 'endoplasmic reticulum (ER) stressors' that correlated with FTD vs. 'mitochondria stressors' in PD or elements of the 'unfolded protein response' and 'ubiquitin proteasome' in FTD vs. 'autophagy' and 'lysosomal' biology in PD [104].

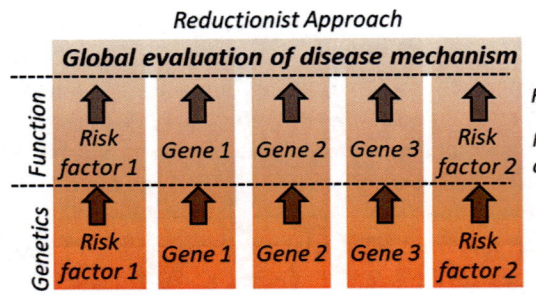 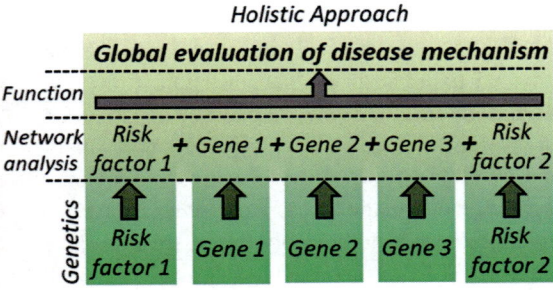

Fig. 2 Reductionist and holistic approach scheme. The 'reductionist' approach studies one gene/risk marker at a time. The 'holistic' approach aims at defining communal functional features across the multiple gene(s)/risk marker(s). Both approaches are important. They are not mutually exclusive but rather incremental and complementary

It is relevant to note that, in parallel with the WGCNA and WPPINA studies and in the context of bridging the biology of fFTDs and sFTDs, additional bioinformatics work showed association of risk variants in sporadic FTD-GWAS with the biology of immune-related disorders [87] or RNA metabolism and cell death pathways to be associated with FTD's language variant syndrome [105] and cell cycle and immune signalling to be associated with tissue-specific expression changes in bvFTD [106].

It must be acknowledged that these are in silico approaches and no practical steps have yet been undertaken to functionally prove the above highlighted risk pathways. Nevertheless, discussions between field professionals (e.g. geneticists, bioinformaticians and functional biologists) on these topics have started and are ongoing, with a focus on FTD models as well. Functional studies will be the next critical step in comparing and understanding disease processes affected in fFTD and sFTD and may subsequently support the development of interventional measures.

Future Directions

The study of FTD – from genetic dissection to disease modelling – will require a significant number of efforts in the years to come. Importantly, the research carried out this far provides us with a solid basis to optimistically look into the future with a clear understanding of the (still) open challenges that will need to be addressed.

FTD genetics will require more powerful and in-depth studies – based on GWAS, fine-mapping and sequencing techniques – to (i) dissect common (i.e. prioritise genes impacted by the genetic risk markers isolated through GWAS), oligogenic and rare genetic factors underpinning disease; (ii) tackle missing heritability; (iii) define the genetic architecture of sFTD with particular focus on the different FTD subtypes (based on both clinical and pathological diagnoses); and (iv) foster meta- and pleiotropy analyses with other closely related neurodegenerative conditions.

In parallel, it will be critical to translate the genetic findings into model systems and molecular mechanisms of disease. More specifically, it will be necessary to implement a paradigm shift from reductionist to holistic approaches to interpret genetics (Fig. 2) and subsequently assist and drive functional studies. This means that precise experimental models (including cell-specificity studies) investigating and validating risk pathways and biological processes that are impacted by genetic variability will (have to) become reality [107, 108].

All this taken together will be instrumental in improving our understanding of the aetiopathogenesis of disease, help stratifying patients for syndrome-specific clinical trials, highlighting efficient endpoints for disease monitoring and therapeutic intervention and deciphering whether and to what extent molecular mechanisms at the basis of fFTD and sFTD are overlapping, convergent or divergent.

Normalising these strategies will be extremely valuable in setting the ground for the development of effective disease management measures in FTD within the frame of precision medicine.

Acknowledgements This work was supported by Alzheimer's Society (grant number 284) to RF.

References

1. Brown RC et al (2005) Neurodegenerative diseases: an overview of environmental risk factors. Environ Health Perspect 113(9):1250–1256
2. Singleton AB (2011) Exome sequencing: a transformative technology. Lancet Neurol 10(10):942–946
3. Manolio TA (2009) Cohort studies and the genetics of complex disease. Nat Genet 41(1):5–6
4. Ciani M et al (2019) Genome wide association study and next generation sequencing: a glimmer of light toward new possible horizons in frontotemporal dementia research. Front Neurosci 13:506
5. Kolber P et al (2019) Gene-environment interaction and Mendelian randomisation. Rev Neurol (Paris) 175(10):597–603
6. Borroni B et al (2009) Revisiting brain reserve hypothesis in frontotemporal dementia: evidence from a brain perfusion study. Dement Geriatr Cogn Disord 28(2):130–135
7. Placek K et al (2016) Cognitive reserve in frontotemporal degeneration: neuroanatomic and neuropsychological evidence. Neurology 87(17):1813–1819
8. Nilsson C et al (2014) Age-related incidence and family history in frontotemporal dementia: data from the Swedish dementia registry. PLoS One 9(4):e94901
9. Cermakova P et al (2015) Cardiovascular diseases in ~30,000 patients in the Swedish dementia registry. J Alzheimers Dis 48(4):949–958
10. Golimstok A et al (2014) Cardiovascular risk factors and frontotemporal dementia: a case-control study. Transl Neurodegener 3:13
11. Rasmussen Eid H et al (2019) Smoking and obesity as risk factors in frontotemporal dementia and Alzheimer's disease: the HUNT study. Dement Geriatr Cogn Disord Extra 9(1):1–10
12. Fenoglio C et al (2018) Role of genetics and epigenetics in the pathogenesis of Alzheimer's disease and frontotemporal dementia. J Alzheimers Dis 62(3):913–932
13. Xylaki M et al (2019) Epigenetics of the synapse in neurodegeneration. Curr Neurol Neurosci Rep 19(10):72
14. Ball N et al (2019) Parkinson's disease and the environment. Front Neurol 10:218
15. LoBue C et al (2020) Beyond the headlines: the actual evidence that traumatic brain injury is a risk factor for later-in-life dementia. Arch Clin Neuropsychol 35(2):123–127
16. Kuter K et al (2010) Increased reactive oxygen species production in the brain after repeated low-dose pesticide paraquat exposure in rats. A comparison with peripheral tissues. Neurochem Res 35(8):1121–1130
17. Bittar A et al (2019) Neurotoxic tau oligomers after single versus repetitive mild traumatic brain injury. Brain Commun 1(1):fcz004
18. Gorno-Tempini ML et al (2011) Classification of primary progressive aphasia and its variants. Neurology 76(11):1006–1014
19. Neary D et al (1998) Frontotemporal lobar degeneration: a consensus on clinical diagnostic criteria. Neurology 51(6):1546–1554
20. Rascovsky K et al (2011) Sensitivity of revised diagnostic criteria for the behavioural variant of frontotemporal dementia. Brain 134(Pt 9):2456–2477
21. Ferrari R et al (2019) Genetics and molecular mechanisms of frontotemporal lobar degeneration: an update and future avenues. Neurobiol Aging 78:98–110
22. Pottier C et al (2016) Genetics of FTLD: overview and what else we can expect from genetic studies. J Neurochem 138(Suppl 1):32–53
23. Forrest SL et al (2019) Heritability in frontotemporal tauopathies. Alzheimers Dement (Amst) 11:115–124
24. Po K et al (2014) Heritability in frontotemporal dementia: more missing pieces? J Neurol 261(11):2170–2177
25. Rohrer JD et al (2009) The heritability and genetics of frontotemporal lobar degeneration. Neurology 73(18):1451–1456
26. Ghetti B et al (2015) Invited review: frontotemporal dementia caused by microtubule-associated protein tau gene (MAPT) mutations: a chameleon for neuropathology and neuroimaging. Neuropathol Appl Neurobiol 41(1):24–46
27. Gijselinck I et al (2008) Granulin mutations associated with frontotemporal lobar degeneration and related disorders: an update. Hum Mutat 29(12):1373–1386
28. DeJesus-Hernandez M et al (2011) Expanded GGGGCC hexanucleotide repeat in noncoding region of C9ORF72 causes chromosome 9p-linked FTD and ALS. Neuron 72(2):245–256
29. van der Zee J et al (2013) A pan-European study of the C9orf72 repeat associated with FTLD: geographic prevalence, genomic instability, and intermediate repeats. Hum Mutat 34(2):363–373
30. Brown J et al (1995) Familial non-specific dementia maps to chromosome 3. Hum Mol Genet 4(9):1625–1628
31. Skibinski G et al (2005) Mutations in the endosomal ESCRTIII-complex subunit CHMP2B in frontotemporal dementia. Nat Genet 37(8):806–808
32. Weihl CC et al (2009) Valosin-containing protein disease: inclusion body myopathy with Paget's disease of the bone and fronto-temporal dementia. Neuromuscul Disord 19(5):308–315

33. Freischmidt A et al (2015) Haploinsufficiency of TBK1 causes familial ALS and fronto-temporal dementia. Nat Neurosci 18(5):631–636

34. Gijselinck I et al (2015) Loss of TBK1 is a frequent cause of frontotemporal dementia in a Belgian cohort. Neurology 85(24):2116–2125

35. Pottier C et al (2015) Whole-genome sequencing reveals important role for TBK1 and OPTN mutations in frontotemporal lobar degeneration without motor neuron disease. Acta Neuropathol 130(1):77–92

36. Momeni P et al (2006) Analysis of IFT74 as a candidate gene for chromosome 9p-linked ALS-FTD. BMC Neurol 6:44

37. Le Ber I et al (2013) SQSTM1 mutations in French patients with frontotemporal dementia or frontotemporal dementia with amyotrophic lateral sclerosis. JAMA Neurol 70(11):1403–1410

38. Synofzik M et al (2012) Screening in ALS and FTD patients reveals 3 novel UBQLN2 mutations outside the PXX domain and a pure FTD phenotype. Neurobiol Aging 33(12):2949 e13–2949 e17

39. Bannwarth S et al (2014) A mitochondrial origin for frontotemporal dementia and amyotrophic lateral sclerosis through CHCHD10 involvement. Brain 137(Pt 8):2329–2345

40. Mackenzie IR et al (2017) TIA1 mutations in amyotrophic lateral sclerosis and frontotemporal dementia promote phase separation and alter stress granule dynamics. Neuron 95(4):808–816 e9

41. Al-Chalabi A et al (2017) Gene discovery in amyotrophic lateral sclerosis: implications for clinical management. Nat Rev Neurol 13(2):96–104

42. Cooper-Knock J et al (2014) The widening spectrum of C9ORF72-related disease; genotype/phenotype correlations and potential modifiers of clinical phenotype. Acta Neuropathol 127(3):333–345

43. Ferrari R et al (2012) Screening for C9ORF72 repeat expansion in FTLD. Neurobiol Aging 33(8):1850 e1–1850 11

44. Galimberti D et al (2014) Incomplete penetrance of the C9ORF72 hexanucleotide repeat expansions: frequency in a cohort of geriatric non-demented subjects. J Alzheimers Dis 39(1):19–22

45. Hensman Moss DJ et al (2014) C9orf72 expansions are the most common genetic cause of Huntington disease phenocopies. Neurology 82(4):292–299

46. Lindquist SG et al (2013) Corticobasal and ataxia syndromes widen the spectrum of C9ORF72 hexanucleotide expansion disease. Clin Genet 83(3):279–283

47. Majounie E et al (2012) Frequency of the C9orf72 hexanucleotide repeat expansion in patients with amyotrophic lateral sclerosis and frontotemporal dementia: a cross-sectional study. Lancet Neurol 11(4):323–330

48. Simon-Sanchez J et al (2012) The clinical and pathological phenotype of C9ORF72 hexanucleotide repeat expansions. Brain 135(Pt 3):723–735

49. Smith BN et al (2013) The C9ORF72 expansion mutation is a common cause of ALS+/-FTD in

Europe and has a single founder. Eur J Hum Genet 21(1):102–108

50. Watts GD et al (2004) Inclusion body myopathy associated with Paget disease of bone and frontotemporal dementia is caused by mutant valosin-containing protein. Nat Genet 36(4):377–381

51. Borroni B et al (2010) TARDBP mutations in frontotemporal lobar degeneration: frequency, clinical features, and disease course. Rejuvenation Res 13(5):509–517

52. Huey ED et al (2012) FUS and TDP43 genetic variability in FTD and CBS. Neurobiol Aging 33(5):1016 e9–1016 17

53. Hardy J et al (2014) Motor neuron disease and frontotemporal dementia: sometimes related, sometimes not. Exp Neurol 262(Pt B):75–83

54. Halliday G et al (2012) Mechanisms of disease in frontotemporal lobar degeneration: gain of function versus loss of function effects. Acta Neuropathol 124(3):373–382

55. Takada LT (2015) The genetics of monogenic frontotemporal dementia. Dement Neuropsychol 9(3)):219–229

56. Pottier C et al (2019) Genome-wide analyses as part of the international FTLD-TDP whole-genome sequencing consortium reveals novel disease risk factors and increases support for immune dysfunction in FTLD. Acta Neuropathol 137(6):879–899

57. Manolio TA et al (2009) Finding the missing heritability of complex diseases. Nature 461(7265):747–753

58. Hasin Y et al (2017) Multi-omics approaches to disease. Genome Biol 18(1):83

59. Van Deerlin VM et al (2010) Common variants at 7p21 are associated with frontotemporal lobar degeneration with TDP-43 inclusions. Nat Genet 42(3):234–239

60. Ferrari R et al (2014) Frontotemporal dementia and its subtypes: a genome-wide association study. Lancet Neurol 13(7):686–699

61. Ferrari R et al (2015) A genome-wide screening and SNPs-to-genes approach to identify novel genetic risk factors associated with frontotemporal dementia. Neurobiol Aging 36(10):2904 e13–2904 e26

62. Pottier C et al (2018) Potential genetic modifiers of disease risk and age at onset in patients with frontotemporal lobar degeneration and GRN mutations: a genome-wide association study. Lancet Neurol 17(6):548–558

63. Zhang M et al (2018) A C6orf10/LOC101929163 locus is associated with age of onset in C9orf72 carriers. Brain 141(10):2895–2907

64. Barbier M et al (2017) Factors influencing the age at onset in familial frontotemporal lobar dementia: important weight of genetics. Neurol Genet 3(6):e203

65. Mackenzie IR et al (2016) Molecular neuropathology of frontotemporal dementia: insights into disease mechanisms from postmortem studies. J Neurochem 138(Suppl 1):54–70

66. Young AI (2019) Solving the missing heritability problem. PLoS Genet 15(6):e1008222
67. Philtjens S et al (2018) Rare nonsynonymous variants in SORT1 are associated with increased risk for frontotemporal dementia. Neurobiol Aging 66:181 e3–181 e10
68. Williams KL et al (2016) CCNF mutations in amyotrophic lateral sclerosis and frontotemporal dementia. Nat Commun 7:11253
69. Kim EJ et al (2018) Analysis of frontotemporal dementia, amyotrophic lateral sclerosis, and other dementia-related genes in 107 Korean patients with frontotemporal dementia. Neurobiol Aging 72:186 e1–186 e7
70. Ng ASL et al (2018) Targeted exome sequencing reveals homozygous TREM2 R47C mutation presenting with behavioral variant frontotemporal dementia without bone involvement. Neurobiol Aging 68:160 e15–160 e19
71. Giannoccaro MP et al (2017) Multiple variants in families with amyotrophic lateral sclerosis and frontotemporal dementia related to C9orf72 repeat expansion: further observations on their oligogenic nature. J Neurol 264(7):1426–1433
72. Kunkle BW et al (2019) Genetic meta-analysis of diagnosed Alzheimer's disease identifies new risk loci and implicates Abeta, tau, immunity and lipid processing. Nat Genet 51(3):414–430
73. Nalls MA et al (2019) Identification of novel risk loci, causal insights, and heritable risk for Parkinson's disease: a meta-analysis of genome-wide association studies. Lancet Neurol 18(12):1091–1102
74. International Schizophrenia, C et al (2009) Common polygenic variation contributes to risk of schizophrenia and bipolar disorder. Nature 460(7256):748–752
75. Karczewski KJ et al (2018) Integrative omics for health and disease. Nat Rev Genet 19(5):299–310
76. Manzoni C et al (2018) Genome, transcriptome and proteome: the rise of omics data and their integration in biomedical sciences. Brief Bioinform 19(2):286–302
77. Manzoni C et al (2020) Network analysis for complex neurodegenerative diseases. Curr Genet Med Rep 8:17–25
78. Liu JZ et al (2010) A versatile gene-based test for genome-wide association studies. Am J Hum Genet 87(1):139–145
79. Pearson TA et al (2008) How to interpret a genome-wide association study. JAMA 299(11):1335–1344
80. Zheng Z et al (2020) QTLbase: an integrative resource for quantitative trait loci across multiple human molecular phenotypes. Nucleic Acids Res 48(D1):D983–D991
81. Chen-Plotkin AS et al (2012) TMEM106B, the risk gene for frontotemporal dementia, is regulated by the microRNA-132/212 cluster and affects progranulin pathways. J Neurosci 32(33):11213–11227
82. Jun MH et al (2015) TMEM106B, a frontotemporal lobar dementia (FTLD) modifier, associates with FTD-3-linked CHMP2B, a complex of ESCRT-III. Mol Brain 8:85
83. Klein ZA et al (2017) Loss of TMEM106B ameliorates lysosomal and frontotemporal dementia-related phenotypes in progranulin-deficient mice. Neuron 95(2):281–296 e6
84. Clayton EL et al (2018) Frontotemporal dementia causative CHMP2B impairs neuronal endolysosomal traffic-rescue by TMEM106B knockdown. Brain 141(12):3428–3442
85. Galimberti D et al (2008) Intrathecal levels of IL-6, IL-11 and LIF in Alzheimer's disease and frontotemporal lobar degeneration. J Neurol 255(4):539–544
86. Sjogren M et al (2004) Increased intrathecal inflammatory activity in frontotemporal dementia: pathophysiological implications. J Neurol Neurosurg Psychiatry 75(8):1107–1111
87. Broce I et al (2018) Immune-related genetic enrichment in frontotemporal dementia: an analysis of genome-wide association studies. PLoS Med 15(1):e1002487
88. Ferrari R et al (2018) Genetic risk factors for sporadic frontotemporal dementia. Springer, Cham, Neurodegener Dis 147–186
89. Hutton M et al (1998) Association of missense and 5'-splice-site mutations in tau with the inherited dementia FTDP-17. Nature 393(6686):702–705
90. Poorkaj P et al (1998) Tau is a candidate gene for chromosome 17 frontotemporal dementia. Ann Neurol 43(6):815–825
91. Malkani R et al (2006) A MAPT mutation in a regulatory element upstream of exon 10 causes frontotemporal dementia. Neurobiol Dis 22(2):401–403
92. Spillantini MG et al (1998) Mutation in the tau gene in familial multiple system tauopathy with presenile dementia. Proc Natl Acad Sci U S A 95(13):7737–7741
93. Rovelet-Lecrux A et al (2010) Frontotemporal dementia phenotype associated with MAPT gene duplication. J Alzheimers Dis 21(3):897–902
94. Rovelet-Lecrux A et al (2009) Partial deletion of the MAPT gene: a novel mechanism of FTDP-17. Hum Mutat 30(4):E591–E602
95. Vandrovcova J et al (2010) Disentangling the role of the tau gene locus in sporadic tauopathies. Curr Alzheimer Res 7(8):726–734
96. Baba Y et al (2005) The effect of tau genotype on clinical features in FTDP-17. Parkinsonism Relat Disord 11(4):205–208
97. Rosen HJ et al (2020) Tracking disease progression in familial and sporadic frontotemporal lobar degeneration: recent findings from ARTFL and LEFFTDS. Alzheimers Dement 16(1):71–78
98. Selkoe DJ et al (2016) The amyloid hypothesis of Alzheimer's disease at 25 years. EMBO Mol Med 8(6):595–608
99. Holtzman DM et al (2016) Tau: from research to clinical development. Alzheimers Dement 12(10):1033–1039

100. Golde TE (2016) Overcoming translational barriers impeding development of Alzheimer's disease modifying therapies. J Neurochem 139(Suppl 2):224–236

101. Langfelder P et al (2008) WGCNA: an R package for weighted correlation network analysis. BMC Bioinforma 9:559

102. Ferrari R et al (2016) Frontotemporal dementia: insights into the biological underpinnings of disease through gene co-expression network analysis. Mol Neurodegener 11:21

103. Ferrari R et al (2017) Weighted protein interaction network analysis of frontotemporal dementia. J Proteome Res 16(2):999–1013

104. Ferrari R et al (2018) Stratification of candidate genes for Parkinson's disease using weighted protein-protein interaction network analysis. BMC Genomics 19(1):452

105. Bonham LW et al (2019) Genetic variation across RNA metabolism and cell death gene networks is implicated in the semantic variant of primary progressive aphasia. Sci Rep 9(1):10854

106. Bonham LW et al (2018) Protein network analysis reveals selectively vulnerable regions and biological processes in FTD. Neurol Genet 4(5):e266

107. Furlong LI (2013) Human diseases through the lens of network biology. Trends Genet 29(3):150–159

108. Skene NG et al (2018) Genetic identification of brain cell types underlying schizophrenia. Nat Genet 50(6):825–833

FTLD Treatment: Current Practice and Future Possibilities

Peter A. Ljubenkov and Adam L. Boxer

Non-pharmacological Management in FTD

Early Education

Many patients and caregivers are unfamiliar with behavioral variant frontotemporal dementia (bvFTD) and primary progressive aphasia (PPA) when these diagnoses are first discussed in clinic. For this reason, early therapeutic invention often involves basic education about the disease. The Association for Frontotemporal Degeneration (AFTD) is a useful reference for patients in North America (www.theaftd.org) and Australia (www. theaftd.org.au), and Rare Dementia Support (www.raredementiasupport.org) offers similar resources in the United Kingdom. These organizations provide basic high-quality information about diagnoses, research opportunities (including clinical trials), and support group services. Patients who are particularly interested in research may also be referred to a local academic center belonging to a large multisite research consortium, such as the Genetic Frontotemporal Dementia Initiative (GENFI, genfi.org.uk) in the United Kingdom and Europe and the ALLFTD

research consortium in the United States (www. allftd.org). Consortia of this kind often provide the best infrastructure to identify and counsel familial FTD cohorts and navigate patients toward relevant clinical trials of interest. Alternatively, patients who carry a strong family history of neurodegenerative disease but who are not interested in research may benefit from an early referral to an independent genetic counselor, particularly when Mendelian forms of FTD are expected.

Initial Safety Evaluations

In bvFTD, like many dementia syndromes, it is important to assess a patient's current level of safety during their early and subsequent evaluations [1]. Patients with features of the behavioral variant of frontotemporal dementia (bvFTD) [2] in particular often lack the capacity to avoid danger, due to disinhibition, apathy, and poor understanding of the internal state of others. Moreover, while patients with bvFTD may occasionally exhibit violent behaviors, they are also at risk of physical or financial victimization due to their impairments in social cognition. Table 1 details a brief list of potential safety concerns and viable intervention strategies in patients with bvFTD. Of these concerns, driving safety is often an early and contentious point of discussion. While a physician's responsibilities may differ by country

P. A. Ljubenkov · A. L. Boxer (✉)
Memory and Aging Center, Department of Neurology, University of California, San Francisco, USA
e-mail: peter.ljubenkov@ucsf.edu;
adam.boxer@ucsf.edu

B. Ghetti et al. (eds.), *Frontotemporal Dementias*, Advances in Experimental Medicine and Biology 1281, https://doi.org/10.1007/978-3-030-51140-1_18

Table 1 Common safety concerns in patients with FTD

Safety concern	Recommended intervention
Firearm and other weapons	Remove all weapons from the home Secure weapons in a locked safe the patient can't access
Driving safety	Report patient to their jurisdictions' relevant authority that controls driving privileges Consider hiding vehicle keys or disabling vehicles if necessary
Medication mismanagement	Recommend caregiver take over medication management Consider securing medication in a locked box or cabinet that the patient can't access Review medication list to limit unnecessary polypharmacy
Poor self-care	Provide early education to caregiver regarding loss of independence in hygiene
Injury using kitchen appliances	Discourage independent access to dangerous kitchen appliances such as the stove or the oven Consider disabling or removing dangerous appliances if the patient must be left unattended
Wandering	Provide early education to caregivers about the need for increased supervision and additional caregiver support Consider ID bracelets, smartphone tracking apps, and/or tracking key fobs Consider door alarms and door locks requiring a key Coordinate with local law enforcement to prepare for potential wandering events
Financial risk/ scams	Consider establishing a durable power of attorney (DPOA) for financial decisions as soon as possible Consider limiting a patient's independent access to bank accounts or credit cards Report any concerns for financial abuse to the local jurisdiction's equivalent of adult protective services

(continued)

Table 1 (continued)

Safety concern	Recommended intervention
Undue approach of strangers	Encourage direct supervision in public places Consider limiting exposure to crowded public places if behaviors are hard to control Consider education of members of the patient's immediate community to promote acceptance and minimize misunderstanding Report any concerns for financial abuse to the local jurisdiction's equivalent of adult protective services
Falls	Educate patient and caregiver about impulsivity if it is present Remove clutter and tripping hazards from walkways and stairs Improve lighting and color contrast on steps Move high-use items to mid-level cabinets that don't require stretching or bending Consider increasing placement of handrails in the home (especially in the shower and by the toilet) Ensure adequate assistance to and from the restroom, particularly at night Consider installing night-lights and a bedside commode If significant parkinsonism is present, consider a weighted walker that defaults to a locked position (such as the U-step)

sive behavior, or any cognitive testing supporting a diagnosis of dementia [3]. Home firearms are also among the most urgent safety concerns in bvFTD and must be removed or secured as soon as a dementia diagnosis is made. Ultimately, if a patient presents persistent safety concerns to themselves or others, a higher level of care may need consideration.

Behavioral Interventions

Caregiver burden is increased in bvFTD relative to Alzheimer's disease dementia [4], due in part to the increase in a variety of difficult and disruptive behaviors. Apathy, disinhibition, compulsive

and jurisdiction, a patient's driving privileges should be reevaluated in the face of caregiver concern, a recent car accident, a recent traffic citation, a recent volitional restriction of the scope of driving, concern for impulsive or aggres-

behaviors, loss of empathy, and dietary changes make up the core clinical features of the bvFTD [2]. These behavioral features are also commonly found in a variety of conditions within the greater clinical spectrum correlating with FTLD pathology on autopsy, especially in advanced stages of disease. In particular, disabling behavioral changes are a well-described phenomenon in FTD with motor neuron disease (FTD-MND) [5], semantic variant primary progressive aphasia (svPPA) [6], non-fluent agrammatic variant primary progressive aphasia (nfvPPA) [7], progressive supranuclear palsy (PSP) [8], and cortical basal syndrome (CBS) [9]. Unfortunately, pharmacotherapy often provides little efficacy in curbing difficult behaviors. For this reason, the core of behavioral management in bvFTD involves caregiver strategies for behavioral redirection and environmental modification. While few studies have sought to test the efficacy of these behavioral interventions in patients with dementia, one potentially viable consensus framework for behavioral intervention was established by Kale et al. in 2014 [10]. This proposed "DICE" model of behavioral intervention advocates careful *d*escription of the circumstances of problem behaviors, thorough *i*nvestigation of potential inciting/contributing factors, *c*reation of an action plan to alleviate exacerbating factors, and follow-up *e*valuation to address the need for implementation of additional interventions. The DICE model advocates three avenues of investigation when considering preventable causes of unwanted behavior: the patient, the caregiver, and the environment. Potentially augmentable patient factors include untreated medical comorbidity (leading to discomfort or delirium), untreated psychiatric comorbidity (including depression and anxiety), untreated pain, untreated sensory deficits, boredom, fear, and poor sleep hygiene. Preventable caregiver factors include a limited understanding of a patient's dementia syndrome, inappropriate expectations for a patient with dementia, a confrontational communication style, and an overly nuanced communication style. Potentially harmful environmental features include unpredictability in the daily routine, a chaotic or uncomfortable physical environment, a poorly lit environment,

an overabundance of distractions or choices, or a lack of recreational distractions. The authors of the DICE model also encourage assessment of safety risk, and while dangerous behaviors require immediate intervention, non-harmful repetitive behaviors may be best managed with acceptance and reframing of expectations.

Speech Therapy

In patients with features of primary progressive aphasia (PPA) [11], treatment typically focuses on referral to a licensed speech and language pathologist (SLP). While clinical trial data is limited, an experienced SLP may offer a variety of interventions and compensatory strategies to patients with PPA. As discussed in a recent review by Volkmer et al. in 2019 [12], PPA interventions commonly tap strategies training individual word retrieval, trained scripts, and compensatory communication methods. A systematic review of 39 studies suggests that word retrieval interventions (e.g., repetitively reading specific words with associated pictures) may transiently help patients with PPA retrieve specific trained words, though these gains may not always be maintained or generalized [13]. Additionally, non-randomized trials in nfvPPA suggest that script training, a common therapy in stroke aphasia, may improve the intelligibility of trained and untrained topics, and these gains may persist for up to a year after treatment [14, 15]. Compensatory communication strategies include communication skills training, including greater implementation of nonverbal gestures, and augmentative and alternative communication (AAC) devices in the form or communication cards, phone apps, or tablet-based devices [12].

Additional Non-pharmacological Interventions

There is some epidemiological data supporting a healthy diet, increased physical activity, increased cognitive engagement, and increased social engagement as mechanisms to prevent all causes of dementia [16]. Many of these lifestyle features are usually hypothesized to modify the risk for vascular dementia rather than the pathophysiology of FTLD. There is, however, relatively new evidence that cognitive activity (e.g., reading or spending time with friends) and physical exercise may be associated with slower rates of clinical decline in familial forms of frontotemporal dementia [17]. While the direction of causality is hard to establish in this early data, increased social engagement and physical exercises tend to be fundamentally positive for quality of life and thus represent low-risk strategies for treatment and prevention of FTD.

Current Pharmacotherapy in FTD

Antidepressants

Selective serotonin reuptake inhibitors (SSRIs) remain the central focus of current pharmacotherapy targeting the behavioral features FTD [18], despite scarce randomized clinical trial support. This practice is supported by early evidence linking disruptive behaviors, such as agitation and aggression, to deficits in serotonergic signaling [19]. Additionally, SSRIs have long been known to induce hyposexuality, which may be a desired side effect in patients with bvFTD [20]. PET studies reflect reduced $5\text{-}HT_{1A}$ receptors throughout multiple frontotemporal regions [21] in bvFTD. Furthermore, postmortem studies in autopsy-confirmed FTLD further support a wide spread of largely postsynaptic deficits involving reduction of $5\text{-}HT_1$ and $5HT_{2A}$ receptors throughout multiple frontal and temporal cortical regions [22–24], as well as 40% loss of serotoninergic neurons in the median raphe nucleus [25]. Consistent with these early pathologic studies, early open-label SSRI studies in FTD suggested benefits in the treatment of depressive symptoms and a variety of core bvFTD features, including disinhibition, compulsions, and dietary changes [26]. Early case data on paroxetine, for instance, suggested benefits in curbing depressive and obsessive symptoms in FTD [20], and in a 14-month open-label trial of paroxetine (20 mg daily), patients experienced improvement

of repetitive behaviors and overall neuropsychiatric index (NPI) score [27]. However, 40 mg daily dosing of paroxetine failed to improve behavior in a follow-up randomized crossover study enrolling ten patients with FTD (and, in fact, patients on active drug performed nonsignificantly worse on cognitive testing [28]). This failure may have been due to the off-target anticholinergic effects of paroxetine. On this note, there is case evidence that anticholinergic tricyclic antidepressants such as clomipramine may also be poorly tolerated in semantic dementia [20]. Current pharmacotherapy trends in bvFTD tend to make greater use of SSRIs with fewer off-target effects, such as citalopram, escitalopram, and sertraline, though trial data is limited in these drugs. Sertraline has so far been shown to improve behaviors in small open-label trials in bvFTD and svPPA [29]. Additionally, a 6-week open-label trial of citalopram (titrated to 40 mg daily) was associated with improvements in depression, disinhibition, and irritability in 15 patients with FTD [30]. Trazodone (a weak SSRI and 5-HT1A, 5-HT1C, and 5-HT2 antagonist) [31] has perhaps the best supported therapeutic rationale in bvFTD, as it yielded significant improvements in depressive symptoms, irritability, agitation, and dietary changes in a randomized, double-blind, placebo-controlled crossover study in ten patients [32]. However, at the dose used in this trial (150 mg daily), patients experienced treatment emergent effects of fatigue, dizziness, hypotension, and cold extremities. Given these and other known side effects of trazodone, it has failed to supplant more typical SSRIs in the standard care of FTD. Additionally, aside from SSRIs, there is a lack of published information supporting or discouraging the use of other depression or anxiety pharmacotherapy in FTD (though there is limited case data supporting the use of mirtazapine for sleep [20] and discouraging the use of buspirone in nvPPA [20]).

Antipsychotics in FTLD

Antipsychotics are commonly used off-label to manage FTD behavioral features, despite a relative paucity of trial data and the high risks associated with these medications. While CSF characterization suggests elevations in dopamine signaling may be associated with increase agitation and aggression in FTD [19], PET studies have revealed overall deficits in dopaminergic receptor binding in striatal [33] and frontal cortical regions [34] in FTD. It is therefore not surprising that patients with FTD may be more susceptible to the extrapyramidal side effects of antipsychotic medications [35]. Additionally, antipsychotics carry a black box warning from the US Food and Drug Administration (FDA) for increased risk of mortality. Moreover, the increased mortality risks of antipsychotics may persist for more than a year after cessation of use [36]. In light of their inherent risks, antipsychotics are used with significant caution in FTD, often as a last resort under a palliative rationale.

Among the atypical antipsychotics, low-dose quetiapine is most often used in bvFTD, given its relatively low rate of extrapyramidal symptoms (EPS) [37, 38] and potentially less severe impact on mortality [39]. So far, limited case data has suggested that quetiapine may have some benefit on agitation [20], but these findings have not been replicated in a clinical trial. Clozapine is also known for its low incidence of EPS [40], but it is seldom used in FTD due to its risk of aplastic anemia. Olanzapine was also found to be helpful in suppressing agitation in an open-label trial with 17 patients with FTD [9], but patients receiving treatment also experienced an increase in EPS [41]. Additionally, olanzapine is associated with increased metabolic syndrome and a relatively high mortality risk compared to quetiapine [39]. Aripiprazole has been used to suppress inappropriate vocalizations in at least one case report [42], but published data on this drug is otherwise limited. Additionally, risperidone has also been effective in stabilizing mood and agitation in case reports [20, 43], but this medication and haloperidol are used relatively infrequently in clinical practice due to their particularly high risk of EPS and morality, even in comparison to the mortality risks of other antipsychotics [39].

Poor Rationale for Alzheimer's Medications in FTD

Previous autopsy studies suggest a relative sparing of the cholinergic system in patients with FTLD pathology on autopsy [23, 24]. For this reason, there is not a firm biological basis for use of cholinesterase inhibitors in FTD clinical syndromes. However, due to the relative paucity of alternative treatments, cholinesterase inhibitors were previously commonly used in FTD after their approval in Alzheimer's disease [44, 45]. Additionally, early trial data appeared to modestly support the use of cholinesterase inhibitors, but these trials contained potential confounding factors that limited their value [46]. For instance, in an early trial, galantamine appeared to stabilize some symptoms in patients with PPA [47], but this modest secondary finding may have been driven by inclusion of patients with logopenic variant PPA (typically associated with Alzheimer's disease pathology). Additionally, a small open-label study of rivastigmine in bvFTD observed a trend of decreased caregiver burden and behavioral impairment after treatment [48], but these measures occasionally spontaneously improve on their own in bvFTD, as apathy overshadows other behaviors. Trials with donepezil have been more definitely discouraging. In a pilot study with 23 patients and separate small open-label trial, donepezil was associated with worsening neuropsychiatric symptoms in FTD syndromes, and this effect improved after cessation of treatment [49, 50]. Given these results, cholinesterase inhibitors are now generally avoided in FTD cohorts. The use of memantine (weak NMDA antagonist) is also generally discouraged, as it failed to improve behavior and potentially worsened cognition in patients with bvFTD and PPA in a multicenter, randomized, double-blind, placebo-controlled trial [51].

Stimulants in FTD

Stimulants are occasionally rationalized as a tool to treat apathy, but they rarely used in the management of bvFTD given fears of increased irritability and disinhibition. There is, however, some limited information that stimulants may occasionally be used for treatment of unwanted behaviors in bvFTD. Methylphenidate (which simulates the release and suppresses the reuptake of dopamine and norepinephrine) appeared to improve withdrawal, apathy, and irritability in an isolated bvFTD case [52]. Moreover, in an eight-patient placebo-controlled crossover trial in bvFTD, a 40 mg daily dose of methylphenidate was associated with decreased risk-taking in a novel testing paradigm [53]. Additionally, dextromethorphan (20 mg daily) appeared to improve apathy and disinhibition relative to Seroquel (20 mg daily) in a small double-blind crossover study [54].

Anticonvulsants

Anticonvulsants like valproate are commonly used to suppress the behavioral features of mania in patients with bipolar disorder, but they are less commonly used in bvFTD due to their potentially unfavorable side effect profiles. Valproate in particular is often avoided due to its risk of encephalopathy [55], hepatotoxicity [56], hyperammonemia [57], parkinsonism [58], and increased mortality [39]. There is, however, some case report data suggesting that valproate may occasionally be used to suppress agitation and hypersexuality [20, 59]. Similarly, case data suggests that carbamazepine can be helpful in suppressing indiscriminate and inappropriate sexual behavior [60]. Additionally, topiramate has been helpful in suppressing compulsive eating and drinking behaviors in a number of case studies [61–64]. While benzodiazepines also occasionally provide a nonspecific tool for sedation, they are seldom used in FTD due to their risk of paradoxical agitation, oversedation, and misuse.

Parkinsonism Medications

Patients with parkinsonism due to FTLD pathology often find little relief from L-dopa. Additionally, when patients with PSP or CBS do respond to L-dopa, the response is typically modest and short-lived [65, 66]. The parkinsonism variant of PSP [8] is, however, occasionally associated with a more measurable and sustained benefit from L-dopa. For this reason, a trial of L-dopa/carbidopa is frequently attempted even in patients with parkinsonism due to suspected underlying FTLD. Direct dopamine agonists, on the other hand, are typically discouraged in patients with suspected FTLD pathology, due to the potential for dysfunctional behaviors from dopamine dysregulation syndrome [67]. Monoamine oxidase (MAO) inhibitors are also infrequently used in clinical syndromes associated with FTLD pathology, but there is limited case data suggesting that selegiline (an MAO-B inhibitor) may improve non-motor symptoms in patients [68].

Future Therapies for FTLD

Future Therapies for Primary Tauopathies

Tau is a microtubule-associated protein that is coded by the *MAPT* gene and is thought to promote microtubule stabilization and axonal transport [69]. Frontotemporal lobar degeneration with tau pathology (FTLD-tau) is characterized by the presence of abnormal tau species, including abnormally misfolded, cleaved, and post-translationally modified (often phosphorylated and acetylated) monomers, oligomers, and filamentous aggregates [70]. Tau proteins can be further subcategorized by the predominance of a subset of six tau isoforms [71], which arise from alternative splicing of mRNA from the *MAPT* gene and chiefly differ in their inclusion or exclusion of exon 10 (which codes for one of four microtubule binding domains). Inclusion of exon 10 results in tau transcripts with four repeated microtubule binding domains (4R tau), while exclusion of exon 10 results in three binding domains (3R tau).

One possible therapeutic strategy in FTLD-tau involves mitigation of toxic loss of microtubule function. TPI-287 (TPI) is a repurposed brain-penetrant, Taxol-related molecule that stabilizes microtubules. TPI-287 was investigated in phase 1 parallel cohort trials enrolling patients with Alzheimer's disease, PSP, and CBS (NCT01966666, NCT02133846). Unfortunately, this drug was poorly tolerated in patients with Alzheimer's disease due to increased anaphylactoid reactions. Increased falls and worsening dementia symptoms were also noted in the PSP/CBD group, and TPI-287 was not pursued in follow-up trials. Additionally, davunetide (AL-108, NAP), a short peptide that was thought to promote microtubule stability, was also not found to be efficacious in a phase 2/3 double-blind placebo-controlled trial in PSP.

Abnormal tau species demonstrate the ability to propagate from neuron to neuron and may induce conformational changes in other tau proteins in a prion-like manner [72]. Given the potential prion-like behavior of tau, anti-tau immunotherapies (including passive and active immunization strategies) are now actively explored as a means to block the interneuronal spread of tau and promote clearance of abnormal tau species [70]. While the majority of anti-tau immunotherapy therapeutic programs intend to target tau in Alzheimer's disease, there is a great interest in the parallel application of these therapies in FTLD-tau clinical trials (ClinicalTrials.gov identifier NCT03068468, NCT03413319, NCT03658135, and NCT04185415). Unfortunately, the ideal epitope for an anti-tau antibody is not clear. It also not clear that the same epitopes should be targeted across differing primary tauopathies. Preclinical studies suggest a multitude of potential epitopes for therapy including monomeric tau, oligomeric tau aggregates, hyper-phosphorylated tau, misfolded forms of tau, the tau N-terminus, the proline-rich regions of tau, microtubule binding domain, and the C-terminal regions of tau [70, 73, 74] (Table 2).

So far, clinical trials in FTLD-tau and Alzheimer's disease have directed the most attention toward passive immunization strategies against N-terminal tau epitopes. This emphasis was encouraged by early preclinical work suggesting that antibodies against the N-terminus may improve cognition in transgenic mice [73], though studies in humans have suggested the most pathogenic species of tau may be truncated at the N-terminus and retain their microtubule binding domains [75]. Unfortunately, in recent clinical trials, antibodies against N-terminal tau epitopes (BIIB092 and ABBV-8E12) have definitively failed to improve the rate of PSP clinical progression in well-powered phase 2 clinical trials (NCT03068468, NCT03413319). Additionally, termination of the BIIB092 PSP trial development program also led to early termination of a parallel phase 1 "basket trial" in patients with primary tauopathies, including CBS, nfvPPA, and pathogenic *MAPT* mutations (NCT03658135). While these events were discouraging, they may have only reflected the limited utility of targeting N-terminal epitopes in FTLD-tau. Additional upcoming trials will seek to target more diverse tau epitopes. Currently, antibodies targeting the midregions of tau, JNJ-63733657 and UCB0107, are being explored in Alzheimer's disease (NCT03375697) and an upcoming phase 1 clinical trial in PSP (NCT04185415), respectively. LY3303560, which targets N-terminal tau but shows preference for tau aggregates, is also currently being explored in a phase 2 trial in Alzheimer's disease (NCT03518073). Additionally, BIIB076, which binds to monomeric and fibrillar forms of tau, is currently being investigated in a phase I clinical trial in Alzheimer's disease (NCT03056729).

Active immunization has received much less attention than passive immunization strategies in previous anti-tau trails. So far, the AADvac1 vaccine (tau peptide aa 294–305/4R coupled to keyhole limpet hemocyanin) was safe and well tolerated in a 72-week open-label trial in patients with Alzheimer's disease [76]. In light of these findings, a phase 1 trial with AADvac1 is currently underway in patients with nfvPPA (NCT03174886). An additional vaccine, ACI-35 (which contains phosphorylated S396 and S404 tau fragments), has also been investigated in a phase 1 trial in Alzheimer's disease but has yet to be investigated in follow-up trials [77].

Antisense oligonucleotides (ASOs) provide an additional promising therapeutic mechanism

Table 2 Potential therapeutics in FTLD

	Mechanism	Indication	Phase	ClinicalTrials.gov identifier	Status
Potential therapies for C9ORF72 expansion					
BIIB078	ASO	ALS-*C9ORF72*	1	NCT03626012	Ongoing
Potential therapies for GRN haploinsufficiency					
Nimodipine	Calcium channel blocker	FTLD-*GRN*	1	NCT01835665	Negative [95]
FRM-0334	HDAC inhibitor	FTLD-*GRN*	2	NCT02149160	Negative
AL001	Anti-sortilin antibody	FTLD-*GRN*	1/2	NCT03987295	Ongoing
PR006	AVV9-based gene therapy	FTLD-*GRN*			Pending
Potential therapies for FTLD-tau					
ABBV-8E12 (C2N-8E12)	Anti-tau antibody (N-terminus)	PSP	2	NCT03413319	Negative
BIIB092 (BMS-986168)	Anti-tau antibody (N-terminus)	PSP	2	NCT03068468	Negative
		CBD, nfvPPA, TES, *MAPT*	1	NCT03658135	Terminated
LY3303560	Anti-tau antibody (N-terminus)	AD	2	NCT03518073	Active
RO 7105705 (RG 6100)	Anti-tau antibody (N-terminus)	AD	2	NCT03289143	Active
UCB0107	Anti-tau antibody (mid-domain)	PSP	1	NCT04185415	Active
JNJ-63733657	Anti-tau antibody (mid-domain)	AD	1	NCT03375697	Unavailable
BIIB076	Anti-tau antibody (monomer and filament)	AD	1	NCT03056729	Active
AADvac1	Tau vaccine	nfvPPA	1	NCT03174886	Active
ACI-35	Tau vaccine	AD	1		Unavailable
Davunetide	Microtubule stabilizations	PSP	2/3	NCT01110720	Negative [101]
TPI-287	Microtubule stabilizations	AD, PSP, CBD	I	NCT01966666, NCT02133846	Negative [102]
ASN001	o-GlcNACase inhibitor	–	1	–	–
Salsalate	Tau acetylation inhibition	PSP	1	NCT02422485	Negative
TRx0237 (LMTx)	Tau aggregation inhibition	bvFTD	3	NCT03446001	Negative
AZP2006	Tau aggregation inhibition	PSP	2	NCT04008355	Active
Lithium Carbonate	Glycogen synthase kinase inhibitor	bvFTD	2	NCT02862210	
Tideglusib	Glycogen synthase kinase inhibitor	PSP	2	NCT01049399	Negative [84]
Young plasma transfusions	Alter peripheral cell signaling	PSP	1	NCT02460731	Negative
Symptomatic FTLD treatments					
Oxytocin	Augmenting social apathy	FTD	2	NCT 01386333	Active
Rivastigmine	Cholinesterase inhibition	PSP	3	NCT02839642	Unknown

(continued)

	Mechanism	Indication	Phase	ClinicalTrials.gov identifier	Status
Suvorexant, zolpidem	Treatment of insomnia	PSP	3	NCT04014387	Active
Transcranial DC stim.	Electrical current stimulation	FTLD-*GRN*	N/A	NCT02999282	Active
Transcranial magnetic stim.	Magnetic field stimulation	PPA, bvFTD	N/A	NCT03406429	Active

ALS-C9orf72 amyotrophic lateral sclerosis due to chromosome 9 open reading frame 72 expansion, *AD* Alzheimer's disease, *bvFTD* behavioral variant frontotemporal dementia, *CBD* corticobasal degeneration, *FTLD* frontotemporal lobar degeneration, *FTLD-GRN* FTLD due to progranulin haploinsufficiency, *MAPT* microtubule-associated protein tau mutation, *nfvPPA* non-fluent variant primary progressive aphasia, *PPA* primary progressive aphasia, *PSP* progressive supranuclear palsy, *TES* traumatic encephalopathy syndrome

for upcoming trials in FTLD. ASOs are short, single-stranded, synthetic oligonucleotides that hybridize with high specificity to complementary pre-messenger RNA (mRNA) or mature mRNA and alter translations in a variety of ways [78]. These drugs require intrathecal delivery, but they offer an attractive diversity of mechanisms to suppress gene expression (mostly by triggering RNAaseH-mediated degradation of target mRNA), increase gene expression (by binding target promoters, suppressing microRNA, or suppressing natural antisense transcripts), or modulate alternative splicing (by forcing the inclusion or exclusion of specific exons). Additionally, previous trials in ASO-based therapies (NCT02193074, NCT00844597, NCT01396239/ NCT01540409, and NCT02255552) have recently resulted in FDA approval of nusinersen [79] for treatment of spinomuscular atrophy (SMA) and eteplirsen [80] for treatment of Duchenne muscular dystrophy (DMD). Currently, BIIB080, an ASO that knocks down tau mRNA expression, is being investigated in patients with mild Alzheimer's disease (NCT03186989) and may have a role in future FTLD-tau trials. While human trial data has yet to be released, intrathecal infusion of BIIB080 resulted in 75% reduction of cortical *MAPT* mRNA and had no dose-limiting side effects in nonhuman primates [81]. An additional ASO-based strategy in FTLD-tau may include manipulation of alternative splicing of exon 10, which is included in the 4R forms of tau that predominate in PSP, CBD, and many pathogenic *MAPT* muta-

tions adjacent to intron 10 or exon 10. So far, ASO-mediated splice alteration has been found to normalize the balance of 3R and 4R tau isoforms in a preclinical model [82] of pathogenic *MAPT* mutations, but this strategy has yet to be implemented in an active clinical trial program.

Multiple therapeutic trials have sought to limit pathogenic posttranslational modification of tau proteins using small molecule therapies. Salsalate is a repurposed small molecule (a nonsteroidal anti-inflammatory typically used to treat pain) that recently gained interest due its inhibition of potentially pathogenic tau acetylation. In preclinical studies, salsalate was found to inhibit tau acetylation via the p300 acetyltransferase and suppress tau accumulation in transgenic mice [83]. Salsalate was recently investigated in a phase 1 trial in PSP (NCT02422485), and while the drug was well tolerated, it failed to show a benefit compared to historic controls. Salsalate is now unlikely to be investigated in follow-up FTLD-tau trials, but a trial in Alzheimer's disease is still ongoing (NCT03277573). Other trial programs have investigated tideglusib and lithium carbonate, which potentially block tau phosphorylation via inhibition of glycogen synthase kinases (GSKs). Tideglusib failed to meet its primary endpoint of efficacy in a phase 2 trial [84]. Additionally, another small molecule therapy, ASN001, has been developed to inhibit O-GlcNAcylation of tau [85] (another potentially pathogenic posttranslational modification) but has yet to transition to an active trial. Several other small molecules have been developed to inhibit tau accumulations directly. LMTM (TRx0237), a proprietary formulation of methyl-thioninium chloride (MTC), is a phenothiazine and perhaps the best clinically studied potential inhibitor of tau aggregation [86]. LMTM has so far failed to show benefits in primary endpoints of efficacy in large, multisite, phase 3 trials in both Alzheimer's disease [87] (NCT01689246) and bvFTD (NCT03446001).

An additional novel strategy for FTLD-tau treatment involves alteration of the extracellular milieu of injured neurons. Studies in aging mice suggest that plasma-derived factors from young mice may improve synaptic health, neurogenesis, and cognitive performance [88, 89]. In light of these findings, a small open-label phase 1 trial in PSP investigated the possible therapeutic benefit of plasma pooled from younger individuals (NCT02460731). This trial failed to show a therapeutic signal relative to historic controls, and whole plasma infusions are unlikely to be investigated in follow-up trials in FTLD-tau.

Future *C9orf72* Expansion Therapies

Pathogenic expansion of the *C9orf72* gene is the single most common genetic mutation causing familial frontotemporal dementia and ALS in North America and Europe [90]. Hexanucleotide expansions of *C9orf72* lead to FTLD pathology via a variety of possible mechanisms, including toxic gain of function from RNA-mediated toxicity and toxic dipeptides (from repeat-associated non-ATG translation) [91]. As discussed in a preceding section on future therapies for primary tauopathies, antisense oligonucleotide (ASO) therapies present a viable mechanism to degrade mRNA targets with high specificity. Intriguingly, in a preclinical rodent model of *C9orf72* expansion, ASOs targeting repeat-containing RNAs were sufficient to decrease toxic mRNA foci, suppress toxic dipeptide production, and improve cognitive performance [92]. Based on the preclinical success of this approach, BIIB078, an intrathecal ASO therapy targeting expanded *C9orf72* RNA, is now being studied in patients with ALS due to *C9orf72* expansion

(NCT03626012). While this trial is not enrolling patients with FTD, any future success of this ALS therapeutic program is likely to translate patients with FTLD due to *C9orf72* expansion.

Progranulin Deficiency Therapies

Haploinsufficiency of the progranulin gene (*GRN*) is associated with an over 50% reduction in plasma and CSF progranulin levels and a high penetrance of FTD [93, 94]. Given the link between low progranulin levels and FTD, several therapeutic trials have sought methods to therapeutically raise progranulin in the blood and CSF. Based on preclinical mouse [95] and cell [96] models of progranulin deficiency, nimodipine (a calcium channel blocker) and histone deacetylase (HDAC) inhibitors were identified as possible oral therapies to increase progranulin levels. In an 8-week, open-label trial, nimodipine failed to raise progranulin levels in participants with *GRN* haploinsufficiency [95] (NCT01835665). Additionally, in a randomized double-blind placebo-controlled phase 2 trial, FRM-0334 (an HDAC inhibitor) also failed to raise plasma progranulin levels in participants with *GRN* haploinsufficiency (NCT02149160). So far more encouraging results have been reported in clinical human trials using AL001, a monoclonal antibody which blocks the sortilin receptor, an important component in the degradation of progranulin [97]. In a phase 1 open-label trial, AL001 successfully raised progranulin levels in healthy volunteers and individuals with *GRN* haploinsufficiency (NCT03636204). Additionally, AL001 appeared to decrease plasma neurofilament light chain levels in mutation carriers, thus providing an early signal of a possible neuroprotective effect [98]. AL001 has subsequently moved on to a phase 2 trial (NCT03987295), with plans for a phase 3 trial currently underway. Another potentially exciting mechanism of progranulin treatment is gene replacement therapy. While the details of these proposed therapeutic mechanisms remain proprietary, at least one potential approach has recently been publically discussed (PR006) which utilizes an AAV9-based vector to deliver a GRN replacement therapy [99].

Symptomatic Treatment Trials

While the majority of recent FTD clinical trial development has focused on disease-modifying interventions, a fair amount of recent trials have alternatively focused on novel strategies of symptom management. Due to encouraging results in a phase 1 study [100], the hormone oxytocin is currently being investigated in a phase 2 trial aimed at improving social apathy in patients with FTD (NCT01386333). Another current symptom management study is comparing the utility of suvorexant (a dual orexin receptor antagonist) to zolpidem (a GABA receptor agonist) in the treatment of insomnia secondary to PSP (NCT04014387). Additionally, a few novel non-pharmacological trials are underway, investigating transcranial electrical and magnetic stimulation techniques as methods of augmenting symptoms in FTD variants (NCT02999282, NCT03406429).

Conclusion

Patients with bvFTD and PPA may be treated with a wide array of current therapies. The current focus of bvFTD and PPA care involves dedicated non-pharmacological therapy, including patient/caregiver education, assessment of safety risks, and behavioral intervention strategies. Among pharmacological interventions, there is the strongest rationale for SSRIs as a means of improving undesired behaviors in bvFTD and PPA. Despite the lack of disease-modifying interventions for the underlying neuropathology of FTLD, there is currently a rich field of therapeutic strategies moving into clinical trials. While early tau immunotherapy trials have failed in FTLD-tau, there is a wide diversity of other therapeutic options available, including ASO therapies, inhibitors of pathogenic posttranslational modification of tau, and additional alternative immunological approaches. Within familial variants of FTD, there is also a growing portfolio of exciting possible therapies tailored to precise mechanisms of pathogenesis. These include possible ASO therapies in *C9orf72* expansion, anti-sortilin immunotherapy in *GRN* deficiency, and possible gene replacement therapy in *GRN* deficiency. Taken together, it is a truly exciting time for the field of FTLD treatment.

References

1. Talerico KA, Evans LK (2001) Responding to safety issues in frontotemporal dementias. Neurology 56(11 SUPPL. 4):S52–S55
2. Rascovsky K, Hodges JR, Knopman D et al (2011) Sensitivity of revised diagnostic criteria for the behavioural variant of frontotemporal dementia. Brain
3. Iverson DJ, Gronseth GS, Reger MA et al (2010) Practice parameter update: evaluation and management of driving risk in dementia: report of the quality standards subcommittee of the American academy of neurology. Neurology 74(16):1316–1324
4. Riedijk SR, De Vugt ME, Duivenvoorden HJ et al (2006) Caregiver burden, health-related quality of life and coping in dementia caregivers: a comparison of frontotemporal dementia and Alzheimer's disease. Dement Geriatr Cogn Disord 22(5–6):405–412
5. Lillo P, Garcin B, Hornberger M et al (2010) Neurobehavioral features in frontotemporal dementia with amyotrophic lateral sclerosis. Arch Neurol 67(7):826–830
6. Perry DC, Brown JA, Possin KL et al (2017) Clinicopathological correlations in behavioural variant frontotemporal dementia. Brain
7. Gómez-Tortosa E, Rigual R, Prieto-Jurczynska C et al (2016) Behavioral evolution of progressive semantic aphasia in comparison with nonfluent aphasia. Dement Geriatr Cogn Disord 41(1–2):1–8
8. Höglinger GU, Respondek G, Stamelou M et al (2017) Clinical diagnosis of progressive supranuclear palsy: the movement disorder society criteria. Mov Disord 32(6):853–864
9. Lee SE, Rabinovici GD, Mayo MC et al (2011) Clinicopathological correlations in corticobasal degeneration. Ann Neurol 70(2):327–340
10. Kales HC, Gitlin LN, Lyketsos CG (2014) Management of neuropsychiatric symptoms of dementia in clinical settings: recommendations from a multidisciplinary expert panel. J Am Geriatr Soc 62(4):762–769
11. Gorno-Tempini M, Hillis A, Weintraub S et al (2011) Classification of primary progressive aphasia and its variants. Neurology 76(11):1006–1014
12. Volkmer A, Rogalski E, Henry M et al (2019) Speech and language therapy approaches to managing primary progressive aphasia. Pract Neurol
13. Jokel R, Graham NL, Rochon E et al (2014) Word retrieval therapies in primary progressive aphasia. Aphasiology 28(8–9):1038–1068
14. Henry ML, Hubbard HI, Grasso SM et al (2018) Retraining speech production and fluency in

non-fluent/agrammatic primary progressive aphasia. Brain 141(6):1799–1814

15. Henry ML, Meese MV, Truong S et al (2013) Treatment for apraxia of speech in nonfluent variant primary progressive aphasia. Behav Neurol 26(1–2):77–88

16. Deckers K, van Boxtel MPJ, Schiepers OJG et al (2015) Target risk factors for dementia prevention: a systematic review and Delphi consensus study on the evidence from observational studies. Int J Geriatr Psychiatry 30(3):234–246

17. Casaletto KB, Staffaroni AM, Wolf A et al (2020) Active lifestyles moderate clinical outcomes in autosomal dominant frontotemporal degeneration. Alzheimers Dement 16(1):91–105

18. Huey ED, Putnam KT, Grafman J (2006) A systematic review of neurotransmitter deficits and treatments in frontotemporal dementia. Neurology 66(1):17–22

19. Engelborghs S, Vloeberghs E, Le Bastard N et al (2008) The dopaminergic neurotransmitter system is associated with aggression and agitation in frontotemporal dementia. Neurochem Int 52(6):1052–1060

20. Chow TW, Mendez MF (2002) Goals in symptomatic pharmacologic management of frontotemporal lobar degeneration. Am J Alzheimers Dis Other Dement 17(5):267–272

21. Lanctôt KL, Herrmann N, Ganjavi H et al (2007) Serotonin-1A receptors in frontotemporal dementia compared with controls. Psychiatry Res Neuroimaging 156(3):247–250

22. Sparks DL, Markesbery WR (1991) Altered serotonergic and cholinergic synaptic markers in Pick's disease. Arch Neurol 48(8):796–799

23. Francis PT, Holmes C, Webster MT et al (1993) Preliminary neurochemical findings in non-Alzheimer dementia due to lobar atrophy. Dementia 4:172–177

24. Procter AW, Qurne M, Francis PT (1999) Neurochemical features of frontotemporal dementia. Dement Geriatr Cogn Disord 10(Suppl 1):80–84

25. Yang Y, Schmitt HP (2001) Frontotemporal dementia: evidence for impairment of ascending serotoninergic but not noradrenergic innervation. Immunocytochemical and quantitative study using a graph method. Acta Neuropathol 101(3):256–270

26. Swartz JR, Miller BL, Lesser IM et al (1997) Frontotemporal dementia: treatment response to serotonin selective reuptake inhibitors. J Clin Psychiatry 58558:212–216

27. Moretti R, Torre P, Antonello RM et al (2003) Frontotemporal dementia: paroxetine as a possible treatment of behavior symptoms a randomized, controlled, open 14-month study. Eur Neurol 49:13–19

28. Deakin JB, Rahman S, Nestor PJ et al (2004) Paroxetine does not improve symptoms and impairs cognition in frontotemporal dementia: a double-blind randomized controlled trial. Psychopharmacology (Berl) 172:400–408

29. Prodan CI, Monnot M, Ross ED (2009) Behavioural abnormalities associated with rapid deterioration of language functions in semantic dementia respond to sertraline. J Neurol Neurosurg Psychiatry 80(12):1416–1417

30. Herrmann N, Black SE, Chow T et al (2012) Serotonergic function and treatment of behavioral and psychological symptoms of frontotemporal dementia. Am J Geriatr Psychiatry 20(9): 789–797

31. The American Psychiatric Publishing Textbook of Psychopharmacology – Alan F. Schatzberg, Charles B. Nemeroff – Google Books. Accessed 2 Feb 2020

32. Lebert F, Stekke W, Hasenbroekx C et al (2004) Frontotemporal dementia: a randomised, controlled trial with trazodone. Dement Geriatr Cogn Disord 17:355–359

33. Rinne JO, Laine M, Kaasinen V et al (2002) Striatal dopamine transporter and extrapyramidal symptoms in frontotemporal dementia. Neurology 58(10):1489–1493

34. Frisoni GB, Pizzolato G, Bianchetti A et al (1994) Single photon emission computed tomography with [99Tc]-HM-PAO and [123I]-IBZM in Alzheimer's disease and dementia of frontal type: preliminary results. Acta Neurol Scand 89(3):199–203

35. Kerrsens CJ, Pijnenburg YAL (2008) Vulnerability to neuroleptic side effects in frontotemporal dementia. Eur J Neurol 15(2):111–112

36. Steinberg M, Lyketsos CG (2012) Atypical antipsychotic use in patients with dementia: managing safety concerns. Am J Psychiatry 169(9):900–906

37. Kurlan R, Cummings J, Raman R et al (2007) Quetiapine for agitation or psychosis in patients with dementia and parkinsonism. Neurology 68(17):1356–1363

38. Komossa K, Rummel-Kluge C, Schmid F et al (2010) Quetiapine versus other atypical antipsychotics for schizophrenia. In: Cochrane database of systematic reviews. John Wiley & Sons

39. Kales HC, Kim HM, Zivin K et al (2012) Risk of mortality among individual antipsychotics in patients with dementia. Am J Psychiatry 169(1):71–79

40. Weiden PJ (2007) EPS profiles: the atypical antipsychotics – are not all the same. J Psychiatr Pract 13(1):13–24

41. Moretti R, Torre P, Antonello RM et al (2003) Olanzapine as a treatment of neuropsychiatric disorders of Alzheimer's disease and other dementias: a 24-month follow-up of 68 patients. Am J Alzheimers Dis Other Dement 18(4):205–214

42. Reeves RR, Perry CL (2013) Aripiprazole for sexually inappropriate vocalizations in frontotemporal dementia. J Clin Psychopharmacol 33(1):145–146

43. Curtis RC, Resch DS (2000) Case of pick's central lobar atrophy with apparent stabilization of cognitive decline after treatment with risperidone. J Clin Psychopharmacol 20(3):384–385

44. Hu B, Ross L, Neuhaus J et al (2010) Off-label medication use in frontotemporal dementia. Am J Alzheimers Dis Other Dement 25(2):128–133

45. Kerchner GA, Tartaglia MC, Boxer AL (2011) Abhorring the vacuum: use of Alzheimer's disease medications in frontotemporal dementia. Expert Rev Neurother 11(5):709–717

46. Lampl Y, Sadeh M, Lorberboym M (2004) Efficacy of acetylcholinesterase inhibitors in frontotemporal dementia. Ann Pharmacother 38(11):1967–1968

47. Kertesz A, Morlog D, Light M et al (2008) Galantamine in frontotemporal dementia and primary progressive aphasia. Dement Geriatr Cogn Disord 25(2):178–185

48. Moretti R, Torre P, Antonello RM et al (2004) Rivastigmine in frontotemporal dementia: an open-label study. Drugs Aging 21(14):931–937

49. Kimura T, Takamatsu J (2013) Pilot study of pharmacological treatment for frontotemporal dementia: risk of donepezil treatment for behavioral and psychological symptoms. Geriatr Gerontol Int 13(2):506–507

50. Mendez MF, Shapira JS, McMurtray A et al (2007) Preliminary findings: behavioral worsening on donepezil in patients with frontotemporal dementia. Am J Geriatr Psychiatry 15(1):84–87

51. Boxer AL, Knopman DS, Kaufer DI et al (2013) Memantine in patients with frontotemporal lobar degeneration: a multicentre, randomised, double-blind, placebo-controlled trial. Lancet Neurol 12(2):149–156

52. Goforth HW, Konopka L, Primeau M et al (2004) Quantitative electroencephalography in frontotemporal dementia with methylphenidate response: a case study. Clin EEG Neurosci 35(2):108–111

53. Rahman S, Robbins TW, Hodges JR et al, Methylphenidate ('Ritalin') can ameliorate abnormal risk-taking behavior in the frontal variant of frontotemporal dementia. Neuropsychopharmacology 31:651–658

54. Huey ED, Garcia C, Wassermann EM et al (2008) Stimulant treatment of frontotemporal dementia in 8 patients HHS public access. J Clin Psychiatry 69(12):1981–1982

55. Rupasinghe J, Jasinarachchi M (2011) Progressive encephalopathy with cerebral oedema and infarctions associated with valproate and diazepam overdose. J Clin Neurosci 18(5):710–711

56. Gram L, Bentsen KD (2009) Valproate: an updated review. Acta Neurol Scand 72(2):129–139

57. Coulter DL, Allen RJ (1980) Secondary hyperammonaemia: a possible mechanism for valproate encephalopathy. Lancet 315(8181):1310–1311

58. Silver M, Factor SA (2013) Valproic acid-induced parkinsonism: levodopa responsiveness with dyskinesia. Parkinsonism Relat Disord 19(8):758–760

59. Gálvez-Andres A, Blasco-Fontecilla H, González-Parra S et al (2007) Secondary bipolar disorder and diogenes syndrome in frontotemporal dementia:

behavioral improvement with quetiapine and sodium valproate. J Clin Psychopharmacol 27(6):722–723

60. Poetter CE, Stewart JT (2012) Treatment of indiscriminate, inappropriate sexual behavior in frontotemporal dementia with carbamazepine. J Clin Psychopharmacol 32(1):137–138

61. Cruz M, Marinho V, Fontenelle LF et al (2008) Topiramate may modulate alcohol abuse but not other compulsive behaviors in frontotemporal dementia: case report. Cogn Behav Neurol 21(2):104–106

62. Nestor PJ (2012) Reversal of abnormal eating and drinking behaviour in a frontotemporal lobar degeneration patient using low-dose topiramate. J Neurol Neurosurg Psychiatry 83(3):349–350

63. Singam C, Walterfang M, Mocellin R et al (2013) Topiramate for abnormal eating behaviour in frontotemporal dementia. Behav Neurol 27(3):285–286

64. Shinagawa S, Tsuno N, Nakayama K (2013) Managing abnormal eating behaviours in frontotemporal lobar degeneration patients with topiramate. Psychogeriatrics 13(1):58–61

65. Litvan I, Grimes DA, Lang AE et al (1999) Clinical features differentiating patients with postmortem confirmed progressive supranuclear palsy and corticobasal degeneration. J Neurol 246(Suppl):1–5

66. van Balken I, Litvan I (2006) Current and future treatments in progressive supranuclear palsy. Curr Treat Options Neurol 8(3):211–223

67. O'Sullivan SS, Evans AH, Lees AJ (2009) Dopamine dysregulation syndrome: an overview of its epidemiology, mechanisms and management. CNS Drugs 23(2):157–170

68. Moretti R, Torre P, Antonello RM et al (2002) Effects of selegiline on fronto-temporal dementia: a neuropsychological evaluation. Int J Geriatr Psychiatry 17(4):391–392

69. Weingarten MD, Lockwood AH, Hwo SY et al (1975) A protein factor essential for microtubule assembly. Proc Natl Acad Sci U S A 72(5):1858–1862

70. Jadhav S, Avila J, Schöll M et al (2019) A walk through tau therapeutic strategies. Acta Neuropathol Commun 7(1):22

71. Buée L, Bussière T, Buée-Scherrer V et al (2000) Tau protein isoforms, phosphorylation and role in neurodegenerative disorders. Brain Res Rev 33(1):95–130

72. Sanders DW, Kaufman SK, DeVos SL et al (2014) Distinct tau prion strains propagate in cells and mice and define different tauopathies. Neuron 82(6):1271–1288

73. Yanamandra K, Kfoury N, Jiang H et al (2013) Anti-tau antibodies that block tau aggregate seeding in vitro markedly decrease pathology and improve cognition in vivo. Neuron 80(2):402–414

74. Castillo-Carranza DL, Sengupta U, Guerrero-Muñoz MJ et al (2014) Passive immunization with tau oligomer monoclonal antibody reverses tauopathy phenotypes without affecting hyperphosphorylated neurofibrillary tangles. J Neurosci 34(12):4260–4272

75. Zhou Y, Shi J, Chu D et al (2018) Relevance of phosphorylation and truncation of tau to the

etiopathogenesis of Alzheimer's disease. Front Aging Neurosci 10

76. Novak P, Schmidt R, Kontsekova E et al (2018) Fundamant: an interventional 72-week phase 1 follow-up study of AADvac1, an active immunotherapy against tau protein pathology in Alzheimer's disease. Alzheimers Res Ther 10(1):108

77. Hung S-Y, Fu W-M (2017) Drug candidates in clinical trials for Alzheimer's disease. J Biomed Sci 24(1):47

78. DeVos SL, Miller TM (2013) Antisense oligonucleotides: treating neurodegeneration at the level of RNA. Neurotherapeutics 10(3):486–497

79. Hoy SM (2017) Nusinersen: first global approval. Drugs 77(4):473–479

80. Lim KRQ, Maruyama R, Yokota T (2017) Eteplirsen in the treatment of Duchenne muscular dystrophy. Drug Des Devel Ther 11:533–545

81. Mignon L, Kordasiewicz H, Lane R et al (2018) Design of the first-in-human study of IONIS-MAPTRx, a tau-lowering antisense oligonucleotide, in patients with Alzheimer disease (S2.006). Neurology 90(15 Supplement)

82. Rodriguez-Martin T, Anthony K, Garcia-Blanco MA et al (2009) Correction of tau mis-splicing caused by FTDP-17 MAPT mutations by spliceosome-mediated RNA trans-splicing. Hum Mol Genet 18(17):3266–3273

83. Min SW, Chen X, Tracy TE et al (2015) Critical role of acetylation in tau-mediated neurodegeneration and cognitive deficits. Nat Med 21(10):1154–1162

84. Tolosa E, Litvan I, Höglinger GU et al (2014) A phase 2 trial of the GSK-3 inhibitor tideglusib in progressive supranuclear palsy. Mov Disord 29(4):470–478

85. Hastings NB, Wang X, Song L et al (2017) Inhibition of O-GlcNAcase leads to elevation of O-GlcNAc tau and reduction of tauopathy and cerebrospinal fluid tau in rTg4510 mice. Mol Neurodegener 12(1):1–16

86. Wischik CM, Edwards PC, Lai RYK et al (1996) Selective inhibition of Alzheimer disease-like tau aggregation by phenothiazines. Proc Natl Acad Sci U S A 93(20):11213–11218

87. Gauthier S, Feldman HH, Schneider LS et al (2016) Efficacy and safety of tau-aggregation inhibitor therapy in patients with mild or moderate Alzheimer's disease: a randomised, controlled, double-blind, parallel-arm, phase 3 trial. Lancet 388(10062):2873–2884

88. Katsimpardi L, Litterman NK, Schein PA et al (2014) Vascular and neurogenic rejuvenation of the aging mouse brain by young systemic factors. Science (80-) 344(6184):630–634

89. Villeda SA, Plambeck KE, Middeldorp J et al (2014) Young blood reverses age-related impairments in

cognitive function and synaptic plasticity in mice. Nat Med 20(6):659–663

90. Greaves CV, Rohrer JD (2019) An update on genetic frontotemporal dementia. J Neurol 266(8):2075–2086

91. Todd TW, Petrucelli L (2016) Insights into the pathogenic mechanisms of chromosome 9 open reading frame 72 (C9orf72) repeat expansions. J Neurochem 138:145–162

92. Jiang J, Zhu Q, Gendron TF et al (2016) Gain of toxicity from ALS/FTD-linked repeat expansions in C9ORF72 is alleviated by antisense oligonucleotides targeting GGGGCC-containing RNAs. Neuron 90(3):535–550

93. Finch N, Baker M, Crook R et al (2009) Plasma progranulin levels predict progranulin mutation status in frontotemporal dementia patients and asymptomatic family members. Brain 132(3):583–591

94. Meeter LHH, Patzke H, Loewen G et al (2016) Progranulin levels in plasma and cerebrospinal fluid in granulin mutation carriers. Dement Geriatr Cogn Dis Extra 6(2):330–340

95. Sha SJ, Miller ZA, won Min S et al (2017) An 8-week, open-label, dose-finding study of nimodipine for the treatment of progranulin insufficiency from GRN gene mutations. Alzheimer's Dement Transl Res Clin Interv 3(4):507–512

96. Cenik B, Sephton CF, Dewey CM et al (2011) Suberoylanilide hydroxamic acid (vorinostat) up-regulates progranulin transcription: rational therapeutic approach to frontotemporal dementia. J Biol Chem 286(18):16101–16108

97. Lee WC, Almeida S, Prudencio M et al (2014) Targeted manipulation of the sortilin-progranulin axis rescues progranulin haploinsufficiency. Hum Mol Genet 23(6):1467–1478

98. Alector Showcases Progress in Immuno-Neurology Clinical Programs and Research Portfolio at R&D Day Nasdaq:ALEC. Accessed 17 Feb 2020

99. Our Programs • Prevail. Accessed 17 Feb 2020

100. Finger EC, MacKinley J, Blair M et al (2015) Oxytocin for frontotemporal dementia: a randomized dose-finding study of safety and tolerability. Neurology 84(2):174–181

101. Boxer AL, Lang AE, Grossman M et al (2014) Davunetide in patients with progressive supranuclear palsy: a randomised, double-blind, placebo-controlled phase 2/3 trial. Lancet Neurol 13(7):676–685

102. Tsai RM, Miller Z, Koestler M et al (2019) Reactions to multiple ascending doses of the microtubule stabilizer TPI-287 in patients with Alzheimer disease, progressive supranuclear palsy, and corticobasal syndrome: a randomized clinical trial. JAMA Neurol

Index

B. Ghetti et al. (eds.), *Frontotemporal Dementias*, Advances in Experimental Medicine and Biology
1281, https://doi.org/10.1007/978-3-030-51140-1

Printed in the United States
By Bookmasters